# MONOLITHIC MICROWAVE INTEGRATED CIRCUITS:

## Technology & Design

## The Artech House Microwave Library

*Analysis, Design, and Applications of Fin Lines*, Bharathi Bhat and Shiban K. Koul

*E-Plane Integrated Circuits*, P. Bhartia and P. Pramanick, eds.

*Filters with Helical and Folded Helical Resonators*, Peter Vizmuller

*GaAs MESFET Circuit Design*, Robert A. Soares, ed.

*Gallium Arsenide Processing Techniques*, Ralph Williams

*Handbook of Microwave Integrated Circuits*, Reinmut K. Hoffmann

*Handbook for the Mechanical Tolerancing of Waveguide Components*, W.B.W. Alison

*Handbook of Microwave Testing*, Thomas S. Laverghetta

*High Power Microwave Sources*, Victor Granatstien and Igor Alexeff, eds.

*Introduction to Microwaves*, Fred E. Gardiol

*LOSLIN: Lossy Line Calculation Software and User's Manual*, Fred. E. Gardiol

*Lossy Transmission Lines*, Fred E. Gardiol

*Materials Handbook for Hybrid Microelectronics*, J.A. King, ed.

*Microstrip Antenna Design*, K.C. Gupta and A. Benalla, eds.

*Microstrip Lines and Slotlines*, K.C. Gupta, R. Garg, and I.J. Bahl

*Microwave Engineer's Handbook*: 2 volume set, Theodore Saad, ed.

*Microwave Filters, Impedance Matching Networks, and Coupling Structures*, G.L. Matthaei, L. Young and E.M.T. Jones

*Microwave Integrated Circuits*, Jeffrey Frey and Kul Bhasin, eds.

*Microwaves Made Simple: Principles and Applications*, Stephen W. Cheung, Frederick H. Levien, et al.

*Microwave and Millimeter Wave Heterostructure Transistors and Applications*, F. Ali, ed.

*Microwave Mixers*, Stephen A. Maas

*Microwave Transition Design*, Jamal S. Izadian and Shahin M. Izadian

*Microwave Transmission Line Filters*, J.A.G. Malherbe

*Microwave Transmission Line Couplers*, J.A.G. Malherbe

*Microwave Tubes*, A.S. Gilmour, Jr.

*MMIC Design: GaAs FETs and HEMTs*, Peter H. Ladbrooke

*Modern Spectrum Analyzer Theory and Applications*, Morris Engelson

*Monolithic Microwave Integrated Circuits: Technology and Design*, Ravender Goyal, et al.

*Nonlinear Microwave Circuits*, Stephen A. Maas

*Terrestrial Digital Microwave Communications*, Ferdo Ivanek, et al.

# MONOLITHIC MICROWAVE INTEGRATED CIRCUITS:

## Technology & Design

**Ravender Goyal,**
**editor**

**Artech House**

Goyal, Ravender
    Monolithic Microwave Integrated Circuit Technology and Design/edited by
Ravender Goyal.

    p.    cm.
    Bibliography: p.
    Includes index.
    ISBN   0-89006-309-5:
    1. Microwave Integrated Circuits—Design and Construction.    2. Gallium
Arsenide Semiconductors—Design and Construction.    I. Goyal, Ravender.
TK7876.m64   1989             621.381′73—dc19                89-327
ISBN   0-89006-309-5                                          CIP


Copyright  ©  1989

**ARTECH HOUSE, INC.**
**685 Canton Street**
**Norwood, MA 02062**

**International Standard Book Number:  0-89006-309-5**
**Library of Congress Catalog Card Number:  89-327**

    **10**        **9**        **8**        **7**        **6**        **5**        **4**        **3**        **2**

*To My Parents,*
*Dr. Kailash Chand Goyal*
*&*
*Mrs. Brij Bala Goyal*

# The Authors

Bert L. Berson, Berson and Associates, Mountain View, California

Tzu H. Chen, TRW, Redondo Beach, California

Jaime Cordero, Litton AMECOM, Maryland

Ian G. Eddison, Plessey Research Caswell, Ltd., Northamptonshire, United Kingdom

Giovanni Ghione, Politecnico de Milano, Italy

Mike Golio, Motorola, Inc., Chandler, Arizona

Ravender Goyal (editor), Anadigics, Inc., Warren, New Jersey

Asu R. Jha, Jha Technical Consulting Services, Cerritos, California

Roger D. Kaul, US Army LAB COM, Adelphi, Maryland

Mahesh Kumar, AEL, Lansdale, Pennsylvania

Barry Manz, Manz Communication, Montville, New Jersey

Steve Miller, Raytheon ESD, Goleta, California

James M. O'Connor, Allied Signal, Columbia, Maryland

Anthony M. Pavio, Texas Instruments, Dallas, Texas

Allan F. Podell, Pacific Monolithics, Sunnyvale, California

Vittorio Rizzoli, University of Bologna, Italy

William J. Roesch, TriQuint Semiconductor, Beaverton, Oregon

Fred Rosenbaum, Washington University, St. Louis, Missouri

Richard A. Sparks, Raytheon Company, Bedford, Massachusetts

Y. Tajima, Raytheon Research Div., Lexington, Massachusetts

Wolfgang Thomann, Technical University of Munich, Federal Republic of Germany

W.S. Titus, Hittite Microwave Corporation, Woburn, Massachusetts

# Contents

FOREWORD                                                                xiii
PREFACE                                                                  xv
Chapter 1 INTRODUCTION                                                    1
  1.1 Introduction                                              1
  1.2 Why MMICs?                                                5
  1.3 Processing                                                7
  1.4 MMIC Performance                                          8
  1.5 MMIC Status                                              15
    1.5.1 GaAs MMIC Reliability                      16
    1.5.2 Yield                                      16
    1.5.3 Cost                                       17
  1.6 Future Developments                                      17
  1.7 MMIC Applications                                        18
    1.7.1 Military Applications                      18
    1.7.2 Commercial Applications                    23
    1.7.3 Consumer Applications                      24
  1.8 Summary                                                  28
Chapter 2 MICROWAVE FUNDAMENTALS                                        31
  2.1 Introduction                                             31
  2.2 Transmission Line Theory                                 35
    2.2.1 ac Solution for Transmission Lines         38
    2.2.2 Propagation Constant, Phase Velocity, and Characteristic
        Impedance                 39
    2.2.3 Losses in Transmission Lines               42
    2.2.4 Reflection Coefficient, Line Impedance, and Conductance    45
    2.2.5 Standing Waves                             49
  2.3 Smith Chart                                              54
    2.3.1 Basics of the Smith Chart                  55
    2.3.2 Changing Characteristic Impedance          57

2.3.3 The Smith Chart and Admittance                                62
2.3.4 Reflection Coefficient, VSWR, and Power                       64
2.3.5 Effects of Frequency Variation                                68
2.3.6 Lossy Circuits                                                70
2.4 Network Parameters                                              70
   2.4.1 Z, Y, and h Parameters                                     70
   2.4.2 ABCD-Parameters                                            72
   2.4.3 Scattering or S-Parameters                                 73
   2.4.4 Properties of S-Parameters                                 76
   2.4.5 Relationship between S-Parameters and Other Network
         Parameters                                                 81
2.5 Noise Parameters                                                82
   2.5.1 Thermal Noise                                              83
   2.5.2 Shot Noise                                                 84
   2.5.3 Noise in Two-Port Network                                  84
   2.5.4 Noise Figure and Smith Chart                               92
   2.5.5 Noise Temperature                                          93
   2.5.6 Noise Figure and Noise Voltage                             94
Chapter 3 MMIC MATERIAL AND MANUFACTURING
          TECHNOLOGY                                                97
3.1 Material Technology and Characterization                        97
   3.1.1 Horizontal Bridgeman Technique                             98
   3.1.2 Liquid Encapsulated Czochralski Technique                  99
   3.1.3 Epitaxial Growth                                          101
   3.1.4 Materials Analysis                                        109
   3.1.5 Advanced Device Structures                                111
3.2 Processing Technology                                          117
   3.2.1 Photolithography                                          117
   3.2.2 Ion Implantation                                          119
   3.2.3 Metalization                                              121
   3.2.4 Dielectric Deposition                                     123
   3.2.5 Etching                                                   124
   3.2.6 Plating                                                   125
3.3 Computer-Aided Process Design                                  125
   3.3.1 Introduction                                              125
   3.3.2 Process Models                                            128
   3.3.3 Device Modeling                                           140
   3.3.4 Calibration                                               146
   3.3.5 Calibration Example Using GATES                           148
3.4 Manufacturing Process for MMICs                                151
   3.4.1 Active Layer Formation and Device Isolation               153
   3.4.2 Thin Film Resistors                                       158

3.4.3 Ohmic Contacts                                             158
3.4.4 Schottky Metalization                                      159
3.4.5 Resistor Etch                                              161
3.4.6 First-Level Metalization                                   162
3.4.7 Dielectric Layer                                           162
3.4.8 Airbridge Metalization                                     163
3.4.9 Final Passivation                                          163
3.4.10 Wafer Thinning and Backside via Etch                      165
3.4.11 Wafer Sawing                                              167
3.5 Process Control and Characterization                         167
3.5.1 Material Characterization                                  169
3.5.2 Process Control Monitors                                   171
3.6 MMIC Foundries: Selection Criteria                           182
3.6.1 Designing MMICs Using Foundries                            187
3.6.2 Processed Wafer Qualification                              190
3.7 Computer-Aided Manufacturing                                 196
Chapter 4 DEVICE MODELING                                        207
4.1 Single-Gate FET                                              207
4.1.1 Introduction                                               207
4.1.2 Basic Operation                                            208
4.1.3 Device Performance Analysis                                211
4.1.4 Characterization                                           215
4.1.5 Equivalent Circuits and Parameter Extraction               221
4.1.6 Device Modeling                                            244
4.1.7 Design Considerations and Applications                     269
4.1.8 Noise Modeling                                             278
4.2 Dual-Gate FET                                                285
4.2.1 Introduction                                               285
4.2.2 dc Characterization and Basic Device Operation             286
4.2.3 High Frequency Lumped Element Equivalent Circuit           289
4.2.4 Applications of Dual-Gate FETs                             299
4.3 Schottky Diodes                                              304
4.3.1 Introduction                                               304
4.3.2 Basic Operation                                            305
4.3.3 Lumped Element Equivalent Circuit                          310
4.3.4 Semidistributed Element Equivalent Circuit                 312
4.3.5 Application and Layout Considerations                      316
4.4 Planar Lumped Elements                                       318
4.4.1 Introduction                                               318
4.4.2 Planar Inductors                                           320
4.5 Planar Capacitors                                            329
4.6 Planar Resistors                                             342

4.7 Transmission Lines 347
  4.7.1 Introduction 347
  4.7.2 Microstrip and Coplanar Lines for MMICs 351
  4.7.3 Line Discontinuities 369
Appendix 4A   Analytical Formulas for Coplanar Lines 373
Chapter 5 MMIC Design Considerations and Amplifier Design 383
  5.1 Introduction 383
  5.2 Design Considerations for MMICs 383
    5.2.1 Chip Size 384
    5.2.2 Thermal Design and Wafer Thickness 384
    5.2.3 Low-Inductance Grounds and Crossovers 385
    5.2.4 Propagation Modes 385
    5.2.5 Other Design Considerations 386
  5.3 Biasing Techniques 387
  5.4 Microwave Amplifier Design 393
    5.4.1 Design Considerations 393
    5.4.2 Procedure for General Design of an Amplifier 405
    5.4.3 Design Examples 406
    5.4.4 2–18 GHz Distributed Amplifier 427
    5.4.5 2–6 GHz Feedback Gain Module 431
    5.4.6 Low-Noise Amplifier Design 440
    5.4.7 Circuit Performance 456
    5.4.8 Combining Techniques for Power Amplifiers 456
  5.5 On-Chip Tuning 462
    5.5.1 Tuning Technique Using an Addition of Elements 463
    5.5.2 Airbridge Removal Technique 463
Chapter 6 MMIC DESIGN: NONLINEAR AND CONTROL CIRCUITS 469
  6.1 Mixer Circuit Design 469
    6.1.1 Introduction 469
    6.1.2 Linearization 470
    6.1.3 Device Models 470
    6.1.4 Distributed Mixer Design 477
  6.2 Phase Shifter Design 489
    6.2.1 Introduction 489
    6.2.2 Design Approach 491
    6.2.3 Design Example 495
    6.2.4 Conclusion 504
  6.3 Double and Single Balanced Mixer Design 508
  6.4 Variable Attenuator and Switch Design 527
    6.4.1 Introduction 527
    6.4.2 Nonlinear FET Operation and Selection 530
    6.4.3 Variable Attenuator Design 532

6.4.4 Measurement *versus* Simulation     540
6.4.5 Switch     540
6.4.6 Measurement *versus* Simulation     546
Chapter 7 MMIC-BASED SUBSYSTEM CASE STUDY:
       THE TVRO CHIP     551
7.1 TVRO     551
7.2 Testing     562
7.3 Sample Wafer Testing     564
7.4 dc Wafer Probing     567
7.5 Designing for Testability     568
7.6 Packaged Unit Testing     568
7.7 Conclusion     571
Chapter 8 DESIGN AUTOMATION TOOLS FOR MMIC DESIGN     573
8.1 Introduction     573
8.1.1 Design Perspective: MMICs *versus* Hybrid     575
8.1.2 Digital-Analog *versus* MMIC Design     576
8.1.3 Typical MMIC Design Cycle     576
8.1.4 Design Approaches     577
8.2 Design Automation Tools     580
8.2.1 Workstation Environment     582
8.2.2 Schematic Capture     591
8.2.3 Circuit Simulation     593
8.2.4 Linear-Frequency Domain Simulation     595
8.2.5 Nonlinear-Time Domain Simulator     596
8.2.6 Mixed-Mode-Harmonic-Balance Simulator     611
8.2.7 Physical Layout     626
8.2.8 Back-End Design Verification     628
8.2.9 Design Languages     634
8.3 Mask Fabrication     636
Appendix 8A     647
Chapter 9 ON-WAFER TESTING OF MMICs     683
9.1 Introduction     683
9.2 Microwave Probes     684
9.3 On-Wafer Calibration     699
9.4 ATE System Architecture     704
9.5 On-Wafer Microwave Testing     707
9.5.1 MMIC Layout Considerations     707
9.5.2 Probe-to-IC Interface Issues     709
9.5.3 Examples of MMIC Wafer Testing     711
9.5.4 Limitations of Microwave Probing     716
9.6 On-Wafer Production Testing     724
9.6.1 Foundry Process Design Support     724

9.6.2  MMIC Product Testing                                                      724
9.7  The Future Status of Microwave Probing                                      731
Chapter 10  MMIC PACKAGING                                                       735
10.1  Introduction                                                               735
10.2  Electrical Design and Measurement                                          738
10.3  Thermal Design                                                             746
10.4  Mechanical Design Considerations                                           752
10.5  Characteristics of Available MMIC Packages                                 758
10.6  MMIC Package Standards and Standardization                                 758
10.7  Conclusions and Acknowledgments                                            763
Appendix 10A   Survey of Available Package Types and Their
                  Properties                                                     767
Chapter 11  MMIC RELIABILITY                                                     781
11.1  Introduction                                                               781
11.2  Circuit Failure Mechanisms                                                 783
11.2.1  Metalization                                                             783
11.2.2  Dielectric                                                               785
11.2.3  Substrate: Gate and Channel                                              785
11.3  Application Failure Mechanisms                                             787
11.3.1  Electrostatic Discharge                                                  787
11.3.2  Assembly                                                                 796
11.3.3  Packaging                                                                797
11.4  Reliability Test Strategy                                                  797
11.4.1  Failure Distributions                                                    798
11.4.2  Acceleration Methods                                                     801
11.4.3  Environmental Testing                                                    805
11.4.4  Reliability Philosophy                                                   807
11.5  Reliability Data                                                           808
11.5.1  Element Studies                                                          808
11.5.2  MMIC Reliability                                                         820
11.6  Radiation Hardening                                                        821
11.6.1  Radiation Hazards                                                        822
11.6.2  Types of Radiation Damage                                               823
11.6.3  Suggestions for Improving Radiation Hardness                             826
11.6.4  Radiation Shielding Guidelines                                           830
INDEX                                                                            837

# Foreword

Solid-state microwave technology has long offered the promise of fulfilling the requirements of nearly all military microwave systems and many commercial ones as well. The largest unsolved problem has been doing so at the right price. We are now at the threshold of a new era, with new levels of capability in the design, manufacture, testing, and use of these types of circuits. However, before the promise can be fulfilled, we know that certain conditions must be met. The products produced must have the performance characteristics, physical characteristics, and reliability required by their customers—the systems designers and developers—all at an acceptable cost.

The field of gallium arsenide microwave technology has gained momentum over the last two decades and is now in a very dynamic, fast-paced state as monolithic microwave circuit (MMIC) techniques are being introduced. It is appropriate to have the information on current and recent developments in this field gathered into one convenient source book available for designers and users alike. It can provide them with knowledge of the techniques, designs, and materials that can be used effectively to produce desired results. Equally important, it can provide information on the limitations to be expected in frequency, power, efficiency, and other parameters.

This book: *Monolithic Microwave: Integrated Circuit Technology and Design,* edited by Ravender Goyal, provides such a source. Each chapter is written by a specialist in that particular technical area. The technical chapters begin with a short introduction to microwave fundamentals and then proceed to detailed information on MMIC processing, device modeling, design considerations including techniques for nonlinear design, image rejection mixer design and MMIC-based subsystem design, automation tools for MMICs, on-wafer testing, packaging, and reliability. Nearly every topic of interest to the prospective MMIC designer, process specialist, and user is treated.

MMICs constitute an opportunity whose time has come. During the 1990s,

MMICs are expected to become a mature, effective technology for use in microwave systems applications on a wide scale. This collection of information will help the efforts to bring MMIC technology to fruition.

*Eliot D. Cohen*                                           *Robert A. Pucel*
*Silver Spring, Maryland*                              *Needham, Massachusetts*

# Preface

The gallium arsenide monolithic microwave integrated circuit (MMIC) is a developing technology that will play a key role in the military and commercial microwave systems of the future. In the microwave industry, microwave integrated circuits (MICs) have been used to refer to circuits where active and passive discrete components such as transistors, and thin- or thick-film chip capacitors, inductors, and resistors are connected on a dielectric substrate. In MMICs, all components, active and passive, are fabricated on a common semi-insulating substrate, thereby eliminating the need for attachment of discrete components. The MMIC offers considerable advantage over its predecessors in many applications. Perhaps the most appealing characteristic of MMICs is their ability to be produced in large volume for very little added cost. MMICs offer the benefits of low cost, improved performance, small size, lower weight, and high reliability.

With MMIC technology maturing rapidly and becoming available through various foundries, there has been a surge in the design of MMICs. However, many of these MMIC designs, layouts, and topologies are based on traditional hybrid techniques. At present, the initial cost of developing a MMIC is considerably higher than that of an equivalent hybrid circuit. To reduce the development cost, it will be necessary to change the design philosophy from that of the hybrid, where circuits can be "tweeked" to obtain the desired performance. Appropriate design tools will play an important part in the success of any MMIC design. This text is developed to discuss all aspects of MMIC design, including material growth, processing, design considerations, design tools, MMIC-based subsystem design, testing, packaging, and reliability. Design automation tools in particular are emphasized at every step of the MMIC design process. Commercially available as well as public domain software tools are discussed wherever applicable.

Chapter 1 sets the background of the book, and provides MMIC application information and its present production status. Chapter 2 describes the conventional microwave fundamentals, but some aspects such as network parameters or noise analysis are discussed, keeping MMIC designs particularly in mind.

Chapter 3 details the material and processing aspects of MMICs and gives an example of a typical MMIC production process with detailed illustrations. Computer-aided process modeling is discussed at such length for the first time. Some guidelines are proposed for choosing a foundry for designing MMICs. In-process testing for yield enhancement is discussed in great detail. Such a chapter not only provides valuable process information for MMIC designers, whereby they can design the MMIC components more intelligently, but also can be helpful to process engineers. The computer-aided MMIC manufacturing environment is also discussed in this chapter.

Chapter 4 details the nonlinear, linear, and noise models—lumped as well as distributed—for both active and passive devices. The characteristics and models of transmission lines are described for their application in MMICs on GaAs as opposed to dielectric substrates. Practical limitations of all the components are described, and parasitics associated with these components are discussed in detail. The most up-to-date information is included in the discussion of single-gate FET modeling with an emphasis on an equivalent circuit model approach directly applicable to circuit simulation software. Design considerations for optimizing the component layout is also discussed. Dual-gate FETs, Schottky diodes, spiral inductors, MIM and interdigitated capacitors, monolithic and thin-film resistors, crossover and via hole modeling are also described. Most of the models discussed are valid up to 25 GHz.

Most of the published work to date in the area of MMICs does not discuss the step-by-step details of MMIC design. This is the first text to discuss design details including computer simulations, sensitivity analysis, and different design trade-offs at length for designing small-signal, power, low-noise, and general-purpose MMICs. Due to the lack of postfabrication tuning capability in MMICs, the importance of accurate device models and appropriate simulation tools is discussed. Chapter 5 starts with a discussion of general design considerations for MMICs and then details different design topologies applicable for designing various types of amplifiers in MMICs. Step-by-step design procedures are described for amplifiers such as 6–18 GHz wideband, 8.5–10.5 GHz power, a general-purpose gain block, low-noise and power combining techniques. Chapter 6 discusses the nonlinear and control circuit designs, providing design examples for the image rejection mixer, distributed mixer, phase shifter, attenuator, and switches. Such examples can serve as a guideline for designing MMICs to a first-time MMIC designer as well as an expert who is designing a particular new component.

Chapter 7 details the example of designing a complete MMIC-based subsystem. Not yet covered elsewhere in the literature, this includes the partitioning of subsystem to circuit level to achieve the optimum overall system performance. Production aspects of a MMIC subsystem from its manufacturing ability and testability points of view are discussed in detail.

Chapter 8 treats the design automation tools for MMIC design. The whole chapter keeps the workstation approach of the design environment in view. In addition to front-end design tools such as schematic capture, circuit simulation, layout, and back-end verification tools are discussed at varying length. The microwave circuit simulator discussion provides an overview of commercial products, as these are well established software tools for designing MICs or MMICs, and the nonlinear simulation section discusses the SPICE program in detail, from the user's point of view rather than dwelling on the algorithms for numerical analysis in the program. Emphasis is on explaining what SPICE can do for microwave MMIC design and what its limitations are. Nonetheless, the harmonic balance technique is described in detail, including the mathematical treatments, as this is the upcoming technique in MMIC and MIC design. Some easy-to-use techniques for making any generic version of SPICE useful for microwave MMIC device modeling and circuit design are described in Appendix 8A. Physical layout techniques, back-end design verification techniques to describe design rules checking, electrical rules checking, and layout *versus* schematic checking are described in detail. The mask fabrication techniques and the technological limitations are also treated in this chapter.

Chapters 9, 10, and 11 address the issues pertaining to the back-end of the MMIC manufacturing process. Chapter 9 discusses on-wafer MMIC testing at microwave frequencies. As this is a new and upcoming technology, a detailed treatment of its implications is provided in this chapter. Advantages of on-wafer testing such as elimination of de-embedding requirements for modeling and expensive labor-intensive assembly and testing of unyielding MMICs is described. MMIC design and layout considerations for on-wafer probing are discussed. Limitations of on-wafer MMIC testing are also covered. Trade-offs between detailed on-wafer testing *versus* final test yields are discussed.

Chapter 10 provides a detailed treatment of packaging, not only for those involved in designing microwave packages, but equally for those who want to use commercially available microwave packages. This chapter, for the first time in print, provides a comprehensive table of commercially available microwave packages, their characteristics, useful frequency range, their die-attach area, and their applications. Cost issues for MMIC packaging are also addressed.

Chapter 11 discusses the least investigated topic of the commercial MMIC environment. Most reliability studies to date have been considered proprietary. This chapter not only discusses the fundamentals of evaluating the reliability of MMICs, but also gives insight into the approach of correlating discrete component reliability with MMIC reliability. This approach is supported by commercial MMIC reliability test data. Electrostatic protection circuits for MMIC protection are discussed as well.

In summary, this book provides detailed information on MMIC technology and design, covering all aspects of the technology with an emphasis on computer-assisted design and processing. This text can be used by the practicing engineers as well as senior or graduate students specializing in microwave integrated circuit design (an appropriate prerequisite would be a course on electromagnetic field theory). This book contains enough background and advanced material for a full one-year or two-semester course with corresponding laboratory for hands-on experience. For a laboratory course, the college or university must be equipped with the circuit simulation, physical layout, and microwave testing capabilities. The first part of the course might consist of Chapters 1, 2, 3, 4, 5 (Sections 5.1 to 5.3), and 8, and the second part would comprise Chapters 5 (Sections 5.4 to 5.5), 6, 9, 10, and 11. Chapter 7 could be used for extra credit or independent study by the students. The first-semester laboratory course might involve learning the circuit simulation, layout tools, and microwave testing for developing device models. The second-semester laboratory course would consist of designing a simple amplifier or control circuit, physical layout of the circuit and its fabrication, either at an internal fabrication facility (if available) or an outside foundry (available at reasonable cost if several students shared the same mask), and final testing of the circuit. (Keep in mind that there is generally a time lag of eight to ten weeks before the fabricated circuits return from an outside foundry. The emphasis of laboratory exercise should be on teaching the students a typical MMIC design cycle rather than achieving the highest current performance levels.

I want to thank most warmly all of the authors who have contributed to this text and spent their valuable time to deliver the manuscript under a tight time schedule. I also thank George Gilbert and Charles Huang of Anadigics, Inc., who permitted me to use Anadigics process and design data.

Additionally, there are numerous experts in the industry and universities who have provided their valuable critical reviews and suggestions in the preparation of this text. I wish to thank all my present and former coworkers and colleagues at Anadigics, Avantek, and Honeywell, and also Ajit Rode (TriQuint Semiconductor), James G. Tenedorio (Harris Microwave), David W. Hughes (Georgia Institute of Technology), Krishan K. Agarwal (E-Systems), M. Mannan (National Semiconductor), Tony Sison (Silvar Lisco), Richard Dougherty (Motorola), John Debolt (PROMIS Systems), Jitendra Goel (TRW), Bob Wiederhold (Cadence), Ulrich Rhode (Compact Software), Peter Parrish and Octavius Pitzalis (EESof), Madhu Gupta (Hughes Research), Eric Strid (Cascade Microtech), Michael Shur (University of Virginia), John K. Lowell (AMD), R. S. Tomar (BNR), Charles Buntschuh (Narda), Ken Kundert (HP), Shashi Karna (IBM), Robert A. Weck (ERADC), Jeff Hantgan (AT&T), Ram Potukucni and H. L. Hung (COMSAT), Greg Boll (Pico Probe), Siegfried G. Knorr (Colby Instruments), S. S. Bharj (M/A COM), Bob Goldwasser (Alpha), V.K. Tripathi (University of Oregon), Martin Stiglitz (Microwave Journal), Barry Gilbert (Mayo Foundation), Satish Puri (AT&T), Bert D. Epler, Ping Yang (Texas Instruments), Jeff Buck (Plessey) and 300 or 400 of those engineers who have attended my seminars on designing MMICs using a foundry, whose feedback and questions have really helped to develop the framework of this book. Thanks to Mark Walsh and Dennis Ricci of Artech House, who have been very patient in accommodating my last-minute changes and additions to the text. Last, but not the least, I thank my parents, my brother Dr. Rakesh Goyal, and my sister Dr. Anita Bansal, who always inspired me to complete this task, and my wife, who has been very patient and supportive throughout the preparation of this book.

RAVENDER GOYAL
BRANCHBURG, NJ

# Chapter 1
# Introduction

*A. R. Jha, R. Goyal and B. Manz*

## 1.1 INTRODUCTION

The *gallium arsenide monolithic microwave integrated circuit* (GaAs MMIC) is a developing technology that will play a key role in the military [1] and commercial microwave systems [2] of the future. MMICs combine various active and passive circuit functions, such as amplification, switching, filtering, *et cetera* on a single substrate.

MMICs differ from *microwave integrated circuits* (MICs) [3]. In the microwave industry, the term *MIC* has been used to refer to planar microwave integrated circuits using one or more different forms of transmission lines, all characterized by their ability to be printed on a dielectric substrate. Active and passive discrete components such as transistors, thin or thick film chip capacitors, and resistors are then attached. This technology is known as hybrid, but more precisely as *hybrid MIC* (HMIC). In the MMIC, all components—passive and active devices—are fabricated on a common semiinsulating substrate, eliminating the need for attachment of discrete components.

Several types of planar transmission line and passive impedance matching circuit MMICs are available [4]. Microstrip circuitry is the type most popular because of its simplicity, relative ease of fabrication, and accessibility. Moreover, it is well described in microwave circuit analysis software. Much work has been done to characterize, analyze, and document various electrical parameters of microstrip circuits using soft and hard substrates. Modern microstrip design practices have become relatively well established and standardized, and they have been incorporated as standard elements in computer-aided microwave circuit design packages.

It is undoubtedly true that many circuit functions now available with MMICs would have been impossible to produce using conventional substrate-based hybrid technology. This is particularly true in the case of circuits requiring many different gate-width FETS. In many cases, however, hybrid integrated circuits could demonstrate concepts to be used in future MMIC designs. For example, a vector modulator using lumped-distributed components can be fabricated on alumina substrates where common-gate and common-source FETs are used in an active power divider, and FETs also are used in pi arrangements for attenuators. Such a hybrid circuit might measure $38 \times 25$ mm. The monolithic version measures $4.5 \times 3$ mm with performance comparable to that of its hybrid counterpart [3].

MMICs offer considerable advantages over their predecessors in many applications [5]. The MMIC is extremely small, compressing many separate functions onto a single substrate of either GaAs or Si. Since devices are not attached to the chip but rather fabricated on it, MMICs substantially reduce the severity of problems caused by interconnect bond wires. GaAs MMIC devices also are 100 to 1000 times more resistant to radiation than silicon devices [6]. Perhaps the most appealing characteristic of the MMIC is its ability to be produced in great quantities for very little added cost in high-volume production [7]. More than any other factor, the MMIC's claim to economical high-volume production has generated wild market projections for GaAs over the years. This heralded the immediate implementation of MMICs in virtually every conceivable microwave system. The most widely published projections were based on the use of MMICs in active antennas and phased-array radars, which would require many thousands of chips.

So far, most of these projections have been too optimistic [8] for the simple reason that technology always has difficulty keeping pace with market projections for it. However, this is not to say that the projections may never be. A historical view of MMIC development during the 1980s shows great strides, and widespread use may still come to pass [9,10]. Several good, high-volume MMIC applications already exist.

A typical example of what can be achieved by MMIC is the solid-state switch, one of the first circuits to be produced in volume by MMIC manufacturers. The MMIC switch uses GaAs FET devices, which have switching speeds less than 2 ns compared to the 10-ns speed of PIN-diode switches. The driver circuit is simple, and the bias power requirements are low compared to PIN-diode switches (350 $\mu$W *versus* 500 mW).

An entire X-band microwave receiver module [6], which includes local oscillator, mixer, low-noise amplifier, and IF amplifier, can be fabricated on a single monolithic chip with dimensions of 0.25 by 0.25 in. This means that 50 such receivers can be fabricated on a single 2 in. diameter wafer. A run of 1000 wafers can be manufactured for $1000 or less per wafer. Thus, the unit cost for our sample receiver will be about \$20 based on a 50% yield.

Missing in this scenario is the design cost, which presently is very high. The cost to run a design through a GaAs foundry is perhaps \$50,000 and allows no room for modification. That is, MMICs are true integrated circuits and, as such, cannot be widely "tweaked" like their hybrid predecessors to meet performance specifications. The design is either right or wrong. If it is wrong, the design must be tweaked, rather than the physical circuit itself, and another pass made through the foundry.

Thus, the design team must produce a circuit that meets the performance specifications the first time through the process. Many variables have to be considered in RF circuits that simply do not exist in their low-frequency analog and digital counterparts. RF circuits always have been difficult to model and predict, and the difficulty in accommodating these peculiarities increases with frequency. However, as more MMIC designs are created and designers increase their database of different MMIC designs, the chances of producing correct designs in few iterations becomes more feasible, ultimately achieving first-time design correctness. Also, the answer, at least in part, lies in the power of the computer-aided design to accurately optimize circuit performance. However, since no computer is any better than the algorithms it uses, first-time design success performance depends heavily on the accuracy of models. And for some microwave circuit elements, accurate models at high frequencies either do not exist or cannot produce good circuits that meet specifications.

On the brighter side, when these difficulties are surmounted, MMICs will bring the microwave industry a revolution in design efficiency, cost, miniaturization, system integration [11], electrical performance, and overall capability never seen in the world of RF [12]. Imagine extremely dense, high-frequency RF circuits on the same chip as digital circuits [13,14], providing complete system functions in the space formerly occupied by a single component. This no doubt will occur, but there is much yet to achieve. The first MMICs were introduced in 1974 employing MESFETS [15]. In 1985, the industry's first 2 to 18 GHz GaAs MMIC distributed amplifier was introduced. More than 15 different types of Si and GaAs MMICs presently are in production.

Successful implementation of an MMIC depends on life-cycle costs, performance requirements, operating frequency, system application, functional capabilities, and reliability-maintainability requirements. Microwave systems operating in UHF and L-band regions favor Si MMIC technology [16] based on the parameters shown in Table 1.1. However, GaAs is the material of choice above these frequencies.

**Table 1.1** Performance Comparison between Si MMIC and GaAs MMIC Device Technologies

| Parameter | Si MMIC | GaAs MMIC |
|---|---|---|
| Typical frequency range | dc to 3 GHz | 2 to 40 GHz |
| Noise figure | Higher | Lower |
| Radiation hardening feature | More susceptible | Less susceptible |
| Gain per stage over same frequency range | Lower | Higher |
| Structure type | Vertical structure device | Planar device |
| Instantaneous bandwidth | One octave or two | Multioctave |
| Optimum performance band | HF, UHF, L-band | S-band and above |
| Tweak-free design | Yes | Yes |
| Type of devices employed | a. Vertical bipolar<br>b. Planar junction FET (JFET) | Planar GaAs MESFET |
| Power added efficiency | a. Much better at UHF, L-band<br>b. Poor at S-band and higher | a. Poor at L-Band<br>b. Higher at 3 GHz and above |

Si MMIC technology is used widely in the design of low-frequency Darlington transistors, high-power silicon bipolar npn transistors, frequency chopper transistors, and other applications in which low noise and high power are principal requirements. GaAs MMICs are used widely in extremely wideband, low-power, high-speed switches (speed less than 2 ns), low-noise, multioctave RF amplifiers, precision electronic control devices, high-power, high-efficiency driver amplifiers, T/R modules for active phased-array radars, stable amplifiers for communication equipment, and high-speed digital circuits. MMICs using high-power FET devices presently yield CW power of 10 W at 10 GHz and 1.5 W at 20 GHz. Integration of GaAs and Si technologies [17] also has been demonstrated, as illustrated in Figure 1.1.

Most early MMIC circuits were extensions of hybrid MIC-based design concepts. As opposed to MICs, MMICs have minimal tuning capability. Monolithic circuits should not be designed as extensions of hybrid technology. Creating interchangeable parts with hybrid elements does not offer a long-term, cost-effective solution to system needs, because hybrid system architectures do not take advantage of the inherent abilities of MMICs.

The use of lumped-element techniques create new opportunities for MMIC designs. This technology has reduced the cost of a simple MMIC circuit from $1000 to $50 (Table 1.1) [18]. Although reduced chip cost is an important benefit of

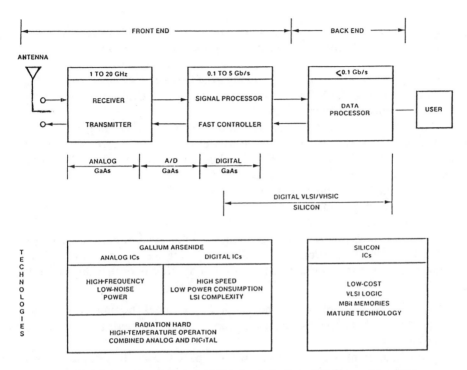

**Fig. 1.1** Optimal implementation of GaAs and Si technologies in an RF–fiber optics system.

lumped-element design, the technique has other advantages. In wideband circuits, it is difficult to realize good chokes with distributed elements. Wideband double-balanced diode mixers are easier to realize with lumped transformers. These transformers require detailed and accurate modeling of parasitic components above about 20 GHz. The lack of models that include the effect of parasitics has been a primary obstacle in using the lumped-element technique. Lumped-element design can be used extensively up to 20 GHz, whereas distributed-element design will be useful at Ku-band frequencies and above. Distributed elements also may be useful in applications requiring the lowest losses; that is, in low noise and power amplifiers where the higher losses of lumped elements may degrade performance.

## 1.2 WHY MMICs?

There are two principal reasons why GaAs is the choice for microwave ICs: The compound has five times greater electron mobility than silicon, and its semiinsulating properties allow natural isolation of individual devices without isolation barriers [19]. Only one-fifth the voltage is needed to speed electrons along in gallium arsenide as in Si. The fivefold greater electron mobility of GaAs results in faster operation with lower power consumption. Since its hole mobility is much lower than its electron mobility, a significant loss of speed would result from the use of complementary designs. The bandgap of GaAs is 1.43 eV at room temperature, compared to 1.12 eV for silicon, which means that GaAs is capable of higher temperature operation than silicon.

The electron mobility and saturated drift velocity of GaAs is higher than in correspondingly doped silicon. Thus, for an electric field of given strength, electrons move more rapidly in GaAs than in silicon, yielding devices that can operate at higher frequencies. However, because of doping levels required in active device channel in GaAs and prominent short-channel effects, its figure of merit is only twice that of silicon.

MMICs offer all the same benefits to the microwave system designers that silicon VLSIs have given to low-frequency digital system designers: low cost, improved electrical performance, small size, lower weight, and high reliability. The main reasons for improved performance of MMICs over their hybrid counterparts are as follows:

- Absence of assembly interconnects greatly reduces the unnecessary parasitics due to bond wires that significantly limit the performance of hybrid circuits.
- Individual components, both active and passive, can be optimally designed to meet the needs of the circuit performance. The designer is not limited to the narrow range of catalog-discrete components.
- Besides the yield dependence on chip size, an MMIC containing many active and passive devices costs no more to fabricate than one containing few. In fact, in MMICs, the active devices are cheaper than the distributed transmission lines, which is quite contrary to hybrids where the cost of the circuit significantly depends on the number of active devices used.
- Bond wires are the biggest reliability hazard in hybrids, which is minimized in MMICs by using on-chip interconnect lines.
- Using on-wafer MMIC testing techniques, the overall testing cost can be reduced significantly compared to that for hybrids. Table 1.2 [20] provides a comparison of the advantages and disadvantages of hybrid vs MMIC technology.

Further advancement in MMIC technology can readily integrate MESFET, MODFET, HEMT, and pseudomorphic devices, depending upon the microwave system performance requirements [21]. The introduction of AlGaAs heterostructures has improved basic MESFET performance. *High-electron-mobility transistors* (HEMTs) and *modulation-doped FETS* (MODFETs) have achieved performance levels thought impossible only a few years ago. Submicron MESFETs (0.25 $\mu$m or smaller), HEMTs, or pseudomorphic HEMTs are suitable for millimeter-wave systems. HEMTs have excellent noise performance and are fine devices for power applications as well. HEMT devices have demonstrated low-noise and medium-power performance from 10 to 100 GHz. A new MMIC development combines MESFETs and HEMTs. These MMICs achieve higher performance levels in critical applications. Alternatively, greater performance margins allow relaxation of process requirements and gate length to increase both producibility and yields.

Implementation of GaAs MMIC technology at millimeter-wave frequencies is critically important for the next-generation of millimeter-wave systems including radars, radiometers, communication equipment, and "smart" weapons [22]. MODFETs, HEMTs, and pseudomorphic HEMTs are the devices that will make it happen.

## 1.3 PROCESSING

The *molecular beam epitaxy* (MBE) processing method has been used for the growth of MESFETs, MODFETs, and pseudomorphic MODFETs for high performance millimeter-wave applications. However, in its current form, MBE [23] is not suitable for volume-production applications because of the high costs of equipment and the limited capability of the machines currently available.

*Metal-organic chemical vapor deposition* (MOCVD) is an epitaxial growth process well-suited for high-rate, low-cost production; 50 to 80 wafers per day can be produced by a single machine, and the capital investment is significantly lower than that for MBE. Microwave and millimeter-wave MODFETs fabricated by using low-pressure MOCVD have excellent gain and noise performance at millimeter-wave frequencies. However, the key to affordable, high-performance MMICs will be the availability of consistently uniform, defect-free AlGaAs-GaAs wafers. The epitaxial growth technique presently is performed by MBE. Other techniques, such as *metal organic molecular beam epitaxy* (MOMBE) and MOCVD are being explored to significantly improve cost and performance in HEMT-based MMICs.

GaAs MMICs depend on the high-resistivity of the semiinsulating substrate material for passive components and the high mobility of the semiconducting layer that is either epitaxially grown onto the SI substrate or implanted into the SI substrate. The active device most commonly used in MMIC is the MESFET. Recently, varactor diodes also have been incorporated into MMICs for VCO applications. A typical MMIC process consists of implanting donors or growing epitaxial layers for the active devices, the former giving improved uniformity of characteristics. Silicon is usually employed for the implant species. Capacitors are formed using either interdigital or *metal-insulator-metal* (MIM) structures, whereas resistors are formed using thin films of metals.

A dielectric layer, such as silicon nitride or polyimide, separates the metal layers. Two methods are commonly employed to produce thick metalization, either to carry current or reduce skin depth losses; namely, plated gold and thick evaporated gold. Interdigital capacitors and spiral inductances are used widely by MMICs to reduce the chip area. Of particular importance in GaAs ICs operating at microwave frequencies is an effective, low-inductance method of connecting components (particularly common-source FETs) to the RF ground. For monolithic circuits at low frequencies, where the FETs can be located near to the chip edges, fine gold mesh has been used to connect the circuit ground areas to the package ground.

There is a definite relationship between microstrip loss and GaAs substrate thickness in MMICs. Thus, if the GaAs substrate is thinned, the microstrip loss will increase, particularly at higher frequencies. Thin GaAs substrates are required to achieve effective heat-sinking of power FET monolithic amplifiers because of the high thermal impedance of the material. Thin substrates in this case are undesirable, however, as high-Q circuit components are needed to transform the low FET impedances to 50 $\Omega$ without excessive losses.

## 1.4 MMIC PERFORMANCE

The power-handling capability of FETs depends on the overall gate periphery, source inductance and resistance, and the ability to dissipate device heat. To achieve the first requirement, multifinger MESFETs are used. As the overall width of the device cell becomes large, it is increasingly difficult to maintain phase and amplitude balance over the device bandwidth. This problem can be alleviated if several MESFET cells are combined, each containing a number of gate stripes. Paralleling cells in a power FET decreases the device impedance to the point at which it becomes difficult to match. Since microstrip impedances range from about 10 to 140 $\Omega$, matching into power FETs is difficult in conventional hybrid circuits, particularly over wide bandwidths. In MMICs, thin-film capacitors are used.

However, low device impedances result in the need for large capacitance values and, thus, large areas and very thin dielectric layers. Both of these conditions are highly undesirable. In these cases, the *miniaturized hybrid MIC* (MHMIC) technique may prove to be of value in permitting off-circuit capacitors to be produced independent of the power FET. Power amplifiers using FETs [24, p. 162] are capable of achieving power-added efficiencies greater than 40%. Even this result leaves almost 60% of the dc power to be dissipated, largely through the FET, underlining the importance of good heat-sinking for power amplifiers.

Table 1.2 [20] compares MMIC and hybrid amplifiers. HMIC components achieving much more than octave bandwidths are not common. However, appli-

Table 1.2 Comparison of Hybrids and MMICs

| | Hybrids | MMICs |
|---|---|---|
| Advantages | Low-cost substrates | Small size and weight |
| | Lower costs in general | Improved reproducibility and reliability |
| | Easily repaired | Broader frequency band performance |
| | Testing and tuning easy | Low cost in large quantity |
| | Wide variety of elements | Circuit design flexibility |
| | Lower transmission line loss | |
| | High Q performance | |
| Disadvantages | Lower reliability | Higher transmission line losses |
| | Limited bandwidth | No circuit tuning |
| | Uncontrolled parasitics | Undesired RF coupling |
| | Limited frequency range | Poor heat sinking |
| | Larger size | High parasitic capacitance to ground |
| | More parts | High materials costs |
| | High assembly costs | Immature process technology |
| | | High equipment costs |
| | | Limited element values |

cation of feedback techniques shows great promise for increased bandwidth performance.

A significant development made possible by the small size of MMICs is the distributed amplifier. This amplifier, which makes use of the MESFET parasitic elements to realize a transmission line structure, produces high gain for the number of FETs used, and is not efficient with respect to consumption of dc power or chip area compared to its hybrid counterpart. However, the performance is still not satisfactory for its application in phased-arrays. Performance comparisons of hybrid and MMIC based amplifiers are shown in Tables 1.3 and 1.4 [3;24]. Current performance data on GaAs MMIC components are summarized in Tables 1.5 through 1.7.

The MMIC generally is able to achieve wider bandwidth and slightly more gain than its HMIC equivalent. However, the hybrid provides higher output power and lower noise with some circuit tuning. A large number of discrete components, wire bonds, and individual substrates are required by the HMIC,

**Table 1.3** Comparison of Hybrid and Monolithic High-Gain Distributed Amplifiers

| *Parameter* | *Details* |
|---|---|
| *Hybrid* | |
| Frequency | 8–18 GHz |
| Noise figure | 4.5 dB typical |
| Gain | 40 ± 2 dB |
| No. of wire bonds | 400 |
| No. of FETs | 16 |
| No. of substrates | 16 |
| No. of carriers | 8 |
| No. of chip resistors (DC and RF) | 32 |
| No. of chip capacitors | 40 |
| Size | Approx. 64 × 8 × 4mm (unpackaged) |
| *Monolithic* (using substrate through via technology and alumina Lange couplers) | |
| Frequency | 7.5–18.5 GHz |
| Noise figure | 5.2 dB typical |
| Gain | 57 ± 1.5 dB |
| No. of wire bonds | 14 |
| No. of FETs | 16 |
| No. of GaAs chips | 5 |
| No. of carriers | 1 |
| Size | 19.5 × 9.5 × 5mm (packaged) |

**Table 1.4** Comparison of Hybrid and Monolithic Amplifier Features

| *Feature* | *HMIC* | *MHMIC** | *MMIC* |
|---|---|---|---|
| Frequency (GHz) | 8–18 | 8–18 | 7.5–18.5 |
| Noise figure (dB) | 4.5 | 4.7 | 5.2 |
| Gain (dB) | 40 ± 2 | — | 57 ± 1.5 |

**Table 1.4** cont.

| Feature | HMIC | MHMIC* | MMIC |
|---------|------|--------|------|
| FETs | 16 | 16 | 16 |
| Substrates | 16 | 9 | 5 |
| Carriers | 8 | 4 | 1 |
| Discrete passive chips | 72 | — | — |
| Wire bonds | 400 | 200 | 14 |
| Fabrication steps | 1 to 2 per substrate | 7–10 | 10–14 |
| Mask expenses | Low | Moderate | High |
| Design iteration times | 1 month | 2–3 months | 3–6 months |
| Manual/mechanical activity | All active and passive chips, and I/Os and tuning | Active chips and I/Os | I/Os only |
| Circuits produced per cycle | 1 | 10s–100s | 100s–1000s |

*Estimated

suggesting a large number of manual, labor-intensive assembly steps. The MMIC reduces this to one substrate and a very small number of wire bonds. The miniature HMIC falls between these two, since the majority of wire bonds and discrete passive chips is eliminated.

MMICs and HMICs require several fabrication and process steps to realize representative amplifiers. The conventional HMIC requires only one or two steps, although this process may have to be repeated for each substrate carrier (16 in the example shown in Table 1.3). The quality and line definition of masks required for HMICs are not demanding or expensive, whereas complete mask sets for MMICs may cost more than $10,000 per iteration.

Due to the specialized equipment needed to produce them, the mask sets often contribute substantially to overall MMIC design iteration times: three to six months compared to about one month for conventional HMICs. These costs underline the importance of accurate device electrical models and computer analysis and optimization routines for MMICs. However, at the completion of the various fabrication steps, the MMIC may yield several thousand complete amplifiers, whereas the HMIC would yield only one.

The cost of any microwave circuit or component depends on five critical factors: yield, size, material, production volume, and automation. Hybrid circuits tend to be cheaper in their initial development but usually undergo expensive

**Table 1.5 Performance Data on Broadband Amplifiers for ECM/ESM Applications**

| Parameters | Broadband Amplifier Number | | | | | | |
|---|---|---|---|---|---|---|---|
| | #1 | #2 | #3 | #4 | #5 | #6 | #7 |
| Frequency range (GHz) | 0.1–12 | 0.050–4 | 2–18 | 6–18 | 2–21 | 2–21 | 3–40 |
| Power output at 1 dB compression (dBm) | 10 | 14 | 13 | 30 | 18 | 20 | 15 |
| Gain (dB) | 16 | 28 | 11 | 30 | 12 | 9 | 6 |
| Noise (dB) | 4.5 | 3.8 | 7.5 | 8.5 | 3.0 | 5 | 4 |
| Application | ESM | ESM | ECM | ECM | ECM | ECM | ECM |
| Device type | Monolithic MESFET | Monolithic MESFET | Monolithic MESFET | Monolithic MESFET | Monolithic HEMT | Monolithic MESFET | Monolithic HEMT |

**Table 1.6** Performance Data on Narrowband GaAs MMIC Power Amplifiers for Satellite and Point-to-Point Communication Applications

| | *Narrowband Amplifier Number* | | |
| --- | --- | --- | --- |
| *Parameter* | *#1* | *#2* | *#3* |
| Frequency range (GHz) | 3.7–4.2 | 10.7–11.7 | 14.0–14.5 |
| Power output (W) | 2 | 12.5 | 12.5 |
| Gain (dB) | 30 | 5.6 | 5.0 |
| Power-added efficiency (%) | 65 | 25 | 21 |
| AM/PM (deg/dB) | 0.25 | 0.50 | 0.75 |

**Table 1.7** Performance Data on GaAs MMIC Microwave Switches for Broadband System Applications

| | *MMIC Switch Number* | |
| --- | --- | --- |
| *Parameter* | *#1* | *#2* |
| Frequency range (GHz) | 2–18 | dc–20 |
| Isolation (dB) | 80 (min) | 35 (min) |
| Insertion loss (dB) | 2.3 (max) | 1.6 (max) |
| Input VSWR | 1.6 : 1 (max) | 1.5 : 1 (max) |

preproduction testing for reliability and reproducibility of performance. The overall cost to produce an MMIC module includes materials, packaging, processing, assembly, and testing-inspection. Relative costs for phase shifters, high-power amplifiers, and low-noise amplifiers using hybrid and MMIC technologies are shown in Figure 1.2 [25].

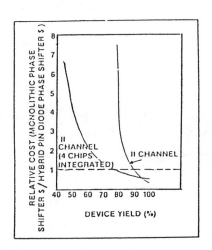

a. Phase Shifter: Device yield of 1.2 mm power FET with a gate length of 1.0–1.2 $\mu$m.

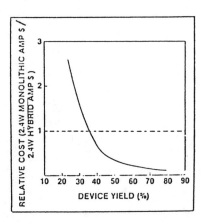

b. High-Power Amplifier: Device yield of 2.4 mm power FET with a gate width of 0.8 $\mu$m.

**Fig. 1.2 Relative costs for microwave components using monolithic and hybrid technologies.**

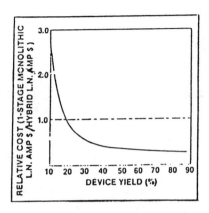

c. Low-Noise Amplifier: Device yield of L.N. 300 $\mu$m FET with a gate length of 0.7 $\mu$m.

## 1.5 MMIC STATUS

A single GaAs power FET is capable of delivering a CW power output of 15 W at 10 GHz and about 1.5 W at 20 GHz. However, power levels are reduced significantly (400 mW from 2 to 18 GHz) in operation over multiple octaves. GaAs devices produced from wafers grown by MBE have demonstrated power-added efficiencies greater than 60% and *maximum available gain* (MAG) greater than 11 dB. Low-noise, submicron GaAs devices now available are capable of yielding overall amplifier noise of less than 5.5 dB from 2 to 18 GHz with amplifier gain as high as 30 dB [26–30].

The noise for an individual device is well below 2 dB from 2 to 18 GHz and over temperature ranges from $-55°$C to 100°C. Low-noise amplifiers using MMIC devices exhibit gain flatness less than $\pm1$ dB, AM/PM conversion less than 0.5° per dB, amplitude tracking of $\pm0.5$ dB, and phase tracking of $\pm5°$. Solid-state switches using GaAs FETs offer superior RF performance over extended frequency ranges. GaAs MMIC RF switches can provide isolation greater than 45 dB, insertion loss less than 1.5 dB, switching speeds less than 2 ns, and VSWR less than 2 : 1 from dc to 20 GHz.

### 1.5.1 GaAs MMIC Reliability

Reliability data on GaAs MMICs are not readily available because their implementation in microwave systems is quite recent. Reliability predictions are based on the thermal efficiency of the substrate, packaging configuration, wire bond, die attachment, and electrostatic-discharge protection. The failure rate of GaAs FET devices (or any other active device) exceeds that of any passive component, such as capacitors or resistors. Reliability analysis of power FETs indicates a MTBF of $10^6$ hours or more at a junction temperature of 150°C [31].

The MTBF for digital GaAs ICs is about $5 \times 10^6$ hours at a junction temperature of 100°C. Reliability data on FETs indicate that the MTBF varies from $2 \times 10^7$ hours for a Si FET to $0.5 \times 10^6$ hours for a low-noise GaAs MESFET at a junction temperature of 125°C.

### 1.5.2 Yield

The term *yield* refers to the number of circuits on a given wafer that deliver acceptable electrical performance. The figure is usually given as a percent of the total devices on the wafer. It is never desirable to compare the yield from one foundry to another, unless the conditions under which the number was generated are precisely given. This is because yield varies dramatically with the level of testing performed. For example, if yield is given after DC test, the value is likely to be quite high, since the test determined only that the device performs its desired function (e.g. amplifies).

If RF testing is performed as well as DC testing, yield will be lower. If testing is conducted after the device is diced and packaged, yield will be lower yet. Consequently, it is not surprising to see yield percentages scattered about the range of 40 to 90 percent, seldom with any qualifier, such as "at the wafer level" attached. In short, when comparing yield, it is important to ask at what stage of production the yield was measured, and what parameters were used to determine it.

Nevertheless, the yield of GaAs MMICs has risen from single-digit levels after RF testing and packaging in the early years, to perhaps 50 percent today. Yield after DC testing is very high, in some cases more than 90 percent, a figure very nearly that of silicon ICs. This is a significant achievement in a short time, since silicon ICs have been produced for far longer than GaAs ICs. The process of improving yield is inherently a long one, with incremental gains achieved after subtle changes are made to the fabrication process. Considering how far GaAs foundries have come already, it is safe to predict that further improvements, while likely to come at a slower rate, will nevertheless continue to appear.

### 1.5.3 Cost

A variety of processing and growth obstacles hinder high-volume production of MMICs. The most challenging problem is that of large-wafer bowing. The prices for wideband MMIC amplifier gain blocks covering 1 to 5 GHz, 2 to 8 GHz, and 6 to 18 GHz are about $30, $45, and $100, respectively, in quantities of 1000.

Overall amplifier cost depends upon output power, gain, flatness, and noise figure requirements. GaAs MMIC switches and electronic variable attenuators capable of operating from dc to 12 GHz with a maximum insertion loss of 1 dB are available at less than $60 each in quantities of 1000 or more.

The price of merchant-market MMICs will unquestionably come down in direct proportion to the number of devices the markets require, and the number of companies that offer them. Perhaps the only comparison that can be made is that of the discrete GaAs FET and GaAs FET amplifier markets, which today are the most fiercely competitive in the industry.

Here, price is significant, and in some cases the only parameter that sets apart one device from another. Consequently, the price of "catalog" FETs and amplifiers has dropped by more than 50 percent in recent years, and in some segments of the market continues to drop. At this writing, these are buyer's markets.

This is not to say the entire MMIC market will follow. However, many MMIC products today compete for market share in the low-noise amplifier market. Performance, price and delivery determine the victor. There are large segments of the MMIC market that are as yet undeveloped, such as phased-array radar, navigation receivers, and various consumer electronics items. As these applications become more clearly focused and the MMICs that serve them appear in larger numbers, it seems likely that a marketplace similar to that for amplifiers will develop. Of course, only time will tell.

### 1.6 FUTURE DEVELOPMENTS

Relatively low thermal efficiency and poor mechanical integrity are the basic limitations of GaAs MMICs. In applications where high mechanical integrity and large power dissipations are the principal requirements, silicon MMICs or GaAs-on-silicon ICs will be found more suitable. GaAs-on-Si shows great promise as a low-cost, high strength, "surrogate" substrate for GaAs devices.

Integration of the element aluminum or indium in the GaAs compound has improved gain and noise performance, particularly at millimeter-wave frequencies. High-electron mobility transistors (HEMTs) yield high gain and low noise over extended frequency ranges and are well suited for ultrabroadband applications due to their large gain-bandwidth product and large noise bandwidth. HEMT device research includes investigation of new materials such as InGaAs and reduction of gate length in the 0.1 to 0.2 $\mu$m range. Future development will probably concentrate on HEMT-based MMIC [32] devices with silicon nitride passivation for scratch protection, titanium-platinum-gold (Ti-Pt-Au) metallization, and advanced fabrication techniques for improved performance, reliability, and thermal aspects.

HEMT and pseudomorphic HEMT MMIC devices (Figure 1.3) yield higher cutoff frequency and transconductance with minimum noise figure in the millimeter-wave region (35 GHz and above). The device configuration is such that it readily can be incorporated into either monolithic or hybrid microwave and millimeter-wave integrated circuits. Computed power-added efficiency and important performance parameters for various millimeter-wave devices are shown in Tables 1.8 and 1.9 [33].

## 1.7 MMIC APPLICATIONS

### 1.7.1 Military Applications

GaAs MMIC technology has significant military applications [34], such as in electronic warfare, radar, phased arrays, and smart munitions, where speed, reliability, and high-frequency operation are essential [35, 36]. The trend in advanced military radar design [37] is toward wider operating bandwidth, rapid and accurate tracking of multiple targets, protection in harsh EW environments [38], radiation hardness, and high reliability. A particularly important approach to meeting these requirements is the active-element phased-array radar employing antennas with thousands of individual *transmit-receive* (T/R) modules [39]. GaAs MMIC technology offers the only path to affordability for such radar systems.

Physical and electrical parameters of X-band T/R modules using various circuit technologies are shown in Table 1.10. Airborne active phased-array radars can employ as many as 3000 T/R modules (Figure 1.4). Each T/R module uses only four MMIC chips and consumes minimum prime power. The number of MMIC chips required per module is a function of RF power output and the level of chip integration. For a typical X-band MMIC chip (for example, a two-stage power amplifier) processed from a 3-inch wafer of GaAs capable of yielding 32 chips using today's technology, a good estimate of wafer processing cost would be $6000 for the wafer. This amounts to $186 per chip.

GaAs MMIC technology also offers significant performance improvements in the components used by communication systems, telecommunication equipment, and high-data-rate fiber optic links. Availability of MMIC amplifiers with low noise, high gain, maximum gain flatness, and excellent AM/PM performance will significantly improve overall performance.

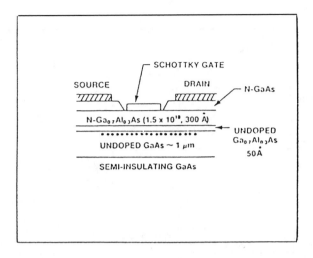

a. HEMT Device.

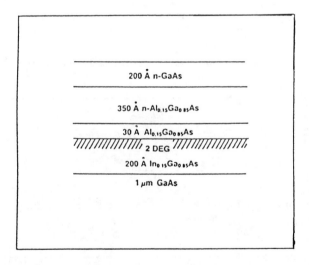

b. Pseudomorphic HEMT Device.

**Fig. 1.3** Epitaxial layer structure for single heterojunction HEMT and pseudomorphic HEMT devices and typical gain and noise performance levels for a HEMT device.

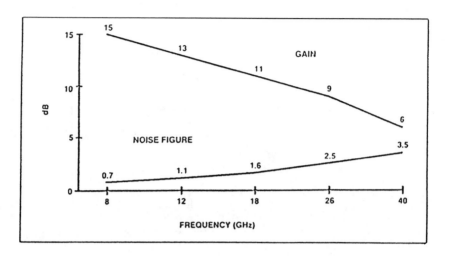

c. Gain and noise figure vs frequency for discrete HEMT device (150 × 0.3 μm gate periphery).

**Fig. 1.3** cont.

**Table 1.8** Power-Added Efficiencies for Various Devices

| | Device Type | | |
| --- | --- | --- | --- |
| Frequency (GHz) | FET | HEMT | Pseudomorphic HEMT |
| 18 | 38 | 44 | 59 |
| 35 | 30 | 34 | 42 |
| 60 | 12 | 16 | 28 |
| 95 | — | 5 | 18 |

**Table 1.9** Performance Comparison for Various MM-Wave Solid-State Devices

| | *Device Type and Gate Length* | | |
| Parameters Assumed | FET<br>(0.25 micron) | HEMT<br>(0.25 micron) | Pseudomorphic HEMT<br>(0.25 micron) |
|---|---|---|---|
| Cutoff frequency $f_T$ (GHz) | 108 | 170 | 230 |
| Minimum noise figure (dB) | | | |
|   18 GHz | 1.3 | 0.91 | 0.85 |
|   35 GHz | 2.40 | 1.75 | 1.55 |
|   60 GHz | 3.55 | 2.54 | 2.35 |
|   94 GHz | 4.90 | 3.50 | 3.30 |
| Maximum available gain (dB) | | | |
|   18 GHz | 15.56 | 19.50 | 22.13 |
|   35 GHz | 9.79 | 13.73 | 16.35 |
|   60 GHz | 5.11 | 9.05 | 11.67 |
|   94 GHz | 1.21 | 5.15 | 7.77 |
|   143 GHz | — | 1.50 | 4.13 |
| Power-added efficiency at 60 GHz (%) | 12 | 16 | 28 (min)<br>38 (max) |
| Saturation electron velocity | $1.7 \times 10^7$ | $2.7 \times 10^7$ | $3.6 \times 10^7$ |

Implementation of GaAs ICs (analog and digital) in fiber optic systems offers data rates greater than 1 Gb/s. High-speed GaAs logic circuits, true time-division multiplexers driven with complementary clocks, and reliable clock-recovery circuits have made significant contributions to fiber-optic communications. Digital processing circuits perform the necessary functions, such as synchronization of data streams, encoding and decoding, and framing of data.

Satellite communication receivers are optimized to give the lowest possible noise figure. A possible configuration in a satellite communication receiver at present would be a discrete MESFET as the first stage in a hybrid MIC, giving low noise with moderate gain, followed by an MMIC high-gain block. The major interest in GaAs ICs for EW systems is in the analog domain. EW systems by

**Table 1.10 Current Performance Parameters on X-band RF Modules for Application in Active Array Radar**

| Parameters | Company A | Company B | Company C | Company D |
|---|---|---|---|---|
| T/R module dimensions (in.) | $3.35 \times 1.57 \times 0.40$ | $4.80 \times 1.33 \times 0.27$ | $4.50 \times 1.50 \times 0.50$ | $4.75 \times 1.65 \times 0.60$ |
| Weight (oz.) | 2.0 | 1.75 | 1.6 | 3.0 |
| Peak power output (W) | 2.5 | 4 | 2.5 | 8.0 |
| Duty (%) | 15 | 50 | 25 | 50 |
| Phase shifter type | 5-bit diode phase shifter | 5-bit diode phase shifter | 5-bit diode phase shifter | 7-bit ferrite phase shifter |
| Circuit technology | Monolithic | Monolithic | Monolithic | Hybrid |

*Remarks*: 1. Switching speeds are in the order of 50 ns for diode phase shifters and 10 to 50 ms for ferrite phase shifters.
2. High power design requires a corporate feed and the array elements are fed by a single high power, high efficiency TWTA.
3. Monolithic circuitry offers substantial reduction in weight and size, but at the expense of performance.

nature are wideband. A typical *electronic support measures* (ESM) radar warning receiver normally covers the frequency band from below 1 GHz up to 18 GHz, with extensions of up to 40 GHz envisaged by about 1990 and to 100 GHz by about 2000.

Monolithic distributed amplifiers using MMIC chips can be cascaded easily for extremely wideband amplifiers with high linear gain. A typical three-stage MMIC amplifier has a nominal gain of 30 dB, maximum noise figure of 8 dB, and minimum output power of 100 mW over a 1–20 GHz range. This type of MMIC amplifier has a low current requirement and offers gain flatness of ±2 dB from 1 to 20 GHz. Such amplifiers require minimum primary power, thereby making them ideal for ECM, ECCM, ESM, ELINT, and expendable jammer [40] applications. For expendable jammers, low cost, small size, minimum weight, and high dc-to-RF efficiency are of paramount importance. High power cascadable GaAs MMICs appear to be best suited for such applications [40].

### 1.7.2 Commercial Applications

GaAs ICs have much to offer high-frequency instrumentation, where wideband operation is demanded. Requirements range from UHF to millimeter-wave frequencies and include analog and digital functions. Instruments that could benefit from the use of GaAs ICs include signal generators, frequency sweep generators, frequency synthesizers, medical equipment, spectrum analyzers, frequency meters, and network analyzers, as well as instruments involving waveform sampling and digitization. Digital ICs providing programmable division, prescaling, and binary counting are finding application in counters and timers.

The RF receiver front end for consumer *direct broadcast satellite* (DBS) receivers implies a large market volume at very low cost. Most current experimental systems and early production versions make use of hybrid MIC technology, but several companies have single-chip MMIC front ends in advanced development. *Very small aperture terminals* (VSATs) for business applications also offer a potentially enormous future market for MMICs.

Advances in lightwave communications include long-distance networks with repeater spacings in excess of 40 km (trunk) and shorter *local area networks* (LANs). The optical fibers feed remote subscriber carrier terminals and remote switches from central offices and provide direct lines to business customers.

There are many prospects for fiber optic interconnections between MMICs: from chip to chip, board to board, and system to system. Examples are interconnection between Si subsystem components at Gb/s data transmission rates using GaAs optoelectronic transceivers, interconnection between GaAs processor chips at 5 Gb/s using GaAs transceivers, and interconnection for analog-to-digital (A/D) conversion. Main circuits based on GaAs MESFET technology designed for fiber optic communication applications are laser driver, transimpedance amplifier, MUX, DMUX, and clock recovery circuits. It is also possible to integrate even the front-end laser diode using GaAs technology.

### 1.7.3 Consumer Applications

The conventional *surface acoustic wave* (SAW) filters currently used in television sets can be replaced by GaAs ICs, which would provide superior electrical performance with minimum cost and power drain. Implementation of low-noise MMIC amplifiers in color televisions will be cost effective, in addition to improving performance, enhancing reliability, and substantially reducing weight and size. GaAs MMICs have the potential of major volume-application in emerging *high definition television* (HDTV) receivers as well. Further, GaAs MMICs have potential applications in cellular radios, marine radar [41], and police radar because of their high reliability, low power consumption, and minimum cost. A summary of GaAs IC applications for present and future systems is provided in Table 1.11.

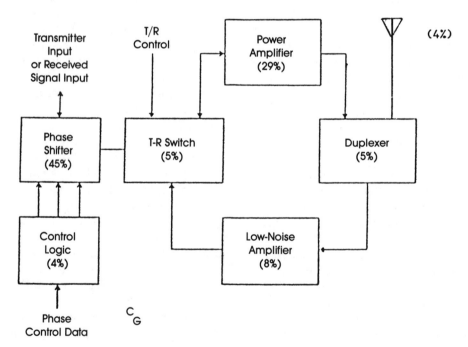

a. General layout of a T/R module with component cost breakdown (%).

**Fig. 1.4** Architecture of a typical module for radar application.

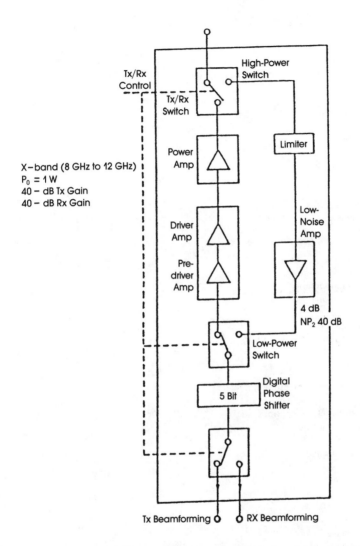

**b. Details of critical elements of X-band module.**

**Fig. 1.4** cont.

**Table 1.11** Selected GaAs IC Applications—Present and Future

Military
* Signal intelligence
* Radar
* Weapons targeting
* Image processing
* Wideband communications
* Antisubmarine search
* Electronic warfare
* Millimeter wave receivers
* Low-noise amplifiers
* Missile seekers
* Expendable decoys
* "Smart" munitions
* Fire control
* Ground-based communications
* Tactical communications
* Navigation
* Training and simulation
* Radiation-hard satellite communications

Communications
* Police radar
* Police radar warning equipment/commercial vehicle
* Cellular mobile-handheld telephones
* Local area networks (LANs)
* Intrusion-identification alarms
* Satellite communications
* Fiber optic repeaters
* Handheld transceivers
* Modems
* Multiplexers
* Switches
* Terrestrial microwave links
* CATV–Pay TV front-ends
* High definition TV receivers

Industrial and Instrumentation
* Computer-aided design & engineering
* Computer-aided testing
* Medical research/imaging
* Seismology
* Nuclear magnetic resonance (NMR)
* R & D simulation
* Robotics
* Machine vision
* Computer-aided tomography

**Table 1.11** cont.

- Plastics research
- Data acquisition subsystems
- Factory automation

Computers & Office
- Optical information storage/retrieval
- Local area networks
- Device/PC board/subassembly interconnection
- Simulation (for CAD-CAE-CAT)
- Image enhancement
- Array processing
- Digital signal processing
- High-resolution–real-time CRT displays
- Aircraft training simulators
- Speech, finger printing recognition (real-time)
- High-speed processing
- Supercomputers
- Minisupercomputers
- Mainframe memory

Consumer
- Automobile radar warning units
- Automobile collision avoidance radars
- TV tuners (completely monolithic)
- Cellular radios (privately owned)
- UHF and microwave amateur radio equipment
- Intrusion alarms (microwave and infrared)
- 3-D electronic cameras (GaAs CCD imagers)
- Voice actuated appliances
- Special purpose personal computers with 3-D imaging
- TVRO receivers (4–6 GHz)
- Direct broadcast satellite (DBS) receivers (11–12 GHz)
- Automotive drive-train, Hall-effect sensors
- Swallowable, low power diagnostic "capsules"
- Advanced heart-stimulation devices (e.g., Pacemakers)
- FM radios (low noise)
- UHF low noise amplifiers (LNAs)-antenna mounted

*Source:* Henderson Ventures.

## 1.8 SUMMARY

True RF integrated circuits have always had the potential for dramatically altering the makeup of the high-frequency electronics industry, as well as the configuration and performance in the systems in which they are used. Although MMICs are increasingly used by microwave subsystem and system manufacturers, the huge volumes envisioned by forecasters have not materialized. However, if the short history of MMICs is a guide to future development, over the next decade they are certain to appear in great quantities. Logic dictates that a day will come soon when the bulk and power consumption of microwave hybrid circuits can no longer be tolerated. And when that day arrives, we can only hope that MMIC manufacturers are ready to deliver the quantities that future generations of systems will need.

# REFERENCES

1. E. D. Cohen, "MMIC: Improving Military Electronic Systems for the 1990s," *Defense Electronics,* November 1987, pp. 87–98.
2. R. E. Lee, "High Volume Microwave Markets Will Demand MMICs over Hybrids," *MSN,* August 1983, pp. 120–130.
3. R. S. Pengelly, "Hybrid versus Monolithic Microwave Circuits—A Matter of Cost," *MSN,* January 1983, pp. 77–114.
4. R. A. Pucel, "Design Considerations for Monolithic Microwave Circuits," *IEEE Trans. on MTTs,* Vol. MMT-29, June 1981, pp. 513–516.
5. G. Summe et al., "Gallium Arsenide: Management Can No Longer Ignore It," *Electronic Business,* February 1987, pp. 118–121.
6. J. W. White, "GaAs: The Bumpy Road," *Microwave J.,* April 1988, pp. 44–48.
7. N. Atherton et al., "7000 MMICs per Week," *Microwaves & RF,* March 1986, pp. 121–123.
8. J. D. Montgomery, "The Elusive GaAs MMIC Market," *Microwave J.,* January 1988, pp. 42–46.
9. H. Emerzian, "The MMIC (R)evolution," *MSN & CT,* March 1986, pp. 61–62.
10. A. F. Podell, "Revolt of the MMICs," *MSN & CT,* March 1986, pp. 54.
11. T. H. Oxley et al., "The System Need for GaAs Devices and Integrated Circuits," *GEC J. Research,* Vol. 4, 1986, pp. 66–70.
12. J. Imperato, "Low-Cost MMICs Replace Hybrids," *Microwave & RF,* March 1986, pp. 115–117.
13. V. G. Gelnovatch, "Is It Time for Microwave VHSIS?" *Microwave J.,* April 1986, pp. 26–30.
14. B. Hoffman et al., "Gallium Arsenide Enhances Digital Signal Processing in Electronic Warfare," *Defense Electronics,* July 1985, pp. 48–56.
15. R. S. Pengelly, "A Decade of GaAs MMICs," IEE Colloquium on GaAs ICs above 12 GHz, 1985.
16. M. R. Stiglitz, "Will Gallium Arsenide Take over Where Silicon Leaves Off?" *Microwave J.,* August 1985, pp. 24–34.
17. D. G. Fisher, "GaAs IC Applications in EW, Radar and Communications Systems," *Microwave J.,* May 1988, pp. 275.
18. Allen Podell et al., "GaAs Real Estate: Making the Most Efficient Use of Semiconductor Surface Area," *Microwave J.,* November 1987, p. 208.
19. K. Wilson, "Monolithic Microwave Integrated Circuits," *GEC J. Research,* Vol. 4, 1986, pp. 140–146.
20. S. M. Bilski, "Microwave Tutorial," *Hybrid Circuit Tech.,* June 1988, pp. 19–25.
21. J. Pustai, "Millimeter-Wave Transistors: The Key to Advanced Systems," *Microwaves & RF,* March 1987, pp. 125–177.
22. T. L. Duffield, "Monolithic mm-Wave ICs for Smart Weapons," *Microwave J.,* April 1988, pp. 91–111.
23. A. Paolella et al., "Advanced mm-Wave Sources by Automated MBE," *Microwave J.,* April 1986, pp. 149–159.
24. R. Douvitt et al., "MMIC Technology for Phased Arrays," *Microwave J.,* March 1988.
25. H. Yamasaki et al., "Hybrid vs Monolithic: Is Monolithic Better?" *Microwave J.,* November 1982, pp. 99.
26. L. C. Upadhyayula et al., "GaAs MMICs Could Carry the Waves of the High-Volume Future," *MSN,* July 1983, pp. 58–82.
27. F. A. Brand, "Microwave Semiconductors: We've Come a Long Way, Baby," *Microwave J.,* September 1988, pp. 197–213.
28. J. J. Spadaro, "The Future Is Now for GaAs ICs," *Electronic Products,* June 1988, pp. 30–33.

29. W. H. Perkins et al., "MMIC Technology: Better Performance at Affordable Cost," *Microwave J.,* April 1988, pp. 135–143.

30. F. H. Eisen, "Gallium Arsenide ICs: A Status Report," *Defense Electronics,* February 1987, pp. 84–90.

31. A. R. Jha, "Improvement in Power Handling Capability and Reliability of X-Band GaAs Power FETs," NAECON, May 1987, pp. 1138.

32. M. R. Stiglitz, "Millimeter-Wave Applications," *Microwave J.,* August 1986, pp. 38–40.

33. A. R. Jha, "Parametric Analysis for an mm-Wave High Electron Mobility Transistor," 12th Int. Conf. Infrared and mm-Waves, December 1987, pp. M-1.1.

34. M. R. Stiglitz, "GaAs Technology and MMIC, 1987," *Microwave J.,* September 1986, pp. 42–63.

35. K. J. Sleger, "Government Systems and GaAs Monolithic Components," *RCA Review,* Vol. 44, December 1983, pp. 507–523.

36. E.H. Gregory, "Applicability, Availability and Affordability of GaAs MMICs to Military Systems," GaAs IC Symp., 1986, pp. 183–186.

37. C. Adricos et al., "GaAs Monolithic ICs Applied to Military-Radar Design," *MSDH,* 1986, pp. 335–347.

38. R. T. Davis, "The Promise of GaAs for EW Systems," *J. Electronic Defense,* March 1988, pp. 51–57.

39. B. Berson, "GaAs MMICs—The Solution for Phased Array?" *MSN & CT,* November 1985, pp. 54.

40. D. W. Wallace, Jr., "Expendable Countermeasures," *J. Electronic Defense,* July 1987, pp. 33–36.

41. M. R. Stiglitz, "The Global Positioning System," *Microwave J.,* April 1986, pp. 34–59.

# Chapter 2
# Microwave Fundamentals

## R. Goyal

## 2.1 INTRODUCTION

Microwaves are electromagnetic waves with short wavelengths. They occupy the range of frequencies from 300 MHz to 300 GHz, which corresponds to wavelengths of 1 m to 1 mm in free space. Most of the frequency bands used for commercial radio and television communication are below 300 MHz. The infrared spectrum begins above 300 GHz. Figure 2.1 shows the complete frequency spectrum from 30 Hz to 300 GHz. The microwave frequency band covers the ultra-high, super-high, and extremely high frequency bands. Microwave frequencies may also be grouped into subbands designated by letters, separated according to application (i.e., commercial, military, *et cetera*). The band designation as used during World War II and adopted in 1969 is shown in Figure 2.2. The other band designations, such as IEEE standard bands and the new band allocation adopted in 1970, also are shown in the figure.

Millimeter waves, as the name signifies, are designated for shorter waves with wavelengths 10 mm or less in free space, which corresponds to 30 GHz or higher frequencies. Figure 2.3 conveniently shows the correspondence between the old and new frequency designations as commonly used today.

From a circuit design point of view, "low frequencies" and "high frequencies" are differentiated on the basis of the circuit design techniques necessary at the frequencies involved. Design techniques are different when the physical size of the components used in the circuit are comparable to the wavelength of the frequencies involved than when the size of the components is much less than the wavelengths. At lower frequencies all the impedance elements are assumed to be lumped elements. In integrated circuits operating at lower frequencies, the interconnect lines are considered to have parasitic lumped resistors and capacitors. In

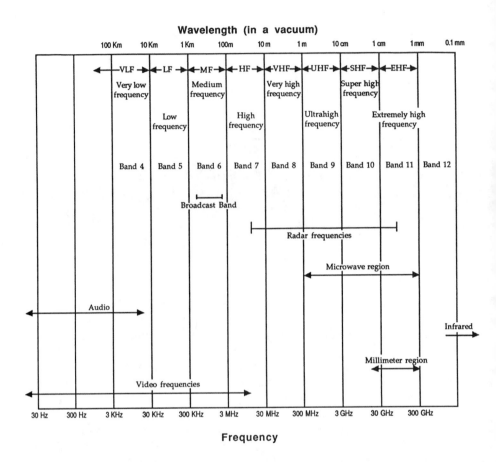

**Fig. 2.1** Complete frequency spectrum.

a more comprehensive analysis the interconnects may be considered in semidistributed form by evaluating the parasitic resistance and capacitance to ground per unit length of the interconnect. However, at microwave frequencies the parasitic resistance, inductance, and capacitance between the interconnect conductors and their surroundings cannot be neglected. Also, as the lengths of interconnect lines increase, these parasitic elements have distributive effects that are not represented in the lumped element approximation. If the physical length of the interconnect line is much smaller than the electrical wavelength of the signal ($\lambda/20$) on a particular substrate, the lumped element equivalent circuit approach is reasonable and the circuit can be analyzed in terms of voltages and currents. However, as the physical length of the interconnect line approaches the signal wavelength, the distributive effects become important. If a voltage wave travels through a

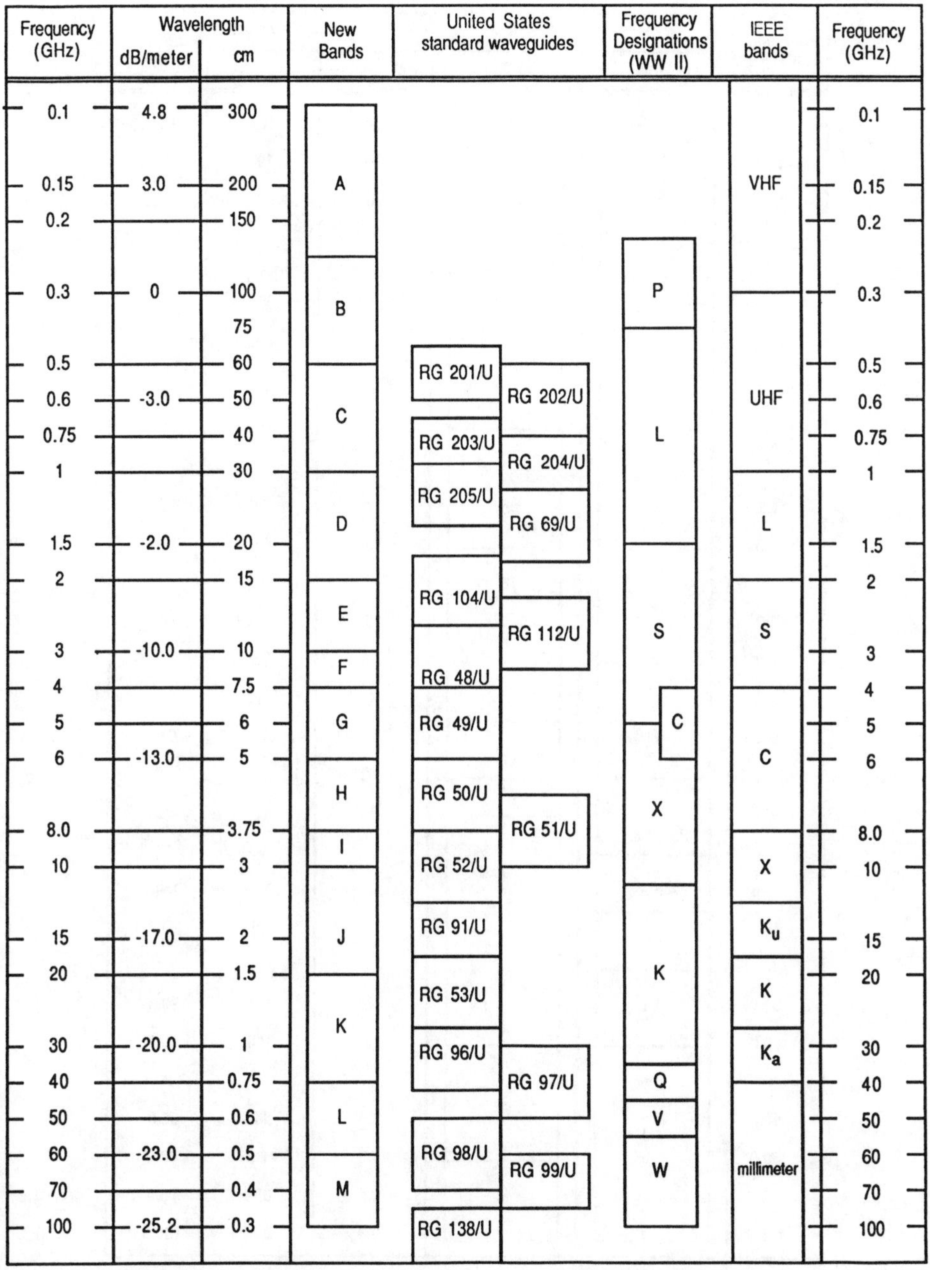

| Frequency (GHz) | Wavelength — dB/meter | Wavelength — cm | New Bands | United States standard waveguides | Frequency Designations (WW II) | IEEE bands | Frequency (GHz) |
|---|---|---|---|---|---|---|---|
| 0.1 | 4.8 | 300 | | | | | 0.1 |
| 0.15 | 3.0 | 200 | A | | | VHF | 0.15 |
| 0.2 | | 150 | | | | | 0.2 |
| 0.3 | 0 | 100 | B | | P | | 0.3 |
| | | 75 | | | | | |
| 0.5 | | 60 | | RG 201/U | | | 0.5 |
| 0.6 | -3.0 | 50 | C | RG 202/U | | UHF | 0.6 |
| 0.75 | | 40 | | RG 203/U | L | | 0.75 |
| 1 | | 30 | | RG 204/U | | | 1 |
| 1.5 | -2.0 | 20 | D | RG 205/U, RG 69/U | | L | 1.5 |
| 2 | | 15 | | | | | 2 |
| | | | E | RG 104/U | | | |
| 3 | -10.0 | 10 | | RG 112/U | S | S | 3 |
| 4 | | 7.5 | F | RG 48/U | | | 4 |
| 5 | | 6 | G | RG 49/U | C | | 5 |
| 6 | -13.0 | 5 | | | | C | 6 |
| | | | H | RG 50/U | X | | |
| 8.0 | | 3.75 | I | RG 51/U | | | 8.0 |
| 10 | | 3 | | RG 52/U | | X | 10 |
| 15 | -17.0 | 2 | J | RG 91/U | | Kᵤ | 15 |
| 20 | | 1.5 | | RG 53/U | K | K | 20 |
| | | | K | | | | |
| 30 | -20.0 | 1 | | RG 96/U | | Kₐ | 30 |
| 40 | | 0.75 | | RG 97/U | Q | | 40 |
| 50 | | 0.6 | L | | V | | 50 |
| 60 | -23.0 | 0.5 | | RG 98/U | W | millimeter | 60 |
| 70 | | 0.4 | M | RG 99/U | | | 70 |
| 100 | -25.2 | 0.3 | | RG 138/U | | | 100 |

Fig. 2.2 IEEE bands, old and new frequency bands.

| Frequency (GHz) | Frequency Designations | | Wavelength (in a Vacuum) | | Frequency (GHz) |
|---|---|---|---|---|---|
| | Previous | Current | centimeters | inches | |
| 0.1 | | | 300. | 118. | 0.1 |
| 0.15 | | A | 200. | 79. | 0.15 |
| 0.2 | VHF | | 150. | 59. | 0.2 |
| 0.3 | | B | 75. | 29.5 | 0.3 |
| 0.5 | | | 60. | 23.6 | 0.5 |
| 0.6 | UHF | | 50. | 19.9 | 0.6 |
| 0.75 | | C | 40 | 15.8 | 0.75 |
| 1.0 | | | 30 | 11.8 | 1.0 |
| 1.5 | L | D | 20 | 7.9 | 1.5 |
| 2.0 | | | 15 | 5.9 | 2.0 |
| | S | E | | | |
| 3.0 | | | 10 | 3.9 | 3.0 |
| 4.0 | | F | 7.5 | 2.9 | 4.0 |
| 5.0 | | G | 6. | 2.4 | 5.0 |
| 6.0 | C | | 5. | 2. | 6.0 |
| 8.0 | | H | 3.75 | 1.5 | 8.0 |
| 10.0 | X | I | 3. | 1.2 | 10.0 |
| 15.0 | Ku | J | 2. | 0.79 | 15.0 |
| 20.0 | K | | 1.5 | 0.59 | 20.0 |
| 30.0 | Ka | K | 1. | 0.4 | 30.0 |
| 40.0 | | | 0.75 | 0.3 | 40.0 |
| 50.0 | mm | L | 0.6 | 0.24 | 50.0 |
| 60.0 | | | 0.5 | 0.2 | 60.0 |
| 75.0 | | M | 0.4 | 0.16 | 75.0 |
| 100.0 | | | 0.3 | 0.12 | 100.0 |

**Fig. 2.3** Commonly used frequency band designations.

component from input to output and the wavelength of the applied signal is comparable to the component's physical dimensions, the instantaneous values of the voltage and current will be different at various points of the component. This leads us to conclude that besides voltage and current, another location dependent parameter is needed to completely define the electrical behavior of such a component. Transmission line theory was developed essentially to handle such problems. Hence, to design circuits that operate at microwave frequencies, it is necessary to have a basic understanding of transmission line theory.

## 2.2 TRANSMISSION LINE THEORY

Just as conducting wires and coaxial cables are used to carry low-frequency signals, waveguides and striplines are used to carry microwave signals. High-quality coaxial cables also are commonly used at microwave frequencies. Several types of transmission structures used at microwave frequencies are shown in Figure 2.4. Let us consider a simple case of two parallel lines separated by one or more dielectric media. The propagation of a microwave signal through this basic transmission line can be analyzed by solving Maxwell's field equations in three space dimensions with time as the fourth dimension. The four Maxwell equations necessary to fully represent the electrical behavior of a simple transmission line— assuming all space is linear, isotropic, and homogeneous—are as follows:

$$\nabla \cdot \mathbf{D} = \rho \qquad \text{(Gaussian Law)}$$

$$\nabla \cdot \mathbf{B} = 0$$

$$\nabla \times \mathbf{E} = -\frac{\partial \mathbf{B}}{\partial t} \qquad \text{(Faraday's Law)} \qquad (2.1)$$

$$\nabla \times \mathbf{H} = \mathbf{J} + \frac{\partial \mathbf{D}}{\partial t}$$

where $\mathbf{D}$ = electric flux density
$\mathbf{B}$ = magnetic flux density
$\mathbf{E}$ = electric field intensity
$\mathbf{H}$ = magnetic field intensity
$\mathbf{J}$ = electric current density
$\rho$ = space charge density

A preliminary observation of these partial differential equations shows that, from the integrated circuit design point of view, use of this approach will be very computationally intensive and impractical when there are large numbers of such

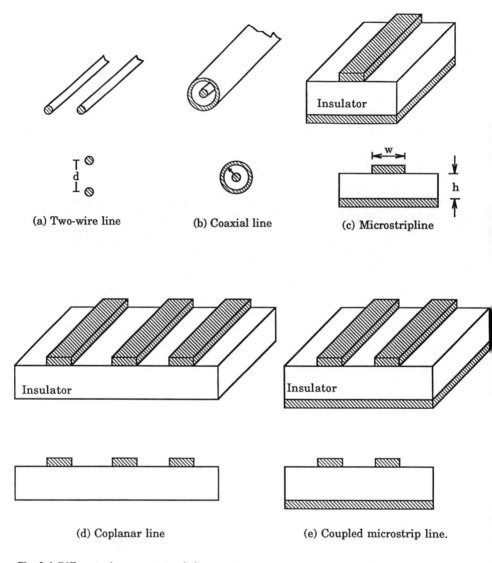

(a) Two-wire line

(b) Coaxial line

(c) Microstripline

(d) Coplanar line

(e) Coupled microstrip line.

**Fig. 2.4** Different microwave transmission structures.

components in a single circuit. A simpler mathematical representation of such a component's electrical behavior is the distributed element equivalent circuit. This approach requires the solution of equations in only one dimension besides the time variable. The transmission line electrical behavior thus can be represented in terms of voltage, current, signal power, and impedances along the line. The laws of circuit theory, Kirchhoff's voltage and current laws, can be used to derive the transmission line equations. In this simplified approach, a chain of closely spaced, discrete resistors, capacitors, and inductors represent a small section of the transmission line. Let us assume a lossy transmission line having a series resistance and inductance of $R$ and $L$ per unit length, respectively, and conductance and capacitance to ground of $G$ and $C$ per unit length, respectively. Considering the whole transmission line as divided into small sections of $x$ length, it can be represented fairly accurately in terms of lumped elements as shown in Figure 2.5. The smaller the sections $x$ are, the better is the lumped element approximation for higher frequencies. Note that the quantities $R$, $L$, $G$, and $C$ are all measured per unit length, because they occur continuously along the line; they thus are distributed throughout the length of the line and we should not assume them to be lumped at any one point.

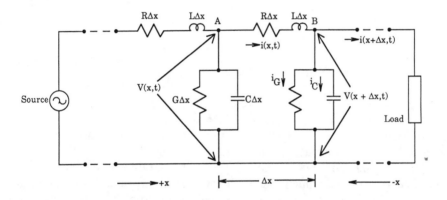

Fig. 2.5 Transmission line equivalent circuit.

By Kirchhoff's voltage law, the summation of the voltages around any loop is given by

$$V(x,t) = i(x,t)R\Delta x + L\Delta x \frac{\partial i(x,t)}{\partial t} + V(x + \Delta x,t)$$

taking our limit as $\Delta x \rightarrow 0$, we get

$$V = iR\Delta x + L\Delta x \frac{\partial i}{\partial t} + V + \frac{\partial V}{\partial x}\Delta x$$

dividing by $\Delta x$:

$$-\frac{\partial V}{\partial x} = iR + L\frac{\partial i}{\partial t} \tag{2.2}$$

By Kirchhoff's current law, the summation of currents at point $B$ in Figure 2.5 is given by

$$i(x,t) = V(x + \Delta x,t)G\Delta x + C\Delta x \frac{\partial V(x + \Delta x,t)}{\partial t} + i(x + \Delta x,t)$$

again taking our limit as $\Delta x \to 0$, we get

$$i = VG\Delta x + \frac{\partial V}{\partial x}\Delta xG\Delta x + C\Delta x \frac{\partial}{\partial t}\left(V + \frac{\partial V}{\partial x}\Delta x\right) + i + \frac{\partial i}{\partial x}\Delta x$$

dividing by $\Delta x$

$$-\frac{\partial i}{\partial x} = GV + C\frac{\partial V}{\partial t} \tag{2.3}$$

### 2.2.1 ac Solution for Transmission Lines

Signals at high frequencies are usually sinusoidal or can be analyzed as though they are composed of sinusoidal waves that are propagating through a component from input to output. Moreover, any signal waveform can be represented in terms of sinusoidal components using Fourier analysis. Considering all signals to be sinusoidal makes the solution of partial differential Equations (2.2) and (2.3) in time domain much easier. Hence, it is appropriate to analyze transmission line equations in steady-state sinusoidal form. Both the voltage and current will be in phasor form and will vary in time and distance. Let $\mathbf{V}$ and $\mathbf{I}$ be the voltage and current phasors, respectively, at an arbitrary point on the transmission line.

From Equations (2.2) and (2.3), the derivatives of the voltage and current phasors can be expressed as follows:

$$-\frac{d\mathbf{V}}{dx} = R\mathbf{I} + L\frac{d\mathbf{I}}{dt} \tag{2.4}$$

$$-\frac{d\mathbf{I}}{dx} = G\mathbf{V} + C\frac{d\mathbf{V}}{dt} \tag{2.5}$$

We can see that a generalized solution of the coupled differential Equations (2.4) and (2.5) can be given by the following:

$$\mathbf{V}(x,t) = (\mathbf{V}^{+}e^{-\gamma x} + \mathbf{V}^{-}e^{+\gamma x})e^{j\omega t} \tag{2.6}$$

$$\mathbf{I}(x,t) = (\mathbf{I}^+e^{-\gamma x} + \mathbf{I}^-e^{+\gamma x})e^{j\omega t} \tag{2.7}$$

where $\gamma = \alpha + j\beta$ is the propagation constant, defined later;
$\quad\quad \mathbf{V}^+, \mathbf{I}^+$ = complex voltage and current amplitudes, respectively; in the positive $x$ direction
$\quad\quad \mathbf{V}^-, \mathbf{I}^-$ = complex voltage and current amplitudes, respectively; in the negative $x$ direction
$\quad\quad \alpha$ = attenuation constant per unit length of the line
$\quad\quad \beta$ = phase constant per unit length of the line

Substituting Equations (2.6) and (2.7) into (2.4) and (2.5) gives

$$\frac{d\mathbf{V}}{dx} = -(R + j\omega L)\mathbf{I} = -\mathbf{Z} \cdot \mathbf{I} \tag{2.8}$$

$$\frac{d\mathbf{I}}{dx} = -(G + j\omega C)\mathbf{V} = -\mathbf{Y} \cdot \mathbf{V} \tag{2.9}$$

where $\mathbf{Z}$ = Impedance phasor per unit length of the transmission line
$\quad\quad \mathbf{Y}$ = Susceptance phasor per unit length of the transmission line

### 2.2.2 Propagation Constant, Phase Velocity, and Characteristic Impedance

Differentiating Equations (2.8) and (2.9) with respect to the distance $x$ results in

$$\frac{d^2\mathbf{V}}{dx^2} = -\mathbf{Z}\frac{d\mathbf{I}}{dx} = \mathbf{Z} \cdot \mathbf{Y} \cdot \mathbf{V} = \gamma^2\mathbf{V} \tag{2.10}$$

$$\frac{d^2\mathbf{I}}{dx^2} = -\mathbf{Y}\frac{d\mathbf{V}}{dx} = \mathbf{Z} \cdot \mathbf{Y} \cdot \mathbf{I} = \gamma^2\mathbf{I} \tag{2.11}$$

where $\gamma = \sqrt{\mathbf{ZY}}$ is the propagation constant

This can be further represented mathematically as follows:

$$\gamma = [(R + j\omega L)(G + j\omega C)]^{1/2}$$

$$= j\omega(LC)^{1/2}\left[\left(1 + \frac{R}{j\omega L}\right)\left(1 + \frac{G}{j\omega C}\right)\right]^{1/2}$$

At high frequencies for a very small section of the transmission line, we can assume that

$$R \ll \omega L \quad \text{and} \quad G \ll \omega C$$

hence
$$\gamma = j\omega(LC)^{1/2}\left[1 + \frac{1}{2}\left(\frac{R}{j\omega L} + \frac{G}{j\omega C}\right)\right]$$

$$= \left[R\left(\frac{C}{L}\right)^{1/2} + G\left(\frac{L}{C}\right)^{1/2}\right] + j\omega(LC)^{1/2}$$

$$= \alpha + j\beta$$

where
$$\alpha = \frac{1}{2}\left[R\left(\frac{C}{L}\right)^{1/2} + G\left(\frac{L}{C}\right)^{1/2}\right]$$

$$\beta = \omega(LC)^{1/2}$$

The phase velocity can be found from the phase constant by the following relationship:

$$v_p = \frac{\omega}{\beta} = \frac{1}{(LC)^{1/2}} \tag{2.12}$$

In free space electromagnetic waves travel at the speed of light; that is, close to 3 $\times 10^8$ m/s. In a transmission line the voltage and current propagations are delayed due to their interaction with the transmission line material. The inductance and capacitance along the line reduce the effective propagation velocity, which is given by Equation 2.12. This, also known as *phase velocity,* is dependent only on the permeability, $\mu$, and the permittivity, $\varepsilon$, of the insulating medium. In free space the phase velocity of a lossless transmission line can also be given by

$$v_p = \frac{1}{\sqrt{\mu_0 \varepsilon_0}} = 3 \times 10^8 \text{ m/s} = \text{velocity of light} \tag{2.13}$$

In an arbitrary medium the phase velocity is given by

$$v_p = \frac{1}{\sqrt{\mu_r \mu_0 \varepsilon_r \varepsilon_0}} \tag{2.14}$$

where $\mu_0$ = permeability of free space
$\varepsilon_0$ = permittivity of free space
$\mu_r$ = relative permeability of the medium
$\varepsilon_r$ = relative permittivity of the medium

The input impedance of the transmission line from Figure 2.5 depends on the values of $R$, $L$, $C$, and $G$, the line length, and the termination at the far end. To simplify the description of such a circuit, a reference impedance, called the *characteristic impedance*, is defined. The characteristic impedance of a transmission line, $Z_0$, is the impedance measured at the input of this line when its length is infinite. Under these conditions the type of termination at the far end has no effect on the characteristics of the transmission line. Let us assume a line of infinite length, as shown in Figure 2.6, the voltage and current at $A - A'$ are $V$ and $i$, respectively; whereas the voltage and current at $B - B'$ are $V'$ and $i'$, respectively. If the line is infinitely long, the input impedance at $A - A'$ will be $Z_0$, the characteristic impedance. At the same time the input impedance at $B - B'$ also will be $Z_0$, for an infinitely long line. Replacing the section of line to the right of $B - B'$ by an impedance of $Z_0$ results in the circuit shown in Figure 2.7.

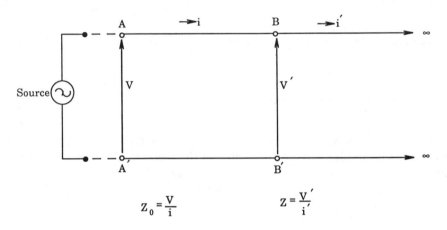

$$Z_0 = \frac{V}{i} \qquad Z = \frac{V'}{i'}$$

**Fig. 2.6** Infinite transmission line.

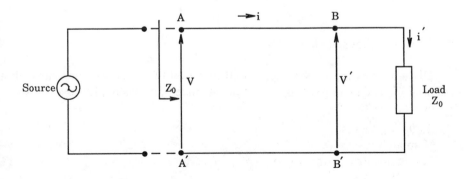

**Fig. 2.7** Properly terminated transmission line.

It follows from the filter theory that the input impedance of such an infinitely long transmission line represented by series and shunt elements as shown in Figure 2.5 is given by

$$Z_0 = \sqrt{\frac{Z}{Y}} = \text{characteristic impedance of the line} \qquad (2.15)$$

where $Z$ = series impedance per section = $(R + j\omega L)$
$Y$ = shunt admittance per section = $(G + j\omega C)$

thus:

$$Z_0 = \sqrt{\frac{R + j\omega L}{G + j\omega C}} \qquad (2.16)$$

At very high frequencies, $R \ll j\omega L$ and $G \ll j\omega C$ and the characteristic impedance is given by

$$Z_0 = \sqrt{\frac{L}{C}} \text{ for a lossless transmission line}$$

Similar to the phase velocity, the characteristic impedance of the line also depends on the permittivity and permeability of the medium, and it is expressed as follows:

$$Z_0 = \sqrt{\frac{\mu_0}{\varepsilon_0}} \approx 377 \ \Omega \text{ for free space}$$

$$Z_0 = 377 \sqrt{\frac{\mu_r}{\varepsilon_r}} \ \Omega \text{ for any other medium}$$

### 2.2.3 Losses in Transmission Lines

There are several reasons for the attenuation of signal traveling through a transmission. The main reasons for the occurrence of losses in physical transmission lines are as follows:

Radiation losses
Mismatch losses
Conductor or resistor losses
Dielectric losses
Hysteresis losses

Radiation losses occur because the signal energy radiates away from the transmission line. With careful shielding of the lines, radiation losses can be minimized. Mismatch losses are caused by part of the signal reflecting toward the source. As a simple example, this could happen because of a mismatch between the load impedance and the characteristic impedance of the transmission line. Signal loss that occurs in this manner is discussed in detail in Section 2.2.4.

Conductor losses are caused by the resistive nature of the conducting medium. At microwave frequencies the signal flow is concentrated in the surface layer of the conductor. The current density is maximum at the surface of the conductor and decreases exponentially with depth into the conductor. The penetration of the current flow is defined by the *skin depth* $\delta$. This is the thickness of the layer of the conductor at which the current density has fallen to $1/e$ (or 30%) of its surface value. Skin depth is a function of frequency and the resistivity of the conducting material. It is given by the following relationship:

$$\delta = \frac{1}{2\pi} \sqrt{\frac{\rho}{f \mu_r}} \tag{2.17}$$

where $\rho$ = specific resistivity of the conductor in $\Omega$-cm
$f$ = frequency in GHz
$\mu_r$ = relative permeability of the conducting material, which is equal to 1 for nonferromagnetic materials.

It can be concluded from this equation of skin depth that $\delta$ increases with resistivity of the conductor and decreases with increasing frequency or with increasing permeability.

If the conductor is perfect—that is, $\rho = 0$ results in $\delta = 0$ and the skin depth is zero, 100% of the incoming signal is reflected and the conducting medium does not absorb any power. If the conductor is lossy, the incoming signal penetrates into the metal by an effective distance, $\delta$, some power is absorbed by the metal, and heat is generated. This power absorption accounts for the attenuation of the signal at microwave frequencies, analogous to the ohmic resistance of an ordinary wire in low-frequency circuits. The skin depth, $\delta$, is plotted as a function of frequency in Figure 2.8 for a number of different metals. The ideal values of specific resistivity of the metals were considered in the computation of these values of skin depth. However, plated metal lines commonly are used in monolithic microwave integrated circuits as interconnects. Caution should be exercised when estimating the skin depth and effective losses of such plated metals. Since plated metals are porous in nature and their surface finish is rough, their effective resistivity is higher than that used in the theoretical calculation of skin depth plotted in Figure 2.8. In order to take into account these imperfections in the resistivity and surface finish of plated metals, it is advisable to have the plated

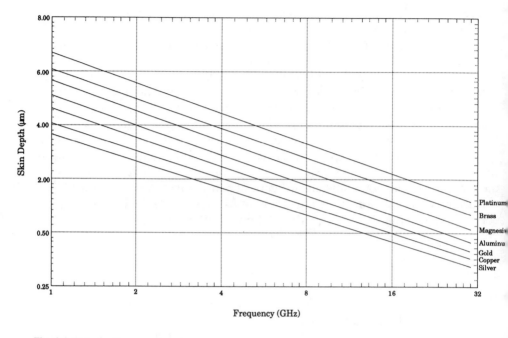

**Fig. 2.8** Skin depth *versus* frequency.

metal thickness at least three–five times the calculated thickness of the skin depth.

Dielectric losses are introduced if the dielectric medium is not a vacuum. This is because the medium will absorb part of the energy from the signal passing through it. These losses can be taken into account by defining the dielectric constant of the material as a complex quantity

$$\varepsilon = \varepsilon' - j\varepsilon''$$

Then the losses of the dielectric material are expressed by the loss tangent, defined as

$$\tan \delta = \frac{\varepsilon''}{\varepsilon'}$$

For most practical cases the loss tangent is a very small number and hence

$$\delta \approx \frac{\varepsilon''}{\varepsilon'}$$

Hysteresis losses occur when ferrite materials are used as a medium for micro-wave signals to propagate. However, these losses can be minimized by plating the ferrite material with a conducting metal.

### 2.2.4 Reflection Coefficient, Line Impedance, and Conductance

Assume that a transmission line of characteristic impedance $Z_0$ and length $l$ is connected to an input source signal $V_{in}$ with impedance $Z_{in}$ and to a load impedance of $Z_L \neq Z_0$ at $x = l$ distance from the input source, as shown in Figure 2.9. The voltage and current at any point $x$ of the transmission line can be given from Equations (2.6) and (2.9):

$$\mathbf{V}(x) = V^+ e^{-\gamma x} + V^- e^{+\gamma x} \tag{2.18}$$

$$\mathbf{I}(x) = I^+ e^{-\gamma x} + I^- e^{+\gamma x} \tag{2.19}$$

$$\mathbf{I}(x) = \frac{\gamma}{Z} (V^+ e^{-\gamma x} - V^- e^{+\gamma x}) \tag{2.20}$$

We assume that the positive $x$ direction is from the source signal toward the load, $V^+$ and $I^+$ are the voltage and current, respectively, in the positive $x$ direction and $V^-$ and $I^-$ are the voltage and current, respectively, in the negative $x$ direction.

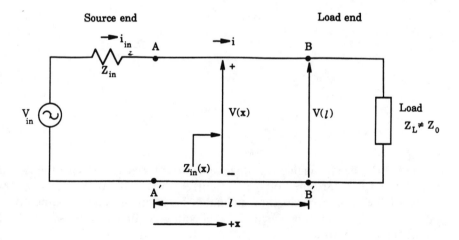

**Fig. 2.9** Transmission line of length *l* terminated in a load impedance.

From Equations (2.18) to (2.20) the input impedance of the line at any point can be represented as

$$Z_{in}(x) = \frac{V(x)}{I(x)} = \frac{Z}{\gamma} \frac{V^+ e^{-\gamma x} + V^- e^{+\gamma x}}{V^+ e^{-\gamma x} - V^- e^{+\gamma x}}$$

$$= \sqrt{\frac{Z}{Y}} \frac{V^+ e^{-\gamma x} + V^- e^{+\gamma x}}{V^+ e^{-\gamma x} - V^- e^{+\gamma x}}$$

(2.21)

When the line is match terminated, the reflected signal is zero; and the input impedance is the same as the characteristic impedance of the transmission line given by Equation (2.15). Combining Equations (2.15) and (2.21) results in

$$Z_0 = \sqrt{\frac{Z}{Y}}$$

Hence,

$$Z_{in}(x) = Z_0 \frac{V^+ e^{-\gamma x} + V^- e^{+\gamma x}}{V^+ e^{-\gamma x} - V^- e^{+\gamma x}}$$

(2.22)

If the transmission line is terminated with a load impedance $Z_L$ other than the characteristic impedance of the line, the voltage and current at the load point ($x = l$) are given as

$$V(l) = V^+ e^{-\gamma l} + V^- e^{\gamma l}$$

(2.23)

$$I(l) = \frac{1}{Z_0} (V^+ e^{-\gamma l} - V^- e^{\gamma l})$$

(2.24)

$$Z_{in}(l) = Z_l = Z_0 \frac{V^+ e^{-\gamma l} + V^- e^{\gamma l}}{V^+ e^{-\gamma l} - V^- e^{\gamma l}}$$

(2.25)

$V^+ e^{-\gamma l}$ and $I^+ e^{-\gamma l}$ are the incident voltage and current, respectively, at the load ($x = l$). $V^- e^{\gamma l}$ and $I^- e^{\gamma l}$ are the voltage and current, respectively, reflected by the load toward the source because of mismatch between the line's characteristic impedance and the load impedance. The reflection coefficient represented by $\Gamma$ can be defined as follows:

$$\Gamma = \text{reflection coefficient} = \frac{\text{reflected signal}}{\text{incident signal}}$$

at $x = 0$, $\Gamma = \Gamma_0$, and $\Gamma_0 = V^-/V^+$. So the reflection coefficient at an arbitrary point can be represented as

$$\Gamma = \Gamma_0\, e^{2\gamma x} \tag{2.26}$$

or,

$$\Gamma = \frac{V^-\, e^{\gamma x}}{V^+\, e^{-\gamma x}} \tag{2.27}$$

This equation shows that the argument of $\Gamma$ goes through $360°$ as $x$ goes through a half-wavelength.

At the load, the reflection coefficient can be represented as

$$\Gamma_l = \frac{V^-\, e^{\gamma l}}{V^+\, e^{-\gamma l}} \tag{2.28}$$

From Equation (2.26):

$$\Gamma_l = \frac{Z_l - Z_0}{Z_l + Z_0} \tag{2.29}$$

Conversely:

$$Z_l = Z_0\, \frac{1 + \Gamma_l}{1 - \Gamma_l} \tag{2.30}$$

These equations, which express the reflection coefficient in terms of impedance and *vice versa,* can be generalized as follows:

$$\Gamma_x = \frac{Z(x) - Z_0}{Z(x) + Z_0} \tag{2.31}$$

$$Z_{in}(x) = Z_0\, \frac{1 + \Gamma_x}{1 - \Gamma_x} \tag{2.32}$$

The transmission coefficient is defined similar to the reflection coefficient in terms of the incident and the transmitted signals as follows:

$$T = \text{transmitted signal} = \frac{\text{transmission coefficient}}{\text{incident signal}}$$

At the load, the transmission coefficient can be given by

$$
T_l = \frac{V^+ \, e^{-\gamma l} + V^- \, e^{\gamma l}}{V^+ \, e^{-\gamma l}}
$$

$$
= 1 + \frac{V^- \, e^{\gamma l}}{V^+ \, e^{-\gamma l}} \tag{2.33}
$$

$$
= 1 + \Gamma_l = 1 + \frac{Z_l - Z_0}{Z_l + Z_0} = \frac{2Z_l}{Z_l + Z_0}
$$

In general the transmission coefficient can be represented at any point along the transmission line by:

$$
T = 1 + \Gamma = 1 + \frac{Z - Z_0}{Z + Z_0} = \frac{2Z}{Z + Z_0} \tag{2.34}
$$

The transmission line impedance, reflection coefficient, and the transmission coefficient may be normalized with respect to the characteristic impedance of the line. Normalized input impedance at any point is expressed as follows:

$$
z_{in} = \frac{Z_{in}(x)}{Z_0} = \frac{1 + \Gamma}{1 - \Gamma} \tag{2.35}
$$

The reflection coefficient is represented as follows:

$$
\Gamma = \frac{z - 1}{z + 1} \tag{2.36}
$$

$$
T = \frac{2z}{z + 1} \tag{2.37}
$$

From Equations (2.18) to (2.20) the input impedance of the transmission line at any point $x$ can be represented as

$$
z_{in}(x) = \frac{V^+ e^{-\gamma x} + V^- e^{\gamma x}}{V^+ e^{-\gamma x} - V^- e^{\gamma x}} \tag{2.38}
$$

The most common transmission line analysis requires the calculation of the input impedance, given the impedance at some other point on the line. As an example assume that we know the impedance at $x = l$ and the load point; then the line impedance at any point on the line is given by

$$
\begin{aligned}
z_{in}(x) &= \frac{V^+(x = l)\,e^{-\gamma x} + V^-(x = l)\,e^{\gamma x}}{V^+(x = l)\,e^{-\gamma x} - V^-(x = l)\,e^{\gamma x}} \\[2mm]
&= \frac{e^{-\gamma x} + \Gamma_l\,e^{\gamma x}}{e^{-\gamma x} - \Gamma_l\,e^{\gamma x}} \\[2mm]
&= \frac{(Z_l + Z_0)\,e^{-\gamma x} + (Z_l - Z_0)\,e^{\gamma x}}{(Z_l + Z_0)\,e^{-\gamma x} - (Z_l - Z_0)\,e^{\gamma x}} \\[2mm]
&= \frac{Z_l\left(\dfrac{e^{-\gamma x} + e^{\gamma x}}{2}\right) - Z_0\left(\dfrac{e^{\gamma x} - e^{-\gamma x}}{2}\right)}{Z_0\left(\dfrac{e^{-\gamma x} + e^{\gamma x}}{2}\right) - Z_l\left(\dfrac{e^{\gamma x} - e^{-\gamma x}}{2}\right)} \\[2mm]
&= \frac{Z_l \cosh \gamma x - Z_0 \sinh \gamma x}{Z_0 \cosh \gamma x - Z_l \sinh \gamma x} \\[2mm]
&= \frac{Z_l - Z_0 \tanh \gamma x}{Z_0 - Z_l \tanh \gamma x}
\end{aligned}
\tag{2.39}
$$

Normalizing the impedances with respect to the characteristic impedance $Z_0$ results in

$$
z_{in}(x) = \frac{z_l - \tanh \gamma x}{1 - z_l \tanh \gamma x}
\tag{2.40}
$$

If the line has small losses (i.e., $\alpha \approx 0$), the normalized input impedance of the line at any point is given by

$$
z_{in}(x) = \frac{z_l - j \tanh \beta}{1 - j z_l \tanh \beta}
\tag{2.41}
$$

### 2.2.5 Standing Waves

When a signal is applied to a transmission line that is terminated in its characteristic impedance or if the length of the line is infinite, all of the signal is absorbed by the load and no power is reflected. A transmission line terminated in such a manner is called a *properly terminated line* or a *nonresonant* or *flat line*. The voltage and current are not attenuated throughout the length of the line if it is lossless or they are reduced exponentially toward the loads if the line is lossy. The only waves present on a lossless line are the voltage and current traveling waves from generator to load. However, if a signal is applied to a finite length of transmission line, terminated in an impedance not equal to its characteristic impedance, part of the signal is absorbed by the load and part is reflected back toward the source. Open and closed circuit loads are extreme cases of unmatched termination. In both the cases, no signal is transmitted to such a load and all of it is reflected back toward the source. If the line is lossless and a signal pulse is sent out from the source and quickly removed, leaving the source side open, then the signal will travel back and forth between source and the load ends, never diminishing. A line terminated in such a manner is called *resonant*. If the termination impedance value is between 0 and $Z_0$ or between $Z_0$ and infinity, part of the incident signal is reflected back and forth between source and load. These sets of traveling waves, going in opposite directions, set up an interference pattern known as *standing waves,* shown in Figure 2.10. Waves having the same length but not necessarily the same magnitude will form this type of interference pattern. The interference pattern shown in Figure 2.11 is the standing waves built as a result of total reflection of the incidence signal at the load. This will be the case when the load is a short circuit or an open circuit.

It must be emphasized that the formation of standing waves as a result of reflection is permanent for a given load and determined by the value of the load and that such waves are truly standing waves. As shown in Figure 2.10, all the nodes are permanently fixed minima; all the antinodes are maxima whose positions are fixed. However, note that standing waves have amplitudes that vary sinusoidally, just like the applied signal. Figure 2.12 shows a standing wave built of a nontotal reflection, which will be the case when $0 < Z < Z_0$ or $Z_0 < Z < \infty$. The reflection coefficient in this case will be less than 1.

The term $V_{max}$ is the sum of the absolute values of the incidence and the reflected voltages:

$$|V_{max}| = |V^+| + |V^-| \tag{2.42}$$

and $V_{min}$ is the difference of the absolute value of incidence voltage and the absolute value of reflected voltage:

$$|V_{min}| = |V^+| - |V^-| \tag{2.43}$$

Then, the standing wave ratio is defined as the ratio of the maximum voltage (current) to the minimum voltage (current) along the transmission line:

$$\text{VSWR} = \left| \frac{V_{max}}{V_{min}} \right|$$

$$= \frac{|V^+| + |V^-|}{|V^+| - |V^-|}$$

$$= \frac{1 + \left| \dfrac{V^-}{V^+} \right|}{1 - \left| \dfrac{V^-}{V^+} \right|} \qquad (2.44)$$

$$= \frac{1 + \Gamma}{1 - \Gamma}$$

where the reflection coefficient is defined using Equation (2.26). Similarly, the reflection coefficient can be expressed in terms of voltage standing wave ratio:

$$\Gamma = \frac{\rho - 1}{\rho + 1} \qquad (2.45)$$

Figure 2.13 shows the relationship between the reflection coefficient and the standing wave ratio.

The power standing wave ratio also can be defined similar to the voltage and current standing wave ratios using

$$\text{PSWR} = \rho^2$$

Standing wave ratio can be expressed in dB as follows:

$$\text{SWR (dB)} = 20 \log\rho$$

The standing wave ratio is a measure of the mismatch between the load impedance and the impedance of the transmission line. Thus, it is one of the most important quantities calculated for a particular load. The standing wave ratio is unity (i.e., $V = V$), a highly desirable situation from the matching point of view as the load is perfectly matched to the impedance of the network. When the line is terminated in a purely resistive load, the standing wave ratio is given by

$$\text{SWR} = Z_0/R_L \quad \text{or} \quad R_L/Z_0 \quad \text{whichever is larger}$$

where $R_L$ = load resistance

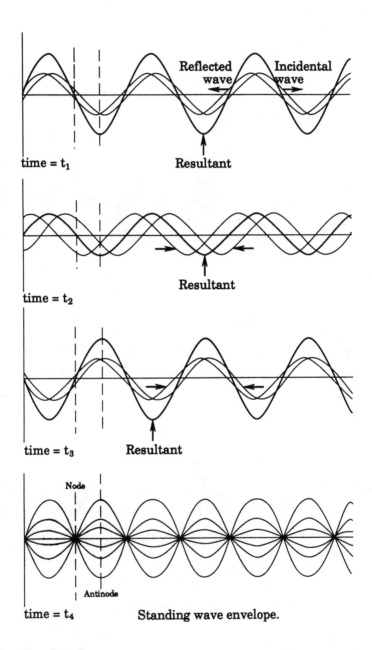

**Fig. 2.10** Formation of standing waves.

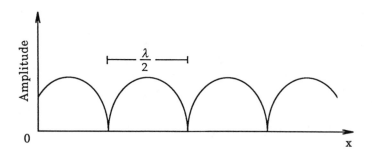

**Fig. 2.11** Standing-wave pattern with total reflection from load.

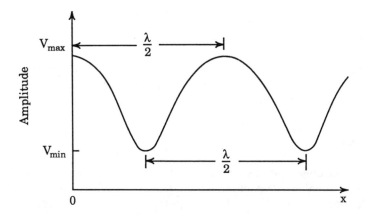

**Fig. 2.12** Standing-wave pattern built of a nontotal reflection.

In the definition of standing wave ratio it is customary to keep a larger quantity in the numerator, so that the ratio is always greater than unity. This does not lead to any confusion, regardless of whether the load resistance is half as small, for example, or twice as large as the line impedance, the ratio of voltage maximum to voltage minimum is $2:1$ and the extent of mismatch is the same in both cases.

If the load is purely reactive, short circuit, or open circuit, the SWR is infinity, since no power is absorbed by the load, and hence the reflected wave has the same magnitude as the forward wave. This results in a voltage minima of zero in the standing wave pattern and, hence, SWR of infinity. A higher value of SWR indicates greater mismatch between the network and the load. A low value of

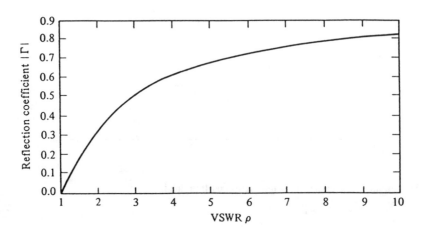

**Fig. 2.13** VSWR *versus* SWR.

SWR is always desirable, except when the transmission line is used as pure reactance or as a tuned circuit. As the value of SWR increases, power loss also increases, and so does the noise in the circuit.

## 2.3 SMITH CHART

Most of the computations required to solve transmission line problems involve complex numbers and complicated mathematical equations, as demonstrated in Section 2.2. One method of easily representing the complex properties of transmission lines is through the use of impedance charts. The most useful charts are graphical representations that provide impedance information at any point on the transmission line for various conditions. By far the most common graphical representation of complex impedance is the Smith chart. As shown in Equation (2.40), to evaluate the line impedance at any point, we must plot the input impedance value for various complex values of $z_l$ and $\gamma_l$. This results in a semiinfinite plane of impedance, even when $\gamma_l$ is held constant in Equation (2.24). Phillip H. Smith provided a graphical solution to this problem by plotting the impedance at any given point on the reflection coefficient plane using the relationship in Equation (2.36). Thus, all impedance values can be mapped within a circle of unit radius since $|\Gamma| \leq 1$. This condition is the basic requirement and restriction of the Smith chart. In certain cases, it is possible for the magnitude of the reflection coefficient to exceed unity. If the value of $\Gamma$ is $> 1$ (i.e., the system resistance is negative), a variation of the Smith chart known as the compressed Smith chart must be used.

## 2.3.1 Basics of the Smith Chart

The Smith chart is a plot of normalized impedance or admittance represent-ing the angle and magnitude of a complex reflection coefficient in a unity circle. The chart is useful for both lossless and lossy transmission lines. For a lossless transmission line, the variation of $\Gamma$ as we move along the line is represented by a rotating vector of constant magnitude in the $\Gamma$ $(x)$ plane. The impedance transfor-mation from input toward load or *vice versa* is determined from the Smith chart by noting the impedances that correspond to the initial and final values of $\Gamma$. To demonstrate how the Smith chart is developed, Figure 2.14 shows the $\Gamma$ plane, representing the real and imaginary parts of $\Gamma$. In general, $\Gamma$ is written in polar form as given in Equation (2.27).

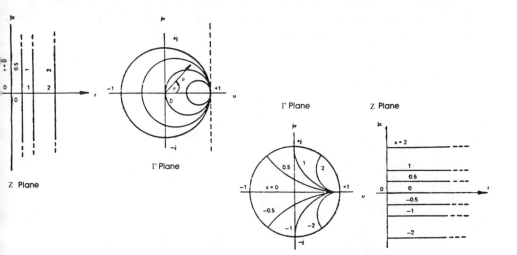

**Fig. 2.14** $\Gamma$ plane representing the real and imaginary parts of $\Gamma$ (from [10], p. 50).

$$\Gamma = \rho e^{j\theta} \tag{2.46}$$

Representing this in the $u$ and $jv$ rectangular system

$$u = \rho \cos\theta$$
$$v = \rho \sin\theta \tag{2.47}$$
$$\rho^2 = u^2 + v^2$$

Substituting Equation (2.47) into (2.46):

$$\Gamma = u + jv \tag{2.48}$$

Substituting Equation (2.47) into (2.48) results in

$$r + jx = \frac{1 + \rho e^{j\theta}}{1 - \rho e^{j\theta}}$$

$$= \frac{1 + u + jv}{1 - (u + jv)}$$

$$= \frac{(1 + u) + jv}{(1 - u) - jv} \qquad (2.49)$$

$$= \frac{1 - u^2 - v^2 + 2jv}{1 + u^2 + v^2 - 2u}$$

To construct the Smith chart it is necessary to know how impedance contours in the Z-plane are mapped onto the $\Gamma$ plane from Equations (2.30) and (2.49). Separating the real and imaginary parts of Equation (2.49):

$$r = \frac{1 - u^2 - v^2}{1 + u^2 + v^2 - 2u}$$

or,

$$\left(u - \frac{r}{r + 1}\right)^2 + v^2 = \frac{1}{(r + 1)^2} \qquad (2.50)$$

and,

$$x = \frac{2v}{1 + u^2 + v^2 - 2u}$$

or,

$$(u - 1)^2 + \left(v - \frac{1}{x}\right)^2 = \frac{1}{x^2} \qquad (2.51)$$

for a fixed value of Equation (2.50) represents a circle in $u - jv$ coordinate system ($\Gamma$ — the reflection coefficient plane). The circle has a radius of $1/(r + 1)$ and is centered at $u = r/(r + 1)$, $jv = 0$. Figure 2.14a shows how these constant resistance contours in the Z-plane are mapped into circles in the $\Gamma$ plane. Similarly Equation (2.51) represents the locus of a circle in $u - jv$ plane for a fixed value of $x$. The radius of the circle is $1/x$ and the center is at $u = 1$, $v = 1/x$.

Figure 2.14b shows how constant reactance contours in the $Z$-plane are mapped into arcs in the $\Gamma$ plane. The combination of parts in Figure 2.14 represent the complete Smith chart—a representation of line impedance in the $\Gamma$ plane as shown in Figure 2.15. The manner in which resistance, capacitance, and inductance are represented on a Smith chart is illustrated in Figure 2.16. Various series and parallel combinations of these components and their corresponding representation on the Smith chart is shown in Figure 2.16 as well. Figure 2.16c shows typical impedance matching circuits and their representation on the Smith chart.

The salient features of the Smith chart can be summarized:

- The horizontal line in Figure 2.15, the only straight line in the figure, represents pure resistive impedance with zero reactance.
- The portion of the chart above the horizontal line represents positive reactive impedance; that is, inductive impedance.
- The portion of the chart below the horizontal line represents negative reactive impedance; that is, capacitive impedance.
- The full circles whose centers lie on the straight line correspond to the normalized resistive part of the impedance, and the arcs correspond to the normalized reactive part of the impedance.
- The circles concentric with the center of the unity circle correspond to the constant reflection coefficient. Since the standing wave ratio is determined by the magnitude of the reflection coefficient, as given in Equation (2.44), these circles also represent the contours of constant standing wave ratio.
- Once around the Smith chart represents one-half wavelength ($\lambda/2$) distance on the line.
- The distances represented in Figure 2.15 are given in wavelengths toward the generator as well as toward the load.

Once the standing wave pattern along the line is established using the Smith chart, we can calculate the magnitude of reflection coefficient, reflected power, transmitted power, and the load impedance. Typical values are shown in Figure 2.17.

### 2.3.2 Changing Characteristic Impedance

In microwave circuit design, it is common to have lengths of transmission lines with different characteristic impedance values $Z_0$. These circuits can be represented on the Smith chart. As an example, let us assume that the circuit shown in Figure 2.18 has a load impedance of $Z_L = 200 + j \times 50$ and is connected to a transmission line with a characteristic impedance of 100 $\Omega$ and a length of $\lambda/8$ at a particular frequency. Next, let us assume that this length of transmission line is connected to another line with a characteristic impedance of 50 $\Omega$ and an

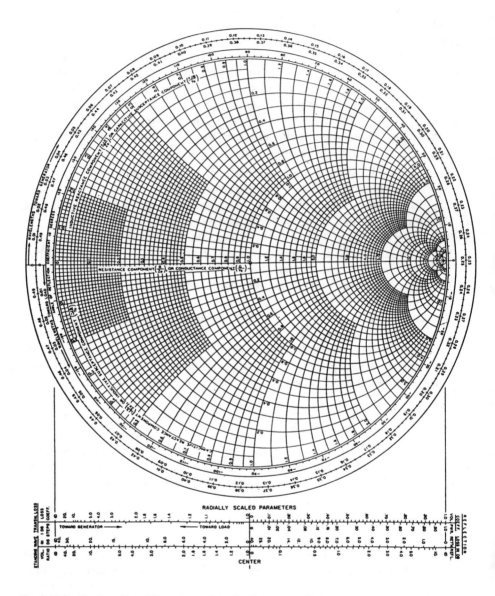

**Fig. 2.15** Smith chart (from [7] courtesy of Analog Instrument Co.).

electrical length of λ/4 at the same frequency. The problem, then, is to calculate the input impedance.

This problem can be solved in the following steps:

1. $Z_L = 200 + j \times 50$ is normalized to a 100 Ω characteristic impedance,

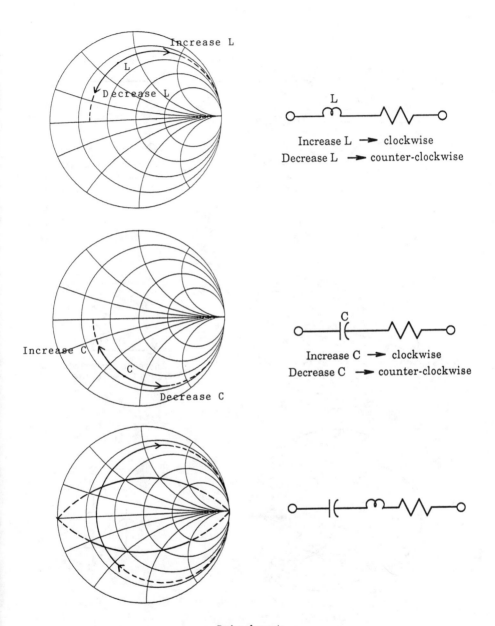

a. Series elements

**Fig. 2.16** Components of the Smith chart.

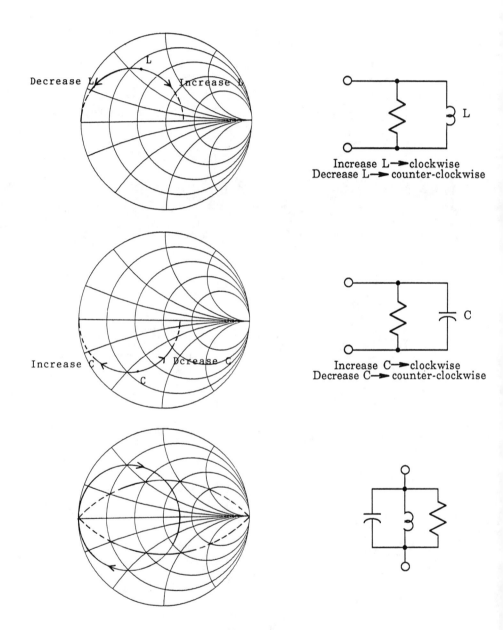

Increase L⟶clockwise
Decrease L⟶ counter-clockwise

Increase C⟶clockwise
Decrease C⟶ counter-clockwise

b. Shunt elements

**Fig. 2.16** cont.

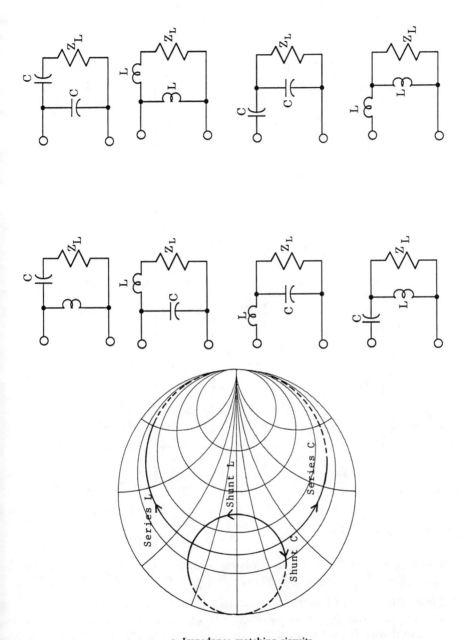

c. Impedance matching circuits

**Fig. 2.16** cont.

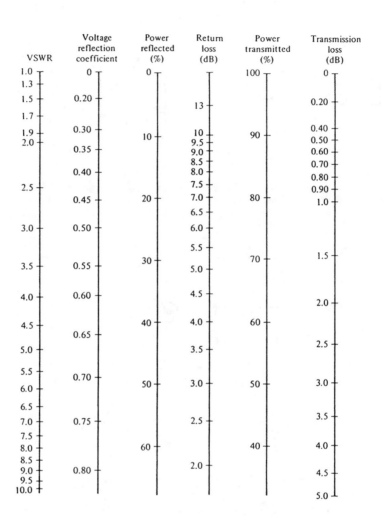

| VSWR | Voltage reflection coefficient | Power reflected (%) | Return loss (dB) | Power transmitted (%) | Transmission loss (dB) |
|------|------|------|------|------|------|
| 1.0 | 0 | 0 | | 100 | 0 |
| 1.3 | | | | | |
| 1.5 | 0.20 | | 13 | | 0.20 |
| 1.7 | | | | | |
| 1.9 | 0.30 | 10 | 10 | 90 | 0.40 |
| 2.0 | | | 9.5 | | 0.50 |
| | 0.35 | | 9.0 | | 0.60 |
| | | | 8.5 | | 0.70 |
| | 0.40 | | 8.0 | | 0.80 |
| 2.5 | | | 7.5 | | 0.90 |
| | 0.45 | 20 | 7.0 | 80 | 1.0 |
| | | | 6.5 | | |
| 3.0 | 0.50 | | 6.0 | | |
| | | | 5.5 | | |
| 3.5 | 0.55 | 30 | 5.0 | 70 | 1.5 |
| 4.0 | 0.60 | | 4.5 | | 2.0 |
| 4.5 | 0.65 | 40 | 4.0 | 60 | |
| 5.0 | | | 3.5 | | 2.5 |
| 5.5 | 0.70 | | | | |
| 6.0 | | 50 | 3.0 | 50 | 3.0 |
| 6.5 | | | | | |
| 7.0 | 0.75 | | 2.5 | | 3.5 |
| 7.5 | | | | | |
| 8.0 | | 60 | | 40 | 4.0 |
| 8.5 | | | 2.0 | | |
| 9.0 | 0.80 | | | | 4.5 |
| 9.5 | | | | | |
| 10.0 | | | | | 5.0 |

**Fig. 2.17** Nomograph of transmission line.

resulting in $Z_L = 2 + j \times 0.5$, and is represented on the Smith chart by point $A$ in Figure 2.19, which is the impedance at point $A - A'$ in Figure 2.18.

2. Point $A$ travels $\lambda/8$ toward the source in the same characteristic impedance circle and moves to point $B$ as shown in Figure 2.19 which corresponds to the point $B - B'$ in Figure 2.18. Normalized impedance at point $B - B'$ is given by $0.95 - j \times 0.75$.

3. Next, point $B$ is normalized to a characteristic impedance of 50 $\Omega$, which is given by $(0.95 - j \times 0.75) \times 100/50 = 1.9 - j \times 1.5$. This corresponds to point $C$ on the Smith chart in Figure 2.19 and point $C - C'$ in Figure 2.18.

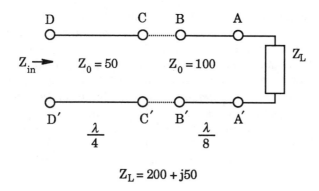

**Fig. 2.18** Circuit for change in characteristic impedance.

4. In order to find $Z_{in}$ at point $D - D'$ in Figure 2.18, the impedance point $C$ in the Smith chart moves toward the generator by $\lambda/4$, which is 180° and shown by point $D$ in Figure 2.19. Hence, the input impedance of the circuit in Figure 2.18 is given by $(0.32 + j \times 0.26) \times 50\ \Omega$, which is equal to a value of $(16.0 + j \times 15.0)\ \Omega$.

### 2.3.3 The Smith Chart and Admittance

When shunt elements are added to a line, it is convenient to work with admittances instead of impedances, since these shunt elements added to a line have their admittance added to the transformed admittance at the point where they are added. It is possible to use the Smith chart to work with admittances. The admittance is obtained by rotating the corresponding impedance by 180°. The relationship between admittance and impedance is $z = 1/y$, where $z$ and $y$ are the normalized impedance and admittance, respectively. As an example, consider impedance $z = 0.5 + j \times 0.5$ as represented by $z$ in Figure 2.20. If we rotate this vector by 180° around the center of the chart, the normalized admittance is obtained. This is represented by point $y$ in the figure:

$$y = 1.0 + j \times 1.0$$

In order to find the actual value of the admittance, the normalized admittance is multiplied by the characteristic admittance. In the example given, the characteristic admittance $y = 1/50 = .02$ is given by

$$y = (1.0 - j \times 1.0) \times .02$$
$$= 0.02 - j \times 0.02$$

This is a convenient method of working simultaneously in the impedance and admittance domains. However, sometimes it is quite inconvenient to rotate the vector by 180° on the Smith chart, particularly if several lengths of transmission lines and parallel elements are to be analyzed. A simpler method is to rotate the Smith chart around the impedance point on a normal Smith chart by 180° to find the admittance point and *vice versa*. Such a *z/y* Smith chart, where the normal Smith chart and another one rotated by 180° are drawn on top of each other is shown in Figure 2.21. The impedance point $z = 0.5 + j \times 0.5$ is plotted as point $z$ on a normal Smith chart in this figure. The same point $z$, when read on the admittance Smith chart provides the corresponding normalized admittance as given by $y = 1.0 - j \times 1.0$ in the figure.

### 2.3.4 Reflection Coefficient, VSWR, and Power

The reflection coefficient of an arbitrary load impedance is expressed as a magnitude and a phase angle. The Smith chart is set up with the reflection coefficient as the radial coordinate, and the circles concentric with the center of the unity circle are circles of constant reflection coefficient. These circles also are the contours of the constant standing wave ratio because the SWR is merely the magnitude of reflection coefficient. Reflection coefficient and VSWR can be found by using the scales at the bottom of the chart shown in Figure 2.15. The magnitude of the reflection coefficient is the distance from the center to the load impedance or admittance on the scale referenced as *Refl.-Coeff./Vol.* The reflection coefficient also can be found in terms of power on the scale labeled *Refl.-Coeff./Pwr*, which provides information about the ratio of power in the reflected wave to the power in the incident wave and is represented as $|\Gamma|^2$.

The VSWR is found in magnitude from the lower left scale, labeled *Standing Wave/Vol.* ratio, and also in dB labeled as *Standing Wave/in dB*. The relative reflected and transmitted powers in a transmission line can be determined from the lower right scale, which is labeled *Reflection/Loss in dB*. Loss in dB/Return, gives the return loss in dB, which is defined as the ratio of the reflected power to the incident power and expressed in terms of the reflection coefficient as

$$\text{return loss} = -20 \log_{10}|\Gamma| \text{ dB}$$

Loss in dB/Reflection provides the reflection loss in dB. This is the mismatch loss and defined as the ratio of power transmitted to the incident power.

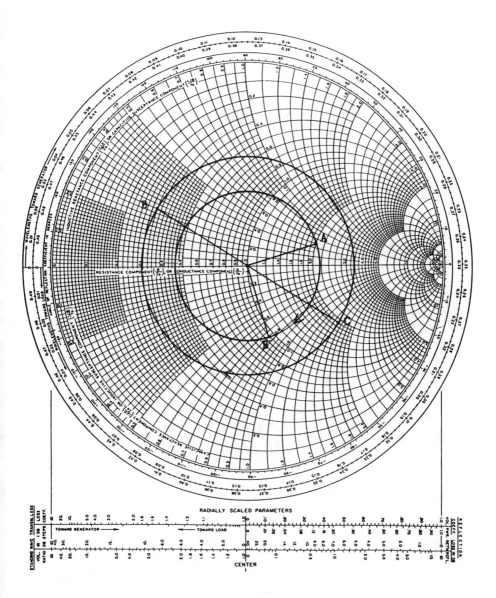

**Fig. 2.19** Smith chart for change in characteristic impedance.

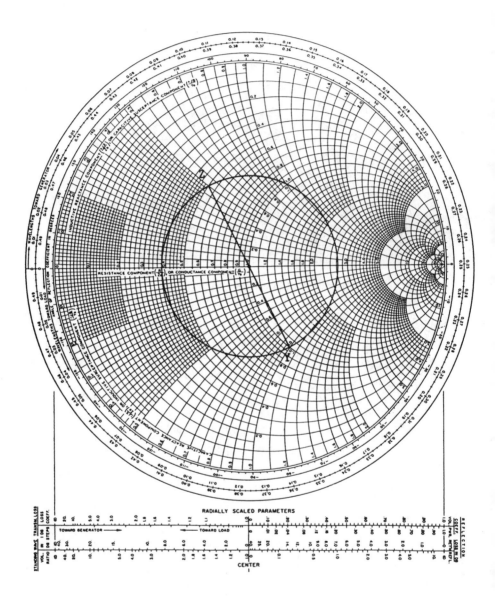

**Fig. 2.20** Rotating the vector by 180° to locate the admittance.

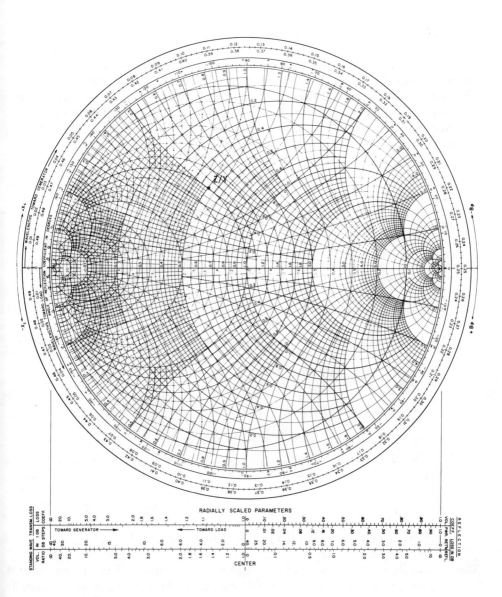

**Fig. 2.21** Rotating the Smith chart to locate the admittance.

### 2.3.5 Effects of Frequency Variation

Almost all practical design problems concern a certain bandwidth or frequency range. The Smith chart often is used for plotting the impedances or admittances of the same circuit at several frequencies. Since the magnitude of reactance depends on frequency, normalized impedance for a range of frequencies is represented by a line on the Smith chart. Assume a simple series resistance and inductance circuit shown in Figure 2.22 with $R = 25\ \Omega$ and $L = 10$ nH.

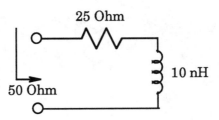

25 Ohm

50 Ohm

10 nH

**Fig. 2.22** Circuit for normalized impedance example.

The normalized impedance in a 50 $\Omega$ system of such a circuit can be given by

$$z_l = \frac{Z_L}{50\Omega} = \frac{25 + j2\pi f 10 \times 10^{-9}}{50}$$

$$= 0.5 + j0.4\pi f \times 10^{-9}$$

$$
\begin{aligned}
\text{for } f &= 1.0 \text{ GHz} & z_l &= 0.5 + j \times 1.26 \\
f &= 2.0 \text{ GHz} & z_l &= 0.5 + j \times 2.51 \\
f &= 3.0 \text{ GHz} & z_l &= 0.5 + j \times 3.77 \\
& \ \vdots \\
f &= 10.0 \text{ GHz} & z_l &= 0.5 + j \times 12.6
\end{aligned}
$$

These normalized impedance points are plotted in Figure 2.23 corresponding to each frequency. In this example the resistive part of the impedance is considered to be independent of frequency. If the resistive part increased with frequency, it would cause the locus of the impedance to shift inward as shown by the dotted line in Figure 2.23.

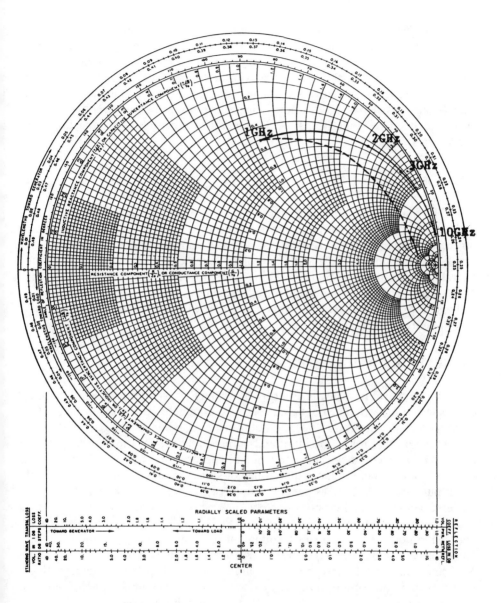

**Fig. 2.23** Smith chart for normalized impedance example.

Plots for resonant circuits can be represented on the Smith chart, too. These plots will be curved and can extend both above and below the horizontal line in the Smith chart. For example, the frequency response for a parallel resonant circuit will be a circle. At dc, the reactive part of the impedance is zero. As frequency increases, the shunt inductive reactance predominates. At resonance, the impedance is pure resistive, and at frequencies above the resonance frequency impedance is capacitive. Finally, as the frequency increases further, the impedance approaches zero.

### 2.3.6 Lossy Circuits

The Smith chart can be used to represent the lossy transmission lines. A line with a constant resistive impedance is represented by a constant resistance circle on the Smith chart. If the resistive part of the impedance of the line varies as a function of frequency, the impedance will be represented by a spiral as opposed to a circle. For example, in Figure 2.23 the solid plot represents the impedance of an $R$, $L$ circuit where $R$ is independent of frequency. If the loss in the circuit at 10 GHz is more than at 1 GHz, the impedance plot will correspond to the dotted plot in this figure.

## 2.4 NETWORK PARAMETERS

An electrical network is composed of many active and passive components connected in a particular configuration. The complete network can be analyzed once each component and its section of transmission line is characterized by a set of parameters that relate the input and output variables for the particular component. At lower frequencies it is convenient to describe electrical networks in terms of the set of parameters that require open or short circuit terminations at the input or output. Such parameters generally involve the network variables voltage and current. The most commonly used network parameters at low frequencies (approximately 30 MHz or less) are $Z$-, $Y$-, $ABCD$-, and $h$- parameters. However, at higher frequencies, because of terminal current and voltage measurement problems, electrical networks generally are represented in terms of scattering or $S$-parameters. All of these network parameters are capable of representing an electrical network of any number of ports, but to maintain simplicity only two-port networks will be considered in the following discussion.

### 2.4.1 Z, Y, and h Parameters

Figure 2.24 shows a general two-port network. The input current and voltage to the network are $I_1$ and $V_1$, respectively, and the output current and voltage are

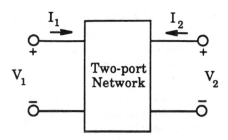

**Fig. 2.24** General two-port network.

$I_2$ and $V_2$ respectively. This two-port network can be represented in Z-, Y-, and h-parameters by the relations given in Equations (2.52–2.54), respectively. The following observations are made from these equations:

1. Z-parameters are defined as in an open circuit environment.
2. Y-parameters are defined as in a short circuit environment.
3. h-parameters are defined as in an open circuit input environment and a short circuit output environment.
4. Z-, Y-, and h-parameters are defined in terms of input and output voltages and currents.
5. Each parameter set is related to a set of four variables associated with the two-port model: two of these variables represent the excitation of the network (independent variables), and the other two represent the response of the network (dependent variables).

$$V_1 = Z_{11}I_1 + Z_{12}I_2$$
$$V_2 = Z_{21}I_1 + Z_{22}I_2 \tag{2.52}$$

$$I_1 = Y_{12}V_1 + Y_{12}V_2$$
$$I_2 = Y_{21}V_1 + Y_{22}V_2 \tag{2.53}$$

$$V_1 = h_{12}I_1 + h_{12}V_2$$
$$I_2 = h_{21}I_1 + h_{22}V_2 \tag{2.54}$$

At lower frequencies it is convenient to describe a network in terms of these parameters because the proper open and short circuit terminations can be realized practically and accurate measurement equipment and techniques are available. As the frequency of operation increases such measurements become increasingly difficult. A broadband open or short circuit is difficult to achieve at higher frequencies; a physical short circuit termination requiring a finite length conductor usually exhibits parasitic inductance, and a physical open circuit termination usually exhibits capacitance due to fringing electric fields. This results in problems measuring voltage and current at high frequencies without disturbing the circuit being measured. Also, active devices very often oscillate and are unstable when terminated in reactive loads such as open or short circuits. Finally, at high frequencies direct total voltage and current measurements are very difficult, if not impossible.

Consequently, there is a clear need for a set of network parameters that is represented in terms of the measurements that can be performed fairly accurately at high frequencies. At high frequencies, incident and reflected wave power measurements are performed comparatively easily. ABCD- and scattering or S-parameters are two examples of network parameters that are widely used at high frequencies to represent a microwave network.

### 2.4.2 ABCD-Parameters

ABCD-parameters, also known as *chain parameters,* are particularly useful to represent cascaded microwave networks. Consider a two-port network as shown in Figure 2.25, where $V_1$ and $I_1$ are the input voltage and current, respectively. $V_2$ and $I_2$ represent the output voltage and current, respectively. It is important to note that $I_2$ is the output current as opposed to the current entering into the output port as in Figure 2.24. The output current $I_2$ becomes the input current to the next two port networks cascaded to the first one. This particular feature of the ABCD-parameters makes them particularly useful for cascaded networks. The ABCD-parameters for such a network are defined in terms of voltage and current as follows:

$$V_1 = AV_2 + BI_2$$
$$I_1 = CV_2 + DI_2 \tag{2.55}$$

or

$$\begin{bmatrix} V_1 \\ I_1 \end{bmatrix} = \begin{bmatrix} A & B \\ C & D \end{bmatrix} \begin{bmatrix} V_2 \\ I_2 \end{bmatrix}$$

For a cascaded network as shown in Figure 2.26 the ABCD-parameters are derived as follows:

$$\begin{bmatrix} V_{11} \\ I_{11} \end{bmatrix} = \begin{bmatrix} A_1 & B_1 \\ C_1 & D_1 \end{bmatrix} \begin{bmatrix} V_{12} \\ I_{12} \end{bmatrix} \tag{2.56}$$

and

$$\begin{bmatrix} V_{21} \\ I_{21} \end{bmatrix} = \begin{bmatrix} A_2 & B_2 \\ C_2 & D_2 \end{bmatrix} \begin{bmatrix} V_{22} \\ I_{22} \end{bmatrix} \tag{2.57}$$

But from Figure 2.26

$$\begin{bmatrix} V_{12} \\ I_{12} \end{bmatrix} = \begin{bmatrix} V_{21} \\ I_{21} \end{bmatrix} \tag{2.58}$$

$$\begin{bmatrix} V_{11} \\ I_{11} \end{bmatrix} = \begin{bmatrix} A_1 & B_1 \\ C_1 & D_1 \end{bmatrix} \begin{bmatrix} A_2 & B_2 \\ C_2 & D_2 \end{bmatrix} \begin{bmatrix} V_{22} \\ I_{22} \end{bmatrix} \tag{2.59}$$

$$\begin{bmatrix} V_{11} \\ I_{11} \end{bmatrix} = \begin{bmatrix} A & B \\ C & D \end{bmatrix} \begin{bmatrix} V_{22} \\ I_{22} \end{bmatrix} \tag{2.60}$$

where, $[ABCD]$ in Equation (2.60) is the product of two matrices $[A_1 \, B_1 \, C_1 \, D_1]$ and $[A_2 \, B_2 \, C_2 \, D_2]$. Thus, a chain of network elements connected in cascade can be represented in [ABCD] parameters by multiplying their individual [ABCD] matrices.

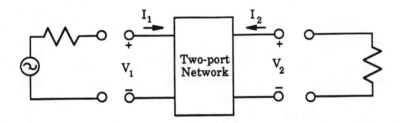

**Fig. 2.25** Two-port network with source and load.

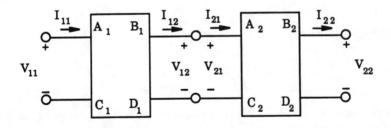

**Fig. 2.26** Cascade of two-port networks.

### 2.4.3 Scattering or S-Parameters

$S$-parameters are a set of network parameters that are used most commonly at microwave frequencies. Consider a transmission line of length $l$, characteristic impedance $Z_0$, and terminated with a load impedance of $Z_L$ as shown in Figure 2.27 (same as Figure 2.9). From Equations (2.23) and (2.24) the voltage and current at any point $x$ on such a transmission line are given by

$$V(x) = V^+e^{-\gamma x} + V^-e^{+\gamma x} \tag{2.61}$$

$$I(x) = \frac{1}{Z_0}\, V^+e^{-\gamma x} - V^-e^{+\gamma x} \tag{2.62}$$

where $V^+$ and $I^+$ are the incident voltage and current, respectively, and $V^-$ and $I^-$ are the reflected voltage and current, respectively. Solving these equations for incident and reflected voltages results in

$$V^+e^{-\gamma x} = \frac{1}{2}\,[V(x) + Z_0 I(x)] \tag{2.63}$$

$$V^-e^{+\gamma x} = \frac{1}{2}\,[V(x) - Z_0 I(x)] \tag{2.64}$$

Both equations are divided by $\sqrt{Z_0}$ and the incident and reflection parameters, $a$ and $b$, are defined at the termination point ($x = l$) by the relations

$$a \equiv \frac{V^+e^{-\gamma l}}{\sqrt{Z_0}} = \frac{1}{2}\left[\frac{V(l)}{\sqrt{Z_0}} + \sqrt{Z_0}I(l)\right] \tag{2.65}$$

$$b \equiv \frac{V^-e^{+\gamma l}}{\sqrt{Z_0}} = \frac{1}{2}\left[\frac{V(l)}{\sqrt{Z_0}} - \sqrt{Z_0}I(l)\right] \tag{2.66}$$

Dividing Equation (2.66) by (2.65) results in

$$b/a = \frac{V^-e^{+\gamma x}}{V^+e^{-\gamma x}} = \frac{V^-}{V^+}\, e^{-2\gamma x} \tag{2.67}$$

The numerator in Equation (2.67) represents the reflected voltage at the termination and the denominator represents the incident voltage at the termination. The ratio $b/a$ therefore represents $\Gamma$, the reflection coefficient at the termination.

Consider a two-port network as shown in Figure 2.28, where $a_1$ and $b_1$ are the incident and reflected parameters at the input port, $a_2$ and $b_2$ are the incident and reflected parameters at the output port, respectively. Extending Equations (2.65) and (2.66), these parameters can be represented as follows:

$$a_1 = \frac{1}{2}\left(\frac{V_1}{Z_0} + \sqrt{Z_0}\, I_1\right)$$

$$a_2 = \frac{1}{2}\left(\frac{V_2}{Z_0} + \sqrt{Z_0}\, I_2\right)$$

(2.68)

$$b_1 = \frac{1}{2}\left(\frac{V_1}{Z_0} - \sqrt{Z_0}\, I_1\right)$$

$$b_2 = \frac{1}{2}\left(\frac{V_2}{Z_0} - \sqrt{Z_0}\, I_2\right)$$

The scattering or S-parameters are defined by

$$b_1 = S_{11}a_1 + S_{12}a_2 \tag{2.69}$$

$$b_2 = S_{21}a_1 + S_{22}a_2 \tag{2.70}$$

The four S-parameters are defined in terms of incident and reflection parameters as follows:

$$S_{11} = \frac{b_1}{a_1}\bigg|_{a_2=0} \tag{2.71}$$

is the input reflection coefficient $\Gamma_i$ with output matched.

$$S_{12} = \frac{b_1}{a_2}\bigg|_{a_1=0} \tag{2.72}$$

is the reverse transmission coefficient with input matched.

$$S_{21} = \frac{b_2}{a_1}\bigg|_{a_2=0} \tag{2.73}$$

is the forward transmission coefficient with input matched.

$$S_{22} = \frac{b_2}{a_2}\bigg|_{a_1=0} \tag{2.74}$$

is the output reflection coefficient $\Gamma_0$ with input matched. The concept of S-parameters just defined for a two-port network can be extended to a generalized $n$-port network.

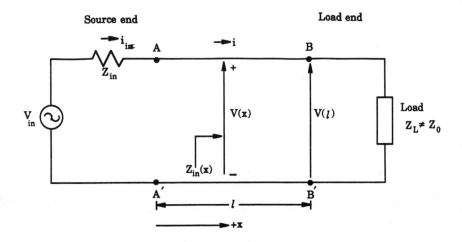

**Fig. 2.27** Transmission line of length *l* terminated in a load impedance.

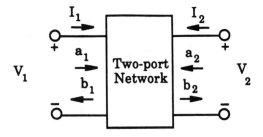

**Fig. 2.28** Two-port network defining *S*-parameters.

### 2.4.4 Properties of S-Parameters

The characteristic impedance $Z_0$ in the definition of the S-parameters is the main conceptual difference between the scattering parameters and other conventional lumped element network parameters such as Z, Y and h. The values of S-parameters for a network will be strongly dependent on the choice of $Z_0$, the characteristic impedance of the measuring system. From Equation (2.65) and (2.66), the port voltage and current are represented as follows:

$$V(l) = (a + b) \sqrt{Z_0} \tag{2.75}$$

$$I(l) = \frac{(a - b)}{\sqrt{Z_0}} \tag{2.76}$$

The power dissipation into the termination load is given by

$$P = \tfrac{1}{2} V(l)I^*(l) = \tfrac{1}{2} (aa^* - bb^*) \tag{2.77}$$

where $a^*$ is the complex conjugate of $a$. This equation demonstrates the conservation of energy, where the power dissipated in the termination is the difference between the incident power $1/2aa^*$ and the reflected power $1/2bb^*$. Extending the concept to a two-port network, the power delivered to port 1 and port 2 is given as follows:

$$P_1 = \tfrac{1}{2} (a_1a_1^* - b_1b_1^*) \tag{2.78}$$

$$P_2 = \tfrac{1}{2} (a_2a_2^* - b_2b_2^*) \tag{2.79}$$

The total power delivered to a two-port network must always be greater than or equal to zero:

$$\begin{aligned} P_{in} &= P_1 + P_2 \\ &= \tfrac{1}{2} (a_1a_1^* - b_1b_1^* + a_2a_2^* - b_2b_2^*) \geq 0 \end{aligned} \tag{2.80}$$

$$= (a^{*T}a - b^{*T}b) \geq 0 \tag{2.81}$$

where $a$ and $b$ are the matrix representations of elements $a_1$, $a_2$ and $b_1$, $b_2$, respectively, and $T$ is the matrix transpose. From Equations (2.71)–(2.74):

$$\mathbf{b} = \mathbf{S} \cdot \mathbf{a} \tag{2.82}$$

and performing the transpose operation on this equation yields

$$\mathbf{b}^T = (\mathbf{S} \cdot \mathbf{a})^T = \mathbf{a}^T \cdot \mathbf{S}^T \tag{2.83}$$

$$\mathbf{b}^{*T} = (\mathbf{S}^* \cdot \mathbf{a}^*)^T = \mathbf{a}^{*T} \cdot \mathbf{S}^{*T} \tag{2.84}$$

Therefore, from Equations (2.81)–(2.84):

$$\begin{aligned} 2\,P_{in} &= (\mathbf{a}^{*T} \cdot \mathbf{a} - \mathbf{a}^{*T} \cdot \mathbf{S}^{*T} \cdot \mathbf{S} \cdot \mathbf{a}) \geq 0 \\ &= \mathbf{a}^{*T} \cdot \mathbf{a} \cdot (\mathbf{I} - \mathbf{S}^{*T} \cdot \mathbf{S}) \geq 0 \end{aligned} \tag{2.85}$$

where $I$ is the identity matrix. Equation (2.85) will be valid only if

$$|\mathbf{I} - \mathbf{S}^{*T} \cdot \mathbf{S}| \geq 0 \tag{2.86}$$

When the network is lossless, the total power dissipated by the network itself is zero, or in other words, no attenuation of the wave is being transmitted through the network. Hence, from Equation (2.81):

$$P_{in} = 0$$

This defines the S-parameters for such a lossless network from Equation (2.86):

$$|\mathbf{I} - \mathbf{S}^{*T} \cdot \mathbf{S}| = 0$$

or

$$\mathbf{S}^{*T} \cdot \mathbf{S} = \mathbf{I} \tag{2.87}$$

For a two-port network, the equation can be written as follows:

$$\begin{vmatrix} S_{11}^* & S_{21}^* \\ S_{12}^* & S_{22}^* \end{vmatrix} \begin{vmatrix} S_{11} & S_{12} \\ S_{21} & S_{22} \end{vmatrix} = \begin{vmatrix} 1 & 0 \\ 0 & 1 \end{vmatrix}$$

A reciprocal network has identical transmission characteristics in the forward as well as in the reverse direction, which implies that the S-parameter matrix is equal to its transpose. In the case of a two-port network, such a condition is expressed by

$$S_{12} = S_{21}$$

In a matched two-port network with no reflection, the reflection coefficients $S_{11}$ and $S_{22}$ will be zero; that is:

$$S_{11} = S_{22} = 0$$

The VSWR of such network ideally is unity at the input as well as the output port.

As mentioned earlier, the scattering parameters depend strongly on the value of the characteristic impedance of the measurement system. It is often useful to convert the S-parameters of the network measured in $Z_0$ characteristic impedance to the corresponding S-parameters in $Z_0'$ characteristic impedance. Figure 2.29a shows an example of a two-port network in which the input port has a source with a characteristic impedance of $Z_0$ and the output port is terminated into $Z_0$ impedance with scattering parameters $S$. Figure 2.29b represents the same two-port network, with an input source impedance of $Z'$ and an output port terminated into $Z_0'$ load impedance, with corresponding scattering parameters $S'$. Then, the $S$

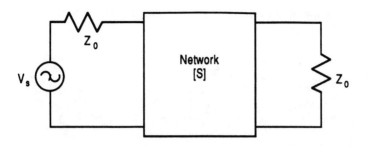

(a) Measured [S] in $Z_0$ impedance system

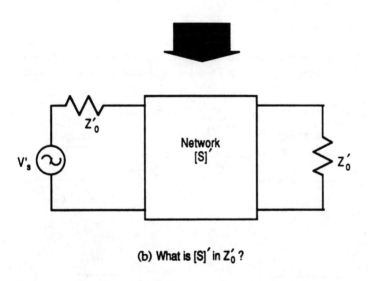

(b) What is $[S]'$ in $Z_0'$ ?

**Fig. 2.29** Change in characteristic impedance.

parameters in the $Z_0'$ impedance system are represented in terms of $S$ parameters in $Z_0$ impedance system by the following relationship:

$$S' = [S - \Gamma U] \cdot [U - \Gamma S]^{-1} \tag{2.88}$$

where

$$\Gamma = \frac{Z_0' - Z_0}{Z_0' + Z_0} \tag{2.89}$$

The reflection coefficient of new base impedance $Z_0'$ with respect to the impedance base $Z_0$ is given by

$$S_{11}' = \frac{(S_{11} - \Gamma)(1 - \Gamma S_{22}) + \Gamma S_{12} S_{21}}{(1 - \Gamma S_{11})(1 - \Gamma S_{22}) + \Gamma^2 S_{12} S_{21}}$$

$$S_{22}' = \frac{(S_{22} - \Gamma)(1 - \Gamma S_{11}) + \Gamma S_{12} S_{21}}{(1 - \Gamma S_{11})(1 - \Gamma S_{22}) + \Gamma^2 S_{12} S_{21}}$$

$$S_{21}' = \frac{S_{21}(1 - \Gamma^2)}{(1 - \Gamma S_{11})(1 - \Gamma S_{22}) + \Gamma^2 S_{12} S_{21}}$$

$$S_{12}' = \frac{S_{12}(1 - \Gamma^2)}{(1 - \Gamma S_{11})(1 - \Gamma S_{22}) + \Gamma^2 S_{12} S_{21}}$$

(2.90)

The properties of S-parameters for different types of electrical networks such as reciprocal, lossy, symmetrical, and lossless are summarized in Figure 2.30.

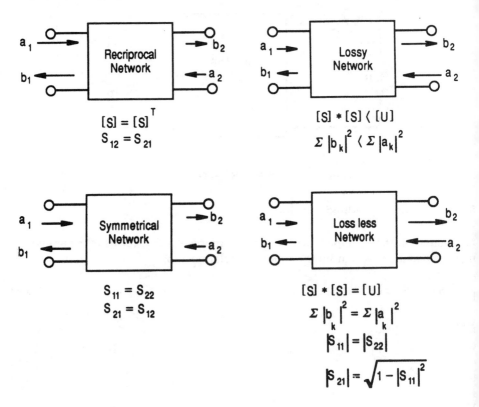

**Fig. 2.30** Properties of S-parameters for different types of networks.

## 2.4.5 Relationship between S-Parameters and Other Network Parameters

Because ABCD-parameters are more suitable to analyze cascade networks, it is sometimes desirable to convert the measured S-parameters into ABCD-parameters. The following equivalency demonstrates the conversion of S-parameters in ABCD-parameters and *vice versa*.

From S-matrix to ABCD-matrix:

$$A = (1 + S_{11} - S_{22} - \Delta S)\ \sqrt{Z_{01}/Z_{02}}\ /\ (2S_{21})$$
$$B = (1 + S_{11} + S_{22} + \Delta S)\ \sqrt{Z_{01}Z_{02}}\ /\ (2S_{21})$$
$$C = (1 - S_{11} - S_{22} + \Delta S)/(2S_{21}\ \sqrt{Z_{01}Z_{02}})$$
$$D = (1 - S_{11} + S_{22} - \Delta S)\ \sqrt{Z_{02}/Z_{01}}\ /\ (2S_{21})$$

where $Z_{01}$ and $Z_{02}$ are normalizing impedances for S-parameters at ports 1 and 2, respectively, and

$$\Delta S = (S_{11}S_{22} - S_{21}S_{12}).$$

Note that for $S_{21} = 0$, ABCD-parameters become indeterminate. The parameter $S_{21}$ represents the forward transmission coefficient and seldom is zero in microwave circuits.

From ABCD-matrix to S-matrix:

$$S_{11} = \frac{AZ_{02} + B - CZ_{01}Z_{02} - DZ_{01}}{AZ_{02} + B + CZ_{01}Z_{02} + DZ_{01}}$$

$$S_{12} = \frac{2(AD - BC)\ \sqrt{Z_{01}Z_{02}}}{AZ_{02} + B + CZ_{01}Z_{02} + DZ_{01}}$$

$$S_{21} = \frac{2\ \sqrt{Z_{01}Z_{02}}}{AZ_{02} + B + CZ_{01}Z_{02} + DZ_{01}}$$

$$S_{22} = \frac{-AZ_{02} + B - CZ_{01}Z_{02} + DZ_{01}}{AZ_{02} + B + CZ_{01}Z_{02} + DZ_{01}}$$

The factor $(AD - BC)$ is equal to unity for reciprocal networks, and therefore $S_{12}$ becomes equal to $S_{21}$.

The port voltage and current are expressed in terms of port incident and reflection parameters by the Equations (2.75) and (2.76). The S-parameters are defined in terms of a and b parameters according to Equations (2.81)–(2.84) and other network parameters such as Z-, Y-, and h-parameters are defined in terms of terminal voltage and current in Equations (2.52)–(2.54). Using the relationships in Equations (2.85) and (2.86), the S-parameters can be represented in terms of Z-, Y-, and h-parameters and *vice versa,* as shown in Table 2.1.

**Table 2.1** Conversion Equations between Z-, Y-, H-, and S-Parameters (Normalized to $Z_0$)

$$S_{11} = \frac{(Z_{11} - 1)(Z_{22} + 1) - Z_{12}Z_{21}}{(Z_{11} + 1)(Z_{22} + 1) - Z_{12}Z_{21}} \qquad Z_{11} = \frac{(1 + S_{11})(1 - S_{22}) + S_{12}S_{21}}{(1 - S_{11})(1 - S_{22}) - S_{12}S_{21}}$$

$$S_{12} = \frac{2Z_{12}}{(Z_{11} + 1)(Z_{22} + 1) - Z_{12}Z_{21}} \qquad Z_{12} = \frac{2S_{12}}{(1 - S_{11})(1 - S_{22}) - S_{12}S_{21}}$$

$$S_{21} = \frac{2Z_{21}}{(Z_{11} + 1)(Z_{22} + 1) - Z_{12}Z_{21}} \qquad Z_{21} = \frac{2S_{21}}{(1 - S_{11})(1 - S_{22}) - S_{12}S_{21}}$$

$$S_{22} = \frac{(Z_{11} + 1)(Z_{22} - 1) - Z_{12}Z_{21}}{(Z_{11} + 1)(Z_{22} + 1) - Z_{12}Z_{21}} \qquad Z_{22} = \frac{(1 + S_{22})(1 - S_{11}) + S_{12}S_{21}}{(1 - S_{11})(1 - S_{22}) - S_{12}S_{21}}$$

$$S_{11} = \frac{(1 - Y_{11})(1 + Y_{22}) + Y_{12}Y_{21}}{(1 + Y_{11})(1 + Y_{22}) - Y_{12}Y_{21}} \qquad Y_{11} = \frac{(1 + S_{22})(1 - S_{11}) + S_{12}S_{21}}{(1 + S_{11})(1 + S_{22}) - S_{12}S_{21}}$$

$$S_{12} = \frac{-2Y_{12}}{(1 + Y_{11})(1 + Y_{22}) - Y_{12}Y_{21}} \qquad Y_{12} = \frac{-2S_{12}}{(1 + S_{11})(1 + S_{22}) - S_{12}S_{21}}$$

$$S_{21} = \frac{-2Y_{21}}{(1 + Y_{11})(1 + Y_{22}) - Y_{12}Y_{21}} \qquad Y_{21} = \frac{-2S_{21}}{(1 + S_{11})(1 + S_{22}) - S_{12}S_{21}}$$

$$S_{22} = \frac{(1 + Y_{11})(1 - Y_{22}) + Y_{21}Y_{12}}{(1 + Y_{11})(1 + Y_{22}) - Y_{12}Y_{21}} \qquad Y_{22} = \frac{(1 + S_{11})(1 - S_{22}) + S_{12}S_{21}}{(1 + S_{22})(1 + S_{11}) - S_{12}S_{21}}$$

$$S_{11} = \frac{(h_{11} - 1)(h_{22} + 1) - h_{12}h_{21}}{(h_{11} + 1)(h_{22} + 1) - h_{12}h_{21}} \qquad h_{11} = \frac{(1 + S_{11})(1 + S_{22}) - S_{12}S_{21}}{(1 - S_{11})(1 + S_{22}) + S_{12}S_{21}}$$

$$S_{12} = \frac{2h_{12}}{(h_{11} + 1)(h_{22} + 1) - h_{12}h_{21}} \qquad h_{12} = \frac{2S_{12}}{(1 - S_{11})(1 + S_{22}) + S_{12}S_{21}}$$

$$S_{21} = \frac{-2h_{21}}{(h_{11} + 1)(h_{22} + 1) - h_{12}h_{21}} \qquad h_{21} = \frac{-2S_{21}}{(1 - S_{11})(1 + S_{22}) + S_{12}S_{21}}$$

$$S_{22} = \frac{(1 + h_{11})(1 - h_{22}) + h_{12}h_{21}}{(h_{11} + 1)(h_{22} + 1) - h_{12}h_{21}} \qquad h_{22} = \frac{(1 - S_{22})(1 - S_{11}) - S_{12}S_{21}}{(1 - S_{11})(1 + S_{22}) + S_{12}S_{21}}$$

## 2.5 NOISE PARAMETERS

In a two-port linear network the noise entering the system at the input port is transmitted to the output in the same way as the input signal. Additionally, noise at the output port includes noise added by the internal components of the network itself. There are two main sources of the noise added by the network components: thermal noise and shot noise.

### 2.5.1 Thermal Noise

Also referred to as *Johnson noise*, thermal noise is associated with random motion of electrons due to the thermal agitation in a conductor. This gives rise to a random ac voltage within the conductor. The added noise power from such a thermal noise source is given by

$$P_n = kTB \text{ W} \tag{2.91}$$

where $k = 1.38 \times 10^{-23}$ J/K is Boltzmann's constant
$T$ = absolute temperature in kelvins
$B$ = noise bandwidth in Hz

An example of a thermal noise source is a resistor. A noise equivalent circuit for a resistor is shown in Figure 2.31a. This circuit consists of a noise voltage source, $V_n$, in series with an ideal noiseless resistor, $R$. When the noise resistor $R$ is equal to the load resistor $R_L$, the maximum noise power is available from $V_n$. The mean square thermal voltage at a temperature $T$ and a frequency bandwidth $B$ can be evaluated using Equation (2.91):

$$\overline{V_n^2} = V_{n\text{rms}}^2 = 4kTBR \tag{2.92}$$

Similarly, the noisy resistor can be represented by a Norton equivalent circuit, consisting of noise current source $I_n$ in parallel with $G = 1/R$, as shown in Figure 2.31b. The mean-square thermal noise current is given by

$$\overline{I_n^2} = I_{n\text{rms}}^2 = 4kTBG \tag{2.93}$$

The maximum available noise power from such a resistor can be written as follows:

$$P_{\text{max}} = \frac{V_n^2}{4R} = kTB \tag{2.94}$$

This concept of noise voltage can be extended to any other device. The noise voltage output from a device can be represented in terms of the equivalent noise resistance given by Equation 2.92. Consider the case of two resistors, of values $R_1$ and $R_2$ both at a temperature, $T$, connected in series. The mean-square voltage of the noise generated by each resistor in bandwidth $B$ is given by

$$\overline{V_{n1}^2} = 4kTBR_1$$
$$\overline{V_{n2}^2} = 4kTBR_2 \tag{2.95}$$

As the thermal noise for both of these resistors are two independent Gaussian random processes, the mean-square voltage measured across the series combination of $R_1$ and $R_2$ can be written as follows:

$$\overline{V_n^2} = \overline{V_{n1}^2} + \overline{V_{n2}^2} = 4kTB(R_1 + R_2) \tag{2.96}$$

This is because, for independent Gaussian random processes, the variance of the sum is equal to the sum of the variances of the individual processes.

The result shown in Equation (2.96) should have been anticipated, since it indicates that the two noisy resistors, $R_1$ and $R_2$, connected in series can be represented by a single noisy resistor, $R = R_1 + R_2$. The same concept can be extended to a more complex network of series and parallel combination of resistors.

### 2.5.2 Shot Noise

Shot noise, also called *Schottky noise*, is generated in active devices due to the random variation in the number of carriers that constitute the current flow in such devices. An example of shot noise is a randomly varying noise current superimposed on the dc current of the drain of a FET. The mean-square shot noise current in a noise bandwidth of $B$ is given by

$$I_s^2 = 2qI_{\text{dc}}B \tag{2.97}$$

where $q = 1.6 \times 10^{-19}$ coulombs is the electron charge and $I_{\text{dc}} = $ dc current through the device.

As shown in Equations (2.92) and (2.97), the noise power is dependent on the system bandwidth irrespective of the center frequency. Such a distribution of noise is called *white noise*.

### 2.5.3 Noise in Two-Port Network

Consider a noisy two-port network with an impedance $Z$, as shown in Figure 2.32a. This generalized two-port network can be represented by the equivalent circuits in Figures 2.32b and 2.32c. The two-port noisy network is replaced by a noise-free network, the noise current sources $i_1$ and $i_2$ and the noise voltage sources $e_1$ and $e_2$ at the input and output, respectively. For the convenience of

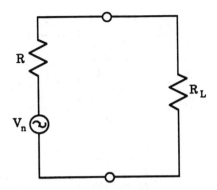

a. Thevenin equivalent circuit

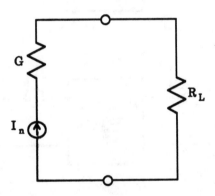

b. Norton equivalent circuit

**Fig. 2.31 Equivalent circuits of noisy resistors.**

analysis, all the noise sources at the output port are referred to at the input port, as shown in Figure 2.32d. The resulting noise sources $e$ and $i$ can be obtained in terms of the network parameters. At the input port in Figure 2.32d:

$$I'_1 = I_1 - i$$
$$V'_1 = V_1 - e$$

(2.98)

Using the definition of Z-parameters from Section 2.4:

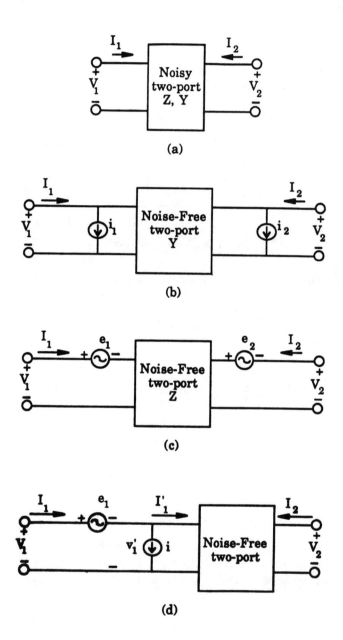

**Fig. 2.32** (a) Two-port network with internal noise sources; (b) equivalent circuit with external noise current sources; (c) equivalent circuit with external noise voltage sources; and (d) equivalent circuit with external noise voltage and current sources transferred to input port.

$$V_1' = Z_{11}I_1' + Z_{12}I_2$$
$$V_2 = Z_{21}I_1' + Z_{22}I_2$$

(2.99)

Solving Equations (2.98) and (2.99):

$$V_1 = Z_{11}(I_1 - i) + Z_{12}I_2 + e$$
$$V_2 = Z_{21}(I_1 - i) + Z_{22}I_2$$

(2.100)

Also, from the two-port network in Figure 2.32c:

$$V_1 = Z_{11}I_1 + Z_{12}I_2 + e_1$$
$$V_2 = Z_{21}I_1 + Z_{22}I_2 + e_2$$

(2.101)

Comparing Equations (2.100) and (2.101) yields

$$i = -e_2/Z_{21}$$
$$e = e_1 - e_2Z_{11}/Z_{21}$$

(2.102)

Now consider the case in which a voltage source $v_s$ of internal impedance $Z_s = R_s + jX_s$ is connected at the input port of the noisy two-port network in Figure 2.32d. Also suppose that the input impedance of the network is $Z = R + jX$, as shown in Figure 2.33. The noise voltage $v_n'$ and $v_s'$ from voltage sources $v_n$ and $v_s$, respectively, are given by

$$v_n' = v_n \frac{Z}{Z + Z_s}$$

(2.103)

$$v_s' = v_s \frac{Z}{Z + Z_s}$$

(2.104)

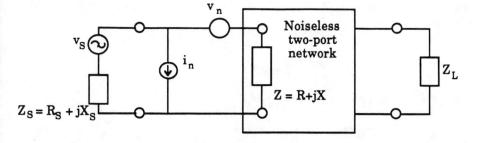

**Fig. 2.33** A noisy two-port network with input signal.

Similarly the noise voltage $v_i'$ across $Z$, from the input noise current source $i_n$, can be written as

$$v_i' = i_n \frac{ZZ_s}{Z + Z_s} \tag{2.105}$$

The noise figure represents the degradation of the signal-to-noise ratio between the input and the output of the two-port network. In other words, the noise figure is defined as the signal-to-noise ratio at the input to the signal-to-noise ratio at the output:

$$NF = \frac{(S/N)_{in}}{(S/N)_{out}} \tag{2.106}$$

If $G$ represents the forward power gain of the network, then

$$NF = \frac{S/N_{in}}{GS/G(N_{in} + N_a)} \tag{2.107}$$

where $N_{in}$ is the input noise power to the network and $N_a$ is the additional noise power added by the network.

Therefore, the noise figure of the network is given by

$$NF = 1 + \frac{N_a}{N_{in}} \tag{2.108}$$

When noise associated with the network is represented in terms of power, it is conventionally called the *noise figure* (NF), whereas when represented in voltage, it is termed the *noise factor* (F). The mean-square input noise power to the two-port network is

$$N_{in} = \overline{v_s'^2} \, Re\left(\frac{1}{Z}\right) \tag{2.109}$$

and the mean-square noise power at the input due to the noise generated by the network itself is

$$N_a = \overline{(v_n' + v_i')^2} \, Re\left(\frac{1}{Z}\right) \tag{2.110}$$

Under matching conditions $Re\left(\frac{1}{Z}\right) = Re\left(\frac{1}{Z_s}\right)$:

$$\frac{N_a}{N_{in}} = \frac{\overline{(v_n' + v_i')^2}}{\overline{v_s'^2}} \tag{2.111}$$

Solving Equations (2.108) and (2.111), the noise figure is given by

$$F = 1 + \frac{\overline{(v_n' + v_i')^2}}{\overline{v_s'^2}} \tag{2.112}$$

Using Equations (2.103)–(2.105):

$$F = 1 + \frac{\overline{(v_n + i_n Z_s)^2}}{v_s^2}$$

$$= 1 + \frac{\overline{(v_n + i_n Z_s)(v_n + i_n Z_s)^*}}{v_s^2}$$

(2.113)

Evaluating the numerator of Equation (2.113):

$$\overline{(v_n + i_n Z_s)(V_n + i_n Z_s)^*} = \overline{v_n^2} + |Z_s|^2\, \overline{i_n^2} + \overline{Z_s i_n v_n^*} + \overline{Z_s^* i_n^* v_n}$$
$$= \overline{v_n^2} + |Z_s|^2\, \overline{i_n^2} + R_s\overline{(i_n v_n^* + i_n^* n_n)}$$
$$+ jX_s\overline{(i_n v_n^* - i_n^* v_n)}$$

(2.114)

There is a correlation between the two noise sources, $v_n$ and $i_n$. The noise voltage $v_n$ can be divided into two parts: one part $v_u$ not correlated to $i_n$ and the second part $(v_n - v_u)$ fully correlated to $i_n$. Defining a correlation impedance $Z_c = R_c + jX_c$:

$$v_n = v_u + Z_c i_n$$

(2.115)

Then

$$\overline{i_n v_n^* + i_n^* v_n} = 2R_c \overline{i_n^2}$$
$$j\overline{(i_n v_n^* - i_n^* v_n)} = 2X_c \overline{i_n^2}$$

(2.116)

Solving Equations (2.115) and (2.116) and rewriting Equation (2.113), the noise figure can be defined as

$$F = 1 + \frac{\overline{v_n^2} + |Z_s|^2 \overline{i_n^2} + 2R_s R_c\, \overline{i_n^2} + 2X_s X_c \overline{i_n^2}}{\overline{v_s^2}}$$

(2.117)

From the definition of equivalent noise resistance given in Section 2.5.1, the mean-square noise associated with the input source $v_s$ and the network noise source $v_n$ and $i_n$ can be written as

$$\overline{v_s^2} = 4kTR_s B$$
$$\overline{v_s^2} = 4kTR_s B$$
$$\overline{v_i^2} = 4kTG_p B$$

(2.118)

where $R_s$ = equivalent input source noise resistance, the real part of $Z_s$
$\qquad R_n$ = equivalent noise resistance of the two-port network
$\qquad G_p$ = equivalent noise conductance of the two-port network

Solving Equations (2.117) and (2.118), the noise factor $F$ is given by

$$F = 1 + \frac{R_n}{R_s} + \frac{G_p}{R_s}(R_s^2 + X_s^2) + 2R_c G_p + 2X_c G_p \frac{X_s}{R_s} \qquad (2.119)$$

We observe from this equation that the noise figure is a function of the source impedance $Z_s$. There is an optimum input impedance, $Z_s = Z_{opt}$, for which the noise factor is minimum, $F_{min}$. By satisfying the conditions of minima in Equation (2.119); that is, $\partial F / \partial X_s = 0$ and $\partial F / \partial R_s = 0$, the value of optimum input impedance, $Z_{opt} = R_{opt} + jX_{opt}$, can be calculated. The optimum values of $R_s$ and $X_s$ at minimum noise factor are given by

$$R_{opt} = \left(\frac{R_n}{G_p} - X_c^2\right)^{1/2} \qquad (2.120)$$

$$X_c^2 \leq \frac{R_n}{G_p} \qquad (2.121)$$

$$X_{opt} = -X_c \qquad (2.122)$$

and the minimum noise factor is expressed as

$$F_{min} = 1 + 2 G_p\left[R_c + \left(\frac{R_n}{G_p} - X_c\right)^{1/2}\right] \qquad (2.123)$$

It is not possible to evaluate the noise figure from Equation (2.119) because $R_n$, $G_p$, $R_c$, and $X_c$ cannot be measured directly. It is important to write the expression for the noise figure in terms of the parameters that can be directly measured, such as $Z_{opt}$ and $F_{min}$. Solving Equations (2.119)–(2.123), the expression for noise figure can be written as

$$\begin{aligned}
F &= F_{min} + \frac{R_n}{R_s(R_{opt}^2 + X_{opt}^2)}[(R_s - R_{opt})^2 + (X_s - X_{opt})^2] \\
&= F_{min} + \frac{R_n}{R_s}\left|\frac{Z_s}{Z_{opt}} - 1\right|^2
\end{aligned} \qquad (2.124)$$

Similarly, the noise factor can be expressed in terms of admittances:

$$F = F_{min} + \frac{R_s}{G_s}[(G_s - G_{opt})^2 + (B_s - B_{opt})^2]$$

$$= F_{min} + \frac{R_n}{G_s}|Y_s - Y_{opt}|^2 \tag{2.125}$$

Using the normalized impedances and admittances:

$$F = F_{min} + \frac{r_n}{r_s}\left|\frac{z_s}{z_{opt}} - 1\right|^2 \tag{2.126}$$

$$F = F_{min} + \frac{r_n}{g_s}[(g_s - g_{opt})^2 + (b_s - b_{opt})^2] \tag{2.127}$$

where $r_n = \dfrac{R_n}{R_0}$

$$y_s = R_0 Y_s = g_s + jb_s$$
$$y_{opt} = R_0 Y_{opt} = g_{opt} + jb_{opt}$$

At high frequencies, it is often convenient to define reflection coefficients instead of impedances, as discussed in Section 2.4. The noise figure can also be expressed in terms of the input reflection coefficient $\Gamma_s$ and the optimum input reflection coefficient $\Gamma_{opt}$. Defining

$$\Gamma_s = \frac{1 - y_s}{1 + y_s} \tag{2.128}$$

$$\Gamma_{opt} = \frac{1 - y_{opt}}{1 + y_{opt}}$$

and solving Equations (2.125) and (2.128), the noise figure can be expressed as

$$F = F_{min} + 4r_n \frac{|\Gamma_s - \Gamma_{opt}|^2}{(1 - |\Gamma_s|^2)(1 + |\Gamma_{opt}|^2)} \tag{2.129}$$

The value of $r_n$ can be evaluated by measuring the noise figure of the network at $\Gamma_s = 0$ (i.e., $Z_s = R_0$). In this condition, the noise factor can be written from Equation (2.129):

$$F_0 = F_{min} + 4r_n \frac{|\Gamma_{opt}|^2}{|1 + \Gamma_{opt}|^2}$$

$$r_n = (F_0 - F_{min})\frac{|1 + \Gamma_{opt}|^2}{4|\Gamma_{opt}|^2} \tag{2.130}$$

### 2.5.4 Noise Figure and Smith Chart

In order to design circuits with noise figure considerations, it is essential to represent noise parameters in terms of matching circuitry. This is possible by representing the noise parameters on the Smith chart. Define a noise figure parameter $N$ as follows:

$$N = \frac{F - F_{min}}{4r_n} |1 + \Gamma_{opt}|^2 \tag{2.131}$$

Solving Equations (2.129) and (2.131):

$$N(1 - |\Gamma_s|^2) = |\Gamma_s - \Gamma_{opt}|^2 = (\Gamma_s - \Gamma_{opt})(\Gamma_s^* - \Gamma_{opt}^*)$$
$$N - N|\Gamma_s|^2 = |\Gamma_s|^2 + |\Gamma_{opt}|^2 - 2 \, \text{Re}(\Gamma_{opt}^* \, \Gamma_s)$$
$$|\Gamma_s|^2(1 + N) + |\Gamma_{opt}|^2 - 2 \, \text{Re}(\Gamma_{opt}^* \, \Gamma_s) = N$$

Multiplying both sides by $(1 + N)$:

$$|\Gamma_s|^2(1 + N)^2 + |\Gamma_{opt}|^2(1 + N) - 2(1 + N) \, \text{Re}(\Gamma_{opt}^* \, \Gamma_s) = N(1 + N)$$
$$|\Gamma_s|^2(1 + N)^2 + |\Gamma_{opt}|^2 - 2(1 + N) \, \text{Re}(\Gamma_{opt}^* \, \Gamma_s) = N^2 + N(1 - |\Gamma_{opt}|^2)$$
$$|\Gamma_s(1 + N) - \Gamma_{opt}|^2 = N^2 + N(1 - |\Gamma_{opt}|^2)$$
$$\left|\Gamma_s - \frac{\Gamma_{opt}}{1 + N}\right|^2 = \frac{N^2 + N(1 - |\Gamma_{opt}|^2)}{(1 + N)^2} \tag{2.132}$$

Equation (2.132) represents a family of circles, where the centers of the circles are $c_F$ on the Smith chart and the radii are $r_F$, given by the following relations:

$$C_F = \frac{\Gamma_{opt}}{1 + N} \tag{2.133}$$

$$r_F = \frac{1}{1 + N} [N^2 + N(1 - |\Gamma_{opt}|^2)] \tag{2.134}$$

For low-noise circuits, the objective is to minimize the value of $N$, which is given by Equation (2.131). For $F = F_{min}$, the value of $N$ is zero and the center of $F_{min}$ circle with zero radius is located at opt, the optimum reflection coefficient on the Smith chart. The centers of other noise figure circles lie along the $\Gamma_{opt}$ vector.

## 2.5.5 Noise Temperature

Noise temperature is an alternative noise representation of the network, closely related to the noise figure. Traditionally, the noise temperature is frequently used to specify the noise in antenna and receivers. Noise temperature, $T_n$, of a circuit is defined as the temperature at which source resistance $R_s$ must be held to a value such that the noise output from the circuit due to $R_s$ equals the noise output due to the circuit itself. Although noise in a network may emanate from many sources, the noise temperature concept describes the resulting noise in terms of thermal noise having the equivalent noise power. In other words, if thermal noise were the only source of noise, the network equivalent noise temperature would be the temperature the network must be maintained at which to produce the same amount of noise power that it generates. The noise figure of a network is given by Equation (2.108), which can be represented in terms of $R_{eq}$, the equivalent noise resistance of the network, as

$$F = 1 + \frac{4kTR_{eq}B}{4kTR_sB} \tag{2.135}$$

where $R_s$ is the source resistance.

The equivalent noise temperature is defined:

$$TR_{eq} = T_n R_s \tag{2.136}$$

then, from Equations (2.135) and (2.136):

$$F = 1 + \frac{T_n}{T} \tag{2.137}$$

where $T_n$ is the equivalent noise temperature of the network.

Similarly, the equivalent noise temperature $T_n$ can be represented in terms of noise figure $F$:

$$T_n = (F - 1)T \tag{2.138}$$

If ideally the network is noiseless, the equivalent noise temperature $T_n$ is absolute zero as $F = 1$ or 0/dB for noiseless network.

## 2.5.6 Noise Figure and Noise Voltage

Conventionally, in microwaves, the noise associated with a network is defined in terms of noise figure in dB; and often it is necessary to estimate the total equivalent input noise voltage of a network as well. It is desirable to develop a relationship between the noise figure and its corresponding value in noise voltage. If the noise figure of the network is $NF$ dB, the corresponding noise factor in terms of voltage is given by

$$F = 10^{NF/20} \tag{2.139}$$

Assume the two-port network in Figure 2.33 to be an amplifier with transfer function $A_0$. The signal-to-noise voltage ratio in the bandwidth $B$ at the input of the network is given by

$$\left(\frac{S}{N}\right)_{\text{in}} = \frac{v_s}{(4kTR_sB)^{1/2}} \tag{2.140}$$

Also, solving Equations (2.111), (2.117), and (2.118) and for simplicity, assuming that the noise sources $v_n$ and $i_n$ are uncorrelated, the signal-to-noise voltage ratio at the output is represented as follows:

$$\left(\frac{S}{N}\right)_{\text{out}} = \frac{A_0 v_s}{(4kTR_sB + v_n^2 + i_n^2 Z_s^2)^{1/2}} \tag{2.141}$$

From Equations (2.140) and (2.141), the noise factor $F$ is represented as follows:

$$\begin{aligned} F &= \frac{(4kTR_sB + v_n^2 + i_n^2 Z_s^2)^{1/2}}{(4kTR_sB)^{1/2}} \\ &= \left(1 + \frac{v_n^2 + i_n^2 Z_s^2}{4kTR_sB}\right)^{1/2} \end{aligned} \tag{2.142}$$

Then, the noise figure in dB is given by

$$\begin{aligned} NF &= 20 \log_{10}\left(1 + \frac{v_n^2 + i_n^2 Z_s^2}{4kTR_sB}\right)^{1/2} \\ NF &= 10 \log_{10}\left(1 + \frac{v_n^2 + i_n^2 Z_s^2}{4kTR_sB}\right) \end{aligned} \tag{2.143}$$

and total root-mean-square equivalent noise voltage at the input is

$$(v_t)_{in} = (4kTR_sB + v_n^2 + i_n^2 Z_s^2)^{1/2}$$
$$= 10^{NF/20}(4kTR_sB)^{1/2} \quad \text{volt} \tag{2.144}$$

If the noise is white in nature, it is conventional to define root-mean-square noise voltage independent of the bandwidth and represented in $V\sqrt{Hz}$

$$v_{in} = 10^{NF/20}(4kTR_sB)^{1/2} \quad V/\sqrt{Hz} \tag{2.145}$$

If the noise voltage is dependent on the frequency, the total noise voltage can be evaluated by integrating Equation (2.145) with frequency. For example, if the noise figure of a network is 0.5 dB, the equivalent root-mean-square noise voltage at 300 K in a 50 $\Omega$ input resistance system is calculated using Equation (2.145):

$$v_{in} = 10^{5/20}(4k \times 300 \times 50)^{1/2}$$
$$= 1.62 \quad mV/\sqrt{Hz}$$

If the bandwidth of the network is 10 GHz, the total noise voltage is given by

$$(v_t)_{in} = 162 \ \mu V$$

# REFERENCES

1. L. N. Dworsky, *Modern Transmission Line Theory and Applications*, John Wiley and Sons, New York, 1979.
2. W. Hilberg, *Electrical Characteristics of Transmission Lines*, Artech House, Dedham, MA, 1979.
3. R. E. Collin, *Foundations for Microwave Engineering*, McGraw-Hill, New York, 1966.
4. J. A. Kong, *Electromagnetic Wave Theory*, John Wiley and Sons, New York, 1986.
5. G. Kennedy and R. W. Tinnel, *Electronic Communication Systems*, McGraw-Hill, New York, 1970.
6. P. H. Smith, *Electronic Application of the Smith Chart*, McGraw-Hill, New York, 1969.
7. P. H. Smith, *Smith Chart Accessories*, Analog Instruments Co., Murray Hill, NJ.
8. H. A. Wheeler, "R. F. Transition Losses," *Electronics*, Vol. 9, January 1936, pp. 26–27, 46.
9. "An Improved Transmission Line Calculator," *Electronics*, Vol. 17, January 1944, pp. 130–133, 318–325.
10. J. F. White, "The Smith Chart: An Endangered Species?" *Microwave J.*, November 1979, pp. 49–54.
11. Robert Thomas, *A Practical Introduction to Impedance Matching*, Artech House, Dedham, MA, 1976.
12. K. C. Gupta, R. Garg, and I. J. Bahl, *Computer-Aided Design of Microwave Circuits*, Artech House, Norwood, MA, 1981.
13. Theodre Saad, ed., *Microwave Engineer's Handbook*, Vol. 2, Artech House, Dedham, MA, 1971.
14. Paul Gray and Robert G. Meyer, *Analysis and Design of Analog Integrated Circuits*, John Wiley and Sons, New York, 1977.
15. Herbert Taub and Donald Schilling, *Principles of Communication Systems*, McGraw-Hill, New York, 1971.
16. John C. Hancock, *An Introduction to the Principles of Communication Theory*, McGraw-Hill, New York, 1961.
17. W. Stephen Cheung and Frederic H. Levien, eds., *Microwaves Made Simple: Principles and Applications*, Artech House, Norwood, MA, 1985.
18. S. J. Algeri, W. S. Cheung, and L. S. Stark, *Microwaves Made Simple—The Workbook*, Artech House, Norwood, MA, 1986.

# Chapter 3
# *MMIC Material and Manufacturing Technology*

*R. Goyal*

## 3.1 MATERIAL TECHNOLOGY AND CHARACTERIZATION

The ultimate performance of a *monolithic microwave integrated circuit* (MMIC) is determined by the quality of the starting material. The best design and processing techniques cannot overcome poor quality material. In addition, designers must be aware of performance limits introduced by the processing technology. Variations in device parameters due to inhomogeneity of starting material and process fluctuations must be taken into account by the circuit designer.

Essentially all MMICs processed today are fabricated on thin wafers of GaAs, which range from 2 to 4 inches in diameter. Due to the relative brittleness of the material, GaAs wafers are thicker than silicon wafers of the same diameter. The wafers first are rough cut from a boule of material and then etched and polished, which can be done on one or both sides. High-quality GaAs substrates are semiinsulating with resistivities in excess of $10^7$ to $10^8$ $\Omega$-cm. This property, along with its excellent electron mobility (up to 8900 compared to 1500 cm$^2$/V-s for silicon at 300 K), makes GaAs an ideal material for MMICs. In order to grow material with these properties, stringent control of stoichiometry and impurity types and levels must be maintained. Because arsenic has a much higher vapor pressure than gallium, particular care must be taken to prevent the loss of arsenic during material growth and in any subsequent high-temperature processing steps.

Two techniques have evolved for the growth of high-quality GaAs. The *horizontal Bridgeman* (HB) process produces ingots by crystallization of molten GaAs in a quartz boat using a slowly moving temperature gradient. *Liquid encapsulated Czochralski* (LEC) growth is a technique, where long, uniformly round GaAs boules are grown under a cap layer of boric oxide.

### 3.1.1 Horizontal Bridgeman Technique

Figure 3.1[1] is a schematic of the apparatus used to growh GaAs using the HB technique. Elemental Ga and As are placed in separately sealed and evacuated ampoules connected by a passageway. The As chamber is heated to 613°C; at this temperature, the solid As sublimes, reaching a vapor pressure of 1 atmosphere. Arsenic vapor is transported through the passageway to the gallium boat, which is held slightly above the melting point of GaAs (1240°C). Once GaAs forms as a liquid, a solid-liquid interface is moved through the boat, either by moving the boat relative to the heater (HB) or, alternatively, by programming the temperature profile in the furnace, the so-called gradient freeze method.

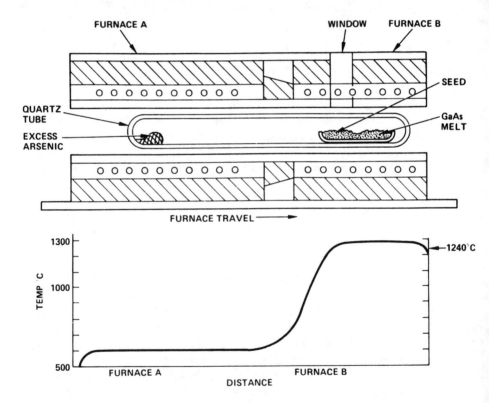

**Fig. 3.1** Horizontal Bridgeman growth apparatus (from [1], p. 26).

GaAs crystals grown by the HB method are nearly dislocation free ($<10^3$ dislocations/cm$^3$) due to the well-controlled stoichiometry afforded by the sealed system and the very uniform thermal gradients that can be maintained. HB material generally results in higher and more uniform activation of ion implanted species during MMIC processing than in material grown by the liquid encapsulated Czochralski method. The best low-noise FETs have been fabricated in buffer layers epitaxially grown on HB GaAs.

Several drawbacks are associated with HB material, however. The resistivity of HB material is not as high as LEC-grown GaAs, possibly due to incorporation of Si from the quartz boat during crystal growth. Small amounts of Cr are added to the melt to compensate for the Si. This Cr is rather mobile in GaAs and has been shown to redistribute during high-temperature processing, particularly during ion-implantation anneal cycles, changing the material properties and sometimes adversely affecting device performance. A second major disadvantage of the HB method is the resulting "D" shape of the GaAs crystal due to the growth from the liquid state in the horizontal boat. These odd shaped wafers are hard to process as most modern processing equipment is designed to handle round wafers. Also, boat-grown methods produce (111) oriented crystals rather than the preferred (100) orientation used for the fabrication of electronic devices. Obtaining the round (100), oriented wafers is costly due to off-axis slicing of the wafers from the boule and the excessive amounts of material lost in grinding to form the round wafers.

### 3.1.2 Liquid Encapsulated Czochralski Technique

The majority of wafers supplied to the industry today are grown by the LEC technique. Figure 3.2[1] shows a cross section of a LEC puller. The LEC technique produces cylindrical boules up to about 5 kg in weight with the preferred (100) crystal orientation. Typical LEC material can have very high resistivity, exceeding $10^8$ $\Omega$-cm, without adding Cr for compensation; however, it suffers from higher dislocation densities ($>10^4$ cm$^{-2}$) than HB material and is less uniform, both radially and from top to bottom of the boule.

As we can see from Figure 3.2, liquid boric oxide, $B_2O_2$, surrounds the GaAs melt, separating the GaAs charge from the crucible and reducing the volatilization of As. An inert gas pressure in excess of the vapor pressure of the melt, on the order of a few atmospheres for low-pressure growth up to 75 atmospheres for high-pressure pullers, prevents the As from volatilizing during growth. Both high- and low-pressure techniques produce GaAs of similar quality.

The initial charge of GaAs can be presynthesized in an HB furnace or synthesized in the puller from elemental Ga and As. Direct synthesis of the GaAs is preferred for high-quality, semiinsulating material. During high pressure synthesis, elemental gallium and arsenic are placed in a high-purity, *pyrolytic boron nitride* (PBN) crucible along with solid $B_2O_2$. For low-pressure growth, gallium

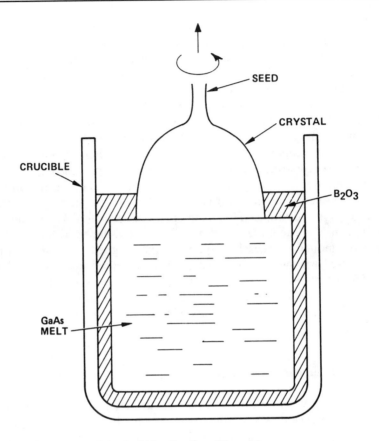

**Fig. 3.2** Liquid encapsulated Czochralski puller (from [1], p. 29).

resides in the crucible while As vapor is injected from a separate source. PBN generally is used, as it is more inert than quartz and does not contribute Si impurities to the melt. The temperature in the puller is slowly raised. At 450°C, the $B_2O_2$ melts and encapsulates the starting materials, preventing As loss. The melt is heated to the melting point of GaAs (1240°C), a seed crystal is introduced through the $B_2O_2$ cap into the melt, and the GaAs crystal is pulled from the melt. Constant rotation of the crystal and close monitoring of the thermal gradients are critical to growing uniform material. A typical growth run lasts between 24 and 48 hours, depending upon the initial charge. The crystal diameter is monitored by crystal weight sensors, and the growth parameters are adjusted continuously to maintain a tight diameter control. High-quality wafers of up to 100 mm diameter have been grown by low pressure LEC techniques.

As noted previously, dislocation densities in excess of $10^4$ cm$^{-2}$ are typical of LEC wafers. The dislocations are distributed in a characteristic W shape pattern across the wafer, with a typical standard deviation of less than 10%. These dislocations are undesirable; although their effect on the microwave performance of devices is not fully understood, digital devices experience a voltage threshold variation that is correlated strongly to dislocation density variations. Attempts at lowering dislocation densities by varying temperature gradients and arsenic overpressure during growth have been unsuccessful so far. However, whole-boule annealing at 800°C for 10–12 hours has been shown to reduce EL2 levels (a deep level directly correlated to dislocation density) by almost a factor of 2.

Ongoing research efforts to reduce dislocations in LEC material include adding a few percent of In to the melt to harden the crystal against dislocation formation [2] and growing material in transverse magnetic fields to reduce fluctuations in melt temperature and associated turbulence. Although essentially zero-dislocation wafers have resulted from adding In to the melt, the inhomogeneous distribution of In throughout the boule due to partitioning of the indium between the solid and molten phases during growth leads to nonuniform properties in wafers cut from different portions of the boule. The addition of In also increases the chances of twinning (boules with sections of material oriented in different crystallographic planes) during crystal growth, limiting the size of single crystal boules that can be grown. The addition of transverse magnetic fields reduces insular inclusions in the crystal but does not appear to improve dislocation density. Improved FET performance due to transverse magnetic field growth has yet to be demonstrated.

### 3.1.3 Epitaxial Growth

Whereas the bulk of GaAs MMICs are fabricated by ion implantation into semiinsulating LEC substrates, a number of applications can be better served using an epitaxial fabrication process. Some of these areas include low-noise FETs, where the addition of a thick, high quality buffer layer can significantly lower device noise figures; power FETs, which benefit from highly tailored, active layer profiles that are difficult to achieve using ion implantation alone; high electron mobility transistors, which feature a GaAs/AlGaAs heterostructure to separate donor ions from mobile electrons in the device channel; and heterojunction bipolar transistors, which require not only heterojunctions but also precise control of base thicknesses and doping. Although epitaxially grown material is more expensive than horizontal Bridgeman or LEC grown material, the increased cost is more than offset by improved device performance. Several epitaxial technologies have evolved, including *liquid phase epitaxy* (LPE), *vapor phase epitaxy* (VPE), *molecular beam epitaxy* (MBE), and *organometallic vapor phase epitaxy* (OMVPE). Table 3.1 outlines the important characteristics of each of these techniques.

**Table 3.1** Epitaxial Technology for MMICs

|  | *LPE* | *VPE* | *MBE* | *OMVPE* |
|---|---|---|---|---|
| Growth rate | 1 μm/min | 1 μm/min | 1 μm/hr | 1 μm/min |
| Interface abruptness (A°) | 100–1000 | 100 | 5 | 10 |
| Uniformity (across wafer) |  |  |  |  |
|   Doping | Good | Good | Excellent | Excellent |
|   Thickness | Poor | Good | Excellent | Excellent |
| Throughput (wafers/run) | 1 | Tens | 1 | Tens |
| AlGaAs/GaAs heterostructures | Yes | No | Yes | Yes |
| Reproducibility (wafer-to-wafer) | Good | Good | Excellent | Excellent |
| Background carrier concentration (cm$^{-3}$) | ~$10^{14}$ | ~$10^{14}$ | ~$10^{14}$ | ~$10^{14}$ |
| GaAs mobility at 77 K (cm$^2$/V-s) | 180 K | 200 K | 140 K | 210 K |

### 3.1.3.1 Liquid Phase Epitaxy

Liquid phase epitaxy is the oldest and simplest technique used to grow epitaxial GaAs. It is used today primarily in the growth of material for optoelectronic components such as LEDs; very little LPE material is used for microwave devices due to the relatively poor surface finish, poor interface abruptness between layers for different doping levels, and lack of doping uniformity, characteristic of LPE growth, over large areas.

Figure 3.3[3] is a schematic of an LPE reactor. GaAs wafers are held face down in a slider that can be moved to bring the wafers in contact with one or more wells of molten material. Both *p*- and *n*-type dopants can be introduced into the molten material; *p-n* junctions are grown by making contact between the wafer and each type well in turn. During growth, the temperature profiles are such that the melts are supercooled to just below their solidification points. When the wafer contacts the melt, material solidifies on the surface.

### 3.1.3.2 Vapor Phase Epitaxy

Vapor phase epitaxy has proven to be a more versatile growth technique for microwave devices than LPE. In the VPE process, gallium and arsenic in the gas

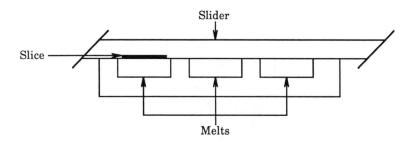

**Fig. 3.3** A liquid phase epitaxy reactor (from [3], p. 43).

phase are mixed together and passed over the GaAs substrate, where the gases react depositing high quality GaAs films. Compared to LPE, VPE offers better control of dopant profiles as well as better surface morphology and abrupt junction formation, at the expense of more costly and sophisticated equipment.

Two main processes can be used in VPE: the halide and the hydride processes. In both cases, gas is injected through a liquid Ga reservoir before moving downstream to the substrate. Figure 3.4 shows a cross-section of a vapor phase epitaxy reactor. The halide process reacts $AsCl_3$ with Ga in the presence of $H_2$ carrier gas producing GaCl, $As_4$, $As_2$, and HCl. These reaction gases are transported to the substrate, which is held at a lower temperature. Once the gases reach the substrate, GaAs films are grown according to the following reaction: $4\,GaCl + As_4 + 2H_2 \rightarrow 4\,GaAs + 4\,HCl$. The hydride process proceeds similarly, except that HCl rather than $AsCl_3$ is passed through the liquid Ga to produce GaCl; and arsenic is provided by injecting arsine, $AsH_3$. In both techniques, layers can be doped by adding suitable doping gases, such as $H_2S$ for $n$-type material.

Both hydride and halide VPE produce high-quality GaAs layers. The best reported low-noise FETs were fabricated on VPE material. Slightly $n$-doped material with background carrier concentrations of $10^{14}$ cm$^{-3}$ and 77 K mobilities approaching 200,000 cm$^2$/V-s have been reported.

However, there are a number of disadvantages to VPE growth. Film growth takes place in a "hot wall" reactor, where the quartz reactor tubes are located inside a clamshell or muffle furnace. The heated walls combine with the highly reactive gases, particularly HCl, to release Si into the gas stream that dopes the epitaxial films, leading to unwanted variations in doping profiles. Gas flow through the reactor is limited by the need to allow $AsCl_3$ or HCl enough time to react sufficiently with the liquid gallium source. Thus, the gas residence time in the reactor is long, and abrupt changes in doping concentrations are difficult. Finally, AlGaAs layers cannot be grown in a hot wall reactor system due to the reaction of Al with any oxygen in the system, a serious limitation as HEMTs and HBTs become increasingly important microwave devices.

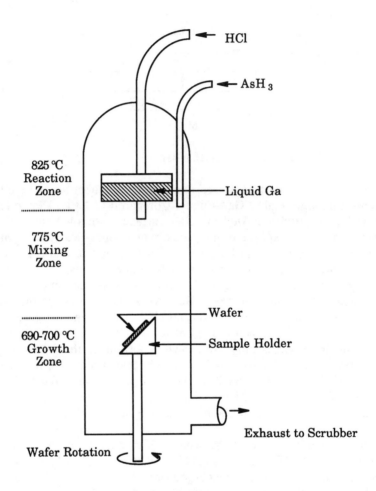

$$x\ Ga + HCl \xrightarrow{825°C} x\ GaCl + (1 - x)\ HCl + x/2\ H_2$$

$$AsH_3 \xrightarrow{800°C} (1 - y)AsH_3 + (1 - z)\ y/4\ As_4 + (zy/2)As_2 + 3/2yH_2$$

$$GaCl + 1/3\,AsH_3 + 1/6As_2 + 1/12As_4 \xrightarrow{\sim700°C} GaAs + HCl$$

**Fig. 3.4** A vapor phase epitaxy reactor.

### 3.1.3.3 Molecular Beam Epitaxy

Molecular beam epitaxy is an excellent growth technique for microwave device materials, offering precise doping control and extremely abrupt junction formation. MBE has been used to grow GaAs, AlGaAs, and other III–V materials in an ultra-high vacuum environment, and high-performance devices have been fabricated on MBE material.

As shown in Figure 3.5[4], an MBE system consists of an ultra-high ($10^{-11}$ torr base pressure) vacuum chamber; in situ analytical equipment to monitor the substrate, vacuum, and source materials; and multiple sources of growth and doping materials. Before growth begins, the sample is heated to approximately 600°C in an arsenic ambient to desorb any residual oxide, leaving an atomically clean surface for epitaxial growth. A typical MBE system has several individually controlled source ovens; one or more ovens each contain gallium, aluminum, arsenic, or dopants such as silicon ($n$-type) or beryllium ($p$-type). A variant of MBE, called organometallic MBE, utilizes gas rather than solid sources. Each source can be individually shuttered to prevent material from impinging on the substrate. To begin growth, the Ga and As shutters are opened. Growth is controlled by the arrival rate of Ga atoms at the substrate; an excessive flux of As is maintained to prevent decomposition of the GaAs. Growth rates of 1 atomic layer per second (1 $\mu$m/hour) are typical.

Dopants can be introduced by opening the shutter in front of the appropriate cell. Dopant levels are set by controlling the flux of atoms arriving at the substrate by varying the temperature of the cell containing the dopant material.

Because of the very clean, high-vacuum environment and the slow growth rate offered by an MBE system, very high-quality materials have been demonstrated with abrupt changes in doping levels, on the order of an atomic layer. However, a major drawback to MBE growth is the limited system throughput due to the slow growth rate and the capacity of growing on only a single wafer at a time. A second drawback is the high cost of an ultra-high vacuum system. Manufacturers of MBE systems are beginning to design machines that can grow on more than one wafer at a time. Unless these efforts are successful, MBE will remain a costly process for epitaxial growth.

### 3.1.3.4 Organometallic Vapor Phase Epitaxy

Perhaps the most versatile and cost-effective epitaxial growth technique for microwave devices is *organometallic vapor phase epitaxy* (OMVPE). This process offers the high throughput of a CVD system as well as the precise doping and thickness control of MBE. Systems that can process up to twenty-five 2-inch GaAs wafers in a single run have been developed.

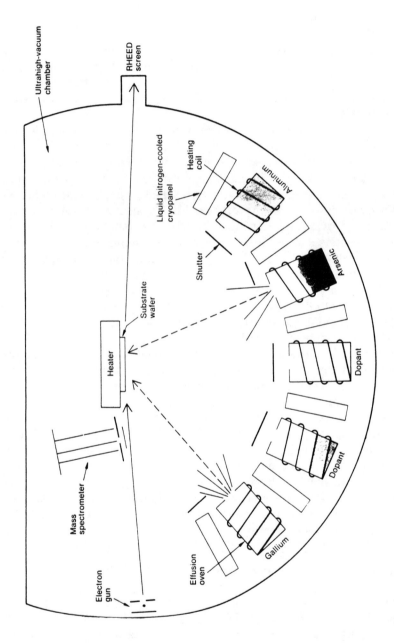

**Fig. 3.5** The growth chamber of a molecular beam epitaxy chamber (from [4], p. 18).

As seen in Figure 3.6, an OMVPE reactor looks much like a VPE reactor, but with several important differences. All of the sources used in an OMVPE system are gases: arsine for As, trimethyl gallium ($Ga(CH_3)_3$) for Ga, and $H_2S$ for an *n*-type dopant are typical, although other organometallics have been used. This eliminates the need for a liquid Ga source as in VPE and allows the gas throughput to be greatly increased. This higher throughput, in turn, reduces the time gas resides in the reactor and facilitates very rapid changes in doping gas concentrations at the GaAs substrate, resulting in very abrupt (nearly as good as MBE) changes in doping concentration in epitaxial films. A second difference between OMVPE and VPE is that the OMVPE reactor is a cold wall reactor. Neither the gases nor the walls of the reactor are heated. The substrate is supported by a susceptor, usually made of graphite, that is heated inductively by an external RF coil. Hence, the reaction of the process gases with the quartzware is greatly reduced, alleviating the problem of unintentional Si doping. Furthermore, since oxygen is not introduced by reaction with the quartz, high-quality AlGaAs films can be grown in an OMVPE reactor.

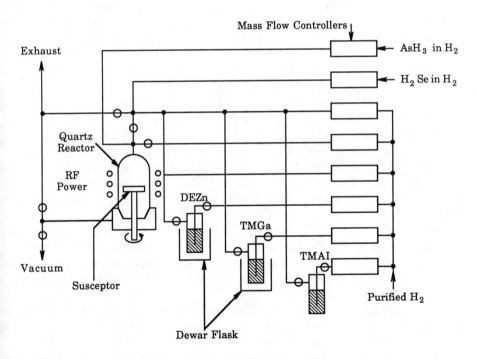

**Fig. 3.6** Cross section of an organometallic vapor phase epitaxy reactor.

### 3.1.3.5 Gallium Arsenide-on-Silicon Material

Although GaAs is the material of choice for high-performance MMICs, several disadvantages are associated with its use. GaAs wafers are more than an order of magnitude more expensive than silicon wafers of comparable diameters and not available in diameters greater than 100 mm. They are more brittle than silicon, more likely to break during processing, decreasing device yields and raising overall costs. And the thermal conductivity of GaAs is much lower than silicon (0.55 *versus* 1.5 W/cm-K), which limits the power handling capability of GaAs MMICs.

If the electrical properties of gallium arsenide could be combined with the mechanical properties of silicon, an almost ideal material for microwave devices and circuits would result. Recently, this combination has been realized by growing thin epitaxial layers of GaAs on silicon substrates using either MBE or OMVPE techniques. Microwave devices were fabricated using standard processing techniques in the GaAs epitaxial layer.

Several problems must be addressed before GaAs-on-Si technology can be used for microwave devices. Although GaAs and Si have the same crystal structure (face centered cubic), there is a lattice constant mismatch of approximately 4% between them. This mismatch causes many types of defects during GaAs epitaxial growth, including edge dislocation, stacking faults, and threading dislocations. The chief problem is the threading dislocations, which are generated at the GaAs-Si interface and propagate up into the GaAs film. Current GaAs-on-silicon substrates have between $10^7$ and $10^8$ dislocations/cm$^2$, far above typical bulk GaAs values of $10^4$/cm$^2$. These dislocations degrade device performance and can contribute to enhanced diffusion of Si into the GaAs epi layers. Several techniques for reducing the number of dislocations are under investigation, including reducing epi growth rates and temperatures, adding interfacial layers of various materials to pin the dislocations at the interface, and postgrowth annealing.

A second serious problem is the large difference in the thermal expansion coefficients of GaAs and Si. Epitaxial growth is carried out at elevated temperatures of at least a few hundred degrees centigrade. When the substrates cool to room temperature, a tensile stress is introduced in the GaAs film, which can lead to bowing of the substrate, making device fabrication difficult. Tensile stresses also can reduce device lifetimes, although long-term device reliability studies have yet to be carried out. Research efforts are directed at lowering the growth temperature to reduce stress accumulation in the GaAs films.

Even with these serious problems, device quality GaAs-on-Si substrates have been demonstrated, and both microwave FETs and lasers have been fabricated successfully with performance comparable to devices fabricated on bulk GaAs. If long-term device reliability can be established, a viable GaAs-on-Si technology could emerge.

### 3.1.4 Materials Analysis

Key to the growth of high-quality semiconductor materials is the capability to rapidly and accurately assess relevant material characteristics either during the growth process or immediately afterwards. Growth conditions can be adjusted to optimize material yield based on test data. A plethora of analysis techniques are available for material evaluation, evaluating the starting raw materials, the boules, and epitaxial layers.

Many of these procedures are of particular benefit to the materials growth. For example, gallium arsenide boules need to be evaluated for stoichiometry, dislocation density, type conversion, impurity levels, and compensation. Wafers are evaluated for surface finish, flatness, resistivity, and mobility. Dopant concentration and distribution, thickness, lattice constant, mobility, and sheet carrier concentration are some of the characteristics important in epitaxial layers. Tests range from relatively straightforward techniques, such as the use of a mechanical profilometer or an optical gauge to measure film thicknesses, to sophisticated techniques, such as Auger or *secondary ion mass spectrometry* (SIMS) analysis for analyzing material at the atomic level, which requires very expensive, dedicated equipment.

It is not the intent in this section to cover the realm of material testing in depth. The reader is referred to several review articles in this field for further information [5–7]. Rather, this section will concentrate on a few, widespread electrical tests that are used to generate data needed by the circuit designer. The relevant parameters needed for accurate modeling and circuit design are resistivity, doping profile, and mobility.

#### 3.1.4.1 Resistivity Measurements

The resistivity of gallium arsenide material varies widely, depending upon the type of material and the dopants present. Semi-insulating gallium arsenide wafers can exhibit resistivities of $10^8$ $\Omega$-cm or higher, whereas highly doped ($10^{18}$ $cm^{-3}$) contact layers can be as low as $10^{-3}$ $\Omega$-cm.

Traditionally, resistivity measurements have been carried out using four contacts on a bar-shaped sample, as shown in Figure 3.7. Current is passed through the outer two contacts, and the corresponding voltage drop across the sample measured between the inner two contacts. Then, if the cross-section and the length of the sample between the two voltage contacts is known, the calculation of the resistivity (in $\Omega$-cm) is straightforward.

In many cases, however, it is not convenient or possible to use a bar-shaped sample, particularly for the thin films or ion-implanted layers encountered in MMIC manufacturing. For films on semiinsulating GaAs substrates, it is possible to etch through the film itself, and then to measure the resistivity as just described.

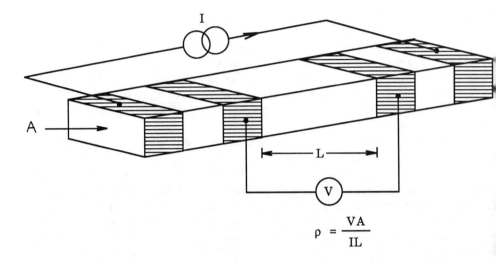

$$\rho = \frac{VA}{IL}$$

**Fig. 3.7** Traditional bar-shaped sample for resistivity measurements.

Such structures are found commonly in *process control monitors* (PCMs) and used to evaluate both ion-implanted and epitaxial layers. A photograph of a four-point pattern used in a PCM can be found later in this chapter, in Figure 3.49.

In materials evaluation, an alternate approach is to form four contacts on the periphery of an arbitrarily shaped sample. Usually, the shape is a square or cloverleaf, as shown in Figure 3.8, with ohmic contacts formed either at the corner of the square or on the leaves of the cloverleaf. Samples of this shape can also be used to measure Hall mobility. To measure resistivity, current is passed between two adjacent contacts and the corresponding voltage recorded between the two remaining contacts.

### 3.1.4.2 Doping Profile

The most effective way to determine the doping level and distribution in a GaAs sample is to measure the *capacitance-voltage* (C-V) characteristics of a reverse-biased diode. This technique is widespread due to many common metals that can be used to form Schottky diodes on GaAs, the accuracy inherent to this technique, and the availability of equipment for measurement.

In C-V profiling, when a reverse bias is applied to the Schottky diode, a depletion layer forms beneath the anode; the thickness of the depletion layer is dependent upon the voltage applied and the free carrier concentration in the material. The capacitance associated with the depletion layer can be determined and in turn the doping density can be calculated. By varying the voltage, a profile can be generated, relating the number of carriers *versus* depth into the sample.

The diode usually is formed by evaporating a thin metal contact onto the sample. An alternate approach, called *electrolytic profiling,* forms a diode between a conductive etching solution and the semiconductor sample. In this technique, the depletion layer capacitance is measured as the sample is etched away, forming a new diode at the electrolyte-semiconductor interface at each step. This technique is particularly useful for structures with both *p*- and *n*-type layers. As each layer is etched through, the doping profiles of the individual layers can be accurately assessed.

### 3.1.4.3 Mobility

The mobility of thin films is the benchmark used to evaluate the quality of epitaxially grown materials, as it is determined strictly by the quality of the material and does not depend upon any other measurement, such as sample thickness. Mobility typically is measured using the Hall effect. This technique is used not only for thin films but also for ion-implanted layers and substrate material.

The Hall effect causes electrons to deflect when moving in a perpendicular magnetic field. Using a cloverleaf or square structure (as described in Figure 3.8), current is passed between two opposing contacts. The sample is placed in a magnetic field so that the field causes the electrons to deflect, generating a Hall voltage that can be measured across the remaining two contacts. At equilibrium, the Hall voltage offsets the effect of the magnetic field. The material mobility and sheet carrier concentration can be determined if the current and Hall voltage are known.

### 3.1.5 Advanced Device Structures

Although the GaAs-based MESFET has been and will continue to be the workhorse of MMIC technology, other devices under development promise higher performance than the conventional MESFET, albeit at increased materials growth and processing complexity. Many of these devices are beginning to emerge as commercially viable alternatives for high-performance MMICs. Some of these devices are AlGaAs/GaAs high electron mobility transistors (HEMTs), heterojunction bipolar transistors (HBTs), and indium phosphide-based devices.

### 3.1.5.1 AlGaAs/GaAs HEMTs

In a conventional MESFET, the charge carriers are confined to a channel within a few thousands Å of the surface. For *n*-type devices, the charge carriers are electrons contributed by donor impurities in the channel region. Electrons

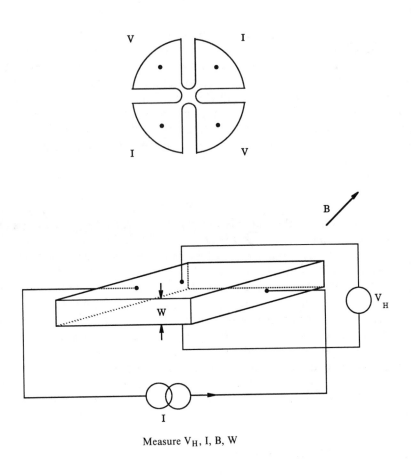

Measure $V_H$, I, B, W

Determine $\mu$, N

**Fig. 3.8** Cloverleaf-shaped Hall sample.

moving from source to drain scatter from these donor atoms and interact with surface states during device operation. These effects limit the frequency response of a MESFET and increase the overall noise associated with this device, primarily as $1/f$ noise.

A HEMT is a transistor that uses heterojunction technology to separate the electrons from the donor atoms and confine the electrons to a very narrow channel (on the order of 50 Å) far away from the surface. Figure 3.9 compares MESFET and HEMT cross sections. In the MESFET, the channel is created by adding dopants, usually silicon atoms, at a level of $10^{17}$ donors/cm³. The HEMT is a multilayer device, consisting of an undoped GaAs conduction layer, a highly

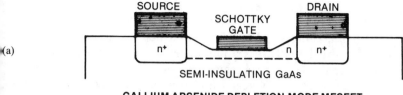

(a)

**GALLIUM ARSENIDE DEPLETION-MODE MESFET**

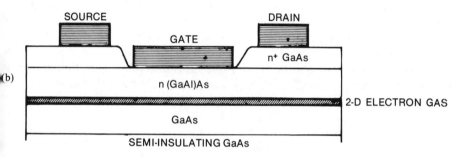

(b)

**HIGH-ELECTRON-MOBILITY TRANSISTOR (HEMT)**

**Fig. 3.9** FET cross sections: (a) standard MESFET; (b) AlGaAs/GaAs high electron mobility transistor.

doped AlGaAs donor layer, and an $n^+$ GaAs cap layer, which passivates the AlGaAs layer and aids in the formation of ohmic contacts. Electrons from the doped AlGaAs are trapped in a potential well formed at the GaAs-AlGaAs interface on the GaAs side. In most cases, a thin (less than 50 Å) layer of undoped AlGaAs is added at the interface to reduce coulombic scattering of the electrons from the nearby doping atoms. Layer thicknesses are chosen so that the AlGaAs layer is depleted completely under the gate electrode for microwave operation. The electrons are confined in a two-dimensional potential well away from the donors.

Compared to MESFETs, HEMT devices offer higher operating frequencies due to reduced scattering. Higher values of transconductance result from the increased saturation velocity of the electron gas and the higher number of electrons that can be introduced because the AlGaAs can be doped more highly than a GaAs layer without reducing the gate-drain breakdown voltage of the device. Recently reported HEMT results include a *noise figure* (NF) of 0.68 dB with associated gain of 9.7 dB at 126 GHz for a 0.25 $\mu$m gate length device, with a corresponding NF and gain of 0.83 dB and 7.7 dB at 18 GHz (Mitsubishi); 1.2 dB NF with 10 dB of gain at 32 GHz, 1.8 dB NF and 6.4 dB of gain at 60 GHz (General Electric); and 1.0 dB NF and 8.2 dB gain at 20 GHz (Fujitsu).

### 3.1.5.2 Heterojunction Bipolar Transistor

A cross section of another advanced device, the heterojunction bipolar transistor, is shown in Figure 3.10[8]. It is a multilayer device, consisting of an $n^+$ GaAs collector contact, a $p^+$ GaAs base region, an AlGaAs wideband gap emitter, and an $n^+$ GaAs emitter contact. Typically, the emitter is graded from 0–0.3 Al composition at the emitter-base and emitter-emitter contact regions.

In an HJBT, the current flow is perpendicular to the device surface rather than horizontal, as in the MESFET and HEMT. As a result, HJBTs offer higher current gain and less trapping. The HJBT threshold is determined by the energy gap discontinuity of the emitter-base junction, determined by the choice of material rather than adjusted by device processing, such as gate recessing, which is common in both HEMT and MESFET technologies. Because the base layer is thin, electrons can travel from the emitter to collector quickly, resulting in a very high (100 GHz) device cutoff frequency and extremely high transconductance (10,000 ms/mm).

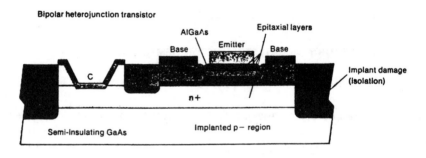

**Fig. 3.10** A heterojunction bipolar transistor (from [8], p. 30).

### 3.1.5.3 Technology Comparison: MESFETs, HEMTs, and HJBTs

As advanced devices emerge into pilot production, designers are forced to choose among several technologies for MMIC applications. In the near future, MESFETs will offer several advantages including a mature, manufacturable technology, a well developed set of CAD design and simulation tools, and multiple foundries with mask-compatible design rules (important for secondary sources). Because of these advantages, the MESFET will be the device of choice for MMICs unless performance requirements of specific applications cannot be met using MESFETs. Table 3.2[9] lists several device parameters and rates the MESFET, HEMT, and HJBT against these criteria.

**Table 3.2** Device Performance Comparison

| Parameter | MESFET | HEMT | HBT | Application |
|---|---|---|---|---|
| Small-signal | | | | |
| $f_T$ | L | M | H | HBT highest speed. |
| $f_{max}$ | M | H | L | HEMT highest frequency. |
| Gain-bandwidth | M | H | H | HEMT best for wideband at HF. |
| Noise figure | M | L | H | HEMT best for LNAs. |
| Phase noise | M | H | L | HBT best for VCOs. |
| $g_m/I_{dss}$ | L | M | H | HBT max. gain for min. power. |
| Threshold voltage uniformity | L | M | H | HBT best for high-precision analog LSI circuits. |
| Hysteresis effects | H | M | L | HBT best for circuits requiring minimal hysteresis in device. |
| Large-signal collector efficiency | M | H | H | |
| Power density | M | M | H | |

*Key:* L = low, M = medium, H = high.

HEMTs have demonstrated performance advantages over MESFETs in two important areas, lower noise figures and higher frequency operation, for a given gate length. Their lower noise is due to the high electron mobility in the two-dimensional electron gas and their lower thermally generated noise. Noise figures are typically better than MESFETs by a factor of 2 or more. High electron mobility transistors are best suited for low-noise applications such as *low-noise amplifiers* (LNAs) in the front ends of equipment. HEMTs are also better suited than MESFETs for high-frequency operation, as they produce more gain than a MESFET with an equivalent gate length.

Heterojunction bipolar transistors, on the other hand, are better suited for applications requiring tight threshold control, high transconductance, and power density. Because device thresholds are set by material, rather than processing, parameters, they can be tightly controlled. This effect leads to HJBT applications in digital circuits and in high-performance sample and hold circuits needed for high-speed A/D converters. Their high transconductance makes them suitable for high dynamic range and very linear amplifier circuits application. And due to the relatively large cross section of the base compared to a MESFET or HEMT channel, HJBTs are potentially high-current devices. For example, HBTs could be used as output drivers for digital chips.

### 3.1.5.4 Future Device Structures

Indium phosphide (InP) and related ternary and quaternary compounds are excellent materials for microwave and millimeter wafer devices, primarily due to the high peak electron velocity in these materials compared to gallium arsenide. The two most widely investigated compounds in this family are InP and InGaAs, which is lattice-matched to InP at composition of 53% In and 47% Ga.

One major barrier to the development of this technology is the relatively low Schottky barrier height (0.3–0.4 eV) between most metals and InP or InGaAs, which limits the gate voltage that can be applied to MESFET-type devices. The addition of insulating layers between the gate and the semiconductor, such as $SiO_2$ or InAlAs, addresses this problem; and several devices with excellent performance have been reported. These MESFET or IGFET devices are analogous to Si MOSFETs except that the native silicon oxide is replaced by a deposited or epitaxially grown film. Another device involves forming a *p-n* junction under the gate (JFET technology). However, long-term device stability needs to be improved in both types of these device structures.

A HEMT is another device structure commonly used for MMIC applications. In the InGaAs/InAlAs HEMT, undoped InGaAs is epitaxially grown on InP and serves as the device channel. A highly doped InAlAs layer on top of the InGaAs is the source of conduction electrons for the device. Recently, 1.3 $\mu$m gate length InGaAs/InAlAs HEMTs were reported with gains over 12.5 dB at 26 GHz and a maximum oscillation frequency ($f_{max}$) greater than 100 GHz [10]. Both the gain and $f_{max}$ are higher than present GaAs devices.

A distinct advantage of the InP-based MMIC systems is that optical devices, which emit light with wavelengths between 1.0 and 1.6 $\mu$m, also can be fabricated on the same substrate. These wavelengths are well matched to the minimum dispersion region for most fiber optics. Integrated circuits combining FETs and lasers also have been developed for higher data rate communications.

## 3.2 PROCESSING TECHNOLOGY

The processing of GaAs MMICs is similar to that for silicon integrated circuits in many respects. This similarity has the fortunate consequence that most of the processing equipment needed for MMIC fabrication has been developed and supported by a much larger silicon community. However, some significantly different processing steps are required, due to the nature of GaAs material and device structures needed for microwave applications.

Gallium arsenide is a compound semiconductor containing both gallium and arsenic. Because the vapor pressure of As is much higher than that of Ga, care has to be taken during high-temperature processing steps to ensure that the arsenic does not volatize and destroy the material. Also, the minimum feature size needed for microwave devices is at or below 1 $\mu$m, whereas the silicon world is just now realizing the *very high speed integrated circuit* (VHSIC) program goal of 1.25 $\mu$m features. Most silicon processing is based on 2.0 $\mu$m or larger design rules. Fortunately, GaAs MMICs do not require the high density of FETs characteristic of most digital silicon designs.

### 3.2.1 Photolithography

Although most of the GaAs MMICs used at X-band (8–12 GHz) and below work well with 1 $\mu$m design rules, there are special applications, particularly for low-noise operation, that benefit from shorter gate length devices. MESFETs with gate lengths as short as 0.1 $\mu$m have been reported, fabricated by advanced (and expensive) lithography techniques. The ultimate resolution of a lithography system is related to the wavelength of radiation used for patterning in photoresist. Optical sources, with wavelengths of approximately 400 nm, can resolve minimum feature sizes down to about 0.8 $\mu$m. Deep ultraviolet sources, close to 300 nm wavelengths, have application for features down to the 0.5–0.3 $\mu$m range. Smaller features require X-ray or electron beam lithography. Whatever the lithography system, it must be able to handle the rather fragile GaAs wafers.

#### 3.2.1.1 Optical Lithography

Two types of optical photolithographic equipment are widely used today for GaAs MMICs: contact and stepper machines. In contact lithography, the photoresist-coated GaAs wafer is brought into intimate contact with a mask, which has been patterned and etched appropriately. Light from the source passes through openings in the mask to expose the photoresist. Contact aligners are relatively inexpensive and have a high throughput, as the entire wafer surface is exposed at one time. Disadvantages of this type of lithography are that the intimate mask-wafer contact can result in broken wafers or masks, particles of photoresist can adhere to the mask (and thus frequent cleaning is required), and the mask lifetime is limited by wear and tear. Proximity alignment techniques, where the mask is separated from the wafer by approximately 1 mil, alleviate many of these concerns, but at a cost of lower resolution. Contact aligners are used widely in GaAs MMIC manufacturing; however, many foundries are beginning to convert to stepper lithography.

In a stepper aligner, light is passed through a reticle up in the optical train and focused on the wafer. Steppers can be either 1 : 1, where the reticle size is the same as the pattern transferred to the wafer, or the image on the wafer can be reduced 5 : 1 or 10 : 1 from the reticle size. In either case, the reticle never contacts the wafer, so reticles do not wear as much as masks and do not need to be cleaned as frequently. Steppers cannot image fields as large as a wafer; rather a smaller area is imaged at one location. A full wafer is patterned by stepping the image across the wafer and carrying out multiple exposures.

### 3.2.1.2 Deep UV Lithography

Many of the contact aligners can be reconfigured as deep UV aligners. Generally, a change in optics is needed to pass the shorter wavelength light. Mask blanks are usually quartz rather than glass for the same reason. Many of the photoresists used in optical lithography are sensitive in the deep UV range, although in some cases exposure times must be increased.

A relatively new deep UV light source is the excimer laser. Using this laser as a source provides higher brightness than conventional high-pressure xenon lamps, reducing exposure times. The high spectral purity of the excimer laser and small spot size also improves the resolution and depth-of-focus of the lithography tool. Deep UV lithography has been used to print features down to 0.3 $\mu$m in small areas and can repeatedly print 0.5 $\mu$m features across the 3-inch wafer.

### 3.2.1.3 X-Ray Lithography

Aligners using x-ray sources can be used in production to print features down to about 0.5 $\mu$m, with an ultimate resolution down to 0.3 $\mu$m. High brightness x-ray sources have been developed that can be used to expose a field on the order of 20 $\times$ 20 mm, with a depth of focus greater than 30 $\mu$m. An x-ray aligner works like an optical stepper in that multiple steps and exposures are required to pattern across a full wafer.

One major drawback of x-ray lithography is the special materials needed for masks and reticles. Standard metal-coated glass or quartz masks cannot be used; most manufacturers use silicon or silicon carbide membranes with patterned tungsten films on these substrates to selectively block x-ray exposure. These masks are more expensive and not widely available. The lack of experience with x-ray lithography and sources along with the costs associated with the masks have slowed the acceptance of this technique by the industry.

### 3.2.1.4 Electron Beam Lithography

Electron beam lithography offers the best resolution of any lithography technique widely used today, down to 0.1 $\mu$m feature size. This technique uses electron beam to write patterns directly on the photoresist-covered wafer. The beam is steered by a pattern generation tape, and patterns can be changed relatively quickly and easily to accommodate design changes. Electron beam lithography equipment also has been used for generating high-quality masks for other lithography tools for several years.

There are two major drawbacks to electron beam lithography: the cost of the equipment and upkeep and the relatively low throughput of the system. Today's e-beam production lithography machines can cost up to $4 million and require full-time maintenance. Unless amortized over a large number of wafers, the cost can be prohibitive. Throughput is limited by the time needed to expose the resist. Advances in beam current density and development of more sensitive resists have reduced the writing times, but these times are still long when compared to optical lithography.

Throughput at this time is not as critical an issue for GaAs MMIC production as it is for very high density silicon digital circuits. Most MMICs use no more than a few MESFETs, the only circuit elements where the high resolution of electron beam lithography is needed. Furthermore, of the lithography steps needed to make a MESFET, only the gate level requires tight resolution and alignment. In MMIC fabrication, electron beam lithography could be used only for this level, with higher throughput techniques used for the less critical levels.

### 3.2.2 Ion Implantation

Ion implantation into semi-insulating GaAs substrates is the method of choice for MMIC fabrication today. Active and contact layers of MESFETs and GaAs resistors are two areas where ion implantation is used. Implanters are available with excellent control of uniformity that can automatically handle GaAs wafers.

Figure 3.11[1] shows a schematic representation of an ion implanter. A beam of atoms for implantation, usually $Si^+$ or $S^+$ ions for $n$-type dopants, is generated

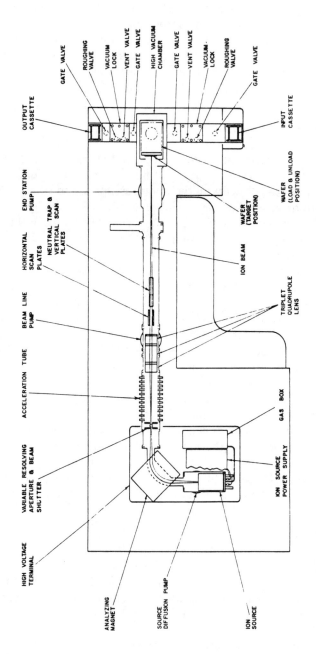

**Fig. 3.11** Cross section of an ion implanter (from [1]).

from a gas or solid source by heating or passing a current through the material. The proper ion is selected by adjusting a magnetic field, and the beam is accelerated down an evacuated tube at high voltage. The wafer is located at the end of the tube, and the beam is raster scanned across it. The number of ions implanted and their distribution in the GaAs lattice is determined by the fluency and energy of the ions reaching the wafer.

Implanted ions follow a Gaussian distribution within the GaAs lattice. As the energies typically used are 200 keV and less, the depth of the peak of the implant profile is less than 3000 A° below the surface of the wafer. Different profiles can be achieved by multiple implants.

Wafers can be blanket implanted, uniformly distributed across the wafer, or selectively implanted, usually using a photoresist or dielectric mask to prevent the ions from reaching the GaAs wafer in selected areas. Both techniques have been used in device fabrication. The arriving ions destroy the crystal lattice, which must be restored to achieve the required electrical properties for device fabrication. This is done by annealing the wafer at an elevated temperature.

Several methods have been developed for implant annealing that prevent the loss of As during the process: capped annealing, where a dielectric cap such as $SiO_2$, AlN, or $Si_3N_4$ is deposited before the anneal and removed afterward but before device fabrication; proximity annealing, where a second wafer is held in close proximity to the device wafer; annealing in an overpressure of arsenic, either in arsine gas or near a solid arsenic source; and rapid thermal annealing, where the temperature is ramped up and down very quickly (on the order of a few seconds) to prevent the loss of arsenic. All of these annealing techniques have been used successfully to activate implants.

Selective ion implantation also can be used to isolate devices from each other. In this case, typically a heavy dose of protons (hydrogen ions), boron, or oxygen ions is implanted to destroy the electrical conductivity of the crystal lattice between devices. This is done after implanting and annealing the active-contact-resistor areas, and annealing is not done after the device isolation implant.

### 3.2.3 Metalization

Most of the metals needed during MMIC fabrication are deposited in a high vacuum environment. The only exceptions are second-level metal lines and the backside ground plane metalization. Both are plated with thick gold to reduce electrical resistance, usually in a gold plating bath, after a thin layer of starting metal is vacuum deposited. Three methods of vacuum deposition commonly are used in the MMIC industry: sputtering, thermal evaporation, and electron beam deposition.

### 3.2.3.1 *Sputtering*

Sputtering is characterized by very uniform deposition over large areas and is well suited for large scale production of MMICs. It is a physical deposition process, during which material is knocked out of a target by high energy ion bombardment and redeposited on a nearby substrate. Samples are placed in a high vacuum chamber, which is evacuated and backfilled with an inert gas, such as argon. Either a dc or (more commonly) an RF discharge ionizes the gas, which is accelerated at a target, typically large diameter disks of the appropriate material. GaAs substrates are positioned within a few centimeters of these targets. The substrates are in direct contact with a grounded metal holder, which can be water cooled to reduce substrate temperatures during the deposition process. In order to reduce surface damage to the GaAs substrates, an external magnet can be used to confine the high-energy gas plasma at the target surface and away from the substrate. This process is called *magnetron sputtering*.

### 3.2.3.2 *Thermal Evaporation*

Metals deposited by thermal evaporation are loaded into refractory metal boats inside a high vacuum chamber. Typical boat materials are tungsten or tantalum. Passing a high current through the boat causes the boat and metal charge in it to heat up to the evaporation point of the metal. The metal vapor impinges on the GaAs substrates, which are located far enough away (typically 30 cm or so) such that the metal vapor arrives perpendicular to the substrate surface. A shutter positioned between the source and substrate controls the arrival of metal vapor at the substrate. The process is carried out at high vacuum, $10^{-6}$ torr or less, so that residual gases are not trapped in the deposited films. Depending on size, thermal evaporators can accommodate multiple sources and wafers to deposit multilayer films at a high throughput rate. A quartz crystal monitor is usually used for in situ determination of film thickness.

Although thermal evaporation is widely used, two main drawbacks are associated with this technique. First, the source temperature is limited to below that of the softening point of the refractory metal boat. Thus, metals such as titanium (used in gates of FET devices) cannot be deposited. Second, in some cases, the metal source and the boat can react, forming unwanted alloys that are deposited on the wafers.

### 3.2.3.3 *Electron Beam Deposition*

Electron beam evaporation differs from thermal evaporation only in the way the source metals are heated. Here, an electron beam is focused on a solid target; local heating causes the metal to evaporate. This method has two main advantages over thermal evaporation. First, the electron beam spot size is very small, and local heating can result in a higher temperature than can be achieved in thermal evaporation. Hence, refractory metals can be deposited by electron beam evaporation. Second, because the heating is localized, the source material does not react with its "boat" (or hearth, in electron beam evaporation technology), and high purity deposition is possible.

### 3.2.4 Dielectric Deposition

The key to successful dielectric deposition on GaAs is not to damage the GaAs surface during the process. Three main techniques are used in MMIC processing: sputtering, chemical vapor deposition, and pyrolytic deposition. The first two are low temperature processes. In pyrolytic deposition, the film is deposited at a higher substrate temperature, but it forms fast enough to prevent the loss of As.

#### 3.2.4.1 Sputtering

In sputtering, atoms to be deposited are physically knocked toward the wafer from a solid target by accelerating gas ions, such as argon or oxygen. The atoms redeposit on the surface of the wafer, which is held in close proximity to the target. Sputtering is a low-temperature process; typical substrate temperatures do not exceed 150°C. Sputtered films have high quality, with good adhesion and step coverage.

One concern about sputtering is the possible damage to the GaAs surface (and subsequent device performance degradation) caused by the presence of energetic ions. Thus, sputtering generally is not used where the device channel is exposed. Sputtering is used extensively to deposit $SiO_2$ caps for ion-implantation annealing, since the subsequent annealing step repairs any surface damage.

#### 3.2.4.2 Chemical Vapor Deposition

In CVD processing, gases containing constituent atoms of the deposited films are mixed together and passed over a heated substrate, where they deposit dielectric films. A typical example is the deposition of $SiO_2$, where silane, $SiH_4$, and nitrous oxide, $N_2O$ react at a heated substrate to form $SiO_2$ by the following reaction: $SiH_4 + 2N_2O \rightarrow SiO_2 + 2H_2$. Other dielectric films such as $Si_3N_4$ have also been deposited by CVD.

The key to the successful deposition of films on GaAs substrates is to keep the substrate temperature low, preferably 350°C or lower. At these temperatures, however, the reaction rates can be slow. One way to improve the deposition rates is to dissociate the gases in the gas phase to speed up the reaction rates. Both optical excitation and plasma-enhanced CVD have been used successfully in dielectric film deposition on GaAs.

### 3.2.4.3 Pyrolytic Deposition

This technique is similar to standard CVD, in that gases are introduced and react with a heated substrate. The substrate is placed on a graphite heater and heated by passing current through the graphite strip. The substrate temperature can be ramped up to over 500°C in a few seconds. At this temperature, the gases react quickly, forming a dielectric film on the substrate that prevents significant arsenic loss. Pyrolytic deposition is particularly suited to the deposition of silicon nitride films, $Si_3N_4$.

### 3.2.5 Etching

Etching is used extensively in GaAs processing. Removing polishing damage from incoming wafers, isolating devices from each other, recessing MESFET channels, forming through-wafer via holes for backside grounding, removing dielectric caps after ion implantation, and cleaning both native oxides and photoresist residue before metal deposition are some steps that require etching.

A large number of wet chemical etches for GaAs and the materials used in processing MMICs have been developed. Etch compositions of acid, peroxide, and water in various proportions have been used to etch GaAs. These etchants work by continuously oxidizing the surface and then dissolving the oxide layer. Etch composition is varied to modify etch rates and selectivity. Other etches, based on HCl or HF, are used to remove oxides and pattern thin-film resistors without damaging the GaAs surface. The wet chemical technique is a relatively inexpensive way to etch GaAs. However, etch rates are sensitive to variations in temperature and concentration, variables that can be difficult to control in large baths when precise control is required.

Gate recess etches, used to control thresholds in GaAs digital circuits and source-drain currents in MMICs, are used where precise control is required and device performance is limited by the control of the etch process. Wet chemical etches do not produce structures with high aspect ratios or vertical walls. Etching closely spaced via holes through 100–150 $\mu$m thick wafers using wet chemical etches is difficult due to the poor aspect ratio.

*Reactive ion etching* (RIE) is one solution to both the fine control and high aspect ratio problems associated with wet chemical etching. In this process, wafers are etched in a low-pressure gas environment. Reactive ions are created by striking a plasma in an appropriate gas, such as oxygen or a chlorine containing compound, and accelerated toward the substrate. Etching occurs at the unmasked regions of the substrate. Because the gas concentrations and substrate temperature can be tightly controlled, excellent control of etch rates results, significantly improving the etching uniformity across a wafer compared to wet chemical process. By choosing the proper combination of gases, temperature, and pressure, highly directional etching of GaAs can be achieved. Via holes with aspect ratios exceeding 20:1 have been demonstrated using RIE.

### 3.2.6 Plating

Due to the relatively high current density carried in RF transmission lines in MMICs, metal thickness in excess of those generally deposited by thermal evaporation are required. The thicker lines also lower the parasitic resistance associated with them, further improving MMIC performance. Moreover, the backside ground plane of an MMIC needs to be plated to form a low resistivity ground, and via holes must be plated to complete low resistance electrical contacts between the topside circuits and the ground plane.

Two types of gold plating processes are used in MMIC fabrication: electrodeless plating and electroplating. In electrodeless plating, the potential difference between a dissimilar metal on the wafer and gold dissolved in a plating bath drives the deposition of gold on the metal; in electroplating, a current passed between the plating bath and the wafer accomplishes the same deposition.

Electrodeless plating is attractive in that no contacts to the wafer or external power supply are necessary. The gold deposits on any exposed metal surface; connections to isolated lines are not required. One major drawback to this process is that the thickness of metal plated is limited to about 2 $\mu$m or less.

Electroplating, on the other hand, is a slightly more complex process in that contacts to the wafer are required. However, this process can be used to deposit an arbitrarily large thickness of gold. Backside gold plating in excess of 25 $\mu$m thicknesses are possible.

## 3.3 COMPUTER-AIDED PROCESS DESIGN

### 3.3.1 Introduction

As the GaAs MESFET technology is maturing toward an era of subsystem-level integration on a single chip, manufacturing methods are needed to improve the uniformity and control of device electrical parameters. This is particularly important for MMIC technology, where low-noise, power, general purpose microwave devices and lower pinch-off voltage or enhancement-depletion mode devices for logic circuits with tight control of device electrical parameters need to be integrated on the same chip. This requires the control and optimization of channel doping profiles for different pinch-off voltage MESFET devices along with the control of other active and passive components. The development of advanced computer-aided design and engineering tools has greatly enhanced the ability of circuit designers to reduce the design and development time of complex integrated circuits. However, very limited simulation software capability is available to address the process development needs.

Process development is an extremely complicated, engineering-intensive effort, more so in GaAs technology because of prominent substrate effects on processed devices. These substrate characteristics are comparatively easy to simulate for Si material because of the high level of purity achieved in Si crystal growth as compared to that in GaAs. Starting from the substrate material several hundreds of processing variables must be tightly controlled, each of which can have profound effects on the final device performance. For example, a slight change in ion-implantation annealing ambience and temperature can change the carrier profiles of the fabricated device and hence the electrical characteristics of the final circuit.

Moreover, trial-and-error methodology to optimize such a complex process no longer is desirable because of the enormous cost and turn-around time. Instead, understanding and formulation of accurate models of the basic physical processes involved are essential. At this time very limited aid is available to a GaAs integrated circuit process engineer to allow anticipating the effects of processing conditions on device performance. With the aid of an ideal computer simulation software tool, the process engineer would be able to quickly investigate the overall impact of process changes on the performance of fabricated devices without spending considerable time and resources in actually doing hands-on fabrication experiments. Such a software tool would be comparable to circuit simulation software for integrated circuit design. Without such circuit simulation software today, it is difficult to comprehend how a designer can successfully design a complicated integrated circuit. At this time an engineer developing a GaAs MMIC process works under conditions similar to a circuit designer, designing circuits without a circuit simulation software. Computer simulation is a cost-effective alternative, not only supplying approximately the right answer for increasingly tight processing windows but also serving as a tool to develop future technologies. When coupled with a device simulation and analysis program, a combined software package calibrated for a particular manufacturing facility can prove to be a powerful process design tool. This is because, with the aid of such a software tool, the process sensitivity to device parameters can be easily extracted by simple changes made to processing conditions in computer program inputs. Current capabilities to predict the changes in the device electrical performance resulting from processing modifications are impeded largely by deficiencies in process models, lack of reproducible GaAs substrate growth and detailed understanding of material characteristics, efficiency of the numerical solvers used in simulation programs, and finally the lack of application of modern software engineering approaches to process simulation tool integration.

SUPREM (Stanford University Process Engineering Models) [1] is one of the widely used process simulation programs, initially developed to address Si processing technology. Version 3.5 of the SUPREM and GATES (Gallium Arsenide Transistor Engineering Models) were developed at Stanford University specifically to address GaAs MESFET technology. (A commercial software package, GATES, supported by Gateway Modeling of Minneapolis, Minnesota, is a combination of process modeling and device simulation.) The overall objective of these programs is to permit a GaAs process or device engineer to simulate accurately GaAs fabrication technology. The input to the program typically is the processing schedule specifying a sequence of time, temperatures, ambient conditions, depositions, implant doses and energies, and so on, along with the starting material properties of GaAs, which play a major role in the final device characteristics. The program output, available after each step in the process sequence is a one-dimensional impurity profile in the bulk GaAs. The structure may incorporate additional material layers above the bulk such as $Si_3N_4$ and $SiO_2$.

The thickness and dopant profiles of each of these layers are calculated by the program during each step. These doping profiles are used to solve the electron transport equations to determine the electrical characteristics of the fabricated device. This integration of process and device simulation software provides a powerful tool to optimize the process and physical structure of the device to achieve best desired electrical performance without making numerous fabrication runs. The following are the most critical issues to the successful development of such a process simulation software to address GaAs technology:

- Development of sophisticated, accurate process models applicable over a broad range of process conditions. The objective is to reduce the amount of experimentation required in process development by building experimentally verified models into the simulation program.

- Understanding and modeling GaAs substrates grown by various methods such as horizontal Bridgeman, undoped liquid encapsulated Czochralski, Cr-doped LEC, liquid phase epitaxy, vapor phase epitaxy, and molecular beam epitaxy.

- Emphasis on the development of analytical models for fast execution of the program; otherwise, application of super fast algorithms for numerical analysis of complex mathematical equations. The objective is to reduce computer simulation time and open the way for a process optimization tool that can automatically iterate to a given solution.

- Development of user-friendly interface for interactive input and graphical output from the program; also, development of a user transparent integration of process, device, and circuit simulation tools (i.e., interface between process simulation and CAD/CAE environment).

• Creation of a large database of one- and two-dimensional process data for model testing.

Figure 3.12 provides an overview of SUPREM 3.5 and GATES programs. The process modeling part of these software packages can model following process steps of GaAs technology:

• Impurity profiles of Se and Si (two main *n*-type dopants for channel and source-drain regions), Be (used for buried *p*-layers), and Mg (used for shallow *p*-type dopants in JFET structures) in one dimension.

• Ion-implantation profile models accounting for wafer tilt and rotation angles and cases where implants are done through caps.

• Diffusion-annealing profiles, examining the annealing effects on the redistribution and activation of implanted species. This includes capped and capless annealing and estimation of channeling tails of ion implant profiles.

• Studies of the effect of material properties on impurity profiles, such as wafer impurities, dislocations, concentration of shallow acceptors and donors EL2, and deep-level acceptor concentrations, like Cr and Mn.

• Profiles of recess etching process.

• Profiles of epitaxial layers.

The device simulation portion of the software shown in Figure 3.12 can provide the following electrical information about the fabricated devices.

• MESFET pinch-off and threshold voltages.
• Resistivity of the implanted layers.
• Source and drain extrinsic resistance.
• I–V characteristics and transconductance of the MESFET.
• Small-signal parameters, such as terminal capacitances and parasitic resistances of the MESFET.
• Piezoelectric and short-channel effects.

### 3.3.2 Process Models

#### 3.3.2.1 Ion Implantation

Four different kinds of one-dimensional profiles are allowed in SUPREM 3.5 and GATES: (1) implants for the channel donor $N_D^-$, which is a comparatively

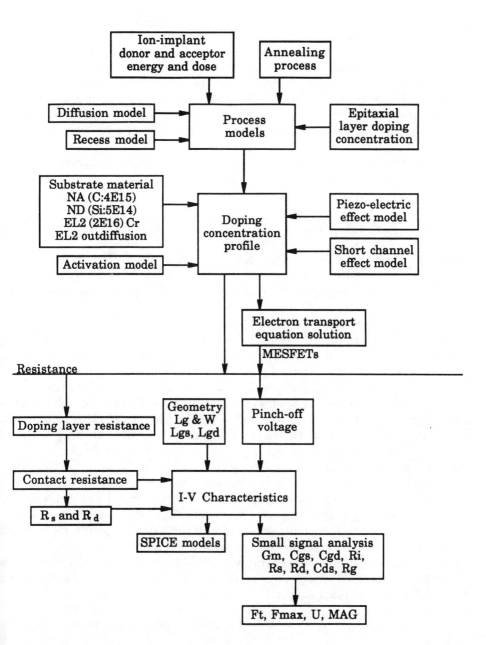

**Fig. 3.12** Overview of SUPREM 3.5 and GATES programs.

low-dose, shallow implant; (2) implants for source and drain regions of the MES-FET $N_D^+$, which are comparatively deeper, large-dose implants; (3) an additional donor implant $N_D'$, which is of intermediate dose; and (4) an acceptor implant $N_A$. In all cases of ion-implantation, particularly at low doses used for channel implants, large exponential ion-channeling tails are observed in profiles of Be, Si, and Se ions implanted into GaAs. Amorphous ion-implantation models such as LSS [2], Boltzmann transport [3], or Monte-Carlo [4], codes that do not account for GaAs crystalline structure, are not capable of accurately representing implantation profiles in GaAs because of the channeling tails [5].

The implantation profiles are modeled by several different methods, such as Pearson IV functions [6], given by the following relationship:

$$\frac{1}{f}\frac{df}{dy} = \frac{a - y}{b_0 + b_1 y + b_2 y^2} \tag{3.1}$$

where the Pearson distribution $f$ is a generalized function with coefficients $a$, $b_0$, $b_1$, and $b_2$ chosen to a fit simulated profile with the measured implant profile. SUPREM 3.5 also uses a Gaussian profile with two exponential tails added to the deep side.

For Si, Se, and Mg, both methods provide similar results and both models are available as a part of SUPREM 3.5. However, for Be, which is a light ion, the Pearson IV profile provides simulations close to the measured results. Parameters for both models are derived empirically from a series of implant profiles obtained for a range of ion implantation energy and dose. The results of measured and simulated implant profiles for Si ions in GaAs are shown [7] in Figure 3.13, where ion energy ranges between 25 keV and 400 keV. The models also include the effects of other implant conditions such as surface coating with $SiO_2$ or $Si_3N_4$ before implantation, effect of capped layer thickness, and the effects of tilt and rotation angles during implant. Implanting at a tilt angle of at least 9° and a rotation angle of 45° from the major flat (011) provides the maximum cross-wafer device uniformity and the narrowest profiles for a given ion energy [8]. For tilt angles greater than 8° and rotation between 30° and 60°, the profiles are nearly independent of angle except for simple projection effects [9]. Due to the ion-implant damage to the crystal, the profiles become narrower for increasing ion doses up to $10^{15}/cm^2$, but there is little effect over the applicable range of doses, which is 1–5 $\times$ $10^{12}/cm^2$ for channel implants and up to 2 $\times$ $10^{13}/cm^2$ for $n^+$ implants for GaAs MESFET source and drain areas.

To model the Be doping profile, when Be is implanted through $SiO_2$ or $Si_3N_4$ encapsulates, the range-scaling method is used [6,10] as given by the following equation:

$$C(x) = C_{PIV}(y) \tag{3.2}$$

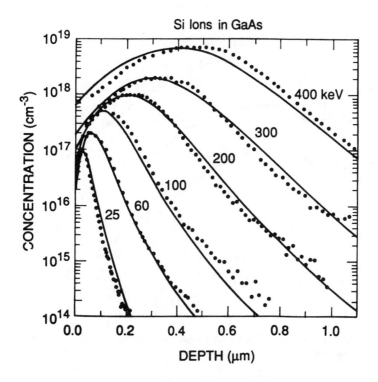

**Fig. 3.13** Measured Si ion-implantation profiles in GaAs compared with Pearson IV fits.

The resulting profile in the GaAs is calculated by shifting the measured profile for implantation into bare wafers toward the surface by an amount $tR_{pS}/R_{pC}$, where $t$ is the cap thickness, and $R_{pS}$ and $R_{pC}$ are the projected ranges in the GaAs substrates and cap [5]. Results of the measured and simulated concentration profiles for 50 keV Be implants into bare and capped GaAs substrate are shown in Figure 3.14. Moments of ion distributions in the amorphous cap are interpolated from tables of values, calculated using the TRIM Monte-Carlo program [4]. The dose in the GaAs is normalized to the incident ion dose reduced by that lost in the cap. Consistent with TRIM results, a Pearson I distribution [6] is used in the cap.

For Si and Se ions, implanting through caps reduces the magnitudes of the channeling tails. This effect is simulated by mixing a shifted Pearson-IV function with a shifted TRIM calculated Gaussian function, using

$$C(x) = (1 - p)C_{Gaus}(y) + pC_{PIV}(y) \tag{3.3}$$

where $y = x - tR_{ps}/R_{pc}$, $C_{Gaus}$, and $C_{PIV}$ are normalized to 1 at $x = R_p$ and $p < 1$. This procedure is illustrated in Figure 3.15. $C(x)$ is normalized to the dose in the

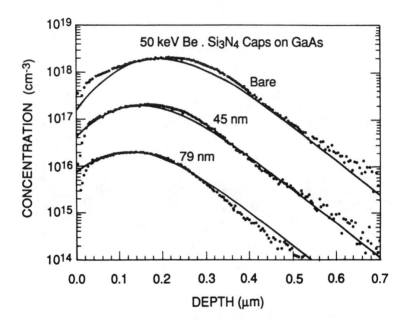

**Fig. 3.14** Measured implantation profiles for Be, implanted onto bare, 45, and 79 nm CVD Si₃N₄ coated GaAs, compared with simulations using range-scaling theory.

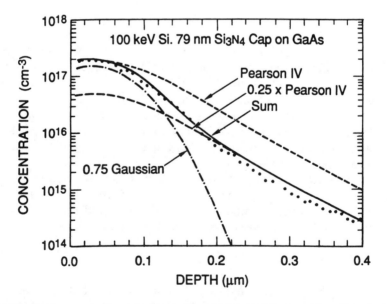

**Fig. 3.15** Measured *versus* simulated doping profiles for 100 keV Si implanted into GaAs through 79 nm CVD Si₃N₄ caps.

GaAs. The mixing factor p ($<1$) depends on the ratio of the cap thickness $t$ to $R_{pS}$ but has been characterized for only a few cases [5].

When ions are implanted through $SiO_2$ and $Si_3N_4$ caps, Si, O, or N ions are knocked into the GaAs, where the Si ions are additional donors, affecting carrier densities. Knock-on atom distributions are calculated using TRIM [4] and are fit to the following expression:

$$C_r(x) = \frac{A}{x^n + b} \exp\left(-(x/a)^2\right) \tag{3.4}$$

where $b = 0.1$ and $x$ is the depth in nm. The parameters $A$, $a$, and $n$ for Si, O, or N atoms can be interpolated from tables as a function of cap ($SiO_2$ or $Si_3N_4$), cap thickness ($0 < t < 1.3R_{pS}$), ion (Si or Se), and ion energy (50–400 keV). The knock-on concentrations are significant only compared to the implant-ion distribution near the surface (where channeling tails are unimportant), and MESFET pinch-off voltages are increased by $<1\%$ (Be), 5–10% (Si), and 20–30% (Se).

### 3.3.2.2 Substrate Material Effects

Four significant impurities are present in either LEC or HB grown GaAs substrates. Table 3.3 gives typical ranges of these impurities present in a commercially available GaAs semi-insulating substrate. In LEC wafers, the Fermi level is pinned to mid-gap by the compensation of the dominant shallow acceptor by the native deep-donor defect EL2. In HB wafers, the dominant donor impurities are compensated by added mid-gap acceptor, Cr. The shallow and deep acceptors both compensate implanted donors; hence, if not controlled, they can lead to threshold voltage and saturated current shifts from wafer to wafer. SUPREM 3.5 allows process engineers to determine the magnitude of these shifts by varying these concentrations and to design processes (e.g., using buried $p$-layers [11]) that reduce the sensitivity to these variations.

Although wafer manufacturers mention properties like resistivity and free carrier mobility in the semi-insulating substrates, the most critical factor affecting device performance fabricated on LEC wafers is the effect of background acceptor impurities, principally carbon. In the substrate material carbon is compensated by EL2. However, in the active device area, carbon is compensated by implanted shallow donors so that higher carbon concentrations lead to lower carrier concentrations and more positive threshold voltages. Though it is important to reduce and control carbon impurities for better processing uniformities, it is undesirable to eliminate them altogether. Due to ion channeling in the crystalline GaAs, the donor implants of Si or Se ions have long tails, which if not compensated by the background acceptors gives broad carrier profiles, and poor device transconductances.

**Table 3.3** Typical Impurity Concentrations ($10^{15}/cm^3$)

| | | LEC Wafers | | HB Wafers | |
|---|---|---|---|---|---|
| Impurity | Ion | Range | Default | Range | Default |
| Shallow acceptor | C | 1–6 | 3 | ≪Si | 0 |
| Shallow donor | Si | ≪C | 0 | 5–15 | 10 |
| Deep donor | EL2 | 5–20 | 20 | ≪Cr | 0 |
| Deep acceptor | Cr | ≪EL2 | 0 | 20–100 | 50 |

The optimum solution is to either tightly control the carbon levels, or to eliminate them altogether, and implant a controlled buried p-layer behind the gate. Figure 3.16 shows what can be done with a buried p-layer for a 60 keV Si ion implant with a target pinch-off voltage of −0.5 V. Without the buried p-layer, the pinch-off voltage varies by nearly 80% over a maximum range of carbon impurity concentrations of $1-6 \times 10^{15}$ cm$^{-cm}$. Implanting 90 keV Be ions with a dose of $2 \times 10^{12}$ cm$^{-3}$ reduces the pinch-off voltage variation to 20%. If we can reduce the range of carbon impurities to $1-2 \times 10^{15}$ cm$^{-3}$, variation of pinch-off voltages can be made almost negligible [12].

### 3.3.2.3 Dislocation Density Effects

In addition to macro nonuniformities cross wafer and wafer to wafer, GaAs devices also exhibit localized nonuniformities such as the pinch-off voltage variations over distances as small as 10–100 $\mu$m. Most of the localized variations are attributed to individual dislocations, on the order of $10^4-10^5$/cm$^2$ in commercial undoped LEC material. Striking evidence for this is the correlation between MESFET pinch-off voltage and distance to the nearest dislocation [13]. Pinch-off voltage of the devices are more negative within a radius of about 30 $\mu$m around dislocations. The physical mechanism behind this decrease may be an increased Si ion activation, due either to increased Ga vacancies [14] or increased As interstitials [15] near dislocations (both of which force Si ions onto Ga sites where they are donors), or a denuded acceptor zone may exist around dislocations [16].

To study the dislocation-induced nonuniformities in the device characteristics, a GATES process and device simulator is used. The physical mechanism

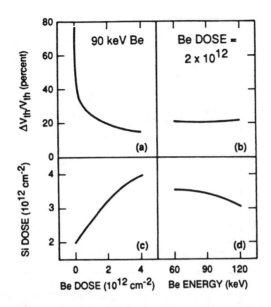

**Fig. 3.16** (a) and (b): The calculated relative change in MESFET threshold voltages when the background C concentration is varied from $1-6 \times 10^{15} \mathrm{cm}^{-3}$, assuming buried $p$-layers are implanted with various doses ((a): constant Be energy = 90 keV) and various Be ion energies ((b): constant dose = $2 \times 10^{12} \mathrm{cm}^{-2}$). The required Si dose needed to obtain a constant threshold voltage of $-0.5$ V with a background C concentration of $3 \times 10^{15} \mathrm{cm}^{-3}$ is shown in (c) and (d).

associated with the dislocation is assumed to be the theory of increased activation. The Monte-Carlo method is used to calculate average and standard deviations of pinch-off voltage and other quantities [14]. The Monte-Carlo calculations consider an area of wafer having dislocations randomly distributed according to some average dislocation density. For each regularly spaced MESFET in this area, the distance to each dislocation is calculated, the effects on the device for all nearby dislocations is summed, and after considering 1,000 to 10,000 FETs, the average and standard deviation of pinch-off voltage are calculated. Figure 3.17 shows these quantities plotted against dislocation density. As the dislocation density increases, the average pinch-off voltage becomes more negative. The standard deviations due to just dislocation effects are small at small dislocation densities, reach a maximum at dislocation densities of about $10^4/\mathrm{cm}^2$, then decrease at large dislocation densities. The decrease in standard deviation with increasing dislocation density above $10^4/\mathrm{cm}^2$ is due to two factors. First, in the increased activation theory assumed here, the activation becomes static near 100%; thereafter, adding more dislocations cannot further affect the device's activation. Second, the counting statistics for the number of dislocations within range of affecting the device improve. The wafer has so many dislocations, that

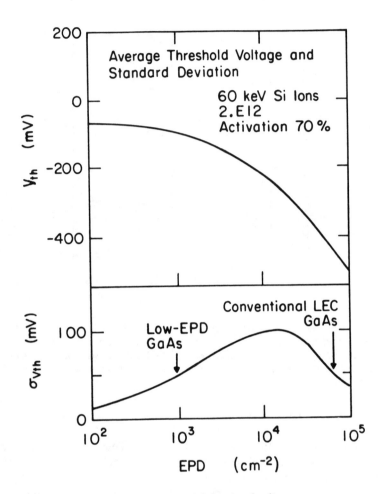

**Fig. 3.17** Pinch-off voltage variation as a function of dislocation density.

each device has about the same probability of being affected; therefore, the differences between device pinch-off voltages decrease. The shape of this curve is independent of the mechanism assumed to affect device characteristics.

This correlation has several other implications. First, the so-called W-patterns of pinch-off voltage *versus* distance across wafer follows directly from the curve in Figure 3.17. The dips of the W-patterns form on the lowest dislocation densities; therefore, the most positive pinch-off voltage and highest standard deviation occur. Second, work at NTT has shown that when we place E- and D-mode MESFETs close together (<20 μm), the pinch-off voltages of the two are closely correlated. This also follows directly from the shape of the pinch-off voltage *versus* dislocation distance curve. If both MESFETs are placed within range of the same dislocation, they will be affected nearly equally by it. Third, to determine whether a dislocation-increased activation mechanism or a denuded acceptor zone mechanism is the correct mechanism, measurements for different doses should be made. The standard deviations should scale as the pinch-off voltage for the activation mechanism but decrease for the denuded-acceptor mechanism, as the influence of acceptors decreases with larger doses. Fourth, another implication of Figure 3.17 is that larger FETs in either gate length or width will give smaller standard deviations—good news for the microwave industry but bad for digital! In essence, at high dislocation densities, the standard deviations fall off with improved counting statistics for the number of MESFETs within an area with a perimeter about 30 $\mu$m around the MESFET gate. If the gate is enlarged the area becomes larger, the counting statistics improve, and standard deviations fall. Finally, we can speculate about the effects of whole-boule annealing. If dislocation-increased activation is due to increased Ga vacancies and the diffusion constants for Ga vacancies are sufficiently large that within a typical boule anneal they can diffuse more than 30 $\mu$m (a large diffusion constant), then the island of increased activation spreads out and decreases in magnitude, leading to smaller standard deviations. Similar explanations can be developed for the denuded-acceptor or increased-As-interstitials mechanisms. Photoluminescence and IR studies indicate islands of optical activity or inactivity near dislocations, and they do spread out with wafer annealing, but their interpretation is not straightforward.

Not all processes result in the same dependence of threshold voltages on distance, so measurements of standard deviations will vary from process to process. The anneal cap is an important factor. $SiO_2$ caps and capless annealing with As overpressure may either allow Ga outdiffusion to increase Ga sites or tie up the As sites, forcing Si ions onto Ga sites, giving large activations. If we can maintain the activation near 100%, dislocations will have little effect on devices. A less well-defined factor is cap porosity and uniformity. While As ion-implants into LOCOS $SiO_2$ show that As does not diffuse in good caps, CVD caps on GaAs may be more porous and are suspected of allowing local As fluctuations, which influences Si-ion activation.

### 3.3.2.4 Diffusion

Redistribution of implanted species occurs during the activation annealing that follows the implant. Generally, there are low diffusivities of Si and Se implants in GaAs during activation anneals. The diffusion depends on implant dose, cap, and cap thickness during the anneal process because of interface stress, material of the cap (i.e., $SiO_2$ or $Si_3N_4$), and arsenic over pressure. The postanneal dopant profile is simulated in SUPREM 3.5 by solving the diffusion equation numerically, and this also can account for the electric field and concentration effects on the diffusion and segregation in the cap. However, more experimental data is needed to establish the relevant diffusivity data needed by such models. GATES simulates the postanneal doping profile by solving Fick's second law of diffusion, assuming no out-diffusion from the substrate, using diffusion coefficients based on the processing parameters. The dopant diffusion during postimplant anneal is calculated using the following equation:

$$C_N(x) = \int_0^\infty C_0(x') \, dx' \left[ \exp\left\{ -\frac{(x - x')^2}{4Dt} \right\} + \exp\left\{ -\frac{(x + x')^2}{4Dt} \right\} \right] \qquad (3.5)$$

where $C_0$ = As-implanted distribution

$\quad\quad C_N$ = distribution of implanted species after anneal

$\quad D(T)$ = concentration independent diffusion coefficient at anneal temperature $T$

$\quad\quad\, t$ = the anneal time

The diffusion constant $D$, for low-concentration Si, Se, and Be ions in GaAs is so small that dopant diffusion during typical furnace anneal is an insignificant factor in MESFET fabrication. MESFET pinch-off voltages are affected when the diffusion length $\sqrt{4Dt}$ exceeds the mean ion range in the semiconductor, $R_{pS} - tR_{pS}/R_{pC}$ [17]. With typical diffusion constants for Se ions at channel concentrations of $10^{17}$ cm$^{-3}$ ($10^{-14}$ cm$^2$/s at 900°C), an anneal time greater than 20 min for bare-wafer implants is needed to affect pinch-off voltage more than a few percent. Si diffusion constants are not well characterized but are much smaller at low concentrations [18–20]. Buried $p$-layer diffusion could also affect pinch-off voltages, but diffusion constants for Be or Mg ions at concentrations $<10^{17}$ cm$^{-3}$ are not well characterized [21–22]. In one case, SIMS measurements detected no diffusion in a Be buried $p$-layer for a typical furnace anneal [22]. It is not well known whether the constant-D assumption, implicit in Equation (3.5) is valid at these low dopant concentrations, but this assumption generally has been used in the extraction of reported diffusion coefficients. Of more concern in MESFET technology is the out-diffusion of the deep donor and acceptor wafer impurities. Cr out-diffusion has been characterized [23], but most manufacturing facilities have switched to LEC wafers, where Cr is not present in significant concentrations.

### 3.3.2.5 Activation

After the implantation and diffusion are modeled, the profiles of the electrically active species must be calculated to determine the electrical behavior of the devices. The activation coefficients of percent active *versus* concentration of implanted and annealed Si and Se provided by Dhiman and Wang [24] presently are implemented in SUPREM 3.5.

Typical Si and Se donor activation coefficients are between 70 and 95%, and at the small concentrations used for buried *p*-layers, near 100% acceptor activation can be expected. For Mg and Be, total activation for anneal temperatures above 700°C and concentrations less than $1 \times 10^{19}$/cm³, as first-order approximation default donor and acceptor activations used by GATES, are 80 and 100%, respectively. Within the expected range of activation achieved with different annealing caps [25], temperature and dose [26], and other factors, varying pinch-off voltages are predicted. The donor activation can be calibrated using a measured pinch-off voltage.

For the high-donor concentrations used for $N_D^+$ implants and sometimes for low-energy, high-dose channel implants, the activation may be concentration or depth dependent [27]. At low concentrations, the electron and donor concentrations differ by a constant factor, but above a certain donor concentration, the electron profile is nearly flat as shown in Figure 3.18. This saturation is modeled by assuming that, for amphoteric Si ions, high concentration activation is controlled by the distribution of Si between Ga and As sites. Assuming equilibrium conditions, this is calculated using [14,28]

$$\frac{[\text{Si}_{\text{As}}^-]}{[\text{Si}_{\text{Ga}}^+]} = \gamma = \frac{1}{K_v} \left[ \frac{\text{Si}}{n_i} \right]^4 \left[ \frac{1 - \gamma}{1 + \gamma} \right]^4 \tag{3.6}$$

where $K_v$ is the equilibrium constant for the reaction between Si on As and Ga sites and Ga and As vacancies, [Si] is the total Si concentration, and $n_i$ is the intrinsic electron concentration at the anneal temperature. This model predicts that $\gamma$ increases from 0 as [Si] increases, therefore the activation coefficient $\eta = (1 - \gamma)/(1 + \gamma)$ decreases from unity. Since the low-concentration activation coefficient $\eta_0$ is less than unity, a minimum value of $\gamma$, $\gamma_0 = (1 - \eta_0)/(1 + \eta_0)$ is used. The model predicts that the saturated electron concentration increases and therefore $n^+$ layer resistivity decreases with higher anneal temperature.

### 3.3.2.6 Recess Etching

Since SUPREM 3.5 and GATES are primarily one-dimensional process simulators, recess etching is modeled by shifting the profiles toward the surface. GATES allows the automatic etching to a desired pinch-off voltage or saturated current.

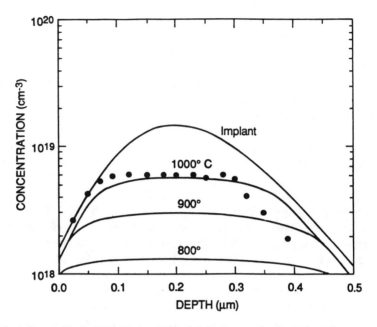

**Fig. 3.18** Implantation profile for 200 keV, $4 \times 10^{12}/cm^2$ Si implant, and estimated carrier concentrations for 800, 900, and 1000°C anneal temperatures. Measured electron concentration after annealing at 1000°C is shown by dots.

### 3.3.2.7 Epitaxial Layers

Competing with ion-implanted MESFET technology is epitaxial-doped-layer growth, best controlled by the molecular beam epitaxy method. SUPREM 3.5 and GATES allow us to input layer thicknesses and doping concentrations. Implantation profiles for the source and drain regions then can be added to these profiles to simulate epitaxial-fabricated MESFETs with implanted source and drain regions.

### 3.3.3 Device Modeling

### 3.3.3.1 Pinch-off Voltage

Impurity profiles in the channel region of the MESFET as derived from the process simulation are used to solve the Poisson's equation in one dimension to predict the electrical $I_{ds} - V_{ds}$ characteristics of the device. The Newton iteration method [29] with the following boundary condition at the surface

$$\frac{dn(x)}{dx} = 0 \tag{3.7}$$

is used to solve the Poisson's equation. This provides the gate-bias-independent electron profile resembling the active donor profile except near the surface and at large depths. At the surface, the boundary condition given by Equation (3.7) forces $n(x)$ to depart from the usually Gaussian doping profile. At larger depths, the substrate junction cuts off the tails in the implant profiles, as shown in Figure 3.19. Using the derived doping profiles, the MESFET pinch-off voltage is calculated using [30]

$$V_p = \frac{q}{\varepsilon_0 \varepsilon_r} \int_0^\infty x n(x) \, dx \tag{3.8}$$

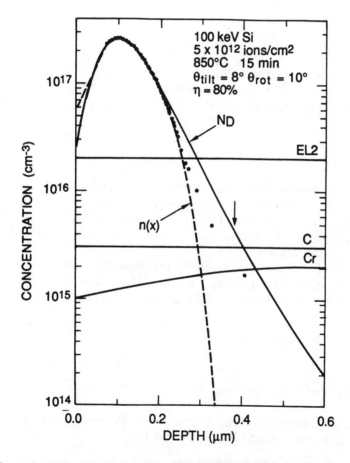

**Fig. 3.19** Comparison of the measured $Si^{29}$ donor profiles (solid line), the calculated electron profile at zero bias (dashed line), and the measured electron profile using the C-V method (dots).

and the threshold voltage then is estimated by

$$V_{th} = V_{bi} - V_P \tag{3.9}$$

where $V_{bi} = 0.78$ V is the Schottky barrier height, which is nearly constant over widely varying process and operating conditions.

### 3.3.3.2 Implanted Layer Resistivity

The conductivity of an implanted layer is given by

$$\sigma = \frac{qW}{L} \int_{x_0}^{\infty} n(x) \mu_m(x) \, dx \tag{3.10}$$

where $W$ and $L$ are the layer width and depth respectively, $\mu_m(x)$ is the electron mobility depending on the carrier concentration and on the compensation ratio.

Similarly, the saturated current through the implanted layers is given by

$$I_{sat} = qW \int_{x_0}^{\infty} n(x) V_s(x) \, dx \tag{3.11}$$

where $V_s$ is the saturated electron velocity, which also depends on the carrier concentration and compensation. Present models for electron mobilities and saturated velocity are derived from [31]. The surface depletion is also taken into account because of the presence of a surface potential $V_{sf}$ on GaAs surfaces which is approximately 0.6 V.

This effect is reflected in Equation (3.11) where the lower limit of the integral is not taken from $x = 0$ but from $x_0$. The value of $x = x_0$ is obtained using abrupt depletion approximation for surface depletion, represented by

$$V_{sf} = \frac{q}{\varepsilon_0 \varepsilon_r} \int_0^{x_0} x n(x) \, dx \tag{3.12}$$

### 3.3.3.3 Source and Drain Resistances

The total resistance between the ohmic metal contact of source or drain and the active channel of a MESFET is given by [32]

$$R_{s,d} = \frac{\sqrt{\rho_c R_{n^+}}}{W} \coth L_c \sqrt{\frac{R_{n^+}}{\rho_c}} + \frac{R_{n^+} L_{n^+}}{W} + \frac{R_{n^-} L_{n^-}}{W} + \frac{R_{n'} L_{n'}}{W} \tag{3.13}$$

where $\rho_c$ is the specific contact resistance, $R_n$ and $L_n$ are the layer resistivities and distances of the regions the electrons traverse between the ohmic contact and gate. To simulate a variety of MESFET topologies, any one of four different structures can be specified, as shown in Figure 3.20. Structure 1 is typical of a depletion-mode MESFET, which may have a recessed channel. Most of resistance in unrecessed depletion-mode devices is in the $N^-$ region. Nonrecessed enhancement-mode MESFETs would have an infinite $R_s$ if fabricated in this way because of surface depletion, so self-aligned processes (structure 2) are used. Since such devices tend to have low breakdown voltages, high drain conductances, and gate-length dependent threshold voltages, a lower energy $N^+$ implant is sometimes used, which can be simulated using structure 4. Structure 3 is a general structure that can be used for a variety of cases, such as those where the gate metal in recess-etched devices do not fill the trench. Source-gate-drain geometry need not be symmetric as shown in Figure 3.20.

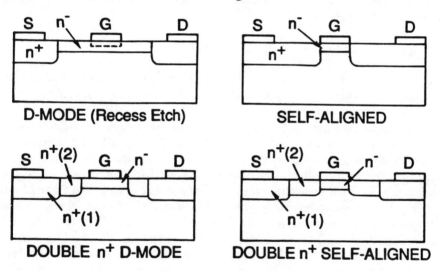

Fig. 3.20 Different MESFET structures.

### 3.3.3.4 Piezoelectric and Short-Channel Effects

Stress from dielectric passivation films located between the gate and source or drain contacts stress the GaAs crystal, inducing piezoelectric charges in the channel [33]. For self-aligned MESFETs (Figure 3.20), lateral straggling and possibly diffusion of the $N_D^+$ implants into channel area results in additional donors beneath the gate. Both factors affect pinch-off voltage and other device characteristics of MESFETs with short gate lengths, $L$. Although the concentrations of additional charges clearly vary considerably across the gate length; because GATES is primarily a one-dimensional simulator, it calculates the concentration of additional carriers at a fixed lateral distance, $fL$, where $f$ is the fractional distance from the drain side. The carrier density is assumed to be constant for all other lateral distances. Poisson's equation in the presence of these charges is given by [34]

$$\frac{d^2\psi}{dx^2} = \frac{q}{\varepsilon_0\varepsilon_r} [N_D^-(x) - n(x) + p(x) - N_A(x)$$

$$+ \rho_{pz}(x,fL) + E(fL)N_D^+(x)] \tag{3.14}$$

where $N_D^-$, $N_D^+$, and $N_A$ are the ionized impurity concentrations and $E$ is the fraction of $N_D^+$ present at the distance $fL$. Assuming the lateral straggling is Gaussian, $E$ is given by [34]

$$E = \text{erfc}[fL/2\sigma_{\text{tot}}] + \text{erfc}[(1 - f)L/2\sigma_{\text{tot}}] \tag{3.15}$$

where $\sigma_{\text{tot}}^2 = \sigma_{\text{lat}}^2 + 2Dt$ and $\sigma_{\text{tot}}$ = the lateral implant straggling parameter calculated using TRIM, $D$ is the Si or Se diffusion constant at the annealing temperature, and $t$ is the diffusion time. The stress on the GaAs at the distance $fL$ comes from the edge of the film on the drain side of the gate minus that from the edge of the film at the drain-metal contact plus a similar difference from the source side. No effect due to the stress of the gate metal is assumed, though this could be accounted for empirically by changing the dielectric film stress. Piezoelectric charges are obtained from the stress following Asbeck, Lee, and Chang [33].

Figure 3.21 shows calculated piezoelectric and short-channel effects on carrier densities. Depending on the orientation of the gate and whether the film stress is compressive or tensile, the piezoelectric charges either add or subtract from the impurity concentrations, leading to either narrower or wider carrier profiles and ultimately to more positive or more negative pinch-off voltages. The $N^+$ encroachment results in more negative pinch-off voltage, because the additional donors always add to the channel carrier concentrations.

### 3.3.3.5 Source-Drain Current, Small-Signal, and Other Model Parameters

Once the free-carrier doping profile is determined in the channel region, many closed-form solutions are available to calculate the source-drain current. It is important to take into account effects such as velocity saturation, substrate conduction, current injection, and drain-barrier lowering. Closed-form analytical equations are preferable over numerical solutions to obtain reasonably accurate

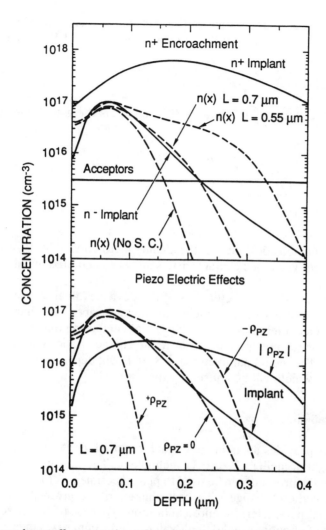

**Fig. 3.21** $N^+$ encroachment effects on carrier profiles for gate lengths $L \gg 1$ $\mu$m (no short-channel effects), $L = 0.55$ $\mu$m and 0.7 $\mu$m. Very deep $N^+$ profiles are used. Piezoelectric effects on the carrier either add or subtract from the net impurities, depending on the sign of stress and orientation of the gate.

values of dc parameters of the device without too much computer time. Such models can be implemented easily in process simulation programs.

As $I_{ds}$ and the depletion region are fully determined for a given gate and drain bias, all of the transconductances and capacitances are calculated analytically and $f_t$, $f_{\max}$, the unilateral and maximum available gains, and the minimum noise figure are computed using standard expressions [32]. The $I_{ds}$ calculations also can include source and drain resistance and gate-current effects. The effective gate and drain biases are modified from the original values $V_{g0}$ and $V_{d0}$ using

$$V_d = V_{d0} - I_{ds}(R_s + R_d)$$
$$V_g = V_{g0} - I_{ds}R_s - I_g(R_s + R_g)$$
(3.16)

where $R_g$ is the gate-metal resistance [32], and $I_g$ is the gate diode current given by:

$$I_g = I_s \exp \frac{V_{g0} - I_g(R_s + R_g)}{nkT}$$

$$I_s = A^*T^2LW \exp - \left(\frac{V_F}{kT}\right)$$
(3.17)

$A^*$ is the Richardson constant, $n$ is the ideality factor, and $V_F = V_{bi}$ is the Schottky forward voltage. The diode characteristics must be calibrated by the user to obtain the best results.

Having dc and rf characteristics of the device evaluated, the SPICE model parameters can be extracted similar to one described in Chapter 6 (Section 6.1.2.2). The characteristics of the device evaluated using the analytical solutions described earlier can be treated similar to the characteristics measured on a fabricated device. The model parameters for SPICE will depend on the model used in that program, which will be chosen by the user.

### 3.3.4 Calibration

GATES provides an estimate of each required input quantity, which is reasonable in light of modern MESFET technology. These estimates are sufficiently accurate that the program can be used to plan experiments or to elucidate physical correlations without changes. To best characterize a given process or device, several crucial parameters require calibration. Table 3.4 lists these parameters, gives ranges of variations and default values, and summarizes measurement techniques to calibrate them. In general, such calibrations require a good set of process control monitoring devices as described in Section 3.5 of this chapter.

The electron carrier activation is the most critical factor and should be calibrated by comparing simulations with a measured MESFET pinch-off voltage, preferably for a long-gate-length FATFET, where short-channel and drain-barrier lowering effects are absent. The surface depletion potential $V_{sf}$ affects $N^-$ layer resistivities and saturated currents and can be calibrated by comparing calculated and measured resistivities or saturated currents. Piezoelectric and $N^+$ encroachment effects may play a role in very short layers, so layers with lengths exceeding $\sim 5$ $\mu$m should be used.

**Table 3.4** Calibration Parameters

| Parameter | Range | Default | Measurement Information |
|---|---|---|---|
| $\eta$: Donor activation | 70–95% | 80% | FET (preferable FATFET) $V_{th}$ |
| $N_A$: LEC acceptor level | $1$–$6 \times 10^{15}$ | $3 \times 10^{15}$ | $V_{th}$ (dose) or tail of $n(x)$ |
| $V_{sf}$: Surface depletion | 0.3–0.8 V | 0.45 V | $N^-$ resistivity |
| $K_v$: Activation saturation | 10–30 | 24 | $N^+$ resistivity, $n(x) > 5 \times 10^{17}$ |
| $r_c, \rho_c$: Contact resistance | $g = 0.1$–10 | $0.2 \ \mu\Omega$–cm$^2$ | TLM data |
| $p$: Pearson-IV mix factor | $<1$ | 1. | SIMS |
| $\sigma_{tot}, \sigma_{lat}$: Short-channel $N^+$ encroachment | $>\sigma_{lat}$ | $\sigma_{lat}$(TRIM) | $V_{th}$ versus $L$ |
| Film stress (piezoelectric) | $0.1$–$5 \times 10^9$ | $1 \times 10^9$ dyn/cm$^2$ | Horizontal-vertical FET pairs |
| $n, V_f, I_s$: Gate current | | | Gate-diode characteristics |

The equilibrium constant describing activation saturation can be calibrated by comparing it with measured $N^+$ layer resistivities, assuming that sufficiently high $N_D^+$ doses are used to achieve saturation. Although the surface depletion effect is less important for $N^+$ layers, $V_{sf}$ should be calibrated first. $K_v$ should be temperature dependent [28]; but in general, a constant value of $K_v$ is found to fit measured resistivities reasonably between 800 and 900°C, suggesting that the steepest temperature dependence is in the $n_i$ factor. Differential resistivity data also may be used to calibrate $K_v$.

Higher wafer acceptor concentrations give narrower electron profiles; hence, the shape of the electron profile, measured using the CV measurement technique at concentrations of about $10^{16}$/cm$^3$, can be used to calibrate $N_A$. Bare wafer implants should be done, and care should be taken in applying resistive corrections to the CV measurements to obtain the correct values near the tail [35]. Also, we should realize that for large reverse biases, needed to probe electron concentrations below $10^{16}$/cm$^3$, the width of the electron profile becomes smaller than the Debye length, so that the approximation upon which the CV method depends breaks down. This produces the tail on the CV derived $n(x)$ profile shown in Figure 3.19.

Resistances of Au-Ge-Ni contacts show great process variations that are not well understood. The specific contact resistance $\rho_c$ is modeled using [36]

$$\rho_c = g\max\left(0.6, 0.2 \times \frac{10^{18}}{\text{Peak}}\right)\mu\Omega - cm^2 \tag{3.18}$$

where Peak is the peak doping (in $cm^{-3}$) of the $N^+$ region, and $g$ is an empirical factor $> 1$. The factor $g$ can be calibrated comparing $\rho_c$ or $r_c$, given by the first term in Equation (3.13), with measurements using the transmission-line method. The contact resistance is not an important component of the total source resistance for non-recess-etched depletion-mode MESFETs, where most of the resistance is from the channel. It can be important in self-aligned or deep-recess devices and for poor processes, where the values of $g$ are large.

The mixing factor $p$ accounting for the channeling reduction that occurs when Si or Se ions are implanted through thick caps is best calibrated using SIMS. CV or other carrier profile data cannot be used reliably, due to wafer acceptor impurity variations.

Dielectric film stress depends on deposition methods, annealing temperatures, and can even change late in the process when second-level dielectrics are deposited or the ohmic metal alloying is done. Pinch-off voltage differences between pairs of short-length MESFETs oriented horizontally and vertically can best calibrate this effect [33], as the added piezoelectric charges have opposite signs in the two orientations as shown in Figure 3.20. Although the MESFETs' pinch-off voltages fluctuate greatly over wafers, pinch-off voltages of FETs with widths $<7$ $\mu$m placed within 5 $\mu$m of one another are tightly correlated [37]. This suggests that random pinch-off voltage variations may be reduced if the horizontal-vertical pairs have small widths and are placed beside one another.

The total lateral straggling parameter can be calibrated by measuring pinch-off voltages *versus* gate lengths for MESFETs where dielectric films are absent. However, we may still obtain a piezoelectric effect due to the gate metal stress, so horizontal-vertical MESFET pairs should be measured and the piezoelectric contribution calibrated. The gate length must be measured accurately; mask values cannot be used because of the variation in gate lines drawn on the mask and that printed and developed on photoresist. Although we would not expect the Si diffusion constant to be sufficiently large to affect $\sigma_{tot}$ [18,19], measurements comparing dependences of pinch-off voltages on gate lengths for varying annealing temperatures and times suggest that diffusion may play a role [34].

### 3.3.5 Calibration Example Using GATES

Figures 3.22–3.25 illustrate the calibration procedures, using data obtained by Magerlein et al. [38] at IBM. Here, $\eta$ and $N_A$ are simultaneously fit using pinch-off voltages obtained from different doses, as shown in Figure 3.22. Magerlein et al. measured pinch-off voltages from E- and D-mode and combined E- and D-mode doses, 1.5, 2.5, and $4 \times 10^{12}$ ions/cm$^2$. Within the range of variations of

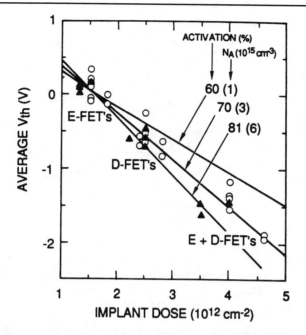

**Fig. 3.22** Measured *versus* calculated pinch-off voltages for various activations and acceptor impurities.

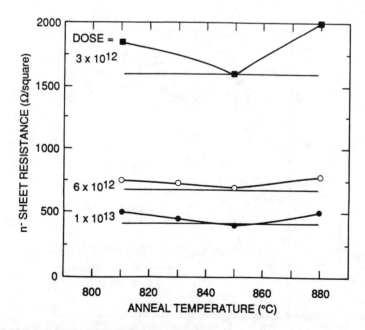

**Fig. 3.23** Measured *versus* simulated sheet resistance of $N^-$ channel layers for various implant dose and anneal temperatures. Surface potential is adjusted to 0.52 V to obtain good agreement at 850°C, the annealing temperature.

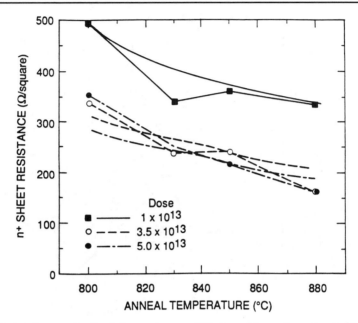

**Fig. 3.24** Measured *versus* simulated sheet resistance of $N^+$ channel layers for various implant doses and annealing temperatures. The activation saturation constant $K_v$ is adjusted to 10 to obtain good agreement with measured data.

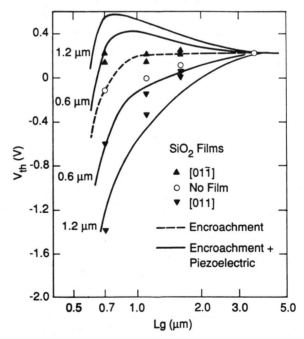

**Fig. 3.25** Measured *versus* simulated pinch-off voltage as a function of gate length for 60 keV, $9 \times 10^{11}/$ cm$^2$ $N^-$ and 175 keV, $1.7 \times 10^{13}/$cm$^2$ $N^+$.

he measured pinch-off voltages, GATES fit the nominal values using $\eta = 70\%$ and $N_A = 3 \times 10^{15}$ cm$^{-3}$. Different wafers give different pinch-off voltages, probably due to different acceptor concentrations. Different activations also can contribute to As/Ga stoichiometries during crystal growth.

The surface depletion potential is calibrated using the $N^-$ resistivity data shown in Figure 3.23. Different resistivities are obtained at different annealing temperatures, probably due to different activation or small wafer-to-wafer acceptor concentration differences. $V_{sf} = 0.6$ V is obtained from the measured pinch-off voltage. The equilibrium constant is obtained from data shown in Figure 3.24. A value of $K_v = 10$ correctly predicts the saturation at a dose of $5 \times 10^{13}$ cm$^{-2}$ for the 50 keV Si implants and the temperature dependence between 800 and 880°C. For the peak $N^+$ doping of $6.4 \times 10^{17}$/cm$^3$, and $R_n^+ = 321$ $\Omega$-mm at the 800°C $N^+$ anneal temperature, $\rho_c = 0.7$ $\Omega$-cm$^2$ is obtained; therefore, $g = 1.17$ is needed.

Figure 3.25 illustrates the calibration of film stress, $\sigma_{tot}$, and $\eta$ by comparing with measurements of Ohnishi et al [39]. In these measurements, horizontally and vertically oriented MESFETs were obtained, then one-half and finally all the dielectric films were removed. A high activation (96%) is needed to fit the measured pinch-off voltage for 3.0 $\mu$m long MESFETs. The short-channel effect is calibrated by using the pinch-off voltage measured for zero-film stress at $L = 0.7$ $\mu$m. Pairs of horizontal and vertical FETs give the same result, indicating that gate-metal stress is negligible here. The fit required $\sigma_{lat} = 84$ nm for the 175 keV Si implants, suggesting either that $D = 7 \times 10^{-14}$ cm$^2$/s for these 15 min anneals and 750°C (a large value) or the lateral straggling is underestimated. The term, $\sigma_{lat} = 190$ nm is reported [40] for 200 keV Si implants, so this large lateral straggling parameter is not unreasonable. However, the data suggest that the short-channel-induced pinch-off voltage roll off is more linear than the model predicts.

The film stress is calibrated using 1.2 $\mu$m films in the (011) direction. The stress coefficient derived depends on the gate-to-drain spacings $L_{gd}$ and $L_{gs}$, which are not reported by Ohnishi et al. [39]. Values of the stress between $2.1 \times 10^9$ dyn/cm$^2$ for $L_{gd} = 0.5$ $\mu$m to $1.4 \times 10^9$ for $L_{gd} = \infty$ are needed. Although the calculated and measured pinch-off voltages of the (011) FETs agree, the agreement for the (011) FETs is not very close. Finally, for the gate current calculations, measurements of the gate-source diode characteristics are used to determine $n$, $V_F$, or $I_s$ in Equation (3.17).

## 3.4 MANUFACTURING PROCESS FOR MMICs

GaAs process technology has been in use since early 1970s to fabricate discrete devices that are used in hybrid microwave circuits. The process technology developed for discrete device fabrication has established a base for manufacturing MMICs designed using MESFET active devices. In fact today's GaAs MMIC technology represents an extension of discrete device process. This coupling between the MESFET and MMIC processing allows the two to coexist in the same fabrication facility, and the data accumulated on discrete devices can be applied to MMIC designs. The fundamental difference in developing the MMIC rather than discrete MESFET devices is the requirement of process repeatability, the components' active and passive reproducibility within specified tolerances, and ability to integrate diverse functions on the same chip (e.g., microwave and digital). At the end of processing, discrete devices can be tested and categorized with different specifications for use in hybrid circuit manufacturing. However, when several devices are used in one MMIC design, there are no means to evaluate the performance of an individual device to adjust for the matching network. The only option is to achieve all individual devices within prespecified performance tolerances. Such an extension of discrete device fabrication process to MMIC manufacturing largely has contributed to several technological innovations, which have been made possible by Silicon technology, such as (1) availability to high-purity, low-defect density, uniform, large-size, and high-resistivity semi-insulating GaAs substrates; (2) techniques for producing a uniform, controlled doping profile for active layer, most commonly using ion-implantation; (3) improved equipment to provide better lithographic and wafer patterning for smaller line widths and uniformity; (4) controlled submicron feature-size etching by wet chemical or dry processes; (5) automated and computer controlled wafer handling equipment for highly accurate and reproducible results with high throughput; and (6) development of in-process electrical testing and material analysis techniques to help evaluate every step of processing for uniformity and control within specifications.

In general, GaAs processing differs from Si processing in several aspects besides the fundamental differences in the materials' properties. Application of GaAs devices for performance driven circuits demands submicron line definitions for the gate of MESFET, low-capacitance metal crossovers requiring airbridge structures and the processing of other analog passive microwave components, such as MIM capacitors and spiral inductors, besides the low-inductance grounding capability. The basic semi-insulating property of the GaAs provides relative ease of device isolation. Typical electrical and mechanical properties of commercially available semi-insulating GaAs starting material are described in Table 3.5. Further advances in GaAs material technology are targeted for lower defect densities, larger wafer sizes (4-inch diameter or more) and also epitaxially grown GaAs-on-Si wafers.

A step-by-step MMIC planar process based on depletion model MESFET technology is shown in Figure 3.26. This represents a typical MMIC process as opposed to the process representing any specific foundry. Most of the MMIC processes based on depletion MESFET, are variations of the process described here. At all steps of photoresist patterning, positive photoresist is commonly used and gold is the interconnect metal of choice. Photoresist-based lift-off processing is used extensively, as opposed to etching process, to define the metalization area.

**Table 3.5 Starting GaAs Semiinsulating Material Properties**

| | |
|---|---|
| Growth method | Undoped LEC is the most commonly used material |
| Resistivity at room temperature | $>10^7$ $\Omega$-cm |
| Mobility at room temperature | $> 4500$ cm²/V-s |
| Size | 3 inch ± 25 mil diameter, 4 inch diameter wafers are also commercially available |
| Thickness | 625 $\mu$m ± 25 $\mu$m |
| Crystal orientation | (100) ± 0.25° |
| Flatness | $<7$ $\mu$m |
| Bow | $<25$ $\mu$m |
| Surface finish | Both sides polished |
| Etch pit density | $<10^7$/cm² |
| Major flat orientation | ($0\bar{1}\bar{1}$) |
| Minor flat orientation | ($0\bar{1}1$) |

Passivation-assisted lift-off processing also is used to define the metal lines where a thin layer of SiN is deposited on the wafer and, along with the photoresist, is used for lifting off the metal. Figure 3.27 shows the key of patterns used in the illustration for different process steps.

## 3.4.1 Active Layer Formation and Device Isolation

LEC undoped semi-insulating GaAs is the starting material most commonly used in MMIC manufacturing. The wafers are polished on at least one side and oriented 2° off (100) toward the nearest (110) plane. Wafers undergo a solvent cleaning to remove any organic residue. A typical cleaning cycle includes a 5-min soak in boiling TCE, acetone, and methanol. Then the polishing damage is removed by etching several microns from the wafer surface using a wet chemical etch. A solution typically used for this purpose is $10:1:1$ of the $H_2SO_4$ : $H_2O : H_2O_2$. The next step is then the formation of active layer or the channel layer. The methods commonly used are blanket $N^-$ or selective $N^-$ ion-implant using Si, a dopant, followed by selective $N^+$ implant to heavily implant the areas where low resistance ohmic contacts are needed, such as the source and drain of the MESFETs and the contacts to the monolithic resistor. Immediately before the implantation, the wafer is dipped in an acid solution typically $10:1$ of $H_2O : HF$ to remove the oxides that might be present on the wafer surface. The peak doping

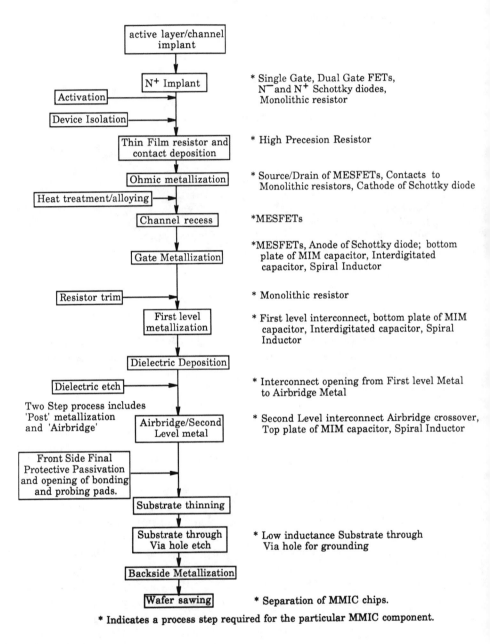

**Fig. 3.26** Flow chart of MESFET-based MMIC process.

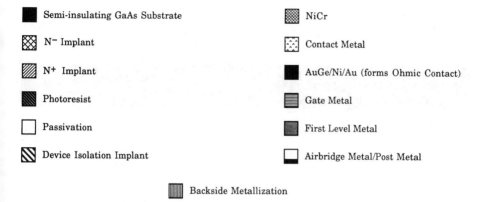

**Fig. 3.27** Pattern reference sheet.

concentration in the active channel layer typically is $10^{16}/cm^3$ to $5 \times 10^{17}/cm^3$ and the depth of the implanted layer is approximately $.15–.4$ $\mu m$ depending on the needed characteristics, such as pinch-off voltage and $I_{dss}$ of the MESFET. Two widely used MMIC manufacturing processes are described here. Figure 3.28 shows the starting semi-insulating wafer with the location of the devices to be fabricated on the wafer in the following steps. Figure 3.29a shows the blanket channel implant all across the wafer, followed by the definition of photoresist for selective $N^+$ implant, shown in Figure 3.29b. Heavy dose and energy implant is performed for $N^+$, the photoresist is removed, and the implanted species are activated, using the annealing process at elevated temperature of about 800–850°C for about 30–40 min resulting in doped layers as shown in Figure 3.29c. To maintain the material stability of GaAs and prevent the out-diffusion of arsenic from GaAs, either an arsine-over pressure technique or a dielectric cap over GaAs is used. The next step is to define the mesa etch area for device isolation, shown in Figure 3.29d. The wet chemical etch technique is commonly used for mesa etch; and the height of mesa is generally 0.6 $\mu m$–0.9 $\mu m$, depending on the depth of the channel implant. The photoresist is then removed and wafer is cleaned, as shown in Figure 3.29e.

The other commonly used process results in a more planar structure, as it uses the selective $N^-$ channel implant instead of a blanket $N^-$ implant and requires no mesa etch for device isolation. As shown in Figure 3.30a, photoresist is used to define the implant area for active channel of MESFETs, monolithic resistors, and Schottky diodes. Also shown is a thin layer of $SiO_2$, which is typically used as a

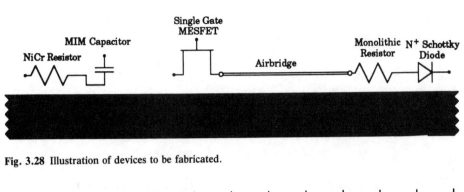

**Fig. 3.28** Illustration of devices to be fabricated.

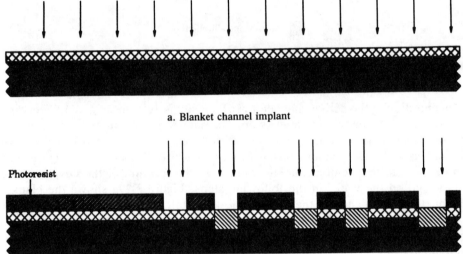

a. Blanket channel implant

b. Selective $N^+$ implant

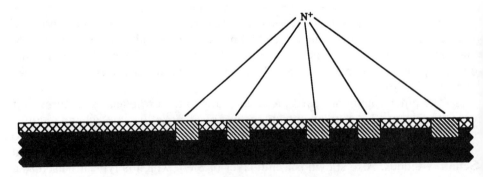

c. Photoresist removed and implant activated

**Fig. 3.29** A widely used MMIC manufacturing process.

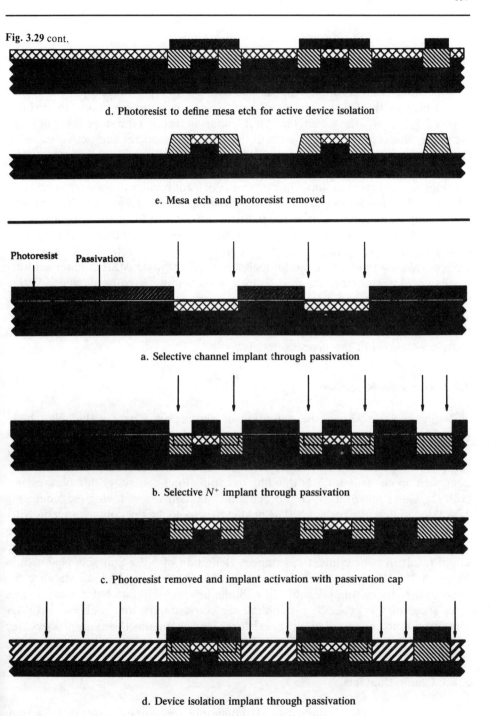

**Fig. 3.29** cont.

d. Photoresist to define mesa etch for active device isolation

e. Mesa etch and photoresist removed

Photoresist    Passivation

a. Selective channel implant through passivation

b. Selective $N^+$ implant through passivation

c. Photoresist removed and implant activation with passivation cap

d. Device isolation implant through passivation

**Fig. 3.30** Other widely used MMIC manufacturing process.

surface protection as well as to control the implant profile. As the implantation results in Gaussian profile, the peak doping is deeper under the surface. Using a $SiO_2$ or other passivation layers, such as SiN, shifts the Gaussian profile toward the surface of the GaAs substrate, resulting in peak doping closer to surface. Figure 3.30b shows the selective $N^+$ implant again through the same layer of $SiO_2$. Next the photoresist is removed and the implanted species are activated using capped anneal method. The surface of GaAs shown in Figure 3.30c is protected by $SiO_2$, which prevents the out-diffusion of arsenic from the GaAs surface. Hence, the implant activation or annealing process does not require arsine over-pressure, which has the disadvantage of using and handling the extremely toxic gas, arsine. Other capping materials used in the industry are SiN, $Al_2O_3$, AlN, *et cetera*. Both the capped and capless annealing processes have been shown to produce uniform, high-activation percentages for $N^-$ and $N^+$ implants. Figure 3.30d shows the device isolation implant, an additional step used by some MMIC manufacturers to ensure better device isolation. Boron is commonly used as device isolation implant. Typical boron implantation at an energy of 200 keV and a dose of $2 \times 10^{13}/cm^2$ is used. Selective $N^-$ planar and mesa etch processes are described up to the channel implant and device isolation steps; however, Figure 3.30c will be used to follow the rest of the process details.

### 3.4.2 Thin Film Resistors

Figure 3.31a shows the definition of photoresist where a thin film, high-precision NiCr resistor is to be fabricated. A thin layer of SiN is deposited on the wafer prior to coating the photoresist. A lift-off process is used to remove NiCr from the undesired area on the wafer. In some processes, thin film resistor is deposited on an insulator SiN thin film to avoid direct contact of thin film resistor with the GaAs material. Such a processing step helps in reducing the sidegating effect due to thin film resistors. The next step is to define the contact area on NiCr resistors to connect them to other components in the MMIC. It is important to make a contact to NiCr just after its deposition, as NiCr oxidizes quickly and could result in poor contact resistance. Definition of NiCr contacts is shown in Figure 3.31b. Heat treatment is done after the NiCr deposition to stabilize the sheet resistivity of thin layer of NiCr. Some processes define NiCr resistors at a later stage of the process, just before dielectric deposition, and use airbridge metalization to connect it to NiCr resistors. This eliminates the extra process step of separate contact metal to NiCr resistors.

### 3.4.3 Ohmic Contacts

Figure 3.32a shows the photoresist definition for ohmic contact metalization.

a. Photoresist to define NiCr resistor and NiCr deposited

b. Photoresist to define NiCr contact and contact metal deposited

**Fig. 3.31** Definition of photoresist on thin films.

AuGe-Ni-Au is a commonly used metal composition for ohmic contacts, which includes 1000 Å of Au-Ge eutectic followed by 330 Å of Ni then a thin layer of gold, followed by the lift-off process to remove the photoresist and ohmic metal from undesired areas and wafer cleaning, as shown in Figure 3.32b. Heat treatment or sintering-alloying is used to achieve low ohmic contact resistance. A rapid thermal annealing system is used widely, in which the temperature is rapidly increased to approximately 400–500°C for 30 s, resulting in the formation of good contact between GaAs and the AuGe alloy. Ni is used as a wetting agent, as Au-Ge eutectic tends to ball up, and is essential for the reliability of ohmic contacts.

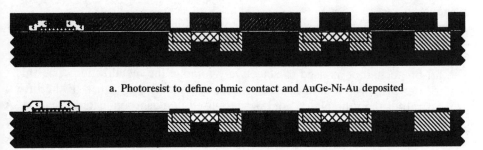

a. Photoresist to define ohmic contact and AuGe-Ni-Au deposited

b. Wafer cleaned and ohmic contact and NiCr heat treated

**Fig. 3.32** Photoresist definition for ohmic contacts.

### 3.4.4 Schottky Metalization

This is the most critical process step in MMIC manufacturing because the submicron line definitions and the device's performance and reliability are heavily dependent on MESFET performance. Wide ranges of metal compositions and process techniques are used in achieving highly reliable, low-resistance, low-leakage current Schottky contacts. Refractory metals, such as Ti, Pt, Pd, and W plated with gold are used widely for improved reliability. Two gate metal structures commonly used are T-type or mushroom gate structures, as shown in Figure 3.33a and lift-off vertical gate structures, as shown in Figure 3.33b. The process steps in the formation of the gate involves

1. Lithographic definition of gate.
2. Channel recess to achieve target drain current ($I_{dss}$) of the FET.
3. Deposition of base metal for Schottky contact.
4. Deposition of thick gold for reduced resistance and reliability.

Mushroom gate formation is generally a two-step process, where the 0.5 $\mu$m (for 0.5 $\mu$m gate length MESFET) gate lines are defined in photoresist; channel recess is performed, commonly using wet chemical etch; and the base metal, which could typically be composed of Ti-W or other refractory metals, is sputtered or evaporated. Aluminum is also used by some manufacturers for Schottky metalization. Next another layer of photoresist is coated on top of the base metal and wider openings (typically 1.0–1.5 $\mu$m) are defined on top of the 0.5 $\mu$m Schottky lines. Gold is plated in the openings, resulting in a mushroom or T-type structured gate. Next, the second layer of photoresist is removed and the base metal is etched followed by the removal of first layer of photoresist. This results in the final Schottky metal structure shown in Figure 3.33a at the gate of the MESFET and the anode of the Schottky diodes. An SEM picture of the mushroom structure gate is shown in Figure 3.33c.

A lift-off vertical gate structure is shown in Figure 3.33b, where the lines are defined in photoresist for Schottky metalization followed by channel recess. The base metal is sputtered or evaporated, and then a thick layer of gold is evaporated. The lift-off technique is used to selectively remove the metalization. The thickness of the evaporated gold in this process is limited, due to use of the lift-off process rather than the plated gold process used in mushroom gate formation. A combination of different photoresists and exposures, such as regular UV and deep UV techniques, is used to achieve thick layers of photoresist without relinquishing use of the lift-off technique.

The typical thickness of Schottky metal-plated gold in mushroom structures is 0.8–1.5 $\mu$m as opposed to 0.5–1.0 $\mu$m in lift-off vertical structures. Also the metal is wider on the top in the mushroom gate structure, which helps the Schottky gate to have more reliability and less resistance, particularly useful if the gate is forward biased. In some processes, Schottky metalization is also used as the first-level interconnect metal and the bottom plate of the MIM capacitor. However, a separate first interconnect metal process is described here.

a. Schottky gate metal deposited and plated with gold

b. Vertical gate structure

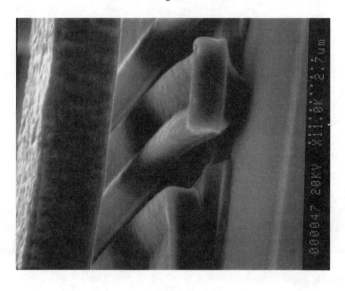

(c)

Fig. 3.33 Gate metal structures in common use.

### 3.4.5 Resistor Etch

To control the sheet resistivity of $N^+$ or $N^-$ monolithic resistors, the channel implant of the $N^+$ implant layer in the resistor area is etched to a specific thickness till the targeted value is achieved. As shown in Figure 3.34, the area where the monolithic resistor is to be etched is patterned in photoresist. A wet chemical etch typically is used for this purpose. However, uniformity across the wafer in the sheet resistivity of the monolithic resistor typically is reduced following this pro-

**Fig. 3.34** Monolithic resistor etch.

cess step due to the loose control of the etching thickness on the active layer. Most often this process step is avoided and the sheet resistivity of $N^-$ or $N^+$ monolithic resistors depend on the channel and $N^+$ implants, respectively. This is not a commonly used process step in manufacturing of MMICs.

### 3.4.6 First-Level Metalization

Figure 3.35 shows the definition of photoresist, where first-level metalization is to be performed. This metal also is typically used to metalize the ohmic contact area on source and drain the FETs' contacts of the monolithic resistors and cathode of Schottky diodes. Further, this metal layer is used as the bottom plate of MIM capacitor. Ti-Pd-Au is typically used for this metalization with 1.0–2.0 $\mu$m metal thickness. We recommend that this layer of metal be deposited on SiN or the insulator, to reduce any potential problems of isolation or backgating.

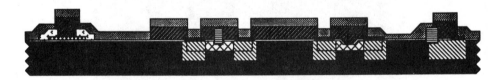

**Fig. 3.35** First-level metal deposited.

### 3.4.7 Dielectric Layer

SiN is deposited on the whole wafer, as shown in Figure 3.36a, typically using a plasma enhanced chemical vapor deposition technique. This protects the active and passive devices and also acts as the insulator for MIM capacitor. Typical thickness of the passivation used is 2000–3000 Å, which results in approximately 350–250 pF/mm$^2$ MIM capacitor. The next step is to define openings in the dielectric layer, as shown in Figure 3.36b, where a connection is made between the first-level metal and the second-level metal. Some processes use other insulated thin films such as $Ta_2O_5$ for MIM capacitors, which provides high density MIM capacitors (typically 600 pF/mm$^2$), however, this introduces an extra process step to define MIM capacitor area.

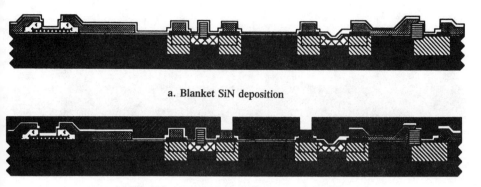

a. Blanket SiN deposition

b. SiN etch to open contact area to first-level metal

**Fig. 3.36** Formation of dielectric layer.

### 3.4.8 Airbridge Metalization

This is the second-level metal, which also is used as the top plate of the MIM capacitor and a low-capacitance crossover for the first-level metal. The process is in two steps. The opening in the first layer of photoresist defines the areas where airbridge metal is supported on the substrate, either on passivation such as in MIM capacitor or on first-level metal where there is an interconnect between first and second level metal. Where there is no opening in the first layer of photoresist, the airbridge metal forms a metal bridge in the air. The base metal is sputtered on top of the first layer of photoresist. Next, the second layer of photoresist is coated and the plating areas are defined by thickly plating it with gold then removing both layers of photoresists, resulting in the structure shown in Figure 3.37a. The typical thickness of airbridge metalization is 1.5–3.0 $\mu$m. The air gap between the air-bridge metal and the substrate is typically 2–4 $\mu$m. Figure 3.37b shows an SEM picture of the section of a MESFET where the sources of the FET are connected using an airbridge metalization and crossing over the gate and the drain of the FET.

### 3.4.9 Final Passivation

To protect the structures on the front side of the wafer, a thick layer of final passivation covers the entire wafer, as shown in Figure 3.38, and openings are etched only in the bonding and probing pad areas. Final passivation improves the overall reliability of the MMIC chip. It is important that the final passivation should not degrade the overall performance of the MMIC.

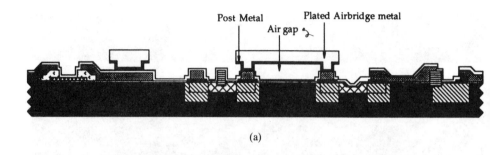

(a)

(b)

**Fig. 3.37** Airbridge metalization.

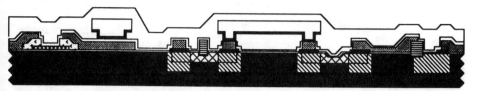

**Fig. 3.38** Final protective passivation.

### 3.4.10 Wafer Thinning and Backside Via Etch

The starting substrate wafer thickness is typically 24 mils. In order to create backside via holes, the wafer is typically thinned down to 4 mil or less thickness. A thin layer of photoresist is spun over the top of the wafer to prevent any accidental damage to the circuits from exposure to the etchants used in the following steps. Using an appropriate wax, the wafer is then mounted from the front side on a carrier. Mechanical lapping initially is used to thin the wafer, followed by chemical etching for a smooth backside finish. Some processes use only the wet chemical etch to thin the wafer from its initial 24 mil thickness to a 4 mil or less thickness. The chemical etchant typically is a $1:3:5:5$ mixture of $HF : HNO_3 : CH_3COOOH : H_2O$, which results approximately 20 $\mu$m/min etch rate at room temperature. A thickness uniformity of within 10 $\mu$m across the wafer is achieved during the thinning operation. The backside of the wafer is coated with photoresist; and the backside via hole mask is aligned to the structures on the front, using an infrared light source to see through the GaAs substrate. Next, the via hole pattern is etched in the photoresist, followed by etching the GaAs material. Via hole etching has proven to be one of the most difficult steps associated with the MMIC process. An etchant must have a moderately fast etch rate and result in a via hole with a reasonable aspect ratio. Several etchants are used for this purpose, including citric acid, sulfuric acid, ammonium hydroxide, and phosphoric acid solutions. A combination of dry reactive-ion etching and wet chemical etching generally results in via holes with a smooth wall surface finish. Formation of via holes is presented in Figure 3.39a. Next the backside metalization is done. This includes deposition of thin layer (typically 100–500 Å) of Cr followed by .5–1 $\mu$m of gold. The wafer is then placed in an electroplating solution and approximately 2–3 $\mu$m of gold is plated onto the backside of the wafer and the via holes are filled with metal, using an optically enhanced plating technique. It is important that the backside metalization fill the substrate via holes and make contact with the frontside metal structure, as shown in Figure 3.39b. The back metal must be suitable for a solder die attach using gold-tin eutectic solder or a silver epoxy die attach. An SEM picture of the substrate via hole is shown in Figure 3.39c. A

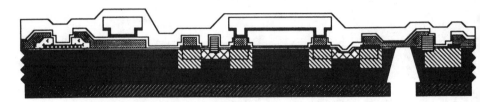

a. Substrate thinning and via hole etching

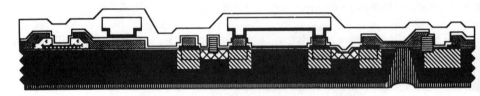

b. Backside metalization

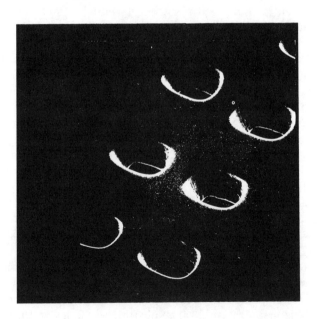

c. Substrate via hole

**Fig. 3.39** Finishing the back side of the wafer.

completely processed wafer with fabricated MMIC discrete devices is shown in Figure 3.40.

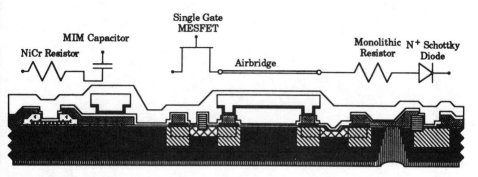

**Fig. 3.40** Process complete.

### 3.4.11 Wafer Sawing

After completing the backside process, the wafer is transferred to adhesive tape to separate the individual chips. A 1–2 mil thick diamond saw is typically used for sawing the GaAs wafers, but a scribing technique using a diamond scribe along scribe lines also may be used for the same purpose. The scribe lines are aligned in the GaAs cleavage planes and the wafer is split along the scribe lines and the dice physically separated from each other by stretching the tape.

Table 3.6 illustrates the correspondence between the processing layer and the devices used in MMICs. The layers essential to fabricating a particular device are indicated by an x in the process layer column. An o in the column indicates an optional layer, which could be used with that particular device depending on other requirements. For example, dielectric etch and airbridge metalization are optional in the MESFET, should it be necessary to either thicken the gold metalization on the source and drain or connecting the source and drain of a multifinger FET.

### 3.5 PROCESS CONTROL AND CHARACTERIZATION

Inherent to any semiconductor IC manufacturing process are the variations in the active and passive device characteristics, both across a wafer and among wafers. Variations in device parameters are directly reflected in the variations of circuit performance. To maintain high circuit yields, it is necessary to control the variation and yield of individual components of the MMIC. This is particularly important in GaAs processing because of nonuniformities in the starting material. The design task is particularly difficult for analog MMICs, where every effort is made to obtain the best gain, gain flatness, noise performance, output power, and isolation. There are two main limitations of yields in MMICs. Material or process defects result in processed devices that have short or open circuits. These defects are randomly distributed across the wafer. Second, there is yield loss because of

**Table 3.6 MMIC Components versus Layers of Process Step Requirements**

| Components | Selective Channel Implant | N⁺ Implant | NiCr (optional) | NiCr Contact Metal (optional) | Ohmic Metalization | Gate Metalization | Resistor Trim (optional) | First Level Metalization (optional) | Dielectric Etch | Airbridge Metalization | Substrate through Via Hole |
|---|---|---|---|---|---|---|---|---|---|---|---|
| Single-gate MESFET | x | x | | | x | x | | x | o | o | |
| Dual-gate MESFET | x | x | | | x | x | | x | o | o | |
| N⁺ Schottky diode | x | x | | | x | x | | x | o | o | |
| N⁻ Schottky diode | x | x | | | x | x | | x | o | o | |
| NiCr resistor | | | x | x | | | | | | | |
| Monolithic resistor | x | x | | | x | | x | o | o | o | |
| MIM capacitor | | | | | | | | x | | x | |
| Interdigitated capacitor | | | | | | | | x | o | x | |
| Transmission line | | | | | | | | x | | x | |
| Spiral inductor | | | | | | | | x | x | x | |
| Probing-bonding pads | | | | | | | | x | x | x | |
| Substrate through via hole | | | | | | | | x | | | x |

limited electrical performance, where ICs do not perform within specifications because of large variations in the performance of individual devices—more variation than is planned in the design cycle.

The MMIC process described in Section 3.4 consists of a large number of interrelated processing steps that influence the performance of both the active and passive devices, and thus the performance of MMICs. The key to lowering the MMIC cost is to target wafers that ultimately will not meet performance specifications and reject them as early as possible or take corrective action early enough in the manufacturing process to ensure compliance with specifications. In-process testing of process control devices is a widely used technique to monitor every processing step, including qualification of the starting material. Fundamental to process control and characterization is the establishment of test structures, test procedures, and specifications that allow rapid assessment of whether a particular process step or process sequence is performed normally. The most difficult task facing the process and test engineer is establishing pass-fail criteria for each process step that will result in reasonable MMIC yield and circuit performance.

### 3.5.1 Material Characterization

Process control begins with the qualification of substrate material. Many of the substrate characterization techniques are destructive and are performed on a sample basis, generally on the seed and tail wafers of the boule and a few wafers from the middle. Typically characterized material parameters are dislocation density, wafer flatness, surface finish, EL2 concentration, resistivity, and mobility.

#### *3.5.1.1 Dislocation Density*

These defects are randomly distributed across the wafer and can result in inoperable MMICs. Molten KOH is a commonly used etchant to determine the etch pit density. Typical etch pit density in LEC-grown, undoped, GaAs semi-insulating substrate wafers is $3 \times 10^4$–$2 \times 10^5$/cm$^2$. This test is generally performed on sample basis on a number of wafers selected from each boule. The upper limits on the acceptable dislocation density depend on the sensitivity of the circuit to the defects and the required yields. These defects are not catastrophic to operation or reliability of GaAs MMICs. A MESFET that is primarily a bulk conduction device is relatively immune to material defects.

#### *3.5.1.2 Flatness*

Total thickness variation across a 3-inch diameter wafer is typically 2–3 $\mu$m.

Larger variations can unfavorably affect several process steps in MMIC fabrication, primarily photolithography and the control of backside processing, such as wafer thinning and substrate via hole etching. Flatness measurement is a nondestructive test and can be performed on every wafer.

### 3.5.1.3 EL2 Concentration

The distribution and concentration of EL2 deep-level donors critically affects the compensation mechanism of undoped LEC GaAs material, which, in turn, affects the control over the variation of MESFET electrical parameters. Tight control of EL2 concentration improves the uniformity of material characteristics, resulting in higher device and circuit yields. EL2-level concentration is measured by focusing one-micron wavelength radiation and projecting it through the wafer normal to its surface. Variations in the detected signal correspond to variations in EL2 concentration. The typical mean EL2 level concentration for a 3-inch diameter substrate is $1-2 \times 10^{16}/cm^3$ with a 2–5% standard deviation across the wafer.

### 3.5.1.4 Resistivity and Mobility

Typical sheet resistivity of LEC grown, undoped, semi-insulating, 3-inch diameter GaAs wafers range between $4-6 \times 10^7$ $\Omega$-cm. This is measured on seed, tail, and selected wafers from the middle of a boule. A common method for measuring resistivity provides average resistivity of the substrate over a larger area, typically 1 mm$^2$. This measurement is performed at five spots in the center and on the edges of the selected wafers. Hall mobility is commonly measured on a sample to determine the quality of substrate material as well. Hall mobility between 6900 and 7100 cm$^2$/V-s is typical for a 3-inch diameter wafer. If the resistivity and mobility measured on the sample wafers meet specifications, the remaining wafers in the boule are typically accepted.

### 3.5.2 Process Control Monitors

*Process control monitors* (PCM) are a set of test structures designed to evaluate each step of the MMIC process [1,2]. Conclusive decisions could be made that a particular process step or sequence had been performed properly and whether the wafer should be processed further. In addition, PCMs contain secondary test structures used to characterize the process steps that do not have primary effect on the performance of the MMIC, but are used to determine process trends and reliability.

The primary PCM test structures, used for pass or fail decisions based on statistical information, are distributed evenly across the wafer. Typical number of PCMs distributed across a 3-inch diameter wafer is 5–40, depending on the process maturity. If the process is mature, fewer PCM structures are required for a decision; however, in the developmental stage or when large process variations are inherent to a step, more sites need to be measured across the wafer. The primary PCMs are evenly distributed across the wafer as a part of the circuit reticle, as shown in Figure 3.41a. A 2 × 2-inch rectangle in this figure indicates the high yielding area of a three inch wafer commonly used to obtain statistical information on PCMs. PCMs outside this area typically are not included in a pass or fail determination of a particular process step. To conserve expensive surface area, the detailed process characterization structures are not distributed across the whole wafer. These structures are provided five locations on the wafer, commonly known as dropouts, as indicated in Figure 3.41b. An alternate location for these dropouts is shown in Figure 3.41c. As the process matures, more test structures are transferred from the primary PCM to dropout locations. A fully mature production process has only four or five locations of in-process test structures.

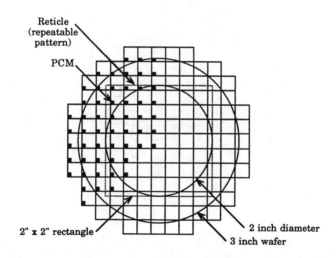

a. Structure evenly distributed

**Fig. 3.41** Distribution of PCMs on a wafer.

The selection of in-process test structures depends on the process. The set of test structures described here is generic in nature and compatible with the GaAs MMIC process described in Section 3.4. Figure 3.42 describes the monitoring

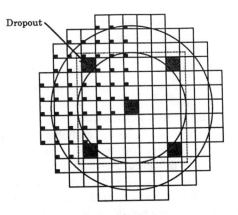

Dropout

b. With dropouts

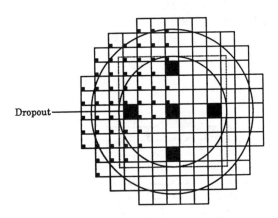

Dropout

c. Another location for dropouts

**Fig. 3.41** cont.

points of the wafer processing where a particular process step is characterized by using a PCM structure. The type of test structure and the objective of the measurements are also detailed in this figure. A set of PCM structures, both primary and secondary, are shown in the block diagram of Figure 3.43.

As shown in Figure 3.42, the check point after substrate material qualification is the channel and $N^+$ implant characterization after implant activation. Mechanical wafers occasionally are used to evaluate the profiles of channel and $N^+$ implants. If the implantation process is found to be drifting with time, proper adjustments can be made to the implant dose and energy or to the equipment itself. After the thin film metal resistor layer is deposited and contacts are made,

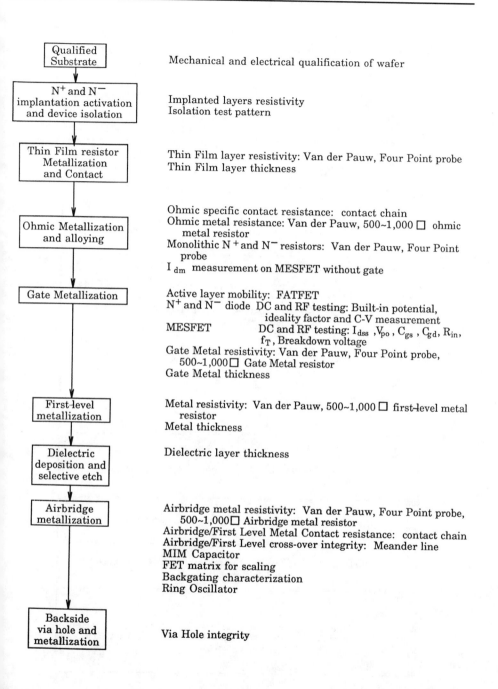

**Fig. 3.42** In-process test points for a typical MMIC process.

| 1 Gate recess control single gate, test FET, 100~200 µm gate width. | 2 FAT FET | 3 Ohmic contact resistance transmission chain | |
|---|---|---|---|
| | | 4 Isolation test pattern | 5 Etch/thickness measurement structure |
| 6 N⁻ resistors 5□, 1□, and 1/5□ | 7 N⁺ resistors 5□, 1□, and 1/5□ | 8 Thin Film resistors 5□, 1□, and 1/5□ | |
| 9 MIM capacitor large value 10~20 pf and small value 0.2~0.5 pf | | 14 Backgating measurement structure | 15 Single gate FET, source and drain connected (N⁻ diode) |

On-wafer microwave probable structures:

| 10 Single-gate MESFET, gate width = 200~500 µm. | 11 Single-gate MESFET, source grounded with substrate via hole | 12 N⁺ diode width ≈ 100~200 µm | 13 N⁻ diode width = 100~200 µm. |
|---|---|---|---|

| 16 Ring Oscillator | Alignment keys |
|---|---|
| 17 Ohmic metal 500~1,000 □ | 18 Gate metal 500~1,000□ |
| 19 First level metal 500~1,000 □ | 20 Airbridge metal 500~1,000□ |

| 21 Meander line First–level metal/Airbridge metal. |
|---|
| 22 First-level metal/Airbridge metal interconnect resistance 200~500 contacts. |
| 23 Van Der Pauw/four point probe structures for N⁺, N⁻, ohmic metal, First level metal, Gate metal, Airbridge metal, and Thin Film resistor metal. |

**Fig. 3.43** PCM block diagram.

the metal layer is evaluated for proper thickness and sheet resistance. A Van der Paw or four-point probe structure [1], as shown in Figure 3.44a and 3.44b, respectively, is used for this purpose.

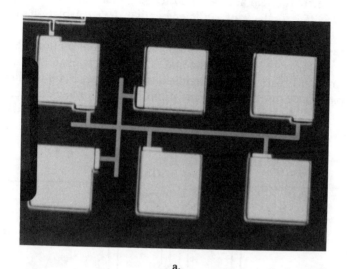

a.

b.

Fig. 3.44 Van der Paw (a) and four-point (b) probe structures.

Next, the device isolation and ohmic contact steps are completed. Device isolation is tested using the pattern shown in Figure 3.45. Leakage current is measured at a particular voltage applied between two $N^-$ islands separated by a certain distance. If the leakage is found to be larger than the specification, the wafers are rejected for lack of sufficient active device isolation. High leakage currents indicate conduction between devices such as MESFETs, diodes, and monolithic resistors through the substrate and hence nonfunctional MMICs. Also the contact resistance between ohmic and $N^+$ implant is evaluated using the test structure shown in Figure 3.46a, which is a transmission chain of ohmic contacts to $N^+$ implanted material. Typical separations between ohmic contact pads in this test structure are 5, 10, 15, and 20 $\mu$m. The resistance, $R$, is measured between the contact pads and the specific contact resistance $R_{SC}$ is evaluated using

**Mesa/selective $N^-$**

**Ohmic**

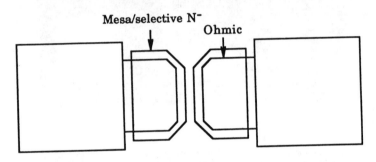

**Fig. 3.45** Device isolation test pattern.

$$R = \frac{2R_{SC} L_t}{W} + \frac{R_s L}{W}$$

where $R$ is the measured resistance between the two contact pads separated b distance $L$, $W$ is the width of the contact, $R_s$ is the resistivity of the $N^+$ implante layer, and $L_t$ is the transfer length.

The contact resistance may also be evaluated graphically by plotting resis tance between different contact pads as a function of pad separation. Interceptio of the extrapolated line to resistance axis, which corresponds to zero spacing provides the value of 2 $\times$ contact resistance [3]. If the mean ohmic contac resistance across a wafer is more than specified, the wafer is rejected for poc ohmic contacts, which eventually will result in poor MESFET, diode, and mon lithic resistor characteristics. Figure 3.46b shows a histogram [4] of ohmic specif contact resistance calculated from PCM test structures across a wafer. $N^+$ and $N$ monolithic resistors are completed at this point and characterized for their resi tivity. Low currents are used in such measurements, as the contact areas need be plated to carry larger current. Also, 5□, 1□, and $\frac{1}{5}$□, aspect ratio $N^+$ and $N$

a.

| Center Value | Percent of Valid Data | | | | | | % IN BIN | CUM % |
|---|---|---|---|---|---|---|---|---|
| | 10 | 20 | 30 | 40 | 50 | 60 | | |
| > MAX LIMIT | | | | | | | 0.0 | 100.0 |
| 0.975 | | | | | | | 0.0 | 100.0 |
| 0.925 | | | | | | | 0.0 | 100.0 |
| 0.875 | | | | | | | 0.0 | 100.0 |
| 0.825 | | | | | | | 0.0 | 100.0 |
| 0.775 | | | | | | | 0.0 | 100.0 |
| 0.725 | | | | | | | 0.0 | 100.0 |
| 0.675 | | | | | | | 0.0 | 100.0 |
| 0.625 | | | | | | | 0.0 | 100.0 |
| 0.575 | | | | | | | 0.0 | 100.0 |
| 0.525 | | | | | | | 0.0 | 100.0 |
| 0.475 | | | | | | | 10.3 | 100.0 |
| 0.425 | | | | | | | 10.3 | 89.7 |
| 0.375 | | | | | | | 10.3 | 79.3 |
| 0.325 | | | | | | | 6.9 | 69.0 |
| 0.275 | | | | | | | 24.1 | 62.1 |
| 0.225 | | | | | | | 27.6 | 37.9 |
| 0.175 | | | | | | | 6.9 | 10.3 |
| 0.125 | | | | | | | 3.4 | 3.4 |
| 0.750E-01 | | | | | | | 0.0 | 0.0 |
| 0.250E-01 | | | | | | | 0.0 | 0.0 |
| < MIN LIMIT | | | | | | | 0.0 | 0.0 |

| Statistics | Total Points | Low Limit | High Limit | Average | Standard Deviation |
|---|---|---|---|---|---|
| Valid Points | 50 | 0 ohm-mm | 1.0 ohm-mm | 0.2 ohm-mm | 0.1 ohm-mm |
| Plotted Points | 50 | 0 ohm-mm | 1.0 ohm-mm | 0.2 ohm-mm | 0.1 ohm-mm |

b. Evaluated from a chain of ohmic contacts

**Fig. 3.46** Test structure (a) and histogram (b) of ohmic contact resistance.

resistors are tested for yield. A histogram of the resistance measured on 100 $\mu$m wide and 100 $\mu$m long $N^-$ monolithic resistors is shown in Figure 3.47.

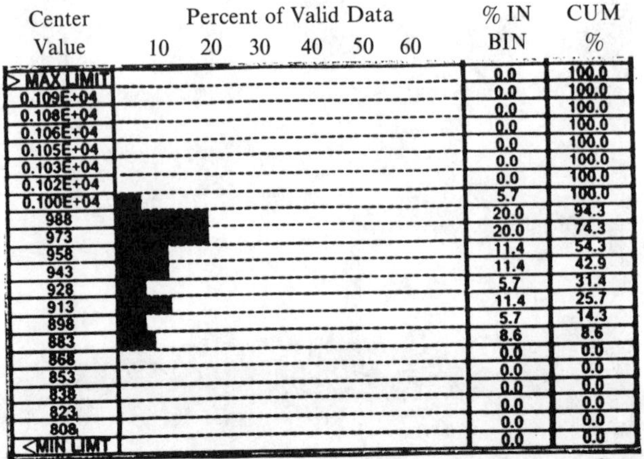

| Center Value | Percent of Valid Data | | | | | | % IN BIN | CUM % |
|---|---|---|---|---|---|---|---|---|
| | 10 | 20 | 30 | 40 | 50 | 60 | | |
| >MAX LIMIT | | | | | | | 0.0 | 100.0 |
| 0.109E+04 | | | | | | | 0.0 | 100.0 |
| 0.108E+04 | | | | | | | 0.0 | 100.0 |
| 0.106E+04 | | | | | | | 0.0 | 100.0 |
| 0.105E+04 | | | | | | | 0.0 | 100.0 |
| 0.103E+04 | | | | | | | 0.0 | 100.0 |
| 0.102E+04 | | | | | | | 0.0 | 100.0 |
| 0.100E+04 | | | | | | | 5.7 | 100.0 |
| 988 | | | | | | | 20.0 | 94.3 |
| 973 | | | | | | | 20.0 | 74.3 |
| 958 | | | | | | | 11.4 | 54.3 |
| 943 | | | | | | | 11.4 | 42.9 |
| 928 | | | | | | | 5.7 | 31.4 |
| 913 | | | | | | | 11.4 | 25.7 |
| 898 | | | | | | | 5.7 | 14.3 |
| 883 | | | | | | | 8.6 | 8.6 |
| 868 | | | | | | | 0.0 | 0.0 |
| 853 | | | | | | | 0.0 | 0.0 |
| 838 | | | | | | | 0.0 | 0.0 |
| 823 | | | | | | | 0.0 | 0.0 |
| 808 | | | | | | | 0.0 | 0.0 |
| <MIN LIMIT | | | | | | | 0.0 | 0.0 |

| Statistics | Total Points | Low Limit | High Limit | Average | Standard Deviation |
|---|---|---|---|---|---|
| Valid Points | 50 | 500 ohm | 1200 ohm | 951.4 ohm | 35.0 |
| Plotted Points | 50 | 700 ohm | 1200 ohm | 951.4 ohm | 35.0 |

**Fig. 3.47** Histogram of monolithic resistor (100 $\mu$m wide, 100 $\mu$m long).

A gateless MESFET structure is also tested at this step. The current through the source-drain of the MESFET without a gate is referred to as the *maximum channel current*. The current is measured on a 100 $\mu$m wide FET, whose source and drain are connected to probing pads by ohmic metal. Such a structure, with large ohmic contacts, alters the device characteristics because of self-gating effects; however, adjustments are made for a corresponding device with appropriate-size source-drain ohmic contacts. A histogram of maximum channel current is shown in Figure 3.48. A large standard deviation in $I_{dm}$ causes a wafer to be rejected for further processing, as the variation in the value of $I_{dm}$ across the wafer corresponds directly to the variation in MESFET drain current.

The gate then is recessed to achieve the desired $I_{dss}$, and the gate metal is deposited. At the completion of this step, MESFET, $N^+$, and $N^-$ Schottky diodes are fully characterized, which includes both dc and microwave measurements. Layout of a 200 $\mu$m single-gate FET is shown in Figure 3.49, which is configured for on-wafer microwave measurements using Cascade Micro Tech probes in ground–signal–ground configuration. A 200 or 300 $\mu$m gate periphery device typically is used for MESFET characterization. Both dc parameters such as $I_{dss}$, pinch-off voltage, transconductance, and breakdown voltage and RF parameters such as $C_{gs}$, $C_{gd}$, and $R_{ds}$ are evaluated. Figure 3.50 shows typical histograms of $I_{dss}$, $V_{po}$ and $C_{gs}$, respectively, measured on a 100 $\mu$m wide single gate FET across the wafer. These histograms take into account all the PCMs across the 3-inch diameter wafer or plots can be made for PCMs within the center 2 $\times$ 2-inch area.

| Center Value | Percent of Valid Data | % IN BIN | CUM % |
|---|---|---|---|
| >MAX LIMIT | | 0.0 | 100.0 |
| 398 | | 0.0 | 100.0 |
| 393 | | 0.0 | 100.0 |
| 388 | | 3.4 | 100.0 |
| 383 | | 6.9 | 96.6 |
| 378 | | 24.1 | 89.7 |
| 373 | | 20.7 | 65.5 |
| 368 | | 17.2 | 44.8 |
| 363 | | 13.8 | 27.6 |
| 358 | | 3.4 | 13.8 |
| 353 | | 0.0 | 10.3 |
| 348 | | 6.9 | 10.3 |
| 343 | | 0.0 | 3.4 |
| 338 | | 3.4 | 3.4 |
| 333 | | 0.0 | 0.0 |
| 328 | | 0.0 | 0.0 |
| 323 | | 0.0 | 0.0 |
| 318 | | 0.0 | 0.0 |
| 312 | | 0.0 | 0.0 |
| 308 | | 0.0 | 0.0 |
| 303 | | 0.0 | 0.0 |
| <MIN LIMIT | | 0.0 | 0.0 |

(Percent of Valid Data axis: 10, 20, 30, 40, 50, 60)

| Statistics | Total Points | Low Limit | High Limit | Average | Standard Deviation |
|---|---|---|---|---|---|
| Valid Points | 50 | 300 mA | 400 mA | 300 mA | 11.5 |
| Plotted Points | 50 | 300 mA | 400 mA | 300 mA | 11.5 |

**Fig. 3.48** Histogram of $I_{dm}$ of a single-gate FET (normalized to a 1 mm gatewidth FET).

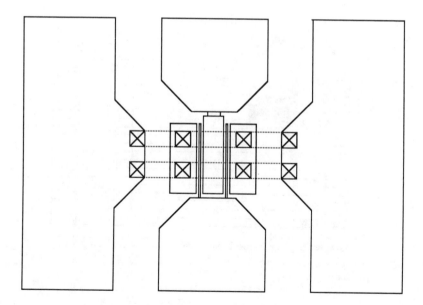

**Fig. 3.49** Layout of a 200 $\mu$m single-gate FET.

| Center Value | Percent of Valid Data 10 20 30 40 50 60 | % IN BIN | CUM % |
|---|---|---|---|
| >MAX LIMIT | | 0.0 | 100.0 |
| 195 | | 0.0 | 100.0 |
| 185 | | 0.0 | 100.0 |
| 175 | | 0.0 | 100.0 |
| 165 | | 0.0 | 100.0 |
| 155 | | 0.0 | 100.0 |
| 145 | | 0.0 | 100.0 |
| 135 | | 4.5 | 100.0 |
| 125 | | 18.2 | 95.5 |
| 115 | | 9.1 | 77.3 |
| 105 | | 31.8 | 68.2 |
| 95.0 | | 18.2 | 36.4 |
| 85.0 | | 9.1 | 18.2 |
| 75.0 | | 4.5 | 9.1 |
| 65.0 | | 4.5 | 4.5 |
| 55.0 | | 0.0 | 0.0 |
| 45.0 | | 0.0 | 0.0 |
| 35.0 | | 0.0 | 0.0 |
| 25.0 | | 0.0 | 0.0 |
| 15.0 | | 0.0 | 0.0 |
| 5.00 | | 0.0 | 0.0 |
| <MIN LIMT | | 0.0 | 0.0 |

| Statistics | Total Points | Low Limit | High Limit | Average | Standard Deviation |
|---|---|---|---|---|---|
| Valid Points | 50 | 0mA | 200 mA | 104.4 mA | 17.0 |
| Plotted Points | 50 | 0mA | 200 mA | 104.4 mA | 17.0 |

a. $I_{dss}$ (normalized to 1 mm gatewidth device)

| Center Value | Percent of Valid Data 10 20 30 40 50 60 | % IN BIN | CUM % |
|---|---|---|---|
| >MAX LIMIT | | 0.0 | 100.0 |
| -0.750E-01 | | 0.0 | 100.0 |
| -0.225 | | 0.0 | 100.0 |
| -0.375 | | 0.0 | 100.0 |
| -0.525 | | 4.3 | 100.0 |
| -0.675 | | 2.1 | 95.7 |
| -0.825 | | 19.1 | 93.6 |
| -0.975 | | 23.4 | 74.5 |
| -1.12 | | 23.4 | 51.1 |
| -1.27 | | 12.8 | 27.7 |
| -1.42 | | 6.4 | 14.9 |
| -1.57 | | 2.1 | 8.5 |
| -1.72 | | 4.3 | 6.4 |
| -1.87 | | 2.1 | 2.1 |
| -2.02 | | 0.0 | 0.0 |
| -2.17 | | 0.0 | 0.0 |
| -2.33 | | 0.0 | 0.0 |
| -2.47 | | 0.0 | 0.0 |
| -2.63 | | 0.0 | 0.0 |
| -2.77 | | 0.0 | 0.0 |
| -2.92 | | 0.0 | 0.0 |
| <MIN LIMT | | 0.0 | 0.0 |

| Statistics | Total Points | Low Limit | High Limit | Average | Standard Deviation |
|---|---|---|---|---|---|
| Valid Points | 50 | -3V | 0V | 1.077V | 0.2 |
| Plotted Points | 50 | -3V | 0V | 1.077V | 0.2 |

b. Pinch-off voltage

**Fig. 3.50** Histograms for a single-gate FET.

| Center Value | Percent of Valid Data 10 20 30 40 50 60 | % IN BIN | CUM % |
|---|---|---|---|
| > MAX LIMIT | | 0.0 | 100.0 |
| 199 | | 0.0 | 100.0 |
| 194 | | 0.0 | 100.0 |
| 189 | | 0.0 | 100.0 |
| 184 | | 0.0 | 100.0 |
| 179 | | 0.0 | 100.0 |
| 174 | | 0.0 | 100.0 |
| 169 | | 0.0 | 100.0 |
| 164 | | 3.4 | 100.0 |
| 159 | | 13.8 | 96.6 |
| 154 | | 10.3 | 82.8 |
| 149 | | 13.8 | 72.4 |
| 144 | | 20.7 | 58.6 |
| 139 | | 13.8 | 37.9 |
| 134 | | 6.9 | 24.1 |
| 129 | | 6.9 | 17.2 |
| 124 | | 3.4 | 10.3 |
| 119 | | 3.4 | 6.9 |
| 114 | | 3.4 | 3.4 |
| 109 | | 0.0 | 0.0 |
| 104 | | 0.0 | 0.0 |
| < MIN LIMIT | | 0.0 | 0.0 |

| Statistics | Total Points | Low Limit | High Limit | Average | Standard Deviation |
|---|---|---|---|---|---|
| Valid Points | 50 | 100 ms | 200 ms | 141.0 ms | 12.0 |
| Plotted Points | 50 | 100 ms | 200 ms | 141.0 ms | 12.0 |

**c. Transconductance (normalized to a 1 mm gatewidth FET)**

**Fig. 3.50** cont.

The choice is between a higher yield penalty or tighter distribution. It is common to expect tighter control over the parameters within the 2 × 2-inch center of the wafer than the full 3-inch wafer.

$N^+$ and $N^-$ diodes are characterized for reverse leakage current, built-in potential, parasitic forward resistance, and ideality factor. A 1 mm FET with source-drain connected is also characterized for gate capacitance and thus estimation of gate length uniformity across the wafer. FATFET [5], which is a single-gate FET with 100 $\mu$m wide and 100 $\mu$m long gate as shown in Figure 3.51, is commonly used to evaluate low field mobility of free carriers in the channel of the MESFET. However, the utility of such a structure is limited for short gate length (<1.0 $\mu$m) devices, where the short channel effects become dominant. The gate metalization resistivity is characterized using a Van der Paw or four-point probe structure similar to that shown in Figure 3.44a and 3.44b, respectively.

After the frontside process is completed, all metal layers—ohmic metal, gate metal, first-level metal, and airbridge metal resistivity—are characterized using a 500–1000 $\square$ structure, as shown in Figure 3.52. Such structures generally are placed in dropout locations. MIM capacitors, backgating test structures, and dual-gate FETs also are characterized at the final step. Integrity of the airbridge cross-over first-level metal is characterized using the meander line structure shown in Figure 3.53. If there is no short circuit between the airbridge and first-level metal,

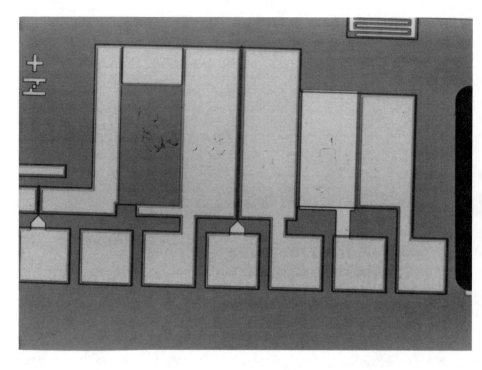

**Fig. 3.51** FATFET.

an open circuit is observed between PAD1 (PAD2) and PAD3 (PAD4). The contact resistance between the first-level metal and airbridge metal is measured using a chain of contacts as shown in Figure 3.54a. The contact resistance depends on the area of contact represented by x in the figure. The total resistance of PAD1–PAD3 is composed of first-level metal lines, airbridge metal lines, and the contacts. Hence, it is necessary to know the first-level metal and airbridge metal resistance before testing such a structure and evaluating the contact resistance. Figure 3.54b is a photograph of contact chains of two different contact areas.

Finally, the scaling effect of the single gate FET is characterized by measuring FETs of different size and gate-drain spacing as shown in Figure 3.55. A summary of commonly used test structures, their purposes, and corresponding tests is provided in Table 3.7.

## 3.6 MMIC FOUNDRIES: SELECTION CRITERIA

The concept of a foundry for semiconductor integrated circuit fabrication originated in the 1970s with the dramatic growth of silicon technology. As the

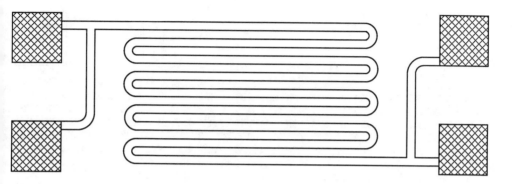

**Fig. 3.52** 500–1000□ resistive pattern.

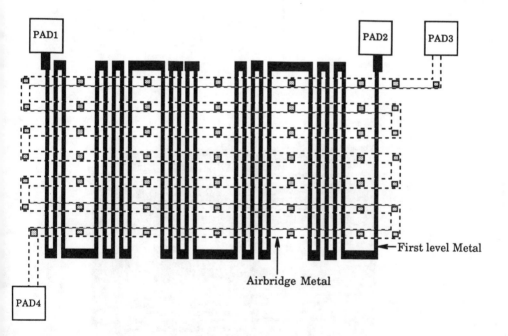

**Fig. 3.53** Airbridge metal–first-level metal meander line.

silicon ICs had progressed to the phase of solid-state integration, there were many small companies with innovative designs, but who were unable to afford the huge investments for the initial setup and maintenance of an IC fabrication facility. At the same time, larger companies with IC manufacturing facilities were running under capacity. The complementary needs of both types of firm evolved into the

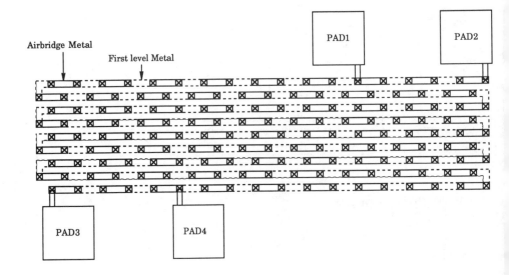

a.

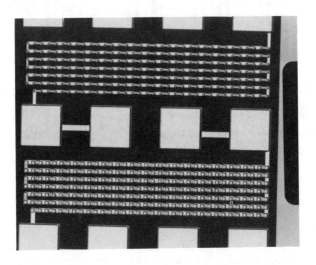

b.

**Fig. 3.54** Chain of first-level metal–airbridge metal (a) and the contact chains of two different areas (b) (courtesy of

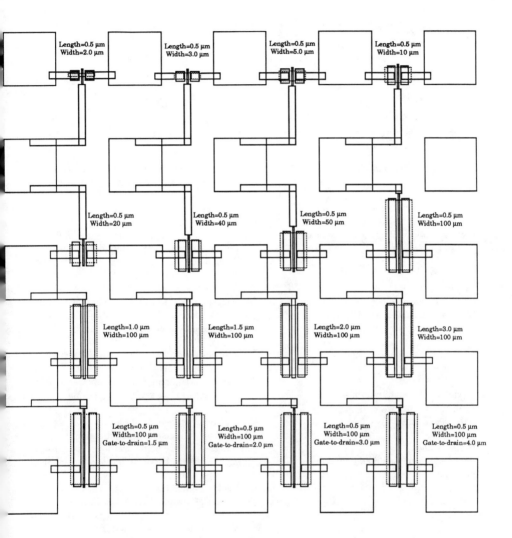

**Fig. 3.55** FET matrix.

IC foundry concept. As the microwave systems are now well into their next generation of monolithic integration, the concept of GaAs IC foundries [1] has become increasingly popular.

By offering convenient and relatively inexpensive manufacturing of customer designed GaAs ICs, these foundries [2,3] aid in the development of new MMICs operating at GHz frequencies. The purpose of the foundries is to provide

**Table 3.7 In-Process Control and Characterization Test Patterns (PCM) and Tests Performed**

| Structure Name | Purpose and Tests Performed |
|---|---|
| Van der Paw, four-point probe | $N^-$ implanted channel layer, $N^+$ implanted layer, sheet resistivity of thin film resistor metal, ohmic metal layer, gate metal layer, first-level metal, airbridge metal |
| Ohmic contact transmission chain | Ohmic contact specific resistance |
| Etch depth–layer thickness measurement structure | Mesa etch depth, monolithic resistor trim depth, thickness of gate metal, thin film resistor metal, ohmic metal, first-level metal, dielectric layer, and airbridge metal |
| Isolation pattern | Device isolation measurement |
| FATFET | C-V measurement, low field mobility calculation |
| 100 $\mu$m wide single-gate FET | In-process channel recess control by measuring $I_{dss}$ |
| Microwave probable single-gate FET | S-parameter measurement and extraction of $C_{gs}$, $g_m$, $C_{gd}$, $I_{ds}$, $C_{ds}$, $V_{po}$, $f_T$, and gate-delay $\tau$ |
| Microwave probable $N^+$ and $N^-$ diodes | S-parameter and dc measurements and extraction of $V_{bi}$, n-ideality factor, leakage current, C-V characteristics, and parasitic resistance |
| Single-gate FET, source-drain connected (1 mm wide) | Measurement of $C_{gate} = C_{gs} + C_{gd}$ during in-process |
| 500–1000 $\square$ metal line | Resistance measurement of ohmic metal, gate metal, first-level interconnect metal, and airbridge metal |
| Meander line | First-level–airbridge metal crossover integrity |
| First-level–airbridge metal contact chain | Contact resistance between first-level metal and airbridge metal |
| Ring oscillator | Gate delay estimation |
| MIM capacitors | Estimation of lower value of capacitor before fringing effects become significant and larger value before yield of MIM capacitor is degraded |
| Single-gate FET with source grounded by via hole | Via hole integrity and estimation of via hole inductance |
| FET matrix | Characterization of MESFET electrical parameters as a function of gate width, gate length, and gate-drain spacing. Parameters like pinch-off voltage, $I_{dss}$, transconductance, and break-down voltage are important. |
| $N^+$, $N^-$, and thin film resistor matrix | Scaling effect on different aspect ratio resistors |
| Backgating structure | Characterize backgating and side-gating effects |
| Microwave probable dual-gate FET | Dual-gate FET characteristics |
| Alignment key | Provides proper alignment between layers during process |

IC processing services based on a well-characterized process and set of design tools. Such services are attractive to system developers whose IC needs may not be sufficient to justify an in-house manufacturing facility. Using an external fabrication facility to process prototype ICs during system development, the system designers can concentrate their efforts on system packaging and integration. As the volume of their manufacturing increases, the systems companies can either continue using the foundry for manufacturing or set up their own facilities.

### 3.6.1 Designing MMICs Using Foundries

The use of GaAs IC foundries is based on the premise that there are accurate process design rules and device models to predict the performance of the circuits over the frequency range of interest. Also, the foundry IC process has to be stable, mature, and reliable enough to produce the modeled characteristics of the devices within the design bounds, repeatedly, over a long period of time. The typical process involved in designing MMICs using foundry is shown in Figure 3.56. As the primary specifications are set up for individual MMICs, the most crucial step is to select a foundry that can fill the primary and secondary specifications of the IC. Several options available from most foundries, such as shared wafer and multiproject chip production, keep the cost of fabricating designs through a foundry relatively inexpensive. However, the extensive time may be required to gain familiarity with a specific foundry's process, design rules, device models, and designing circuits, and this would have to be repeated if the initial choice of foundry is wrong. The most important criterion in selecting a foundry is the performance of the active devices and the availability of different passive elements and their characterization.

Having the device model library available for use in conjunction with commercial circuit analysis software packages is helpful in evaluating the suitability of the foundry without any commitment to a process. It also allows analysis of the secondary specifications of the MMIC. A broad range of foundry processes are availability for different types of MESFET devices, such as high-power, low-noise, and general purpose devices. Integration of several types of active devices on a single chip, including E/D and microwave depletion mode devices, also is an important consideration. Most MMIC foundries offer depletion-mode MESFET based process technology with 0.5 $\mu$m, 1.0, or more $\mu$m gate lengths. A few foundries offer the HEMT based MMIC process, and TriQuint is the only foundry offering the most extensive product line, which includes the E/D process, the 0.25 $\mu$m gate length MESFET process for millimeter wave applications, and also the integration of microwave and digital process that is very useful for further integration of microwave and digital control functions on the same IC.

As important as the foundry's process suitability is its customer support and the availability of a detailed device characterization data base. A detailed design

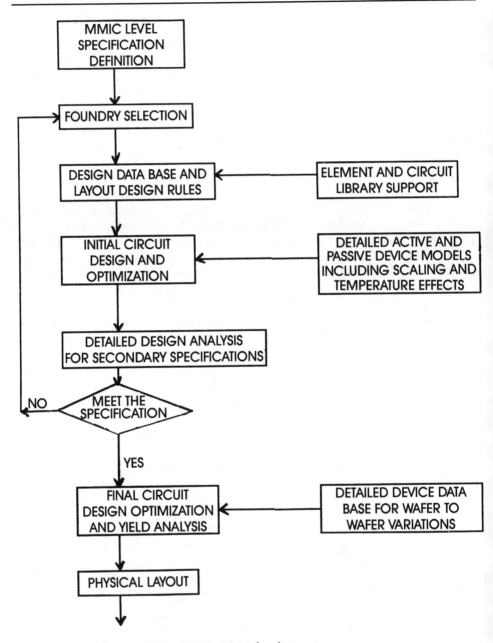

**Fig. 3.56** Typical flow of designing MMICs using a foundry.

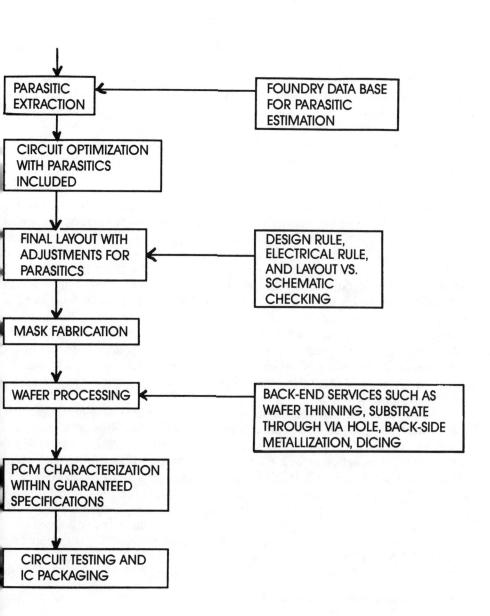

**Fig. 3.56** cont.

data base is needed for yield analysis. Availability of detailed device models and their compatibility with commercial circuit design software (e.g., SPICE, Touchstone™, Supercompact™, and harmonic balance programs) are the crucial selection criteria.

When the circuit is designed in schematic form, we must convert the schematic to layout. The foundry can make this conversion easier by offering layout component and cell libraries. The next step is to extract the parasitics, such as interconnects or coupling between the active and passive components. Besides the experience of an individual circuit designer, the foundry should provide basic design rules for estimating these parasitics; parasitic extraction software is more desirable, but at least there should be a service to provide such estimates of parasitics.

Once the layout is finalized, it is important to check the physical design for violations of any design or electrical rules, such as minimum interconnect width and spacing of interconnect lines. The physical layout must always be compared to the schematic to identify connectivity errors, a service that must be provided by the foundry. A final layout is sent to the foundry for mask generation and finally wafer fabrication. As the wafers are processed, the foundry must provide detailed PCM measurements of active and passive devices to make sure that the processed wafers have devices within the guaranteed specifications. Tail-end processing services such as wafer thinning, substrate through via holes, backside metallization, and wafer dicing also are important. Circuit testing, both dc and on-wafer microwave probing, should be available from the foundry if such capabilities are not accessible in-house. Availability of a suitable microwave package from the foundry is also helpful.

Table 3.8 provides simple guidelines for evaluating the suitability of a particular foundry for a specific circuit design. Accurate knowledge of a foundry's ability to meet these parameters is essential for predictable MMIC design. The 1.0 $\mu$m gate length depletion mode FET-based technology is useful for circuits up to 10 GHz, 0.5 $\mu$m gate length MESFET process is applicable to 20–25 GHz, 0.25 $\mu$m gate length process is useful between 35–60 GHz, and HEMT-based process is useful up to 60 GHz. High precision thin film resistors, spiral inductors, and MIM capacitors are useful for precise biasing and matching network design. Two-layer interconnect metalization with airbridge crossover capability is essential for any MMIC design.

### 3.6.2 Processed Wafer Qualification

As the foundry concept, semicustom and standard cell design procedures are becoming popular in GaAs MMICs, it is important to set criteria for acceptable processed wafers. JEDEC recently published an initial set of standards for

**Table 3.8** Representative Electrical and Physical Design Rules

*Technology:*
Wafer diameter
Chip thickness
Limitation on chip or reticle size
Ion implanted channel–epitaxially grown active layer
Depletion-enhancement-E/D MESFET devices
Availability of components:
  Single- and dual-gate FET minimum gate length, mixing various gate length devices on the same chip
  Schottky diode, varactor diode, PIN diode
  Monolithic resistor
  Thin film resistor
  MIM capacitor
  Airbridge crossover
Substrate via hole
Highest usable frequency of process for amplifier and oscillator design
Measurement of processed device performance
Performance parameters to qualify a wafer
Ability to tailor process parameters to meet specific requirements (e.g., doping profiles, insulator thickness, resistivity of monolithic and thin film resistors)

*Single-Gate FET Performance:*
Saturation current $I_{dss}$ and pinch-off voltage $V_{po}$, gate-drain breakdown voltage $BV_{gd}$
Unity current gain bandwidth $f_T$
$G_m/I_{dss}$ ratio; a measure of gain-power dissipation
$G_m \cdot R_{ds}$ product; a measure of voltage gain
Power output (W/mm)
dc $\rightarrow$ RF conversion efficiency (%)
Maximum stable gain
Forward gain S21
Noise figure
Minimum noise figure bias point
Maximum power output bias point

*Sheet Resistances:*
$N^-$ layer ($\Omega/\square$)
$N^+$ layer ($\Omega/\square$)
Thin film resistor ($\Omega/\square$)
Ohmic metal ($\Omega/\square$)
Gate metal ($\Omega/\square$)
First-level metal ($\Omega/\square$)
Airbridge metal ($\Omega/\square$)

**Table 3.8** cont.

*Thickness of Layers:*
Gate metal ($\mu$m)
First-level metal ($\mu$m)
Airbridge metal ($\mu$m)
MIM capacitor dielectric ($\mu$m)
Airgap between airbridge metal and substrate

*dc Current Carrying Capability:*
Gate metal (mA/$\mu$m width)
First-level metal (mA/$\mu$m width)
Airbridge metal (mA/$\mu$m width)
Ohmic metal (mA/$\mu$m width)
Thin film resistor (mA/$\mu$m width)
Monolithic resistor (mA/$\mu$m width)

*Minimum Line Width:*
Gate metal ($\mu$m)
First-level metal ($\mu$m)
Airbridge metal ($\mu$m)
Ohmic metal ($\mu$m)
Thin film resistor ($\mu$m)
Monolithic resistor ($\mu$m)

*Minimum Line Spacing:*
Gate metal to gate metal ($\mu$m)
First-level metal to first-level metal ($\mu$m)
Airbridge metal to airbridge metal ($\mu$m)
Ohmic metal to ohmic metal ($\mu$m)
Thin film resistor to thin film resistor layer ($\mu$m)
First-level metal to airbridge metal ($\mu$m)
Thin film resistor to gate metal–first-level metal–airbridge metal ($\mu$m)
Gate metal to airbridge metal ($\mu$m)
Ohmic metal to airbridge metal ($\mu$m)

*Passive Components:*
MIM capacitance (pF/mm$^2$), minimum and maximum range
Quality factor ($Q$) of MIM capacitors and spiral inductors
Range of standard inductors (nH)
Inductance/Area (nH/mil$^2$)
Inductance/dc Resistance (nH/$\Omega$)
Resonance frequency of inductors
Schottky diode cut-off frequency and ideality factor

**Table 3.8** cont.

---

*Design-Model Support:*
Linear and nonlinear models for single- and dual-gate FETs compatible with small-signal RF analysis and time-domain or harmonic balance programs
Accuracy of models in subthreshold region
Device scaling models
Schottky diode; linear and nonlinear models
Design models for passive components:
    Monolithic and thin film resistors, MIM capacitors fringing capacitance model, interdigitated capacitor, and spiral inductor

*Statistical Data Base:*
Across the wafer and wafer-to-wafer process variations for active and passive devices

*User Interface:*
Layout input-output format
Standard cell, component library
Design verification; design rule, electrical rule and layout *versus* schematic checking

*Backend services:*
Backside metalization
Wafer sawing
On-wafer microwave testing of circuits, including power and noise measurements
Assembly of finished circuits
Packaging availability
Reliability testing

---

accepting processed wafers in GaAs MESFET technology. Wafers to be evaluated must have completed all the fabrication processes prior to microcircuit wafer evaluation and wafer lot acceptance. Measurement of the *process monitor* (PM), verifying that the identified parameters are within process limits, are required from each wafer lot.

Process monitors are defined as the collection of test structures that provide data on wafer acceptability. The PM may be either stepped onto every wafer in dedicated drop-in locations, incorporated into kerf locations, or located on each die, such that they can be probed at the conclusion of processing up to and including final topside metalization and passivation where applicable. Table 3.9 presents a list of minimum PM test structures that are applicable for each process technology. The manufacturer sets PM parametric limits as called for by design or process rules. The performance of each wafer is evaluated individually and independent of the performance of other wafers in the lot. This wafer acceptance procedure is based on physical testing, visual inspection, and electrical testing of suitable PMs. Wafers failing any process specifications, with the exception of acceptable rework instances, are removed from further processing. This method is restricted to a well characterized (controlled) and baseline process. *Well characterized* means that the fabrication line has been adequately documented in relation to the capabilities of the process. *Baseline* refers to the existence of a well-defined process parameter with associated variances (based on characterization data) against which the actual wafer-to-wafer process data is measured.

**Table 3.9 Minimum Suggested Set of Structures Used in a PM**

N-channel transistors for measuring transistor parameters (minimum and maximum geometries)
P-channel transistors for measuring transistor parameters (minimum and maximum geometries)
Threshold–pinch-off voltage
Leakage current
Sheet resistance
E-mode transistor parameters
D-mode transistor parameters
Dielectric integrity
Via hole–airbridge resistance
Contact resistance
Step coverage
Alignment verniers
Functional blocks characteristics
Line width
Diode parameters
Backgating-isolation
Backside via hole metalization
Doping profile structure
FATFET
Thin film resistor characteristics
Capacitance value measurements

### 3.6.2.1 Wafer Acceptance Criteria

There are four steps to wafer acceptance:

1. Processing to specifications in accordance with the manufacturer's established baseline documents.
2. PM measurement and evaluation.
3. Visual-SEM inspection.

4. Physical testing as required in MIL-STD-976.

Wafers are rejected for failing any test, with the exception of acceptable rework instances in accordance with MIL-STD-976 or applicable rework documents.

### 3.6.2.2 PM Evaluation

Prior to qualification, the type of PM structures utilized are submitted for approval to the qualifying activity. Acceptance is made on a wafer-by-wafer basis, depending upon the information derived from PM room temperature testing. If drop-in PMs are utilized, each wafer shall have at least five PMs; one stepped in the center and the others in each of the four quadrants. For kerf PMs and for PMs on individual dies, the five probed PMs shall be located in the center and in each of the four quadrants. Quadrant PMs shall lie at least two-thirds of a radius away from the wafer center. Wafer acceptance shall be based on four out of five PMs or 80% of the tested PMs passing, if more than five PMs are routinely tested as per submitted documentation to qualifying activity.

### 3.6.2.3 Visual-SEM Inspection

Inspection is performed at critical process steps during wafer fabrication. Inspections may include patterns, alignment verniers, and critical dimension measurements. Defective wafers are removed from the lot for scrap or rework.

### 3.6.2.4 Data Reporting

When required by the applicable document and for qualification, the following data is made available for each wafer submitted:
- Results of each test conducted, initial and any resubmissions.
- Number of wafers accepted or rejected per lot.
- For pattern failures, failure analysis data and failure mode of each rejected wafer (if applicable) and the associated mechanism for catastrophic failures for each rejected device.
- Number of reworked wafers and reason for rework.
- Measurements and record of data for specified PM electrical parameter.

### 3.6.2.5 Defective Wafers

All wafers that fail any test criteria are removed at the time of observation or at the conclusion of the test in which the failure is observed. Rejected wafers may be subjected to approval rework operations as detailed in the baseline operations. Once rejected and verified as unreworkable, no wafer may be retested for acceptance.

## 3.7 COMPUTER-AIDED MANUFACTURING

A typical monolithic microwave integrated circuit manufacturing process is composed of many individual steps as described in Section 3.4. The success of any integrated circuit manufacturing facility depends on its ability to execute these steps with reasonable and repeatable yields. Yield is one of the most important factors in determining the ability of a manufacturing facility to produce integrated circuits in large quantities, cost effectively. Yield depends on many factors throughout the manufacturing process such as design engineering; particle count and controlled environment in the fabrication area; proper operation of equipment; quality of materials, skills, and training of the manufacturing personnel; and the quality of each step of the fabrication process.

As the product technologies and manufacturing processes have become increasingly complex, the amount of data involved in controlling and analyzing the process has also increased substantially [1,2]. Moreover, for this data to be useful in characterizing a manufacturing process, it must be readily accessible to the user and the data must be represented in a consistent and meaningful graphical format. *Computer-aided manufacturing* (CAM) software systems [3–6] have been very useful for data management in tracking the IC manufacturing process from early product planning stages to design, in-process data collection and process control, assembly, and on-time final delivery. A typical CAM software [7] system for integrated circuit fabrication contains modules that address the following issues:

- Lot tracking
- Production control and reporting
- Statistical process control
- Engineering data analysis
- Inventory management
- Automation and equipment management
- Priority scheduling
- Quality control
- Cost accounting–cost model
- Capacity planning and reporting
- Maintenance management
- Facility environment monitor-control

The features and capabilities of a typical CAM system are given in Table 3.10.

**Table 3.10 Typical CAM System Features**

| *Features* | *Capabilities* |
| --- | --- |
| Process specification management | Store and maintain detailed process specifications with engineering chance control and up-to-date display to the operating staff |
| Work-in-process tracking | Track work-in-process, yields and cycle time to the desired operational detail |
| Inventory control | Keep records of available materials and inform user when to reorder, depending on delivery time of the material |
| Work planning and factory floor scheduling | Plan production activity schedules on daily, weekly, and monthly basis; set priorities on work |
| Engineering data collection | Collect and store data in a specified format for each lot and every wafer in a lot. Data includes electrical and operational data and is base for statistical process control |
| Equipment related data collection | Collect global data, not associated with a particular lot, such as equipment status and environmental status. This provides base for preventative maintenance scheduling |
| Production data collection | Collect detailed data related to production such as process yields, cycle times, processing costs. This feature provides for extensive production reporting |
| Factory automation and equip-ment control | Direct process, data collection, and material transport equipment to perform downloaded process programs for error-free, paperless production |

Lot tracking is the core of the CAM software package in the [8,9] semiconductor IC manufacturing environment as shown in Figure 3.57. This function typically is performed by using a slip of paper as a lot traveler in manual operations. The lot tracking module of the CAM software is a powerful tool for the manufacturing manager to control and obtain information about the movement of work in process. Such a module is extremely useful in the production area as well as the engineering wafer fabrication facility, where it is important to track processing differences from lot to lot and wafer to wafer. The lot tracking module spans the entire manufacturing cycle from bulk material to assembly and testing of the final product. Because the manufacturing cycle consists of many work stations, this module requires entry of many bits of data and a data-base system to store and report these data.

Inventory management tracks and reports the stock quantity of the materials throughout the process and the finished goods. This includes the raw material, chemicals, gases, and other consumable items used in the fabrication process. The manufacturing data base, built from lot tracking events and inventory disposition, provides the necessary information for operational planning such as scheduling, cost information, yields, capacity planning, and engineering data analysis for product design. The cost accounting module tracks the manufacturing cost associated with a particular product and, using an appropriate cost model, can project the price of a product.

Several computer-aided manufacturing software programs are commercially available to address the needs of semiconductor IC manufacturing needs. Table 3.11 provides a summary of the software packages most commonly used. Most of these programs run in real time on medium-size computers, except CAMEO®, which is a personal computer–based program. IC-10® is one of the most widely used CAM software system in small volume wafer fabrication facilities. (The commercial support for this package is not available from Hewlett-Packard as of this writing.) Aside from these commercial packages, many large manufacturing companies have developed their own CAM programs, which further emphasizes the importance of computer controlled manufacturing.

**Table 3.11 Summary of Commonly Used CAM Software Packages**

| Company Name | CAM Software |
|---|---|
| Cameo Systems | CAMEO |
| Consilium | COMETS |
| Promis Systems | PROMIS |
| Qronos Technology | Advantage |
| Hewlett-Packard | IC-10 |

The manufacturing issues in an MMIC processing facility are different from those of its counterpart in silicon technology. Because of the demand for a high

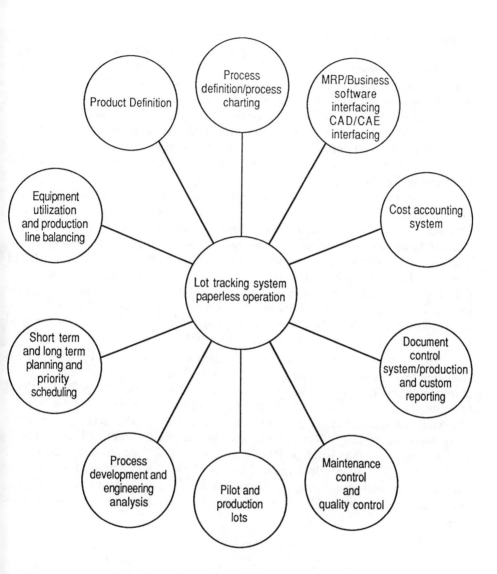

**Fig. 3.57 Computer-aided manufacturing system.**

volume of silicon ICs at low prices, the manufacturing issues in silicon processing focus on efficient resource planning, capacity planning, inventory management, scheduling, and process floor control in addition to yield tracking and improvement. For facility processing GaAs MMICs, the volume requirements are not as high. Because of the inherent material nonuniformity of GaAs and the nature of the MMICs itself, the manufacturing issues focus more on in-process data collection and statistical analysis, engineering data-base management, lot tracking and traceability, and minimized rework and scrap. In order to track process yield, tests are used to collect large amounts of data at every step of wafer fabrication as described in Section 3.5.

One of the most important requirements in such a data collection environment is to select a computer-aided manufacturing system that is easily interfaced with the in-line test equipment [10] to minimize manual data entry. Such automatic data collection interface not only improves the throughput but also avoids any operator error in recording the data. One such an easily interfaced CAM system is PROMIS, which uses its SCOPE module for automated interfaces to data collection and process equipment. This module formalizes the automation interface away from custom software and provides a standardized style for process test data entry into the manufacturing data base. Table 3.12 shows a typical equipment interface developed as a part of SCOPE.

The CAM system must have a user friendly engineering data base in order to store and analyze the process data. Several generic data-base programs are available commercially that may be customized to fit the individual needs. INGRES from Relational Technology and ORACLE from Oracle Corp. are two examples of commonly used data-base programs for management of data generated in wafer fabrication and assembly. These data bases can be interfaced effectively to statistical analysis packages, such as RS/1™ from BBN Software Products Corp. and SAS from SAS Institute for detailed analysis of process variables in developing, characterizing and monitoring wafer processing. These statistical analysis programs run on a variety of hardware including personal and medium-sized computers.

**Table 3.12** Typical SCOPE Interfaces

*Process:*

| | |
|---|---|
| Thermco TMX 9000 | Diffusion furnace |
| Eaton Nova NV 10/80/160 | Ion implanter |
| Varian 80/120/160-10 | Ion implanter |
| Eaton 6200A Implanter | Ion implanter |
| Canon Stepper | Photolith stepper |
| AME 8100 | Plasma etcher |
| GCA 6300B | Stepper |
| SVG System 88 | PhotoLith spin track |

*Metrology:*

| | |
|---|---|
| ITP80 | Line width measurement |
| Nanoline V | Line width measurement |
| Nanospec AFTXXX Series | Film thickness |
| Rudolph AutoEllipsometer EL-IV | Film thickness |
| Rudolph AutoEllipsometer EL-III | Film thickness |
| OSI DN-15 | Optical wafer inspection |
| Four Dimension 180 | Resistivity measurement |
| Leitz CD Scope | Line width measurement |
| Leitz FTA | Film thickness |
| Prometrix Four-Point Probe | Resistivity measurement |

*Electrical Test:*

| | |
|---|---|
| Accutest | Parametric tester |
| Keithley | Parametric tester |

*Cell:*

| | |
|---|---|
| Cell Interface Message Set | To cell controllers |
| Cell Controller | Diffusion cell |

*Data Collection:*

Burr-Brown Data Collection Products

## ACKNOWLEDGEMENTS

The editor wishes to thank the following for their contributions to this chapter: Dr. James M. O'Connor of Allied Signal, Columbia, Maryland for contributing Sections 3.1 and 3.2; Dr. Robert Anholt of Gateway Modeling, Minneapolis, Minnesota for critically reviewing Section 3.3 and providing the original figures for this section; Dr. Ajit Rode of TriQuint Semiconductor, Beaverton, Oregon for critically reviewing Sections 3.4 and 3.5.

# REFERENCES

## *Section 3.1*

1. J. V. DiLorenzo and D. D. Khandelwal, *GaAs FET Principles and Technology*, Artech House, Norwood, MA, 1982.
2. C. Jacob et al., "Dislocation-Free GaAs and InP Crystals by Isoelectronic Doping," *J. Crystal Growth*, Vol. 61, 1983, pp. 417–424.
3. R. E. Williams, *Gallium Arsenide Processing Techniques*, Artech House, Norwood, MA, 1984. This is an excellent review of GaAs processing.
4. M. G. Panish and A. Y. Cho, "Molecular Beam Epitaxy," *IEEE Spectrum*, April 1980.
5. H. F. Matare, "The Electronic Properties of Epitaxial Layers," *CRC Critical Reviews of Solid State Science*, November 1975.
6. C. A. Evans, Jr., and M. D. Strathman, "RBS Technique Exposes Surface Properties of Electronic Materials," *Industrial Research and Development*, December 1983.
7. M. A. Kelley, "More Powerful ESCA Makes Solving Surface Problems a Lot Easier," *Research and Development*, January 1984.
8. R. C. Eden, A. R. Livingston, and B. M. Welch, "Integrated Circuits: The Case for Gallium Arsenide," *IEEE Spectrum*, December 1983.
9. J. Berenz and B. Dunbridge, "MMIC Device Technology for Microwave Signal Processing Systems," *Microwave J.*, April 1988.
10. A. Fathimulla, H. Hier, and J. Abrahams, "Microwave Performance of Pulsed-Doped-Heterostructure GaInAs MESFETs," *Electronics Letters*, Vol. 24, January 1988, p. 93.


## *Section 3.2*

1. Eaton Corporation, 200 MC-A Ion Implanter product literature.
2. M. R. Stiglitz, "GaAs Foundry Operations 1988," *Microwave J.*, March 1988, p. 40.


## *Section 3.3*

1. D. Antoniadis, S. E. Hansen, R. W. Dutton, and A. G. Gonzales "Models for Computer Simulation of IC Fabrication Processes," *IEEE Trans. Electron Devices*, April 1979, pp. 490–500.
2. J. F. Gibbons, W. S. Johnson, and S. W. Mylroie, *Projected Range Statistics*, Stroudsburg, PA, Dowden, Hutchinson and Ross, 1975.
3. L. A. Cristel, J. F. Gibbons, and S. W. Mylroie, "Application of Boltzmann Transport Methods to Ion Implantation," *J. Applied Physics*, 1980, p. 6176.
4. J. P. Biersack and L. G. Haggmark, "A Monte-Carlo Program for the Transport of Ions in Amorphous Targets," *Nuclear Instrumentation Methods*, 1980, pp. 257–269.
5. R. Anholt, P. Balasingam, S. Chou, T. W. Sigmon, and M. D. Deal, "Ion Implantation into GaAs," *J. Applied Physics*, 1989.
6. H. Ryssel and J. P. Biersack, in *Process and Device Modeling*, ed. W. L. Engl, Amsterdam, North-Holland Press, 1986.

7. R. Anholt, T. W. Sigmon, and M. D. Deal, "Process Devices Models for GaAs MESFET Technology," *GaAs IC Symposium Digest,* 1987, pp. 53–55.
8. R. Anholt and T. W. Sigmon, "A Process and Device Model for GaAs MESFETs," *IEEE Trans. Computer Aided Design,* April 1989.
9. D. H. Rosenblatt, W. R. Hitchens, R. Anholt, and T. W. Sigmon, "GaAs MESFET Device Dependences on Ion-Implant Tilt and Rotation Angles," *IEEE Electron Device Letters,* 1988, p. 139.
10. H. Ryssel and I. Ruge, *Ion Implantation,* John Wiley and Sons, New York, 1986.
11. R. Anholt and T. W. Sigmon, "Substrate Impurities Effects on GaAs MESFET's," *J. Electronic Materials,* 1988, p. 5.
12. R. Anholt, "Modeling Materials and Process Effects on the Uniformity and Manufacturability of GaAs MESFET Integrated Circuits," unpublished report.
13. S. Miyazawa and K. Wada, "Mechanism for Threshold Voltage Shifts of a GaAs FET around Dislocations," *Applied Physics Letters,* 1986, p. 905.
14. R. Anholt and T. W. Sigmon, "Model of Threshold Voltage Fluctuations in GaAs MESFETs," *IEEE Electron Device Letters,* 1987, p. 16.
15. T. Egawa, Y. Sano, H. Nakanura, and K. Kaminishi, "Influence of Annealing Methods on Correlations between Threshold Voltages of GaAs MESFETs and Dislocation," *Japanese J. Applied Physics,* 1986, p. L973.
16. K. Watanabe, F. Hyuga, H. Nakaniski, and K. Hoshikawa, "Inhomogeneity of Electrical Properties around Dislocations," *1985 GaAs and Related Compounds Conf.* (Inst. of Physics Conf. Series No. 79, 1985), p. 277.
17. R. Anholt and T. W. Sigmon, "Ion-Implant Effects on GaAs MESFETs," *IEEE Trans. Electron Devices,* 1988.
18. M. D. Deal et al., "SUPREM 3.5, Process Modeling of GaAs," *Proc. 1987 IEEE IEDM Conf.*
19. H. Kanber, M. Feng, V. K. Eu, R. C. Rush, and W. B. Henderson, "Correlation between Chemical and Electrical Profiles in Si, Se and S Implanted GaAs," *J. Electron Materials,* 1982, p. 1083.
20. Based on SIMS comparisons of As-implanted and annealed Si profiles using $SiO_2$ caps, we discount the CV measurements of diffusion constants of T. Ohnuma, T. Hirao, and T. Sugawa "Study of Encapsulants for Annealing Si-Implanted GaAs," *J. Electrochemical Soc.* 1982, p. 837.
21. W. V. McLevige, M. J. Helix, K. V. Vaidyanathan, and B. G. Streetman, "Electrical Profiling and Optical Activation Studies of Be-Implanted GaAs," *J. Applied Physics,* 1977, p. 3342.
22. H. M. Macksey, G. E. Brehm, and S. E. Matteson, "Improved GaAs Power FET Performance Using Co-Be Implantation," *IEEE Electron Device Letters,* 1987, p. 116.
23. M. D. Deal and D. A. Stevenson, "Diffusion of Cr in GaAs," *J. Applied Physics,* 1986, p. 2398.
24. J. K. Dhiman and K. L. Wang, *J. Electrochemical Soc.,* 131, 1984, p. 2957.
25. F. Hyuga, K. Watanabe, J. Osaka, and K. Hoshikawa, "Activation Mechanism of Si Implanted into Semi-insulating GaAs," *J. Applied Physics Letters,* 1986, p. 1072.
26. Y. K. Yeo, R. L. Hengehold, Y. Y. Kim, A. Ezis, Y. S. Park, and J. H. Ehret, "Substrate Dependent Electrical Properties of Low-Dose Si Implants in GaAs," *J. Applied Physics,* 1985, p. 4083.
27. D. E. Davies, P. J. McNally, J. P. Lorenzo, and M. Julien, "Incoherent Annealing of Implanted Layers in GaAs," *IEEE Electron Device Letters,* 1982, p. 102.
28. M. Greiner, "Silicon Diffusion in GaAs Using Rapid Thermal Processing," Ph.D. thesis, Stanford University, 1984.
29. R. W. Klopfenstein and C. P. Wu, "Computer Solution to One-Dimensional Poisson's Equation," *IEEE Trans. Electron Devices,* June 1975, pp. 329–333.
30. C. H. Chen, A. Peczalski, M. S. Shur, and H. K. Chung, "Orientation and Ion-Implanted

Transverse Effects in Self-Aligned GaAs MESFETs," *IEEE Trans. Electron Devices,* July 1987, pp. 1470–1481.

31. C. M. Maziar and M. S. Lundstran, "Caughey-Thomas Parameters for Electron Mobilities in GaAs," *Electronics Letters,* 1986, p. 565.

32. H. Fukui, "Determination of the Basic Device Parameters of a GaAs MESFET," *Bell Systems Technical J.,* 1979, p. 771.

33. P. M. Asbeck, C. P. Lee, and M. C. F. Chang, "Piezoelectric Effects in GaAs FETs," *IEEE Trans. Electron Devices,* October 1984, pp. 1377–1380.

34. T. Ohnishi, Y. Yamauchi, T. Onodera, N. Yokoyama, and H. Nishi, "Experimental and Theoretical Studies on Short-Channel Effects in Lamp-Annealed Self-Aligned MESFET's," *Abstracts, 16th Conf. Solid State Devices and Materials* (Kobe, Japan), 1984, p. 391.

35. R. E. Williams, *Gallium Arsenide Processing Techniques,* Artech House, Dedham, MA, 1984, p. 368.

36. W. Dingten and K. Heime, "New Explanation of the $Nd^{-1}$ Dependence of the Specific Contact Resistance," *Electron Letters,* 1982, p. 940.

37. M. Hirayama, M. Togashi, N. Kato, M. Suzuki, Y. Matsuoka, and Y. Kawasaki, "A GaAs 16 kbit SRAM Using Dislocation Free Crystals," *IEEE Trans. Electron Devices,* 1986, p. 104.

38. J. H. Magerlein et al., "Characterization of GaAs Self-Aligned MESFET IC's," *J. Applied Physics,* 1987, p. 3080.

39. T. Ohnishi, T. Onodera, N. Yokoyama, and H. Nishi, "Comparison of the Orientation Effect of $SiO_2$ and $Si_3N_4$ Encapsulated GaAs MESFET's," *IEEE Electron Device Letters,* 1985, p. 172.

40. R. G. Wilson, *In GaAs and Related Compounds,* Bristol, 1984.

## Section 3.5

1. A. Gupta et al., "Yield Considerations for Ion-Implanted GaAs MMICs," *IEEE Trans. Microwave Theory Tech.,* MTT Vol. 31, January 1983, pp. 16–20.

2. R. M. Welch et al., "LSI Processing Technology for Planar GaAs Integrated Circuits," *IEEE Trans. Electron Devices,* Vol. ED-27, June 1980, pp. 1116–1124.

3. R. E. Williams, *Gallium Arsenide Processing Techniques,* Artech House, Dedham, MA, 1984.

4. R. Goyal et al., "High Performance 0.5 $\mu$m Manufacturing Process for MMICs: Yield Perspective," *Microwave J.* August 1987.

5. A. A. Immorlica, Jr., et al., "A Diagnostic Pattern for GaAs FET Material Development and Process Monitor," *IEEE Trans. Electron Devices,* Vol. ED-27, December 1980, pp. 2285–2291.

## Section 3.6

1. M. R. Stiglitz, "GaAs Foundry Operations 1988," *Microwave J.,* March 1988, p. 38.

2. A. G. Rode et al., "GaAs Custom MMIC Foundry: The Right Tool for the Times," *MSN & CT,* November 1985, p. 61.

3. R. Schneiderman, "GaAs Foundries Key," *Microwaves & RF,* March 1988, pp. 8–15.

## Section 3.7

1.  I. Krause and C. R. Suchors, "Design Automation: A Strategic Necessity," *Electronic Business*, April 1987, pp. 122–126.
2.  T. J. Sanders and J. W. Boarman, "Application of Management Tools for Improving Integrated Circuit Manufacturing," *Solid-State Tech.*, May 1987, pp. 105–110.
3.  John H. Powers, Jr., "Automating Electronic Manufacturing," *IEEE Circuits and Devices*, March 1987, pp. 21–32.
4.  Mehran Sepehri, "Integrated Data Base for Computer-Integrated Manufacturing," *IEEE Circuits and Devices*, March 1987, ｜ . 48–54.
5.  Brian Moore, "What Is Computer Integrated Manufacturing," *Test and Measurement World*, October 1988, pp. 30–37.
6.  "CIM Modernizes Manufacturing," *Digital Review*, November 1987, pp. 93–97.
7.  H. A. Watts, "CAM: Lot Tracking and More," *Semiconductor Int.*, July 1987, pp. 66–68.
8.  P. Burggraaf, "CAM Software–Part 1: Choices and Capabilities," *Semiconductor Int.*, June 1987.
9.  P. Burggraaf, "CAM Software—Part 2: Implementation and Expectations," *Semiconductor Int.*, July 1987.
10. W. Bottoms, "Interfacing for Automated Wafer Fabrication," *Proc. of the Advanced Semiconductor Equipment Exhibition*, January 1985.

# Chapter 4
# Device Modeling

*R. Goyal, M. Golio and W. Thomann*

## 4.1 SINGLE – GATE FET

### 4.1.1 Introduction

The GaAs MESFET has become one of the most highly utilized devices in the microwave industry. A major reason for its importance is its potential for exploitation in both power and low-noise applications. Over the range from about 1 GHz to millimeter-wave frequencies, the MESFET offers several advantages over competing device technologies such as bipolar transistors, IMPATTS, *pin* diodes, or transferred electron devices. Applications where the MESFET has seen significant utilization include low-noise [1] and high-power [2] amplifiers, microwave oscillators [3], mixers [4], low-power consumption and high-isolation RF switches [5], and high-speed digital circuitry [6]. Recently, interest in the MESFET has been furthered by the continued use of monolithic circuits for microwave, millimeter-wave and high-speed digital subsystems.

One key to the enduring success of microwave MESFET circuitry has been the continuing improvements in the high-frequency performance characteristics achieved since fabrication of the first MESFET in 1966 [7]. These improvements have been attributable in large part to reduction in device size (scaling). Typical gate length dimensions of GaAs MESFETs produced in research labs have been decreased by an order of magnitude (to approximately 0.1 $\mu$m) from 1968 to the present [8]. During that time corresponding frequency capabilities have been extended upward ($f_T$ values greater than 100 GHz) by approximately the same factor.

The success of monolithic circuits and of device scaling has brought with it new challenges in device modeling. In addition, monolithic technology does not allow for easy postfabrication tuning, as is often possible with hybrid circuit technology. The cost of monolithic circuit fabrication places greater pressure on the designer to achieve first-time success. Thus, the accuracy of the device model is more critical for monolithic circuit development. Unique device modeling challenges also have been posed by the scaling to submicron dimensions. These small gate length devices often exhibit anomalous characteristics termed *short channel effects*. Such characteristics affect device performance in first-order ways and therefore cannot be neglected by device models.

### 4.1.2 Basic Operation

The cross section of a MESFET device is illustrated in Figure 4.1. Three metal electrodes labeled *gate, source,* and *drain* connect to a thin semiconductor active channel layer. The active channel is created by ion implantation of donor atoms into semi-insulating material or by growing doped material using epitaxial growth techniques. Contact to both the source and drain is accomplished with ohmic metalization. For reasonably low biases the current through the source-drain contacts is related linearly to the voltage across the contact junction. Thus, for small biases, the source-drain terminals behave like a linear resistor. For higher biases, the semiconductor material itself will tend to limit the maximum carrier velocity, causing current saturation to occur.

The gate contact in a MESFET device is accomplished using Schottky

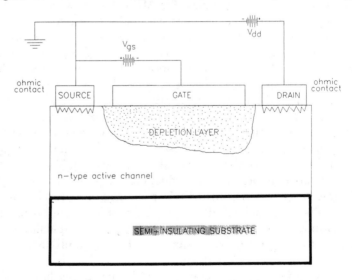

**Fig. 4.1** Cross section of a MESFET device.

metal. The Schottky barrier contact creates a depletion layer beneath the gate that is completely depleted of free charge carriers. Because no free-carriers exist in this depletion layer, no current can flow through it, thereby reducing the available cross sectional area for current flow between the source and drain. As reverse bias is applied to the gate contact, the depletion layer penetrates deeper into the active channel and current is further reduced. The gate bias, then, acts as a mechanism for limiting the maximum flow of source-drain current. With or without gate bias, the source-to-drain current in the device does not continue to increase linearly as a function of applied drain voltage. Instead, the carriers reach a limiting velocity at some critical electric field value, and current saturates. This saturation effect occurs in all semiconductors.

Figure 4.2a shows the current-voltage relationship expected from an ideal MESFET, as just described. Actual MESFET I–V characteristics are similar to the ideal characteristics, with the important exception that the slope of the curves remains slightly positive even after semiconductor-limited velocity has been reached. The slope of the $I_d$-$V_{ds}$ curve represents the output conductance of the device, and actual devices do exhibit finite output conductance even in saturation. A number of factors contribute to this finite conductance including charge domain formation, charge injection into the nonideal semi-insulating substrate, and conduction mechanisms via surface and channel-substrate interface states.

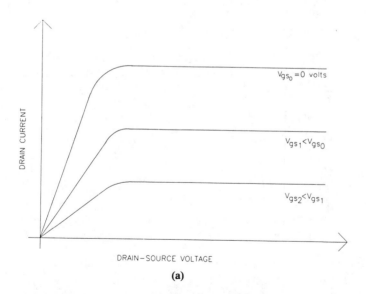

(a)

**Fig. 4.2** dc current-voltage characteristics of a MESFET: (a) ideal $I_d - V_{ds}$ characteristics; (b) measured $I_d - V_{ds}$ characteristics for two devices with identical doping and geometry except for gate length; (c) measured $I_d - V_{gs}$ characteristics for the same two devices.

Figure 4.2b presents measured $I_d$-$V_{ds}$ characteristics for two GaAs MES-FETs that are similar except for gate length. Although the curves of Figures 4.2a and 4.2b are similar in many respects, the differences between the characteristics have a significant effect on device performance. It is apparent that not only do the actual measured characteristics exhibit finite output conductance, but the conductance is greater for the 0.5 μm gate device than for the 1.0 μm gate device. Figure 4.2c presents the drain current dependence on gate-source voltage and illustrates that the device pinch-off voltage is increased in magnitude for the shorter gate device. These short channel effects often are more pronounced in devices with small gate lengths (typically with gate lengths below 0.5 μm) and sometimes lead to severely degraded device performance. Small gate length to channel thickness ratios and carrier injection into the semiinsulating substrate are two causes of short channel effects. To complicate matters further, some of the characteristics of the GaAs MESFET, such as output conductance, change dramatically as a function of frequency between dc and around 100 kHz. The corner frequency of 10 Hz–10 MHz where the changes occur also strongly depends on how the semiinsulating GaAs material is grown and thus varies from vendor to vendor. Nonideal characteristics of MESFETs such as these represent challenges to the device modeler.

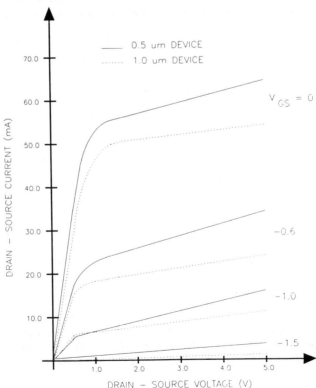

**Fig. 4.2 (b)**

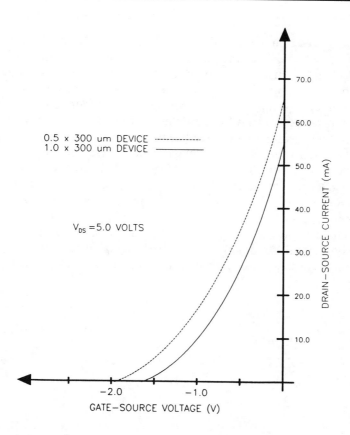

**Fig. 4.2 (c)**

### 4.1.3 Device Performance Analysis

A number of procedures related to modeling circuits containing active devices must be complete before performance predictions can be obtained. A device model can be applied only once model parameters have been extracted to describe the particular device of interest. Likewise parameter extraction may be accomplished only after appropriate device characterization has been performed. These processes of characterization, parameter extraction, and device modeling are clearly related in a number of important ways. The accuracy of any device model ultimately is limited by how accurately the model parameters are determined. The parameters of a model might include equivalent circuit element values, empirical constants, or physical characteristics of the device. The determination of required model parameters (parameter extraction) is dependent on the type and accuracy of available device characterization data. The merits of the device model itself are partially determined by the amount and type of characterization required to use the model. A device model that involves only a few parameters, which are determined from a limited number of simple measurements, has tremendous time saving and accuracy advantages over device models that require that a large number of parameters be determined from a number of tedious measurements.

Several different methods have been used for the parameter extraction process [9–11], and a wide variety of device models have been investigated [12–16] by the microwave research community. Figure 4.3 illustrates the general relationship between device characterization, parameter extraction, device modeling, and circuit simulation for one approach to the microwave circuit design problem. Several other approaches involving different characterization and parameter extraction schemes have also been used. The processes described by Figure 4.3 apply to both small- and large-signal modeling applications (the small-signal models being determined as an intermediate step toward large-signal model determination).

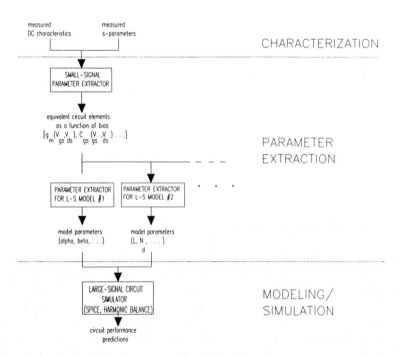

**Fig. 4.3** Diagram of the relationship between device characterization, parameter extraction, and modeling.

The lower portion of Figure 4.3 refers to *modeling and simulation*. Simulation deals with analysis that leads to circuit performance predictions. Computer-aided design packages such as Supercompact™, SPICE™, Touchstone™, and harmonic balance programs typically are used by the microwave community to perform circuit simulation tasks.

A device model can be composed of a set of equivalent circuit element values in a particular circuit topology or an equation or set of equations that, when evaluated, predict device performance. Small-signal models are capable of predicting device performance when the device is operated in a linear application, whereas large-signal models are applicable for both linear and nonlinear applications.

The primary advantages offered by empirical device models is the simplicity and efficiency of their most basic forms. The engineer with essentially no knowledge of device physics can utilize the most commonly available empirical models to obtain qualitative results in a very short period of time. These more basic empirical models require as few as four modeling parameters to define first-order device behavior. Such models, however, provide only qualitative performance information and often fail even to show characteristic behavior, particularly of short channel devices. Typically, empirical models capable of providing accurate performance predictions are much more complex. Determination of as many as 25 different empirical parameters may be required to define the more accurate models. A number of tedious measurements also may be needed to determine these parameters. Such models preclude performing simultaneous device and circuit optimization studies because model parameters cannot be directly related to device physics.

Both physically based and empirical models have been exploited for use in circuit simulations. Physically based models rely on physical parameters that describe the device dimensions and physical properties of the material in order to obtain their predictions. Such parameters will typically include gate length, gate width, channel thickness, and doping density. Other parameters that might be required by a physically based model include terminal spacings, low field mobility and saturation velocity of the GaAs material, deep-level density of the bulk material, or surface state density. In contrast, empirical models rely solely on parameters that are empirically determined by fitting model predictions to measured data. These parameters often possess no particular physical significance.

There are a number of reasons that physically based models are preferred over empirical models for MMIC applications. One advantage over empirical models is that a physical model will allow studies of the effects of process variation on ultimate device performance before the device is fabricated. A second advantage of physically based over empirical modeling is that a physically based model allows optimization of the active device suitable for a particular type of circuit (e.g., low-noise, power). Simultaneous device-circuit optimization offers a significant potential advantage over standard hybrid design practices. Fully exploiting this opportunity, however, requires that the device be described in terms of the physical properties of the FET that can be altered during the fabrication process. Devices utilized in unique circuit topologies to perform specialized functions could be improved significantly utilizing such physically based models. Unfortunately, physically based models often are more complex computationally than the simplest empirical models. This complexity can lead to models too inefficient for practical design applications in simulators. Furthermore, the geometric dimensions and material properties required to describe the device can be more difficult to obtain than the electrical characteristics required to define empirical models.

The portion of Figure 4.3 labeled *parameter extraction* refers to the process of determining parameter values for a model. For an equivalent circuit model, parameter extraction is the process used to determine the values equivalent circuit elements must take to accurately describe device performance. When a physically based model is utilized, the parameter extraction process involves determining the physical dimensions and material properties of the device. The parameter extraction of gate length, for example, might be accomplished by SEM measurements of finished devices. Determination of other physical properties, such as doping density or electron mobility in the channel, can be much more difficult to extract. Nonphysically based models require that empirical parameter values be determined as part of the parameter extraction process. Typically this is accomplished by adjusting model parameter values until an acceptable agreement is obtained between measured and modeled characteristics. The determination of which characteristics (i.e., dc, S-parameters, load-pull contours, *et cetera*) should be compared to model predictions, however, is not always a trivial process.

To determine appropriate parameter values for a device model, characterization is required. This is illustrated in the upper portion of Figure 4.3. Characterization can involve dc measurements, small signal RF measurements, large-signal measurements, noise measurements, or materials and dimensional measurements. The type of characterization required for a particular application depends on the type of model and parameter extraction process to be used.

For the particular scheme illustrated in Figure 4.3, both dc and RF measured data is entered into a small-signal parameter extractor routine. When small-signal S-parameter data are used for parameter extraction of large-signal models, characterization is performed on the same device at several different bias levels. The results of this process are then used in conjunction with a large-signal parameter extraction routine. An alternate technique could be used that utilizes only large-signal S-parameter or load-pull data. This extraction process results directly in large-signal model parameter values. Typically, the output of such a process will be an equivalent circuit whose behavior closely approximates the measured device characteristics. The primary advantage of using only dc and small-signal characterization data as opposed to large-signal data is that the former is considerably easier to obtain while providing equally good modeling accuracy [17]. The resulting model is also compatible with the widely used circuit simulation programs. The parameters obtained by the parameter extractor may be used in a large-signal circuit simulation that has the model of interest incorporated within the code.

### 4.1.4 Characterization

The various types of data that might be considered for use in the device analysis process include (1) dc current-voltage characteristics, (2) microwave S-parameters, (3) large-signal S-parameters or load-pull characteristics, (4) noise parameters, and (5) physical characteristics of the device, such as geometry and material properties. Each of these types of data offers certain attractive features, and each type is also associated with certain difficulties and limitations.

#### *4.1.4.1 dc Measurements*

The primary advantage of utilizing dc data for model parameter extraction is ease of performance. Unfortunately, dc data fails to accurately describe the RF characteristics of the device. The output resistance of GaAs MESFETs, for example, is a strong function of frequency [18–20]. The output conductance inferred from dc characteristics such as those of Figure 4.2b, therefore, is completely inadequate for determining microwave device performance. Differences between dc and microwave output conductance for GaAs MESFETs of more than an order of magnitude are often observed. Figure 4.4 illustrates measured output resistance characteristics of a $0.8 \times 400$ $\mu$m ion-implanted MESFET device between the frequencies of 20 Hz and 100 kHz. The behavior described by these curves has serious implications regarding parameter extraction processes that utilize only dc data. Recent investigations into such processes have shown them to produce unacceptable results [9]. In addition to the output conductance shifts at lower frequencies, device transconductance has also been observed to change by a lesser amount (approximately 5–30%) [18].

Although dc data alone are inadequate for complete model parameter determination [9,11], some extremely useful information can be obtained from such measurements. Figure 4.5 presents measured drain-source current as a function of both applied drain-source and gate-source bias levels. By definition, the output conductance of the device should be given by the derivative of the drain-source current with respect to drain-source voltage:

$$g_{ds} = 1/r_{ds} = \frac{dI_{ds}}{dV_{ds}}\bigg|_{V_{gs} = \text{constant}} \tag{4.1}$$

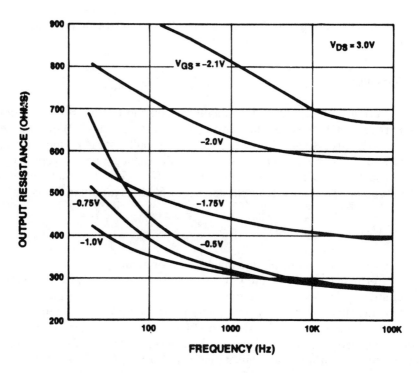

**Fig. 4.4** Measured output resistance of a 0.8 × 400 μm MMIC device as a function of frequency between 20 Hz and 100 kHz.

Likewise, transconductance can be expressed as

$$g_m = \frac{dI_{ds}}{dV_{gs}}\bigg|_{V_{ds} = \text{constant}} \tag{4.2}$$

Because of the frequency dependence of these characteristics, the curves of Figure 4.5 alone cannot provide adequate characterization for use in parameter extraction. We should point out that, although the derivative defined by equation (4.1) varies significantly as frequency is increased from dc to approximately 1 MHz, the actual measured currents and the transconductance expression given by (4.2) varies by a relatively small amount [20]. The data presented in Figure 4.5, therefore, may be useful as first-order estimates of device performance characteristics.

Parasitic resistance values, $R_s$, $R_d$, and $R_g$, illustrated in the MESFET equivalent circuits of Figures 4.6 can be estimated using three separate dc current-voltage measurements. The three simple measurements are illustrated in Figure

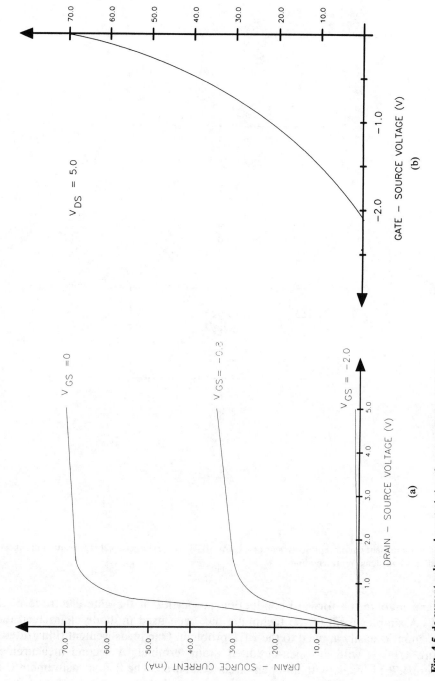

**Fig. 4.5** dc current-voltage characteristics of a 0.8 × 400 μm MMIC device: (a) drain-source current as a function of drain-source bias; (b) drain-source current as a function of gate-source bias.

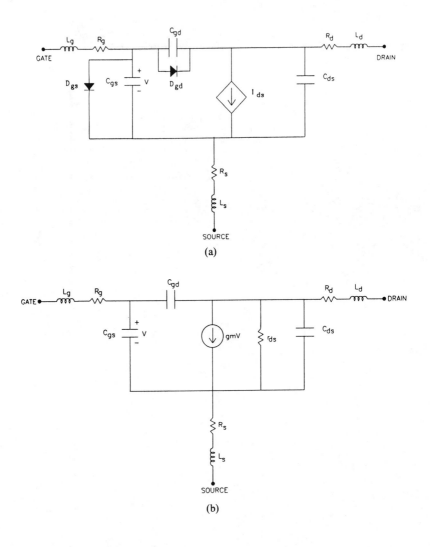

**Fig. 4.6** Equivalent circuit representations of a GaAs MESFET: (a) large-signal representation; (b) corresponding small-signal representation.

4.7 and involve the forward conduction properties of the gate electrode of the FET. A simpler form of the technique has been used in diode characterization [21], and it is easily applied to the FET problem. One measurement (illustrated in Figure 4.7a) is with an open-circuited drain terminal. A second measurement (Figure 4.7b) is made with an open-circuited source. The final measurement (Figure 4.7c) is made with the drain and source grounded. Any of these three forward

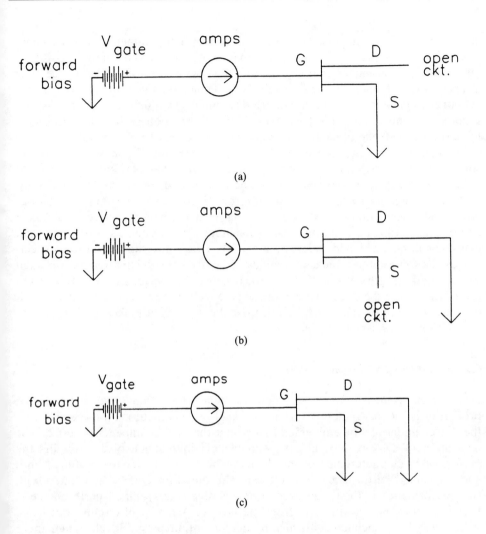

**Fig. 4.7** Equipment and device configurations for forward bias dc measurements.

bias measurements also can be used to extract the parameters required to describe the gate-drain and gate-source diodes of Figure 4.6a.

### 4.1.4.2 RF Measurements

The most common RF measurements utilized for the characterization of MESFETs for parameter extraction purposes are microwave S-parameter measurements. Automated measurement equipment to perform these measurements is readily available in most laboratories. Standard S-parameter measurements characterize only the small-signal performance of a device. To obtain large-signal information, S-parameter characterization must be performed on the device over a large range of operating bias levels [17]. Typically, the S-parameters corresponding to each bias level are used to determine a unique set of values for the small-signal equivalent circuit elements. This information is then used to determine the dependence of each equivalent circuit element to bias conditions.

Measurements at low frequencies (e.g., 1 MHz) can be used to determine some device characteristics. Output resistance, for example, can be measured directly using a sine wave generator and ac voltage and current meters. The ratio of the ac voltage across the drain-source terminals to the ac current into the output port is the device output resistance. A similar measurement of device capacitance can be made using an ac bridge. The capacitance measured this way is the total gate capacitance, which includes the sum of the gate-source and gate-drain capacitances. This capacitance measurement can be extremely difficult to make on small devices (gate lengths of less than 1 $\mu$m), because the capacitance of these structures is small (on the order of 1 pF/mm of gate width). The use of large gate width devices also often is limited by device oscillation, which makes such measurements almost impossible.

### 4.1.4.3 Large-Signal Measurements

An early measurement technique designed to obtain large-signal device properties is the load-pull measurement. Load-pull characterization requires that the device under test be embedded in a circuit that can be impedance tuned. The measurement system must also be capable of simultaneously determining the performance characteristics of the device. The device under test is tuned until some specified performance is achieved. The tuned impedance associated with that performance is then measured. For example, the specified performance is typically an output power level from the device. A family of circuit impedance values will be associated with any particular output power level. When these impedance values are plotted on a Smith chart they are referred to as *output power contours* and can be used in the design process of large-signal circuits. Note that this technique need not be limited to large-signal characterization, although such an elaborate technique would be extremely inefficient compared to standard small-signal characterization methods. By determining a number of different load impedance values associated with one or several performance specifications, a map of performance as a function of circuit impedance can be produced. A characterization performed in this manner applies only for the RF power levels used in the measurement. Measurements must be made at a number of different incident power levels in order to make the characterization process more complete.

The most common use of load pull measurements has been for the design of power amplifiers [22]. The technique is tedious but can be automated [23]. The typical implementation of the technique, however, does not describe the harmonic content of the device. Thus, power saturation characteristics are predicted using this method, whereas third-order intercept is not. A load-pull system capable of measuring the harmonic content of the RF output signal as a function of device load can be realized [24]. Although, in principle, this system could be used to completely determine large signal device characteristics, the amount of equipment required to perform these measurements is prohibitive for most applications.

Large-signal device characterization has also been performed using measurements similar to S-parameter measurements. The technique, termed *large-signal S-parameter characterization,* is distinct from standard small-signal S-parameter measurements primarily in the incident power used. The data obtained is applicable only at the power level of the experiment; measurements performed at multiple signal levels are required for complete characterization. Both power amplifiers [25] and oscillators [26] may be designed using large-scale S-parameters. Large-signal S-parameter methods and load-pull measurements provide similar information concerning the device and are related to each other mathematically. For that reason, these two techniques also share many of the same limitations. Neither technique is easily employed to determine a complete large-signal device model that can be utilized in a general circuit simulator. Because these techniques (as they are typically implemented) do not provide harmonic information about the measured device, they cannot provide the data required to describe all aspects of a nonlinear device model.

### 4.1.5 Equivalent Circuits and Parameter Extraction

General circuit simulation packages can be used for device predictions when an equivalent circuit model can describe the device behavior. Figure 4.6a illustrates fairly standard equivalent circuit representations of the GaAs MESFET. Figure 4.6a shows a large-signal circuit representation of the FET and Figure 4.6b illustrates a corresponding small-signal circuit. Although more sophisticated models (i.e., models that utilize more equivalent circuit elements) often are required to describe high-frequency or broadband device characteristics, the circuits of Figure 4.6 are adequate for many applications to a frequency of about 18–20 GHz, can be expanded easily by the addition of extra elements, and serve as excellent illustrative tools.

The process of obtaining appropriate equivalent circuit element values from measured S-parameters can be accomplished in several ways. Each element can be assigned an arbitrary value and optimized to minimize the error between simulated and measured data. This method is possible because of the availability of efficient optimizer routines, which are built into small-signal circuit simulation packages. The solution found to this error minimization process, however, is not unique [27]. Instead, a large number of numerically equivalent solutions can be determined, depending on the initial estimates of element values. If the application of interest involves only small-signal operation over the range of frequencies actually measured, then such a technique may produce adequate predictions. For many applications, however, a more systematic approach is preferred.

Equivalent circuit models have an advantage over directly measured S-parameters of devices because circuitry can be designed to operate beyond the frequency range of the available device measurements. For this application, the model topology and element values are first determined so that model predictions match the available measurements. Because the equivalent circuit model can be evaluated at any frequency, it can be used to extrapolate the available measurement data. The validity of such extrapolations is determined largely by how causally the equivalent circuit is related to the actual device performance. Therefore, the equivalent circuit element values need reflect not only the measured performance but also the physical processes responsible for device behavior. The equivalent circuit approach also is useful for scaling the device sizes.

When small-signal equivalent circuit information is to be utilized for large-signal parameter extraction, it is also important that a more systematic approach to optimization be utilized. As in the frequency extension application, the relationship between physical phenomena and element values is important for large-signal parameter extraction problems.

### 4.1.5.1 Extraction of Parameters from dc Data

The dc forward bias data described in Section 4.4 can be plotted on a semi-logarithmic graph, as shown in Figure 4.8. For small amounts of forward bias, the gate junction should act as an ideal diode as described by the following equation:

$$I = I_s \left[ \exp(qV/nkT) - 1 \right] \tag{4.3}$$

or equivalently:

$$V = \frac{nkT}{q} \ln[I/I_s + 1] \tag{4.4}$$

where $I_s$ is the reverse saturation current of the Schottky junction, $q$ is the electronic charge, $V$ is the applied forward potential, $n$ is the ideality factor, $k$ is Boltzmann's constant, and $T$ is the device temperature. As the drive level increases, the series resistance of the gate diode begins to affect the I–V character-

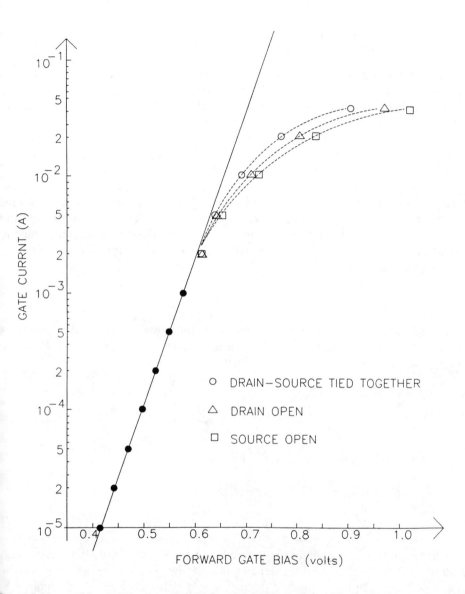

**Fig. 4.8** Measured forward bias characteristics for a 0.8 × 400 μm device. The three curves correspond to the three configurations shown in Figure 4.7.

istics. The complete equation describing the data will actually be given by

$$V = \frac{nkT}{q} \ln[I/I_s] + IR \tag{4.5}$$

where $R$ represents the resistance in series with the gate electrode. The deviation of the curve from a straight line on a semilogarithmic plot is caused by the voltage drop across the resistance in series with the gate diode. Both the linear characteristics and the digression from linear characteristics are illustrated in Figure 4.8. For the measurement of Figure 4.7a, this resistance is the sum of $R_g + R_s$. For the measurement of Figure 4.7b, the series resistance is $R_g + R_d$; and for the final measurement, the series resistance is the sum $R_g + R_s \| R_d$. The individual parasitic series resistance values are determined in the following manner. First, fit the linear portion of the data to a straight line. Extend the line beyond the linear current region. For an arbitrary current value, $I_0$ (see Figure 4.9), determine the difference between the actual voltage applied to achieve this current, and the voltage corresponding to the same current lying on the straight line data fit. The series resistance is then given by

$$R_{\text{series}} = \Delta V / I_0 \tag{4.6}$$

Figure 4.9 illustrates this process for one of the measurements. By computing this series resistance for all three cases described, three separate equations for the three parasitic resistances are determined:

$$R_a = R_g + R_s$$
$$R_b = R_g + R_d$$
$$R_c = R_g + (R_s \times R_d)/(R_s + R_d)$$

The solution of this set of equations is given by

$$R_g = R_c - \{[R_c^2 - R_c(R_a + R_b) + R_a^* R_b]\}^{1/2}$$
$$R_d = R_a - R_g \tag{4.7}$$
$$R_s = R_b - R_g$$

The parasitic resistance values determined in this way are slightly bias dependent. Figure 4.10 illustrates the bias dependence of the parasitic resistances for the $0.8 \times 400 \ \mu m$ gate ion-implanted device, whose data is presented in Figure 4.8. The only significant changes in $R_s$ and $R_d$ at low forward bias are due primarily to experimental accuracy. The value of $R_g$, however, does vary from about 1 to 2 $\Omega$.

One criticism of the method just described is that this measurement of resistance represents the resistance under forward bias conditions. For many FET

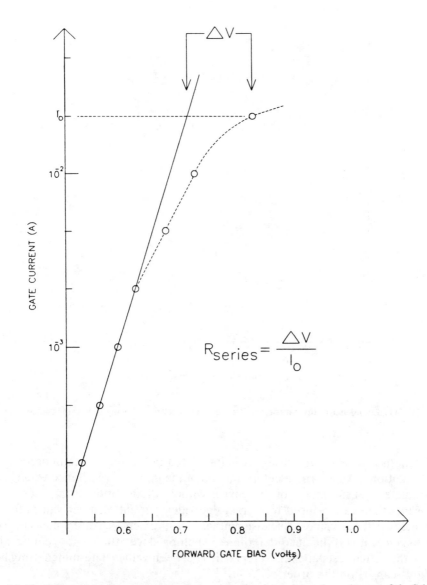

**Fig. 4.9** Illustration of the measurement analysis required to determine series resistance of the Schottky gate circuit using forward bias characteristics.

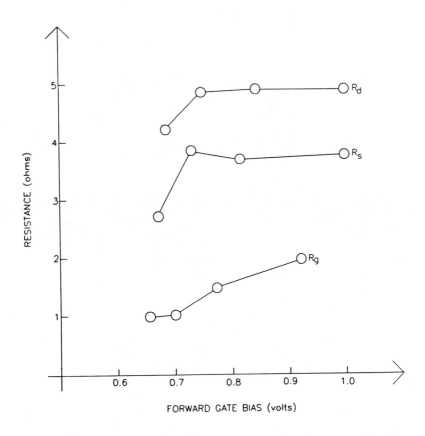

**Fig. 4.10** Bias dependence of the parasitic resistance values determined using forward bias characteristics of the gate of the device.

applications, however, the designer is interested in the resistance under reverse bias conditions. An alternate determination of the parasitic resistance values uses S-parameter measurements of the device without drain-source bias. These measurements then can be used to extract equivalent circuit element values for the parasitic resistance values. Because neither measurement is a direct measurement of resistance, it is difficult to determine which produces the most accurate measure of the actual resistance. Discrepancies between values determined using both methods can be on the order of 25%.

The linear portion of the forward bias characteristics shown in Figure 4.8 can also be utilized to obtain equivalent circuit information. These characteristics can describe the large-signal forward bias properties of the gate. This behavior is approximated well using the ideal diode equation as expressed by (4.3). The gate-source and gate-drain diodes described in this manner are illustrated in the circuit presented in Figure 4.6a. The parameters required to specify these diodes are the ideality factor, $n$, and the reverse saturation current, $I_s$. To determine the ideality factor it is first necessary to find the slope of the linear portion of the Figure 4.8 curve. This can be expressed as

$$m = [\log(I_2/I_1)]/[V_2 - V_1] \tag{4.8}$$

where $I_2$ and $I_1$ are the measured currents in the straight line region of Figure 4.8 corresponding to arbitrarily chosen applied voltage levels $V_2$ and $V_1$, respectively. The ideality factor then is determined from

$$n = \frac{q \ln(e)}{mkT} \tag{4.9}$$

Once the ideality factor is determined, saturation current is found by substitution of specific current and voltage values into Equation (4.3).

### 4.1.5.2 Small-Signal Equivalent Circuits

A preferred approach to the determination of equivalent circuit element values is one that provides a unique solution for the values. Such a technique can be employed for an appropriate equivalent circuit topology at any given frequency or possibly over a narrow band of frequencies [28]. Refinements on the equivalent circuit topology representing the device eventually may allow this approach to be utilized for general applications.

A realizable compromise to this approach is to determine initial estimates of circuit element values from a direct calculation that yields a unique solution, and then to optimize the solution from this starting point. During the optimization, circuit element values can be limited to a narrow range, thus assuring that variations from the systematic estimates are not great. A simplified equivalent circuit model for the GaAs MESFET that neglects parasitic elements is presented in Figure 4.11. This circuit model is representative of the MESFET operated to approximately 10 GHz [29].

Using simple circuit analysis techniques, the $y$-parameters for the simplified circuit model can be derived and expressed as

$$y_{11} = \omega^2 C_{gs}^2 R_i/E + j\omega(C_{gs}/E + C_{gd}) \tag{4.10a}$$

$$y_{22} = 1/R_{ds} + j\omega(C_{gd} + C_{ds}) \tag{4.10b}$$

$$y_{21} = (g_{m1} - g_{m2}\omega C_{gs}R_i)/E - j[(g_{m1}\omega C_{gs}R_i + g_{m2})/E + \omega C_{ds}] \tag{4.10c}$$

$$y_{12} = -j\omega C_{gd} \tag{4.10d}$$

where $\quad E = 1 + (\omega C_{gs} R_i)^2$

$$g_{m1} - g_{m2} = g_m[\exp - j\omega\tau]$$

From these equations, the seven element values required to define the equivalent circuit of Figure 4.11 are given by

$$C_{ds} = [\text{Im } y_{22}]/\omega - C_{gd} \tag{4.11a}$$
$$C_{gs} = [\text{Im } y_{11}]/\omega - C_{gd} \tag{4.11b}$$
$$C_{gd} = -[\text{Im } y_{12}]/\omega \tag{4.11c}$$
$$R_{ds} = 1/\text{Re } y_{22} \tag{4.11d}$$
$$R_i = [\text{Re } y_{11}]/(\omega C_{gs})^2 \tag{4.11e}$$
$$g_m = \text{Re } y_{21} \tag{4.11f}$$
$$\tau = (-[\text{Im } y_{21}]/\omega - g_m R_i C_{gs} - C_{gd})/g_m \tag{4.11g}$$

There are certain limitations to this analysis. The element $R_i$ is the most difficult parameter to extract using this approach. Equations (4.11b), (4.11e), and (4.11g) strictly apply only at frequencies where $(\omega R_i C_{gs})^2 \ll 1$ [29]. The value obtained for the charging resistance, $R_i$, is sensitive to measurement errors and should not be evaluated at frequencies where the magnitude of $s_{11}$ approaches unity. If parasitic resistance values of the device are high, the model of Figure 4.11 is less accurate.

For typical microwave applications, S-parameters and not $y$-parameters are available. The $y$-parameters for any linear two-port, however, can be computed directly from S-parameter data as described in Chapter 2. Once S-parameters are converted to $y$-parameters, equations (4.11) are used to determine equivalent circuit element values.

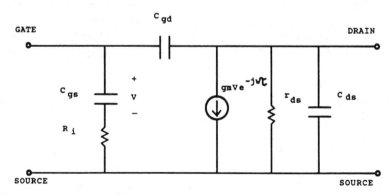

**Fig. 4.11** Equivalent circuit of the MESFET used in conjunction with Equations (4.10) and (4.11).

Calibration and de-embedding accuracy is critical to the accurate determination of equivalent circuit element values using this technique. Therefore, on-wafer S-parameter measurements typically are preferred. The accuracy of the element value determinations also is highly dependent on the frequency at which the measurements are made. Frequencies of tens of MHz to several GHz are better for determining all of the parameters except for $R_i$ and $\tau$. Higher frequency measurements (approximately 10–26 GHz) are preferred for determining these two parameter values.

The extraction process results in multiple values of equivalent circuit elements. To determine a single value, independent of frequency, a mean of the multiple values for each element can then be taken. Typically, S-parameters taken at 5 to 20 frequencies between dc and 10 GHz are sufficient to accurately determine element values. By combining the dc parameter extraction techniques described in this section with this method, an initial estimate of all the equivalent circuit parameters illustrated in Figure 4.6 (with the exception of parasitic inductances) can be found. Standard circuit optimization routines then can be utilized to refine these estimates until excellent agreement is reached. In practice, these techniques have been found to produce estimates that have values within 5–10% of the final values obtained through optimization. Typically, not all of the element values can be varied during the optimization process. Parasitic resistances, $R_s$, $R_d$, and $R_g$ for example, are often held constant at the value determined from the forward bias measurements. Similarly, if a circuit element value has been determined from an independent process such as a low frequency ac measurement, it may be desirable to fix that element value during the optimization process.

Figure 4.12 presents measured and modeled S-parameters from 1 to 10 GHz for a $0.8 \times 400$ $\mu$m $n$-channel ion-implanted GaAs MESFET biased at $V_{ds} = 3.0$ V and $V_{gs} = -2.0$ V. The equivalent circuit element values used to achieve this match are presented in Table 4.1. Also shown in the table are the initial estimates for each parameter determined from the dc and RF techniques described earlier. For this optimization process, initial parasitic inductance values were all assumed to be 0.1 nH. In addition, the parameters required to describe the forward Schottky gate characteristics are found to be $n = 1.33$ and $I_s = 14.17$ pA.

The gate bias level used to obtain the characteristics presented in Figure 4.12a is very near the device pinch-off voltage (low-drain current), whereas the drain bias level is well into the saturation region. This bias level is typical of common source, gate-driven MESFET mixer applications but is not typical of standard amplifier designs. The agreement obtainable between measured and modeled characteristics is significantly better at higher current bias levels. Agreement can also be improved through the use of more elaborate equivalent circuit topologies.

A process similar to the one just described can be accomplished with commercially available software [30]. More involved equivalent circuit configurations

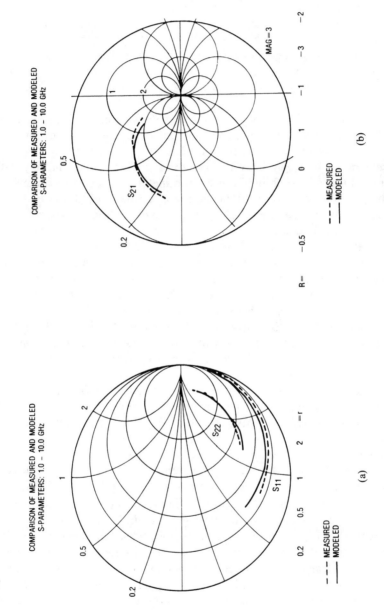

**Fig. 4.12** A comparison of measured and modeled S-parameters for a $0.8 \times 400$ $\mu$m MMIC device. Bias levels are $V_{ds} = 3.0$ V and $V_{gs} = -2.0$ V (a) $S_{11}$ and $S_{22}$; (b) $S_{21}$; (c) $S_{12}$.

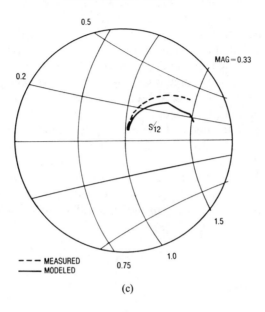

COMPARISON OF MEASURED TO MODELED
S-PARAMETERS: 1.0 – 10.0 GHz

(c)

**Fig. 4.12** cont.

**Table 4.1** Initial and Final Element Values for the $0.8 \times 400$ $\mu$m Device (bias conditions are $V_{ds} = 3.0$ V, $V_{gs} = -2.0$ V; the initial estimates are made from the dc and RF measurement techniques described in the text)

| Equivalent Circuit Element | Initial Estimate | Final Value |
| --- | --- | --- |
| $R_g$ | 1.35 Ω | 2.00 Ω |
| $R_s$ | 3.50 Ω | 3.60 Ω |
| $R_d$ | 4.75 Ω | 4.89 Ω |
| $L_g$ | 0.1 nH | 0.14 nH |
| $L_s$ | 0.1 nH | 0.17 nH |
| $L_d$ | 0.1 nH | 0.12 nH |
| $r_{ds}$ | 452 Ω | 492 Ω |
| $g_m$ | 20.5 mS | 20.4 mS |
| $C_{gs}$ | 0.28 pF | 0.26 pF |
| $C_{gd}$ | 0.074 pF | 0.071 pF |
| $C_{ds}$ | 0.11 pF | 0.11 pF |

such as the circuit illustrated in Figure 4.13 can be used with this software. This equivalent circuit includes three element values not included in the circuit of Figure 4.6b: $R_{gs}$, $R_i$, and a time constant, $T_i$, associated with the device transconductance. The addition of these elements greatly enhances the ability of the model to match measured device characteristics over a broad frequency range. A unique solution for all of the model element values can be determined provided measured S-parameter data is available to 26 GHz and the value of $R_g$ is first determined independently. Without an independent determination of $R_g$, this element value cannot be uniquely distinguished from $R_i$. The determination of $R_g$ can be accomplished using the dc forward biased measurement techniques described in the previous section. The circuit element value $R_{gs}$ shown in Figure 4.13 is important to model predictions only when the device is forward biased and significant gate current is flowing. Through the use of appropriate optimization techniques, this type of software provides consistent element values independent of starting circuit values.

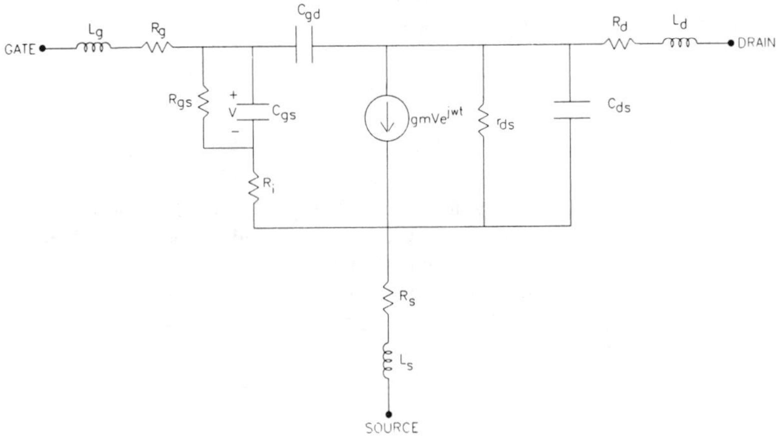

**Fig. 4.13** An equivalent circuit model of the MESFET that can be used to describe device performance to approximately 26 GHz.

Figure 4.14 illustrates the agreement between measured and modeled S-parameters from 45 MHz to 22 GHz, which is obtained using FETFITTER and the associated equivalent circuit of Figure 4.13. The device represented by the characteristics shown in Figure 4.14 is a 0.5 $\mu$m gate length, ion-implanted device. The equivalent circuit element values used to obtain this agreement are presented in Table 4.2. This agreement is seen to be excellent and the resulting model is adequate for most small-signal modeling applications.

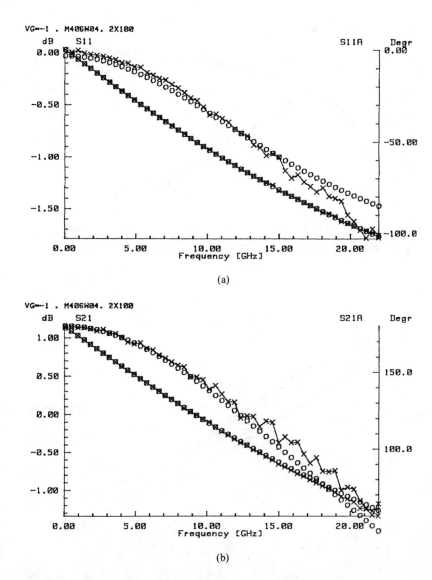

**Fig. 4.14** A comparison of measured and modeled S-parameters for 0.5 $\mu$m ion-implanted device. Bias levels are typical of amplifier operation. The equivalent circuit of Figure 4.13 is used to obtain the match: (a) S11; (b) S21; (c) S12; (d) S22.

When performing MMIC design tasks, it is often desirable to scale the gate width of active devices. Active devices used as current sources, for example, typically must have narrower gate-width dimensions than the RF device they

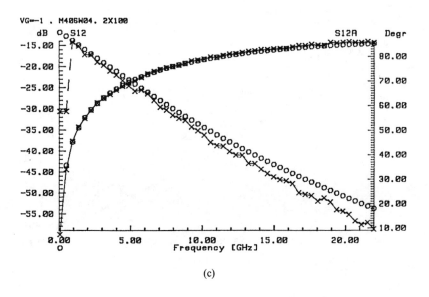

(c)

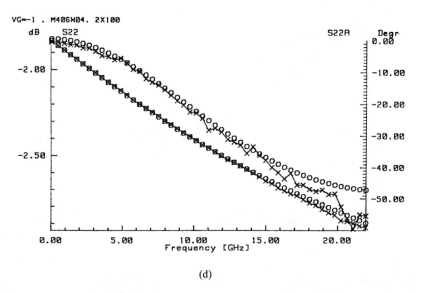

(d)

**Fig. 4.14** cont.

drive. Gate-width scaling has proven extremely useful in the design of distributed amplifiers [31]. Tremendous benefits are similarly realized by the freedom to choose devices of various gate widths in the design of broadband microwave

**Table 4.2** Element Values for a 0.5 $\mu$m Gate Length Device at a Bias Condition Typical of Amplifier Operation (the equivalent circuit is shown in Figure 4.13 and agreement between measured and modeled characteristics is illustrated in Figure 4.14)

| Equivalent Circuit Element | Value |
|---|---|
| $R_g$ | 1.00 $\Omega$ |
| $R_s$ | 0.58 $\Omega$ |
| $R_d$ | 0.00 $\Omega$ |
| $L_g$ | 0.053 nH |
| $L_s$ | 0.0 nH |
| $L_d$ | 0.0 nH |
| $r_{ds}$ | 474 $\Omega$ |
| $g_m$ | 12.7 mS |
| $C_{gs}$ | 0.12 pF |
| $C_{gd}$ | 0.024 pF |
| $C_{ds}$ | 0.053 pF |
| $T_t$ | 4.36 pS |
| $R_{gs}$ | 27 $\Omega$ |
| $R_i$ | 1.94 $\Omega$ |

switches or voltage controlled attenuators. For this reason, it is desirable to produce an equivalent circuit model that can predict the performance of similar devices with arbitrary gate widths. In practice, either the gate width of individual gate fingers, $Z$, or the number of equivalent gate fingers, $n_f$, can be scaled. The total gate width of the device can be expressed as

$$W = Z \cdot n_f$$

Some of the equivalent circuit element values of microwave devices scale not only with total gate width but also with the number of equivalent gate fingers on the device.

The most accurate method for establishing an equivalent circuit element model that is gate width scalable is to measure several devices with identical physical properties except for gate width. A similar family of devices that alter only the number of gate fingers can also be measured. This sort of FET matrix is described in Chapter 3.5. Once an equivalent circuit has been determined for each device, the element values can be plotted as a function of gate-width and number of fingers. Finally, an expression can be determined using linear regression that describes the plotted data.

An alternative to this technique, which is sufficiently accurate for many applications, is to characterize only one device of known gate width and then to assume the following scaling relationships:

1. $R_s$, $R_d$, $R_g$ values are inversely proportional to gate width. For example, $R_s$ is given by

$$R_s = R_{s0}/(Z \cdot n_f)$$

where $R_{s0}$ is the source resistance per unit gate width. Similar expressions are assumed for the other parasitic resistance values. In practice this expression often does not work well for gate resistance, $R_g$. The actual gate parasitic resistance value is a complex function of the number of gate fingers and the gate electrode layout. For gate resistance, a more accurate expression used is $R_g = (R_{g0} \cdot Z)/n_f$ where, $R_{g0}$ is the gate metal resistance per unit length of gate finger. This expression represents a first-order approximation to the gate resistance scaling relationship.

2. $C_{gs}$, $C_{gd}$, $C_{ds}$, $g_m$, and $g_{ds}$ values are directly proportional to gate width. As an example, the expression to define gate-source capacitance would be

$$C_{gs} = C_{gs0} \cdot Z \cdot n_f$$

Again, similar relationships should be assumed for the other parameters mentioned. Also note that the element $g_{ds}$ is the output conductance, given by the inverse of the output resistance $r_{ds}$ as defined in Equation (4.1).

Using these first-order scaling relationships will not predict some effects caused by more subtle gate-width scaling properties. Gate resistance and parasitic inductances, in particular, are not directly scalable with gate width. Likewise, device capacitance values often will be associated with a small amount of fringing capacitance (due to airbridge crossovers, for example), which does not scale with gate width. This is discussed further in Section 4.1.7. Some caution also must be taken to limit individual gate finger width. If this dimension becomes too long with respect to wave length of the signal (on the order of 10% of the wave length), the gate fingers will act as transmission line segments. For a signal frequency of 18 GHz the individual finger length should be limited to approximately 100 $\mu$m or less.

Some discussion of physically based models is appropriate at this point

Purely physically based models require no electrical measurement data in order to produce device performance predictions, which is often presented as an advantage of physically based modeling over empirical modeling. Such an argument is misleading because the physical parameters required of a device model may not be known to the circuit designer. Furthermore, the accuracy required of some physical parameters in order to achieve acceptable device performance predictions is beyond standard fabrication and measurement tolerances. For example, a change of as little as a few one-hundredths of a micron in the simulated gate length or active channel thickness of a device can have significant effects on performance predictions. It is very difficult, however, to measure the dimensions of a finished device to this accuracy. Material parameters such as doping densities or mobility can be even more difficult to determine. Multiple, tedious measurements requiring considerable analysis can be needed to determine physical parameters of a finished device [32].

There is a technique that enjoys many of the advantages offered by physically based models while avoiding the difficulties associated with determining the exact physical properties of the device. The required physical parameters such as gate length, electron mobility, and channel doping density profile can be taken as semiempirical parameters and varied until simulated results match measured performance to within a specified error tolerance. The technique becomes very similar to that just described for empirical models, with the exception that the values of these physical parameters are varied instead of the values of equivalent circuit elements. Although the physical parameters required of the model may not be known with required accuracy prior to the optimization process, such parameters usually are known within fairly small tolerances. Thus, the optimization portion of the parameter extraction process often is more efficient than the analogous process for empirical models. The resulting model also is directly scalable in all dimensions. Furthermore, using these models with a minimum amount of information concerning fabrication process tolerances allows the effect of processing fluctuations to be studied.

### 4.1.5.3 Large-Signal Parameter Extraction

The calculation of the electrical characteristics of equivalent circuits is not the only method used to model device behavior within large-signal circuit simulation packages. Purely mathematical models have been used that express terminal currents directly in terms of terminal voltages and derivatives of voltages with respect to time [16,33]. These mathematical expressions for the relationships between time and the terminal voltages and currents cannot be represented exactly by standard circuit elements configured in a reasonable circuit topology.

The goals of nonlinear circuit simulation are considerably more varied and challenging than linear circuit simulation. Whereas linear circuit simulation is concerned primarily with gain, phase, and noise *versus* frequency, nonlinear circuit performance objectives include gain compression, saturated power, efficiency, AM-to-PM and PM-to-AM conversion, harmonic distortion-generation, and *multitone intermodulation distortion* (IMD) products. Simulation of these nonlinear performance characteristics places new requirements on active device models.

The operating range of active devices is much more varied in nonlinear circuits than in typical linear circuits. For linear circuit applications, the drain source voltage is typically 2.5–4 V and, depending on the application, the gate-source voltage is adjusted to produce a drain-source current 20–80% of $I_{dss}$. Modeling the active device at one such dc operating point is all that is required, because the amplitude of the RF signals are much smaller than the dc component and do not perturb the dc operating point. For nonlinear circuit applications, the RF signals are comparable to the dc component; and during an RF cycle, the instantaneous operating point can range from pinch-off and gate-drain breakdown to "ohmic," low drain voltage region below the "knee," and even into forward gate conduction.

Although the equivalent circuit is not required for many models, the discussion that follows will center on equivalent circuit element behavior as a function of bias. The equivalent circuit element values provide a basis for comparison of model results with measured characteristics and also provide a basis for parameter extraction.

Figures 4.15 through 4.18 present the measured behavior with bias level of the four nonlinear equivalent circuit elements—$g_m$, $g_{ds}$, $C_{gs}$, and $C_{gd}$—for the device utilized in Table 4.1. Note that direct measurements of these characteristics would be difficult or impossible to make. The curves were determined using RF measurements as described in Section 4.4. Measurement of S-parameters was made under 12 different bias conditions. The equivalent circuit assumed to apply for this problem is shown in Figure 4.6a, and the parasitic resistance and inductance values along with the drain-source capacitance are assumed not to vary with bias conditions.

The general trends observed in the characteristics are typical of those found for other microwave MESFET devices. Studies indicate that when large-signal device models accurately describe equivalent circuit element behavior, such as that shown in Figures 4.15 through 4.18, accurate RF performance predictions are obtained [17]. This suggests that large-signal parameters can be extracted by choosing model parameters that minimize the error between modeled and measured equivalent circuit element values. To initiate this parameter extraction process, it is first necessary to define an error function to be minimized. The preceding discussion suggests that an appropriate error function might be expressed as

$$E^2 = E_{gm}^2 + E_{rd}^2 + E_{C_{gs}}^2 + E_{C_{gd}}^2$$

(4.1)

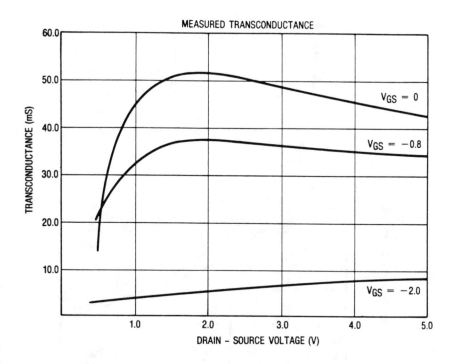

**Fig. 4.15** Transconductance as a function of bias determined for a 0.8 × 400 μm MMIC device.

where

$$E_{gm}^2 = \frac{1}{N} \sum_{i=1}^{N} \frac{A}{g_{m_i}^2} [g_{mmeas_i} - g_{m_i}]^2 \tag{4.13a}$$

$$E_{rd}^2 = \frac{1}{N} \sum_{i=1}^{N} \frac{B}{r_{ds_i}^2} [r_{dsmeas_i} - r_{ds_i}]^2 \tag{4.13b}$$

$$E_{Cgs}^2 = \frac{1}{N} \sum_{i=1}^{N} \frac{C}{C_{gs_i}^2} [C_{gsmeas_i} - C_{gs_i}]^2 \tag{4.13c}$$

$$E_{Cgd}^2 = \frac{1}{N} \sum_{i=1}^{N} \frac{D}{C_{gd_i}^2} [C_{gdmeas_i} - C_{gd_i}]^2 \tag{4.13d}$$

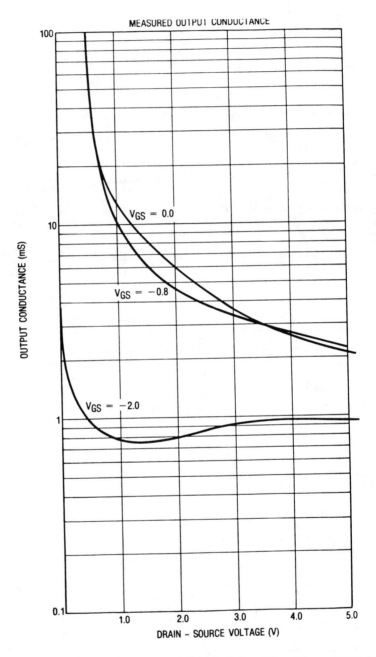

Fig. 4.16 Output conductance as a function of bias determined for a 0.8 × 400 μm MMIC device.

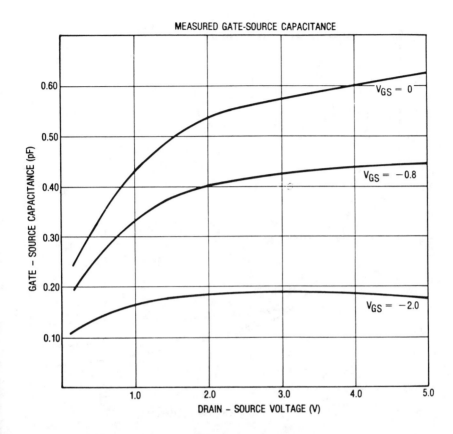

MEASURED GATE-SOURCE CAPACITANCE

$V_{GS} = 0$

$V_{GS} = -0.8$

$V_{GS} = -2.0$

GATE – SOURCE CAPACITANCE (pF)

DRAIN – SOURCE VOLTAGE (V)

**Fig. 4.17** Gate-source capacitance as a function of bias determined for a 0.8 × 400 μm MMIC device.

In Equations (4.12) and (4.13), $N$ is equal to the number of data points considered (i.e., the number of different bias points where data was measured). The term $g_{m_i}$ represents the modeled transconductance at the $i$th bias point and the term $g_{mmeas_i}$ is the measured transconductance at the $i$th bias point. Similar definitions apply to Equations (4.13b) through (4.13d). The parameters $A$, $B$, $C$, and $D$ represent optical weighting functions. A numerical parameter extraction technique that minimizes the error function expressed by Equation (4.12) can be applied to any device modeling application.

The functional form of the device model determines how small the error term $E^2$ can be made. It is not possible, in general, for a model to match all the data considered. A model with few parameters typically will be associated with a higher value for minimum error, $E^2$, than a more elaborate model with several parameters and therefore more flexibility.

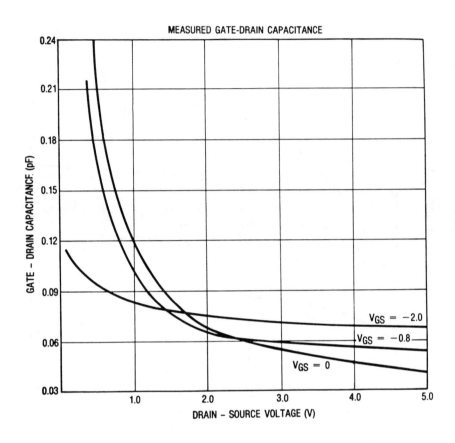

**Fig. 4.18** Gate-drain capacitance as a function of bias determined for a 0.8 × 400 μm MMIC device.

The error function expressed by Equations (4.12) and (4.13) completely neglects the dc characteristics of the device. It is possible, therefore, for a model whose parameters have been chosen by this process to produce accurate predictions of RF performance but produce inaccurate dc current-voltage predictions. Although such a model is acceptable for many applications, it is not desirable for others. For example, MMIC biasing is often accomplished using FETs as current sources. The design of such a circuit requires that accurate dc as well as RF predictions be established. To accommodate such situations, an additional term can be added to Equation (4.12). The new equation becomes

$$E^2 = E_{gm}^2 + E_{rd}^2 + E_{Cgs}^2 + E_{Cgd}^2 + E_{Id}^2 \tag{4.14}$$

where

$$E_{Id}^2 = \frac{1}{N} \sum_{i=1}^{N} \frac{F}{I_{d_i}^2} [I_{d\text{meas}_i} - I_{d_i}]^2 \qquad (4.15)$$

and with $I_{d_i}$ equal to the modeled and $I_{d\text{meas}_i}$ equal to the measured drain current at the $i$th bias point.

Model parameters also can be optimized by using the public domain software SUXES (Stanford University extractor model parameters). The SUXES program was written primarily to optimize the model parameters for CMOS device models (Doganis et al. [34,35]). However, as the source code of the program is available to its user, with simple modifications in the model equations, the same program can be extended for modeling GaAs MESFETs. SUXES is modular in structure, and a separate subroutine contains the device model equations. The user needs to develop a subroutine similar to that for CMOS device model equations or make a simple modification in the existing routine to implement GaAs MESFET model equations. The optimization part of the program uses the Levenberg-Marquardt method with modifications to incorporate linear constraints and prevent unexpected deviations of the model parameter values. The basis of the SUXES program is the nonlinear least squares fit to the experimental data by means of a modified version of the Marquardt algorithm operating in a multivariable space. The objective function to be minimized can be written as

$$\|f(p)\|^2 = \sum_i f_i(p)^2 \equiv \sum_i \left[\frac{I_i(p) - I_i^*}{\max(I_i^*, I_0)}\right]^2$$

where $p$ is the parameter vector to be optimized and $f(p)$ is the error vector between the calculated drain current, transconductance or output conductance, $I_i$, and the measured value $I_i^*$ for a given set of biases. User input $I_0$ determines whether a relative or absolute (weighted by $I_0$) error is taken.

The program extracts the parameters in a user-specified constrained region of operation: $f(p)$ is calculated at each iteration, and $p$ is adjusted to reduce the norm of the error. The process converges or terminates when any of the following conditions is met:

1. On two successive iterations, the parameter estimates agree to a specified number of significant digits.
2. The number of function evaluations is limited when convergence is not otherwise achieved.

3. The Euclidean norm of the approximate gradient is less than a specified small value, which indicates that the solution is close to optimum.
4. The current value of all outputs from the function simulations agrees with their respective values within some specified tolerance.
5. In exceeding the upper bound of the Marquardt parameter, λ, the search for a descent direction is abandoned.

The program displays the termination or convergence criterion and the final parameter vector. The user may determine whether this vector is suitable or the extraction may be repeated with other initial conditions. A block diagram in Figure 4.19 shows the operation of SUXES program in automated mode, where data files having the measurement results and the optimization strategy information are used. SUXES has been successfully used for implementing and optimizing model parameters for the Curtice model, modified Curtice model, Statz model, and other user-defined models for GaAs MESFETs down to 0.5 μm gate length devices.

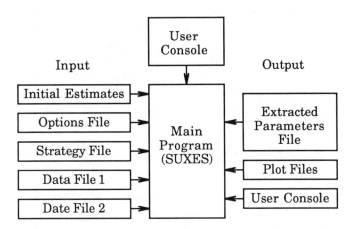

**Fig. 4.19** Block diagram of the SUXES computer program operating in the automated mode (from Doganis *et al.* [34]).

### 4.1.6 Device Modeling

A large-signal device model is an expression or set of expressions that describe device performance with respect to transient electrical stimulation. These expressions can take many different forms. Madjar and Rosenbaum [33] have described the total terminal currents of the device using explicit functions of the gate-source voltage, $V_{gs}$; the drain-source voltage, $V_{ds}$; and their derivatives. The model predicts instantaneous terminal currents for each set of instantaneous voltages from equations of the forms

$$I_d = I_{con} + D_{vsg} \frac{dV_{gs}}{dt} + D_{vds} \frac{dV_{ds}}{dt} \tag{4.16}$$

$$I_g = G_{vsg} \frac{dV_{gs}}{dt} + G_{vds} \frac{dV_{ds}}{dt} \tag{4.17}$$

where $I_{con}$, $D_{vsg}$, $D_{vds}$, $G_{vsg}$, and $G_{vds}$ all are functions of the terminal voltages.

An alternative approach to device modeling is to describe device behavior using an equivalent circuit. For large-signal models, at least some of the equivalent circuit elements must take on values that depend on voltage or current. Often this approach involves independent equations to determine the current-voltage characteristics and the capacitance-voltage relationships. When these relationships can be treated independently, the parameter extraction processes also can be independent of each other. This often will simplify the error minimization process described in Section 4.1.5.

One more commonly available empirical current-voltage description of GaAs MESFETs is the hyperbolic tangent model made popular by Curtice and Ettenberg [15]. The basis of this model can be understood by examining measured I–V characteristics of actual devices. Figure 4.5b shows an approximately quadratic dependence of drain-source current to gate-source voltage for voltages greater than some threshold voltage, $V_{T0}$. Likewise, the $I_d - V_{ds}$ characteristics of Figure 4.5a resemble a hyperbolic tangent function in form. Finally, the characteristics of devices operating in current saturation exhibit a finite output conductance. Combining these facts leads to a description of the I–V characteristics of the MESFET given by

$$I_d = \beta(V_{gs} - V_{T0})^2 \tanh(\alpha V_{ds})(1 + \lambda V_{ds}) \tag{4.18}$$

In Equation (4.18), $\alpha$, $\beta$, and $\lambda$ are arbitrary empirical parameters whose value must be chosen such that evaluation of (4.18) is consistent with measured characteristics. Values for transconductance and output conductance can be determined by direct application of Equations (4.1) and (4.2).

Table 4.3 presents values for $\alpha$, $\beta$, $\lambda$, and $V_{T0}$ determined for the $0.8 \times 400$ $\mu$m device characteristics presented in Figures 4.15 through 4.18. The minimization of Equation (4.14) was used as a basis for making the parameter choices. Comparison of measured and modeled parameters are presented in Figure 4.20. This figure also presents data obtained from a physically based model, which will be discussed later in this section. From the figure we can see that fairly good agreement with measured data is obtained, although transconductance values obtained by the model tend to considerably overestimate actual measured RF transconductance values for many bias levels.

**Table 4.3** Hyperbolic Tangent Model Parameters for the 0.8 × 400 μm Device Whose Equivalent Circuit Element Characteristics Are Presented in Figures 4.15 through 4.18

| Model Parameter | Value |
|---|---|
| $\alpha$ | 1.87 |
| $\beta$ | $14.4e^{-3}$ |
| $\lambda$ | $55.73e^{-3}$ |
| $V_{T0}$ | $-2.23$ |

The weighting factors for the error terms of Equations (4.13) and (4.15) were all equal for the parameter extraction process implemented. Note that it is not possible for the hyperbolic tangent model of Equation (4.18) to describe both the dc current-voltage curves of Figure 4.20a and the RF transconductance and output conductance characteristics of Figure 4.20b and 4.20c. The low frequency dispersion of the output resistance for this particular device results in an order of magnitude resistance drop from 5000 Ω at dc to approximately 400 Ω at 10 kHz. The equal weighting factors applied to the dc and RF data causes the parameter extractor to compromise on the match for both characteristics. Excellent match can be obtained for either characteristic individually, if the weighting factors for the other characteristic is set to zero.

The model described by Equation (4.18) often is found to provide inadequate predictions of actual device characteristics. At best, the model parameter values can be chosen so that accurate predictions are obtained over a limited range of bias levels. Improvements in the model can be implemented easily by slight alterations in the equation. For example, the $I_{ds}$-$V_{gs}$ characteristics of GaAs FET devices are seldom exactly quadratic in nature. This suggests that the exponent of the ($V_{gs} - V_{T0}$) argument might be chosen to have a value other than 2 in equation (4.18). Such an approach creates another model parameter, the exponent $V_{gexp}$, which needs to be determined. The resulting equation is given by

$$I_d = \beta(V_{gs} - V_{T0})^{V_{gexp}} \tanh(\alpha V_{ds})(1 + \lambda V_{ds}) \tag{4.19}$$

Another useful modification to this model can better describe the threshold voltage. Careful observation of measured device characteristics indicates that threshold voltage is actually slightly dependent on drain-source bias. To account for this dependence, the argument ($V_{gs} - V_{T0}$) in Equation (4.19) can be replaced with the argument [$V_{gs} - (V_{T0} + \gamma V_{ds})$], where the empirical parameter $\gamma$ has been added to the model. In a similar fashion, gate-source voltage dependence can be

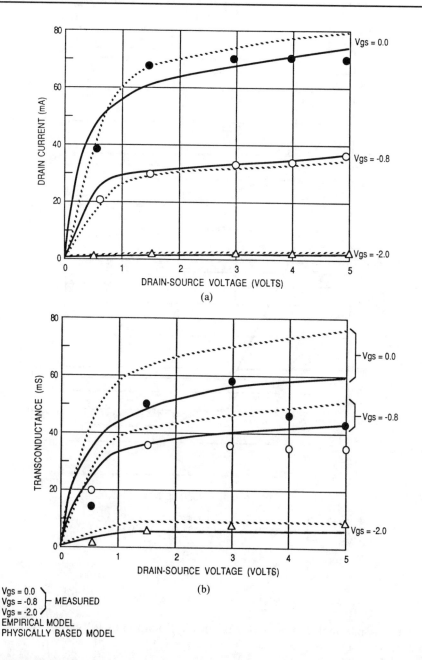

Fig. 4.20 Measured and modeled characteristics for a 0.8 × 400 μm MMIC device. The solid line represents results from the physically based model of Equation (4.21). The dashed lines are calculations from the empirical model of Equation (4.18). Parameter choices for the models are presented in Tables 4.5 and 4.3, respectively. Points are measured values: (a) dc current-voltage characteristics; (b) RF transconductance; (c) RF output conductance.

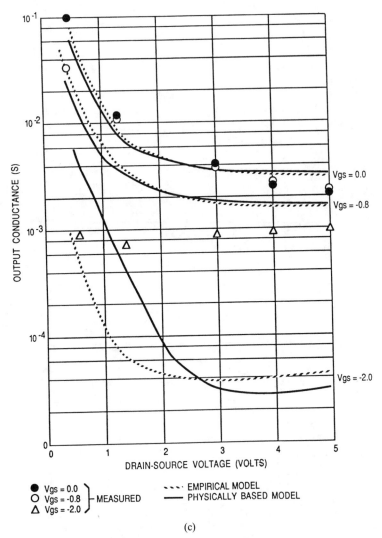

Fig. 4.20 cont.

added to the parameter beta by replacing $\beta$ in Equation (4.19) with $\beta'$ where $\beta'$ is given by

$$\beta' = \frac{\beta}{(1 + b(V_{gs} - V_{T0}))}$$

and the empirical parameter $b$ has been added to the model.

The techniques described in the preceding paragraphs, or modifications similar to them, are required to accurately predict behavior of devices with gate length geometries of 0.5 $\mu$m or less. Table 4.4 presents values for $\alpha$, $\beta$, $\lambda$, $V_{T0}$, $\gamma$, $b$, and $V_{gexp}$ determined for a 0.5 $\times$ 300 $\mu$m MMIC device. As was done with the 0.8 $\mu$m device, the minimization of Equation (4.14) was used as a basis for determining the parameters. Equal weighting factors as defined in Equations (4.13) and (4.15) were used for all error terms. A comparison between measured and modeled characteristics is shown in Figure 4.21. As in the case of the 0.8 $\mu$m device, better agreement can be obtained for any particular characteristic at the expense of the agreement with other characteristics. This is accomplished by adjusting the weighting function values of the error terms in Equations (4.13) and (4.15).

**Table 4.4** Modified Hyperbolic Tangent Model Parameters for a 0.5 $\times$ 300 $\mu$m MMIC Device

| Model Parameter | Value |
| --- | --- |
| $\alpha$ | 1.93 |
| $\beta$ | 22.7e$^{-3}$ |
| $\lambda$ | 12.49e$^{-3}$ |
| $V_{T0}$ | $-1.86$ |
| $\gamma$ | $-55.99$e$^{-3}$ |
| $b$ | 51.16e$^{-3}$ |
| $V_{gexp}$ | 1.920 |

Figure 4.20c illustrates a problem associated with most MESFET device models. For gate-source voltage levels near the threshold voltage, many device models predict output conductance values that are much smaller than those actually observed. Although for many applications, this underestimation does not significantly affect performance predictions, it can be important for circuitry such as RF switches.

When the model of Equation (4.18) or (4.19) is evaluated within a computer routine, a significant portion of the computation time is devoted to evaluating the hyperbolic tangent function. To improve the computational efficiency of the model, the hyperbolic tangent function can be approximated using only the first few terms of a series expansion.

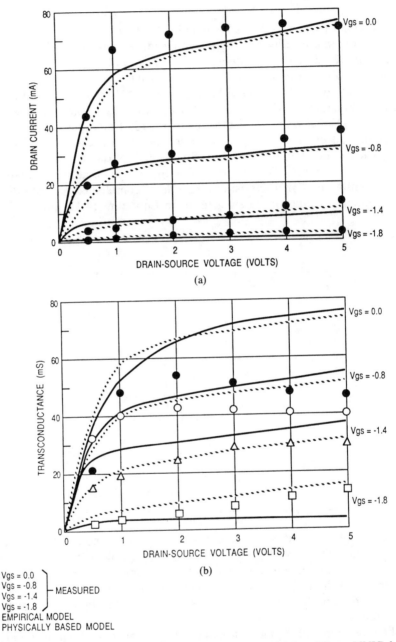

**Fig. 4.21** Measured and modeled characteristics for a short channel $0.5 \times 300 \ \mu$m MMIC device. The solid line represents results from the physically based model of Equation (4.21). The dashed lines are calculations from the empirical model of Equation (4.19), including the modifications discussed in Section 4.6. Parameter choices for the models are presented in Tables 4.4 and 4.6, respectively. Points are measured values: (a) dc current-voltage characteristics; (b) RF transconductance; (c) RF output conductance.

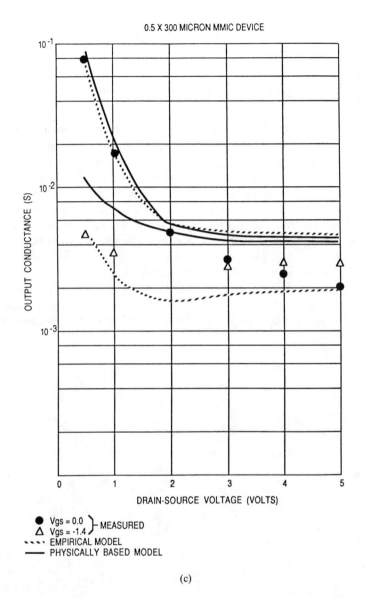

0.5 X 300 MICRON MMIC DEVICE

● Vgs = 0.0 ⎫ MEASURED
△ Vgs = -1.4 ⎭
▸▸▸▸ EMPIRICAL MODEL
—— PHYSICALLY BASED MODEL

(c)

**Fig. 4.21** cont.

Statz et al. [14] have combined these last two modifications to the hyperbolic tangent model to arrive at a modeling expression that appears quite different from that of equation (4.19). The resulting equations to describe device performance are given by

$$I_d = \frac{\beta(V_{gs} - V_{T0})^2}{1 + b(V_{gs} - V_{T0})}\left[1 - \left(1 - \frac{\alpha V_{ds}}{3}\right)^3\right](1 + \lambda V_{ds}) \text{ for } 0 < V_{ds} < 3/\alpha \tag{4.20}$$

$$I_d = \left[\frac{\beta(V_{gs} - V_{T0})^2}{1 + b(V_{gs} - V_{T0})}\right](1 + \lambda V_{ds}) \text{ for } V_{ds} > 3/\alpha$$

Because of the availability of the large-signal circuit simulation program SPICE, a different class of empirical models enjoys certain advantages over those described in the preceding paragraphs. These models utilize a basic device model, which is built into the circuit simulator (such as the JFET model in SPICE) in conjunction with other circuit elements available within the circuit simulator. Thus, the models may be implemented in standard simulators without revising the computer code. Figure 4.22 illustrates a circuit topology which is capable of adequately describing GaAs MESFET performance for many applications. In the figure, $J1$ represents a device model that is built into the simulator, such as a JFET model, but that may not adequately describe MESFET device performance. The voltage control voltage source, $E_{vgs}$, the diodes, $D1$ and $D2$, and the external resistors modify the characteristics of the built-in device model so that the predicted behavior more nearly approximates GaAs MESFET performance.

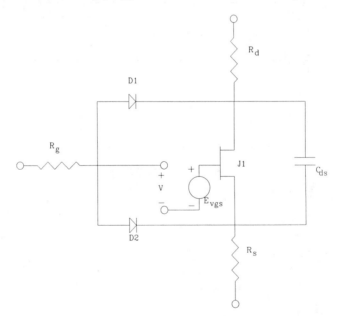

**Fig. 4.22** Large-signal empirical model for a GaAs MESFET using device model and circuit elements built into commercially available software.

The advantage of such a model is its ready availability without the need for specialized software or modification of the computer code. A large number of parameters must be determined, however, before the model can be used. In addition to all the parameters required by the built-in model, the model of Figure 4.22 also requires two sets of diode parameters and three resistor values.

Although almost any degree of accuracy can be obtained by continuing modifications to existing empirical models, such improvements still fail to address issues related to device process tolerances and simultaneous device and circuit optimization. These issues can be addressed only by using physically based models. Such models are almost always more complex than empirical models in appearance and computational effort required for evaluation. As engineering computer capabilities continue to increase, however, the advantages of physically based models rapidly begin to outweigh the computational disadvantages.

A modified version of the model of Lehovec and Zuleeg [36], is one example of a physically based model that has been implemented successfully as a large-signal microwave simulation tool [11,12]. The current-voltage description that defines this model is given by

$$I_d = 4k_t(V_{T0} - V_{bi})^2[(p^2 - s^2) - 2/3(p^3 - s^3)] \qquad (4.21)$$

with

$$k_t = \frac{1/3\, u_c(\varepsilon\, W/a L)}{1 + v(p^2 - s^2)}$$

$$v = \frac{u_c V_{T0}}{v_{sat} L}$$

$$s = \left(\frac{V_{bi} - V_{gs}}{V_{T0}}\right)^{1/2}$$

$$p = \left[\frac{V_{bi} - (v_{gs} - V_{ds})}{V_{T0}}\right]^{1/2}$$

In Equation (4.21), $L$ is the device gate length, $W$ is the device gate width, $a$ is the thickness of the active channel, $u_c$ is the low field mobility of the electrons in the active layer, $v_{sat}$ is the saturated velocity of the carriers, $\varepsilon$ is the dielectric constant of the material, and $V_{bi}$ is the built-in potential of the Schottky gate. As with the empirical hyperbolic tangent model, improvements can be made to physically based models by the incorporation of appropriate terms into the modeling equations. Such improvements have been made to the model defined by Equation (4.21) [12,37]. A parameter, $\lambda$, is used to more accurately describe output conductance behavior of the device; whereas a parameter, $\alpha$, is used to describe the behavior of the device in the subthreshold region of operation.

Table 4.5 presents the physically based parameters that describe the device whose characteristics are presented in Figures 4.15 through 4.18. The geometry and material parameters were allowed to vary slightly from the nominal fabrication values in order to minimize the error function defined by Equation (4.14). A comparison of measured and modeled device characteristics is given in Figure 4.20. The agreement of model predictions with measured performance is very good. The previous discussion regarding weighting functions with respect to agreement between measured and modeled characteristics also applies for this physically based model.

**Table 4.5** Physically Based Model after Lehovec and Zuleeg for the 0.8 × 400 μm Device Whose Equivalent Circuit Element Characteristics Are Presented in Figures 4.15 through 4.18

| Model Parameter | Physical Significance | Value |
|---|---|---|
| $L$ | Gate length | 0.794 μm |
| $W$ | Gate width | 400 μm |
| $V_{T0}$ | Threshold $V$ | −2.05 V |
| $a$ | Epi-thickness | 0.211 μm |
| $V_{bi}$ | Built-in potential | 0.85 V |
| $u_c$ | Mobility | 4008 cm²/Vs |
| $v_{sat}$ | Saturation velocity | $2.09 \times 10^5$ m/s |
| $\alpha$ | Subthreshold parameter | 0.603 |
| $\lambda$ | Conductance parameter | 0.824 |

Table 4.6 presents the physically based model parameters determined for the 0.5 × 300 μm MMIC device used to produce Figure 4.21. The error function minimization process and error term weighting was identical to those used in the previous cases. A comparison between measured performance and modeled characteristics of the short channel device obtained using the physical model also is shown in Figure 4.21. We see that the agreement is comparable or better than that observed using the empirical modified hyperbolic tangent model.

**Table 4.6** Physically Based Model after Lehovec and Zuleeg for a 0.5 × 300 $\mu$m MMIC Device

| Model Parameter | Physical Significance | Value |
|---|---|---|
| $L$ | Gate length | 0.540 $\mu$m |
| $W$ | Gate width | 300 $\mu$m |
| $V_{T0}$ | Threshold $V$ | −1.82 V |
| $a$ | Epi-thickness | 0.228 $\mu$m |
| $V_{bi}$ | Built-in potential | 0.85 V |
| $u_c$ | Mobility | 4290 cm$^2$/Vs |
| $v_{sat}$ | Saturation velocity | 4.33 × 10$^5$ m/s |
| $\alpha$ | Subthreshold parameter | 0.322 |
| $\lambda$ | Conductance parameter | 0.581 |

To complete the device models of Equations (4.18), (4.19), (4.20) or (4.21), it also is necessary to describe capacitance voltage characteristics of the device. An expression for gate-source and gate-drain capacitance can be developed from first-order $p$-$n$ junction theory applied to a two-terminal Schottky diode structure. The expression for gate-source capacitance that results is expressed as

$$C_{gs} = \frac{C_{gs0}}{(1 - V_{gs}/V_{bi})^m} \tag{4.22a}$$

where $C_{gs0}$ is the zero bias gate-source capacitance and $m$ is an empirical parameter equal to 0.5 for an abrupt junction diode. A similar expression often used for the gate-drain capacitance is

$$C_{gd} = \frac{C_{gd0}}{(1 - V_{gd}/V_{bi})^n} \tag{4.22b}$$

A major limitation to the capacitance model described by Equation (4.22) is that each device capacitance is assumed to be dependent on only one voltage. The measured gate-source capacitance presented in Figure 4.17, however, is clearly a function of not only gate-source voltage but also of drain-source voltage. Although less obvious, a similar comment is applicable to the gate-drain capacitance illustrated in Figure 4.18. The gate-drain capacitance does not depend solely on gate-drain voltage. The zero bias capacitances, $C_{gs0}$ and $C_{gd0}$, therefore are ambiguously defined. The effect of this problem is an inaccurate estimate of device capacitance for many applications. Table 4.7 presents a reasonable choice for capacitance parameters required by Equation (4.22) to describe the FET whose characteristics are shown in Figures 4.15 through 4.18. Figure 4.23 presents a comparison of measured device capacitance characteristics to model predictions. The failure of the model to adequately describe gate-source and gate-drain capacitance for low drain-source voltage levels is evident.

**Table 4.7** Parameters for the *p-n* Junction Capacitance Model of the 0.8 × 400 μm Device Whose Equivalent Circuit Element Characteristics Are Presented in Figures 4.15 through 4.18

| Model Parameter | Value |
|---|---|
| $C_{gs0}$ | 0.62 pF |
| $C_{gd0}$ | 0.24 pF |
| $m$ | 0.7 |
| $V_{bi}$ | 0.8 V |

Capacitance parameter choices for the 0.5 × 300 μm device whose current, transconductance, and output conductance characteristics are shown in Figure 4.20 are presented in Table 4.8. Comparison to measured values are presented in Figure 4.24. The limitations of this model are also seen in the 0.5 μm device.

**Table 4.8** Parameters for the *p-n* Junction Capacitance Model of the 0.5 × 300 μm Device

| Model Parameter | Value |
|---|---|
| $C_{gs0}$ | 0.60 pF |
| $C_{gd0}$ | 0.23 pF |
| $m$ | 0.7 |
| $V_{bi}$ | 0.85 V |

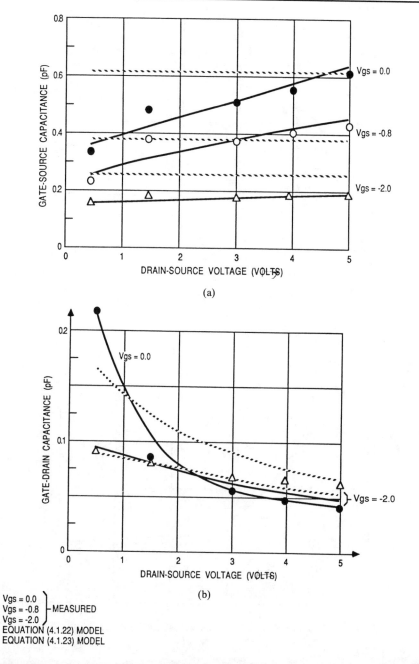

**Fig. 4.23** Measured and modeled capacitance characteristics for a $0.8 \times 400 \ \mu$m MMIC device. The solid line represents results from the semiempirical model of Equations (4.23) and (4.24). The dashed lines are calculations from the empirical model of equation (4.22). Parameter choices for the models are presented in Tables 4.7 and 4.9, respectively. Points are measured values: (a) gate-source capacitance characteristics; (b) gate-drain capacitance characteristics.

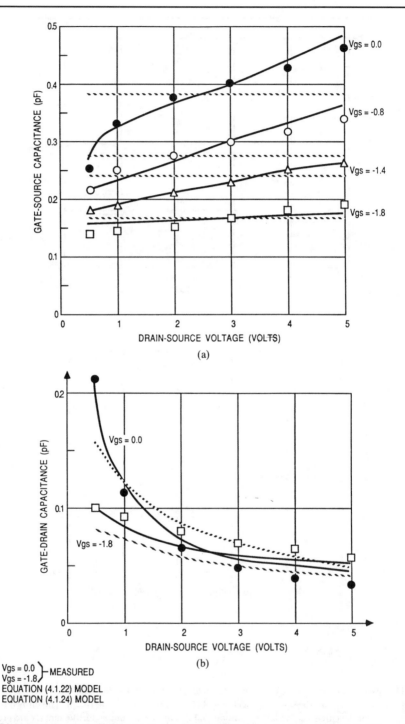

**Fig. 4.24** Measured and modeled capacitance characteristics for a short channel $0.5 \times 300 \ \mu$m MMIC device. The solid line represents results from the semiempirical model of equations (4.23) and (4.24). The dashed lines are calculations from the empirical model of equation (4.22). Parameter choices for the models are presented in Tables 4.10 and 4.8, respectively. Points are measured values: (a) gate-source capacitance characteristics; (b) gate-drain capacitance characteristics.

The model of Statz et al. [14] addresses the problem of dual voltage dependence of the device capacitances. Three empirical parameters are utilized in this model to provide significantly improved predictions over the *p-n* junction model described by Equation (4.22). A simple modification in capacitance equations, given by (4.22), can increase the capacitance model accuracy. Such model equations are

$$C_{gs} = C_{gs1} + \frac{C_{gs0}}{\left(1 - \frac{V_{gs}}{V_{bi}}\right)^m} + C_{gs2} \cdot V_{ds} \qquad \text{(for 4.22a)}$$

$$C_{gd} = C_{gd1} + \frac{C_{gd0}}{\left(1 - \frac{V_{gd}}{V_{bi}}\right)^n} \qquad \text{(for 4.22b)}$$

These equations can accurately model the capacitances in the saturation region of MESFET operation but fail to accurately represent the measured data in linear region when $V_{ds}$ is less than knee voltage. Another suggestion to improve the accuracy of the capacitance model equation is to use a tanh function of $V_{ds}$.

Currently available capacitance models that depend on purely physical data rarely provide sufficient accuracy for microwave applications. Semiempirical models, however, are capable of providing extreme accuracy while including many of the appropriate physical characteristics of the device. The expressions (4.23) and (4.24), for example, have been utilized to match device characteristics with extreme accuracy [11,12]. The gate-source capacitance is described well by

$$C_{gs} = W C_{GS0} + L_{GS1}(\varepsilon W L/a) \left(\frac{V_{gs} - V_{T0}}{V_{bi} - V_{T0}}\right)^m f_c(V_{ds}) \qquad (4.23)$$

where

$$f_c = 1 + L_{GS2}\left(\frac{V_{bi} - V_{T0}}{V_{gs} - V_{T0}}\right)^{1/m} (V_{ds} - V_{dsat}) \quad \text{for } V_{ds} > V_{dsat}$$

$$f_c = (L_{GS2}V_{dsat} + B_{gs} - 1)\left(\frac{V_{ds}}{V_{dsat}}\right)^2$$

$$+ (2 - L_{GS2}V_{dsat} - 2B_{gs})\frac{V_{ds}}{V_{dsat}} + B_{gs} \quad \text{for } V_{ds} < V_{dsat}$$

Evaluation of equation (4.23) requires that five empirical parameters be determined: $C_{gs0}$, $L_{gs1}$, $L_{gs2}$, $m$, and $B_{gs}$. The variables $W$, $L$, and $a$ correspond to the physical dimensions of the device as defined in Table 4.5.

Although the expression (4.23) appears complex as compared to equation (4.22a), it can be evaluated within a circuit simulation routine very quickly and efficiently. Some additional care must be taken when such expressions are used within a circuit simulation routine to ensure charge conservation is maintained. Once such concerns have been addressed, however, the expression offers physical scaling information and considerably more accuracy than can be achieved using Equation (4.22a).

A corresponding semiempirical expression for gate-drain capacitance is given by

$$C_{gd} = WC_{GDI} \left[ 1 - \left( \frac{L_{GD1}V_{gs} - V_{ds}}{L_{GD1}V_{bi}} \right)^r \right]^{-1} (1 - L_{GD2}V_{gs}) + W_{CGD0} \tag{4.24}$$

As with Equation (4.23), evaluation of Equation (4.24) requires determination of five empirical constants: $C_{GDI}$, $L_{GD1}$, $L_{GD2}$, $r$, and $C_{GD0}$.

Table 4.9 presents semiempirical parameter values required for evaluation of Equations (4.23) and (4.24), which apply for the device characterized by Figures 4.15 through 4.18. A comparison of modeled and measured capacitance characteristics for the device is shown in Figure 4.23. The figure also presents a comparison of the CV model described by Equations (4.23) and (4.24) with the model of Equation (4.22). The figure clearly illustrates the accuracy advantage of the more complex expressions.

**Table 4.9** Capacitance Model Parameters for the 0.8 × 400 $\mu$m Device Whose Equivalent Circuit Element Characteristics Are Presented in Figures 4.15 through 4.18 (the model is defined by Equations (4.23) and (4.24); physical parameters are identical to those given in Table 4.5)

| Model Parameter | Value |
|---|---|
| $C_{GS0}$ | $36.1 \times 10^{-11}$ F/m |
| $L_{GS1}$ | 2.27 |
| $L_{GS2}$ | 0.185 |
| $m$ | 1.33 |
| $B_{gs}$ | 0.542 |
| $C_{GD0}$ | $83.2 \times 10^{-12}$ F/m |
| $C_{GDI}$ | $13.8 \times 10^{-10}$ F/m |
| $L_{GD1}$ | 0.941 |
| $L_{GD2}$ | 0.591 |
| $r$ | 2.05 |

The capacitance characteristics of the $0.5 \times 300$ $\mu$m MMIC device also have been measured and parameters of the semiempirical capacitance model were extracted. Table 4.10 presents the capacitance parameters determined for the short channel device. A comparison between measured and modeled characteristics for both the semiempirical model and the simpler model of equation (4.22) is presented in Figures 4.24. When implementing a complicated model for capacitances $C_{gs}$ and $C_{gd}$ in a nonlinear circuit simulation program, we must be careful about charge conservation issues and discontinuities in the model equations, because these can result in nonconvergence problems in some programs, such as SPICE.

**Table 4.10** Capacitance Model Parameters for the $0.5 \times 300$ $\mu$m Device (the model is defined by Equations (4.23) and (4.24); physical parameters are identical to those given in Table 4.6)

| Model Parameter | Value |
| --- | --- |
| $C_{GS0}$ | $51.6 \times 10^{-11}$ F/m |
| $L_{GS1}$ | 3.08 |
| $L_{GS2}$ | 0.214 |
| $m$ | 1.31 |
| $B_{gs}$ | 0.150 |
| $C_{GD0}$ | $13.9 \times 10^{-11}$ F/m |
| $C_{GDI}$ | $15.7 \times 10^{-10}$ F/m |
| $L_{GD1}$ | 1.186 |
| $L_{GD2}$ | 1.340 |
| $r$ | 2.57 |

When a GaAs MESFET is used as an element in an RF switch or as a variable attenuator, it typically is operated at a dc drain-source bias level of zero volts. For a device whose gate electrode is symmetrically located between the drain and source contact, the gate-source and gate-drain capacitance values are expected to be nearly identical for this bias condition. In practice, some slight difference between the two values is expected because of nonsymmetries in the terminal pad and airbridge layouts. Such behavior was not considered in the parameter extraction process that led to the values presented in Tables 4.7–4.10. For circuit applications requiring low drain-source bias levels, these symmetry constraints should be considered in performing parameter extraction. Specialized symmetric device models also can be considered for these applications.

### 4.1.6.1 Low-Frequency Anomalies in GaAs MESFETs

As mentioned during our discussion of dc measurements, several of the GaAs MESFET's electrical characteristics shift significantly in value at low frequencies. Output resistance [13,18,19,20,48], transconductance [13,18,49,52], and device capacitances [52] have all been observed to shift. Output resistance has been reported to drop by as much as a factor of 5 to 10 as frequency increases above dc. The dispersion in this characteristic has the greatest effect on device performance. The characteristic frequencies at which these decreases occur vary from less than 100 Hz [20] to greater than 100 kHz [48]. Decreases in transconductance tend to be less significant (on the order of 5–10% of the dc value), but also occur at widely varying frequencies [18,49,52] and can have an important effect on device performance.

A wide variety of measurement techniques have been used by a number of researchers to observe the low frequency dispersion of GaAs MESFET characteristics. Pulsed current-voltage measurements can be used to distinguish dc from RF current-voltage behavior. More elaborate measurements involving significant amounts of equipment can produce similar data.

A wide range of physical phenomena have been identified as the cause for low frequency dispersion. Electron trapping at the channel-substrate interface has been indicated as the mechanism responsible for output resistance dispersion [19,48,51], whereas other work indicates that surface state occupation plays a major role in that behavior. Other explanations involve hole traps [49] or hole injection [50] from the channel-substrate interface and backgating phenomena [13].

A number of device models have described low frequency dispersion in MESFETs, one of the earliest modeling efforts being a simple resistor-capacitor network placed in series with the device's standard output resistance [51]. This approach successfully predicts both dc and microwave output resistance behavior of a device at one bias level. The model is also easy to incorporate into most circuit simulation routines. Major shortfalls of this simple model include its inability to predict the device characteristics in the transition region from high to low resistance, failure to predict the corresponding observed device transconductance dispersion, and inability to predict characteristics at more than one bias level. The model also requires an extremely large capacitance (on the order of microfarads) to obtain agreement with measured characteristics. Such a high capacitance value is clearly nonphysical for a microwave MESFET structure.

Larson [48] has suggested the use of a standard large-signal GaAs MESFET model in conjunction with an added resistor-capacitor network and two MOSFET devices. Such a model is easy to implement in large-signal circuit simulations and offers considerable versatility in matching device characteristics. The model accounts for some of the bias characteristics of low frequency dispersion, but the model is difficult to use because of its complexity and computational effort.

Another model uses an added resistor, capacitor, and dependent current source. This model successfully predicts the existence of frequency dispersion in both output resistance and transconductance. In addition, the model's required capacitance assumes a physically realistic value. The is easy to implement in circuit simulation and has many desirable characteristics for a model that describes low frequency behavior. A more detailed description of this model is therefore appropriate.

Low-frequency anomalies observed in GaAs MESFETs have been referred to as *drain lag* [38], *frequency dependence of the output resistance* [20], and *hysteresis*. Drain lag manifests itself as a long settling tail in a linear amplifier, as shown in Figure 4.25a, which is a GaAs MESFET inverter circuit biased for linear operation. When an input wave form of Figure 4.25b is applied to the input of this linear amplifier, the initial rise time of the output is short, but a second time constant on the order of milliseconds appears, produces the long settling tail referred to as *drain lag*.

The gain of the circuit shown in Figure 4.25a, found from the small-signal equivalent circuit, is given by

$$\text{gain} = gm_1(R_{DS1}\|R_{DS2}) \tag{4.25}$$

where $gm_1$ is the transconductance of $M1$ and $R_{DS1}$ and $R_{DS2}$ are the small-signal output resistances of $M1$ and $M2$, respectively. The measured gain of the amplifier as a function of frequency is shown in Figure 4.25c. The low-frequency rolloff in gain can be accounted for by the frequency dependent term in $R_{DS}$ of Figure 4.25a or Equation (4.25). Direct measurement of $R_{DS}$ of the MESFET at low frequencies, from 1 Hz to 100 MHz, show that $R_{DS}$ varies as a function of frequency and, thus, can account for the gain curve shown in Figure 4.25c.

Such a low-frequency anomaly can be accurately modeled by considering MESFET as a four terminal device [13]. Figure 4.26 shows the basic physical structure of the MESFET and indicates the rationale for including a fourth terminal. Even though the GaAs substrate used in manufacturing MMICs is a semiinsulating substrate, its conductivity and dielectric introduce parasitic resistance $R_{SP}$ and capacitance $C_{SP}$. Voltage changes at the drain can modulate the backgate depletion layer through the substrate, spreading resistance and capacitance time constant. Such parasitic capacitance and resistance produce prominent effects in device performance manufactured using Cr doped semiinsulating substrates. Intentional Cr doping introduces deep-level traps that can respond to low-frequency signals and, thus, large time constants. However, because of the existence of such traps close to the middle of the energy bandgap, they are incapable of responding to high frequency signals; hence, the $R_{DS}$ does not change after certain frequency. When an undoped starting substrate is used for fabrication, such change in $R_{DS}$ with frequency is less prominent.

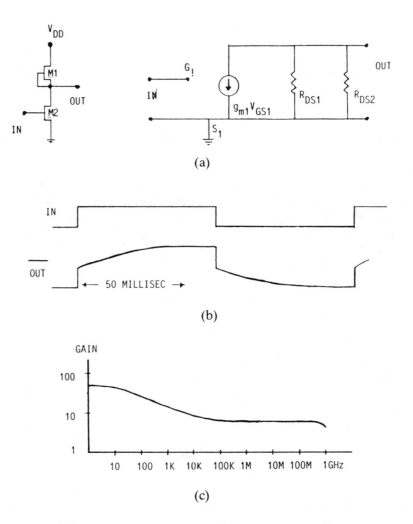

**Fig. 4.25** Low-frequency anomaly in a GaAs linear inverter: (a) inverter and small-signal equivalent circuit; (b) step response; (c) frequency response (© IEEE, reproduced by permission).

To simplify the MESFET model, such parasitic resistance and capacitance can be reduced to a lumped equivalent circuit, as shown in Figure 4.27. A nonlinear model can be developed to account for this effect. For the sake of simplicity, we start with a simple [39] drain current equation generally used for MESFET:

$$I_{DS} = \beta(V_{GS} - V_T)^2(1 + \lambda V_{DS}) \tanh \alpha V_{DS} \qquad (4.26)$$

The effect of the back gate can be introduced into this equation by considering it to have the same effect on drain current as the front gate but with much lower

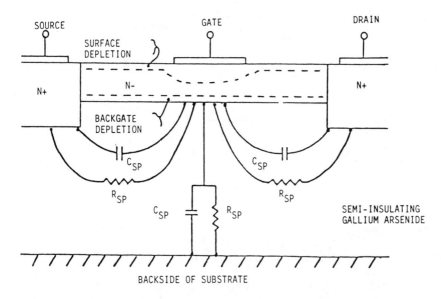

**Fig. 4.26** Model of a MESFET including the backgate parasitic elements (© IEEE, reproduced by permission).

magnitude, resulting in the following drain current equation:

$$I_{DS} = \beta(V_{GS} + \gamma V_{BS} - V_T)^2(1 + \lambda V_{DS}) \tanh \alpha V_{DS} \qquad (4.27)$$

where $V_{BS}$ = backgate voltage
$\gamma$ = backgate transconductance parameter

The small-signal equivalent circuit for such a four-terminal model is presented in Figure 4.27b.

The small-signal parameters, such as $g_m$, $g_{mb}$, and $R_{DS}$ can be evaluated from Equation (4.27) by differentiating $I_{DS}$:

$$g_{m^-} = \frac{dI_{DS}}{dV_{GS}} \approx 2\beta(V_{GS} + \gamma V_{BS} - V_T) \quad \text{for } V_{DS} > 2/\alpha \qquad (4.28)$$

$$g_{mb} = \frac{dI_{DS}}{dV_{BS}} \approx 2\beta\gamma(V_{GS} + \gamma V_{BS} - V_T) \quad \text{for } V_{DS} > 2/\alpha \qquad (4.29)$$

$$1/R_{DS} = \frac{dI_{DS}}{dV_{DS}} \approx \beta\gamma(V_{GS} + \gamma V_{BS} - V_T)^2 \quad \text{for } V_{DS} > 2/\alpha \qquad (4.30)$$

Finally, the input impedance of the FET can be evaluated from Figure 4.27b, represented as

$$Z_{DS}(j\omega) = \frac{R_{DS}}{1 + g_{mb}R_{DS}} \left\{ \frac{j\omega + 1/(R_{BS}C_{BD})}{j\omega + 1/[R_{BS}C_{BD}(1 + g_{mb}R_{DS})]} \right\} \tag{4.31}$$

In this derivation, it is assumed that $R_{BS} \gg R_{DS}$, which is practically true.

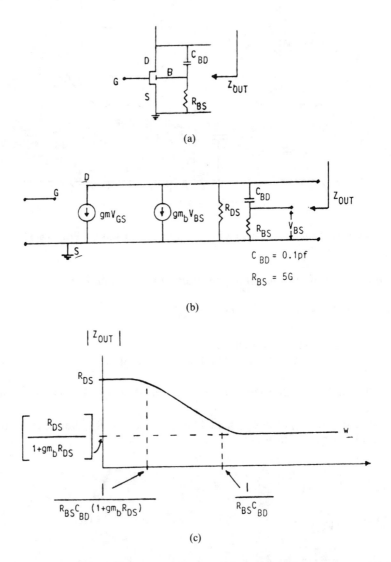

(a)

(b)

(c)

**Fig. 4.27** Four-terminal MESFET model: (a) MESFET model; (b) small-signed equivalent circuit; (c) $Z_{out}$ *versus* frequency, calculated from the circuit (© IEEE, reproduced by permission).

Equation (4.31) is plotted in Figure 4.27c and indicates the frequency cut-off points and their relationship with equivalent circuit element values when the output impedance changes with frequency. Such relationships can be used to estimate the values of $R_{BS}$ and $C_{BD}$. $R_{BS}$ commonly is set to a large value, so that $R_{BS} \gg R_{DS}$, generally several hundred megaohms and $C_{BD}$ is evaluated by setting the term $1/2\pi R_{BS}C_{BD}$ equal to the frequency at which the low-frequency $R_{DS}$ change ends.

It is important to include the fourth terminal as a part of the device. A simple RC circuit [40] without introducing the bulk terminal, as shown in Figure 4.27b, can be used to predict the $R_{DS}$ variation with frequency, but it has two major problems. First, when the MESFET is in the off state, the RC network still appears from drain-to-source and, thus, makes the device conduct, despite the off state. Second, in order to match $R_{DS}$ as a function of frequency, capacitance $C_{BD}$ has values on the order of microfarads, which is totally nonphysical. By introducing the fourth terminal, the device can be turned off completely in the off state; because of the backgate $gm$, a large resistance or practically open circuit is in parallel with $C_{BD}$ and $R_{BS}$. In the four-terminal model, $C_{BD}$ is very small, practically an open circuit, and thus has minimal conductance; hence the off state.

Figure 4.28(a) shows the curve tracer plot of I$_{DS}$ versus V$_{DS}$ of a GaAs MESFET and also the simulated MESFET using SPICE, with input V$_{DS}$ voltage varying, as shown in Figure 4.28(a), to simulate the curve tracer voltage at V$_{DS}$. Figure 4.28(b) is a plot of R$_{DS}$ as a function of frequency as calculated using the above model and as measured for a 0.5 $\mu$m gate length MESFET.

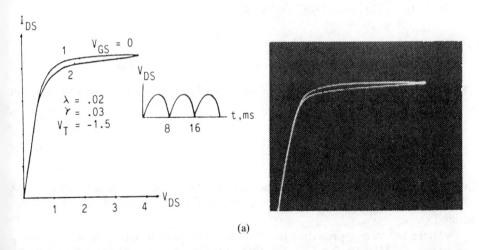

(a)

**Fig. 4.28** Simulated and measured data for a GaAs MESFET: (a) simulated curve tracer measurement using the transient analysis option and the four-terminal MESFET model in SPICE, and an actual curve tracer measurement; (b) a comparison of $R_{DS}$ versus frequency for measured and simulated data. (© IEEE, reproduced by permission).

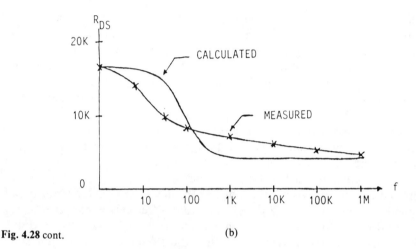

**Fig. 4.28** cont.                                    (b)

The model parameters in Equation (4.26) can be estimated easily. The output impedance $R_{dc}$ at low frequencies is given by Equation (4.31) at $j\omega = 0$:

$$R_{dc} = R_{DS}$$

At high frequencies, beyond low-frequency anomalies, the output impedance is calculated from Equation (4.31), when $j\omega \to \infty$:

$$R_{ac} = \frac{1}{g_{mb}} \parallel R_{DS}$$

The backgate transconductance parameter $\gamma$ can be evaluated by measuring drain impedance at dc and high frequency:

$$\gamma = 2\beta \frac{1}{(V_{GS} - V_T)} \frac{R_{dc} - R_{ac}}{R_{ac}R_{dc}} \tag{4.32}$$

Such a concept of introducing the backgate model in a drain current equation for MESFET can be extended to more sophisticated and detailed models, as described earlier.

There are two approaches to introducing the fourth terminal in an existing three-terminal MESFET model. First, the SPICE source code can be changed to introduce the four-terminal model, which means major modifications in the program. Second, the fourth terminal can be introduced using the subcircuit approach in SPICE. To illustrate, the second approach, Equation (4.27), can be written as

$$I_{DS} = A[(V_{GS} - V_T)^2 + (\gamma V_{BS})^2 + 2(V_{GS} - VT)(\gamma V_{BS})]$$

where

$$A = \beta(1 + \lambda V_{DS}) \tanh \alpha V_{DS}$$

Using nonlinear voltage controlled current source components in SPICE, we can use the basic three-terminal MESFET model in SPICE and generate the backgate voltage dependent terms using a subcircuit approach. These techniques are described in detail in Appendix A8. However, such an approach is difficult to implement for very complicated drain current equations.

## 4.1.7 Design Considerations and Applications

In many MMIC applications, it is desirable to design not only the passive matching circuitry but also the active device. The freedom to perform this simultaneous device-circuit optimization previously was not practical for the microwave engineer using hybrid technology. Device design, however, can provide a powerful tool to the microwave engineer, leading to significantly improved circuit performance. The physical characteristics of the optimum MESFET will be determined by the circuit function, the signal power levels, bias levels, and the frequency range of interest. Simultaneous device and circuit design can be useful in a wide range of applications. Gate width scaling, for example, has been shown to offer significant advantages in the design of distributed amplifiers [31]. In this application, scaling the gate periphery is an effective way to compensate for series capacitance effects. Scaling of other physical device characteristics for distributed amplifier design has not been studied in detail.

Active devices in MMIC circuitry are beginning to find more nonstandard applications. For example, circuit biasing often is provided through a current source device, as shown in Figure 4.29. In this application the optimum device to be used as the current source may not share either gate length or gate width dimensions with the RF device. Other geometric and material properties of the optimum current source device may also differ from that of the RF device. Another example of an FET utilized in a nonstandard manner is as an active inductor [31]. Figure 4.30 illustrates a useful configuration for such a circuit element. The optimum device for this unique application may bear no resemblance to the well-designed FET structures that are used in amplifier applications.

In addition to the device's intended application, the power and frequency range of interest also affect its design, even when MESFET devices are to be used as amplifier elements; for example, the optimum device for a low-noise, high-frequency amplifier will be quite different than the optimum power device. In general a low-noise, high-frequency device will have a thinner active channel

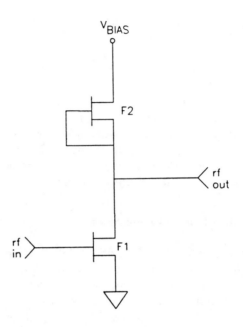

**Fig. 4.29** A common-source FET amplifier topology. Device F1 is the amplifying device, while device F2 is used as a current source. Input and output matching circuitry are not shown.

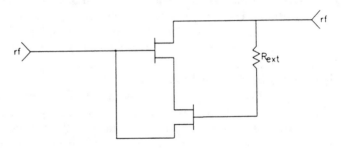

**Fig. 4.30** A circuit topology used as an active inductor.

dimension with higher doping density than a power device. Likewise, the gate width of power devices tend to be greater than those of low-noise devices.

Layout considerations also are important in designing MESFETs for various applications. When laying out a multifinger FET to achieve larger gate widths, airbridge crossover connections are commonly used to connect source, drain, and gate fingers. Such connections can be achieved in various ways, as shown in Figure 4.31. In Figure 4.31a, all gate fingers are connected with airbridge or first-level metal interconnect. The sources and drains are connected by airbridge crossing over gate fingers and drains sources. Such a layout configuration introduces the crossover capacitances and increases $C_{ds}$ significantly. The quantitative increase in $C_{ds}$ depends on the airbridge width. Increase in $C_{gs}$ and $C_{gd}$ is minimal because of small gate lengths and, thus, minimal crossover areas between airbridge and gate metal. An alternative method to connect drains using either airbridge or first-level interconnect metal is shown in Figure 4.31b. In this case, the crossover to connect drains are avoided, which helps reduce $C_{ds}$. If $C_{ds}$ is a critical parameter in the particular application, this configuration definitely is preferable to the preceding one. In Figure 4.31c, the only crossover is between the line connecting the drains and that for sources. This configuration results in a minimal $C_{gs}$ and $C_{gd}$ and very little parasitic $C_{ds}$ because of the crossover between drain and source lines. The next approach is shown in Figure 4.31d. Here, the gate fingers are connected by the first-level interconnect metal, drains are connected by either an airbridge or the first-level metal, and the sources are connected by an airbridge crossing over the gate finger connections. In this configuration $C_{ds}$ is minimized; however, $C_{gs}$ is increased slightly because of the crossover of source airbridge connections over the gate finger connection. The increase in the value of $C_{gs}$ will depend on the crossover area of gate and source connection line widths. The fifth configuration is shown in Figure 4.31e, which is similar to Figure 4.31d except that the drain connection airbridge crosses over the gate connection, thus minimizing $C_{gs}$ and increasing $C_{gd}$ slightly. It is difficult to calculate exact parasitic capacitances in any of these five configurations. Table 4.11 shows a qualitative comparison of the parasitic capacitances due to crossover connections in multifinger FETs. Depending on the configuration of the FET layout for which detailed characterization data base is available (say, for a single 100 $\mu$m finger FET), we can choose one of the four configurations to achieve a multifinger FET, minimizing one or another parasitic capacitance due to crossover connections ($C_{gs}$, $C_{gd}$, or $C_{ds}$) and have accurate estimates of the other by simple linear scaling with the width of the device. This will be particularly practical in circuit designs using a distributed elements approach, as in the distributed amplifier, which requires precise estimates of one or another ($C_{gs}$, $C_{gd}$, $C_{ds}$) capacitance of the MESFET.

One important figure of merit for microwave devices is the maximum stable gain, $G_{max}$. This figure represents the maximum power gain a device can produce at a given frequency under stable operating conditions. The maximum stable gain of a device can be computed directly from the device's S-parameters and typically is an optional output parameter of commercially available circuit simulation packages. One possible goal for device scaling is to maximize this quantity.

The choice of all dimension and material parameters also is affected by the frequency range of interest. The parameters listed in Tables 4.5 and 4.6, for example, are typical of those for a medium-power microwave device, with frequencies up to about 10 and 18 GHz, respectively. The doping density in the active channel for these devices is approximately $10^{17}$ cm$^{-3}$. To scale the device

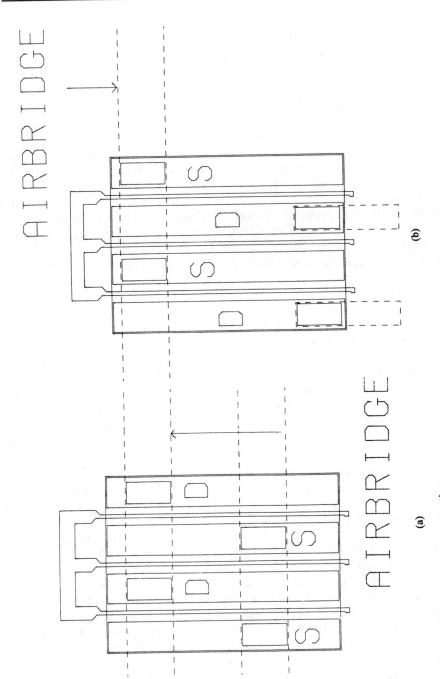

(b)

(a)

**Fig. 4.31** Airbridge crossover connections.

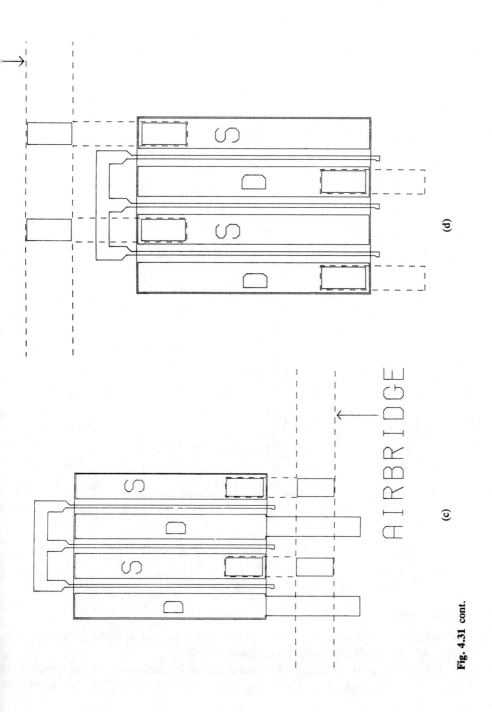

AIRBRIDGE

(c)

(d)

**Fig. 4.31** cont.

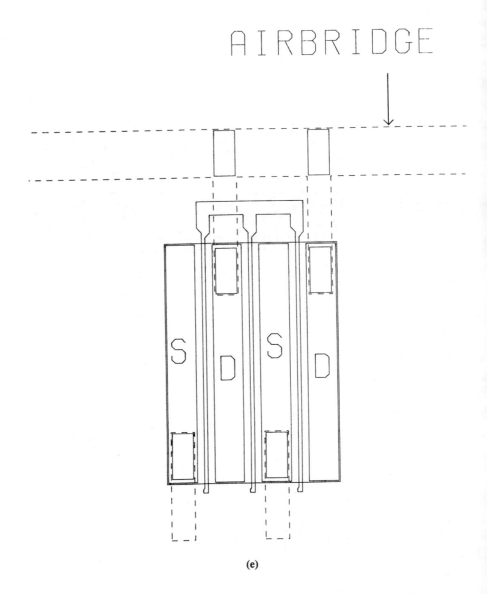

(e)

**Fig. 4.31** cont.

for higher frequency applications, gate length, gate width, and epi-thickness are reduced, and doping density is increased.

Another figure of merit and indicator of device frequency performance is the gain-bandwidth product, $f_T$. This represents the frequency at which the magnitude

**Table 4.11** Parasitic Capacitances of Configurations in Figure 4.31

| | Configuration | | | | |
|---|---|---|---|---|---|
| Parasitic Capacitance | (a) | (b) | (c) | (d) | (e) |
| $C_{gs}$ | Moderate | Moderate | Minimum | Moderate | Minimum |
| $C_{gd}$ | Moderate | Minimum | Minimum | Minimum | Moderate |
| $C_{ds}$ | Maximum | Moderate | Moderate | Minimum | Minimum |

of the short-circuit current gain of the device drops to unity. As a first-order estimate, the $f_T$ of an FET can be approximated as

$$f_T = \frac{g_m}{2\pi(C_{gs} + C_{gd})}$$

The primary physical parameter of the FET affecting this figure of merit is the device gate length. To estimate, the gate-source capacitance, which dominates the denominator of the expression for $f_T$, decreases linearly with decreased gate length. The transconductance of the device will increase approximately linearly as the gate-length dimension is scaled down.

Although it is difficult to prescribe precise rules for scaling down GaAs MESFETs, some guidelines relating various physical parameters of the FET to each other can be determined from an analysis of previously fabricated high-frequency amplifier-oscillator devices [8]. The ratio of gate length to epi-thickness, for example, remains nearly constant at a value of 3 to 5 regardless of other dimensional scaling:

$$3 < L/a < 5 \tag{4.33}$$

The relationship between doping density and gate length is well approximated by the following expression:

$$N_d = 1.5 \times 10^{17} \, L^{-1.43} \tag{4.34}$$

The obtainable gate length dimensions for MESFET devices continue to decrease as each year passes. The minimum gate length that can be utilized by the microwave engineer depends on the processing limits of the fabrication facility to be used. Ultimate limits, however, are rapidly being approached [8].

The channel doping density and epi-thickness are related to device pinch-off voltage by the following expression:

$$V_{bi} - V_{T0} = \frac{qN_d a^2}{2\varepsilon} \tag{4.35}$$

The actual threshold voltage of the device can vary over a range of 1 to 2 V for most low-noise devices or to approximately 5 to 10 V for power devices.

The guidelines defined by expressions (4.33) through (4.35) provide good first-order estimates for microwave device design but have definite limitations. The device rules specified apply only to devices that might be utilized in amplifier, oscillator, or mixer circuitry. As mentioned previously in this section, however, little work has been done to optimize such devices for other applications, such as RF switches, voltage controlled attenuators, current sources, and active inductors. Futhermore, MMIC devices are fabricated primarily through ion implantation, a process that does not create a constant doping density in the active channel. Instead, dopant ions are distributed in the material, as illustrated by the solid line in Figure 4.32. The doping density and epi-thickness values therefore are defined somewhat ambiguously. Although most device models do not take this information into account, some successful modeling has utilized this carrier density *versus* depth information [16,41].

Second-order phenomena such as short channel effects, which are not accounted for in expressions (4.33) through (4.35), can have a significant effect on actual device geometries and doping. For example, Equation (4.35) predicts device threshold voltage to be independent of gate length. Figure 4.2c, however, clearly illustrates that, for short-channel devices, gate length can affect the measured threshold voltage. A corresponding shift in transconductance also takes place as gate length is reduced. Hauser [42] identified one cause of this observed shift to be small gate length to epi-thickness ratios. A model has been developed which partially explains the measured data.

For many microwave and millimeter wave applications, the driving force in the design is circuit efficiency rather than improved frequency or power performance. For such applications, a reasonable figure of merit is the maximum stable gain efficiency, defined as

$$E_G = \frac{G_{max}}{I_d}$$

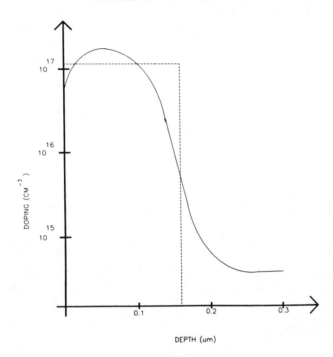

CHANNEL DOPING vs DEPTH

DEPTH (um)

**Fig. 4.32** Doping density profile of a typical ion-implanted channel. Solid line represents actual doping density and dashed line represents a device modeling approximation.

Table 4.12 presents the computed $f_T$, $G_{max}$, and $E_G$ for both the 0.8 × 400 $\mu$m device and the 0.5 × 300 $\mu$m device that have been used to produce the data presented in this section. The maximum stable gain and gain efficiency are computed for a frequency of 12 GHz. By 18 GHz, the 0.8 $\mu$m gate length device cannot produce power gain, and the $G_{max}$ value of the 0.5 $\mu$m device is 8.9 dB. Note that both devices are capable of providing power gain beyond the $f_T$ frequency. The gain-bandwidth product serves only as a basis of comparison between devices and indicates that the device with the higher $f_T$ value will tend to perform at higher frequencies. Gain efficiency is seen to be much greater for the 0.5 $\mu$m device than for the 0.8 $\mu$m device. This trend, however, is not generally observed and devices of similar geometry can exhibit drastically different gain efficiencies.

Clearly, device design and simultaneous device and circuit optimization are easier to perform when access to a physically based model is available. First-order device optimization utilizing such a model can be performed by altering the physical device parameters of the model until optimum performance predictions are

**Table 4.12** A Comparison of Figures-of-Merit for the 0.8 × 400 $\mu$m and the 0.5 × 300 $\mu$m FETs Whose Characteristics are Presented in Chapter 4 (the maximum stable gain and gain efficiency are computed at a frequency of 12 GHz; the figures apply to the two devices when biased at $\frac{1}{2} I_{dss}$)

| Device | $f_T$ (GHz) | $G_{max}$ at 12 GHz (db) | $E_G$ at 12 GHz (dB/mA) |
|--------|-------------|--------------------------|-------------------------|
| 0.5 × 300 | 16.4 | 10.5 | 0.42 |
| 0.8 × 400 | 9.1 | 5.2 | 0.17 |

achieved. Scaling the device gate width is possible even when only empirical models are available through procedures similar to those described in Section 4.5.

### 4.1.8 Noise Modeling

Modeling noise mechanisms in solid-state devices is more complex than modeling other electrical behavior. Although empirical models have been developed [43], the noise figure measurements required for these models are extremely tedious, even when automated. Physically based noise models [44,45], although accurate, require knowledge of a large number of theoretical parameters, which generally are not known or easily determined.

Podel developed an analysis approach that limits the number of measurements to obtain noise figure predictions [46]. One noise measurement and a small-signal equivalent circuit model for the FET are sufficient for the complete noise analysis. More recently, Gupta et al. [47] described a model that further simplifies the measurement requirements while improving the noise figure predictions. This model also eliminates assumptions required of the earlier work related to circuit losses, noise correlations, and the functional form of the output noise current.

A key factor influencing the noise figure of an amplifier circuit is the generator admittance, $Y_g = G_g + jB_g$, connected at the input port of the amplifier. The effect this has on noise figure is given by

$$F(Y_g) = F_{min} + \frac{R_n}{G_n}[(G_g - G_{gop})^2 + (B_{gop})^2] \tag{4.36}$$

where $F_{min}$ = the minimum value of $F$ with respect to $Y_g$

$Y_{gop} = G_{gop} + jB_{gop}$ = the admittance value at which $F = F_{min}$

$R_n$ = a quantity that measures the sensitivity of $F$ to $Y_g$

Figure 4.33 illustrates the equivalent circuit models used in the most general form of this analysis. The FET device is represented by a five-element equivalent circuit model using gate-source capacitance, $C_{gs}$; input resistance, $R_T = R_g + R_s + R_i$; transconductance, $g_m$; output resistance, $R_o$; and a white noise current source of spectral density $S_{i0}$. The first four values may be determined using S-parameter measurements and the small-signal analysis described in Section 4.5. Evaluation of the spectral density of the white noise current source requires one noise measurement. This measurement may be made directly using a low noise receiver on the output of the device with the gate-source terminals of the device short circuited. The receiver measures the noise power, $P_{out}$, delivered by the device to the input resistance, $R_{in}$, of the narrow-band receiver in an effective noise bandwidth, $B$. The power spectral density of the device then is calculated from the following relationship:

$$S_{i0}(f_L) = \frac{P_{out}}{B} \frac{(R_{in} + R_o)^2}{R_o^2 R_{in}} \tag{4.37}$$

Because the power spectral density, $S_{i0}$, is frequency independent in the microwave frequency range, measurement needs to be carried out at only one frequency, $f_L$. An alternate approach to determining the value of $S_{i0}$ is to compute this value from a single measurement of minimum noise figure at some microwave frequency.

The analysis is simplified by assuming that the operating frequency, $f_0$, is significantly below the gain-bandwidth product of the device:

$$\omega_0^2 C_{gs}^2 R_T^2 \ll 1$$

and that the gate leakage current is small.

Using the equivalent circuit element values determined for the device, the input conductance of the device can be computed as

$$G_{in} = \frac{\omega^2 C_{gs}^2 R_T}{1 + \omega^2 C_{gs}^2 R_T^2} \tag{4.38}$$

Likewise, the optimum input circuit susceptance is given as $\tag{4.39}$

$$B_{opt} = -\omega C_{gs}$$

*280*

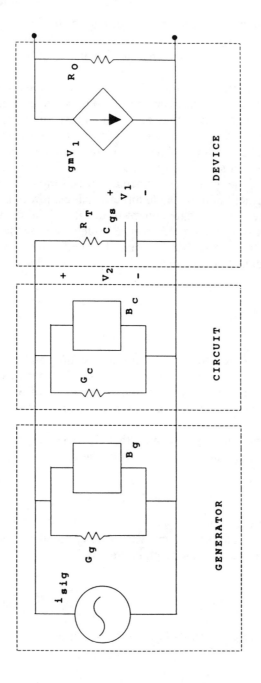

(a)

**Fig. 4.33** Equivalent circuits used for noise and analysis: (a) idealized linear amplifier circuit; (b) noise equivalent circuit of the amplifier; (c) simplified noise equivalent circuit.

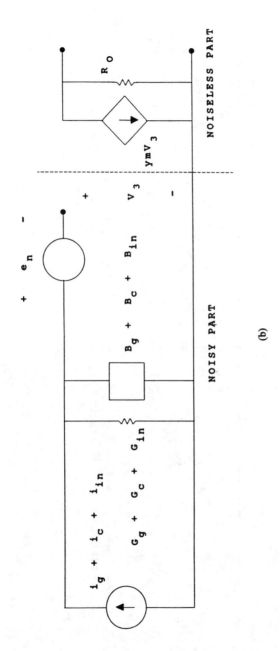

**Fig. 4.33 cont.**

282

cont.

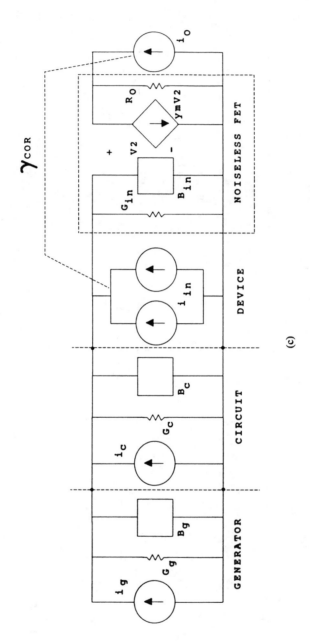

Fig. 4.33 cont.

Using both the equivalent circuit element values and the power spectral density computed in Equation (4.37), the quantity $R_n$ can be evaluated from

$$R_n = S_{i0}/(4kT)\frac{1 + \omega^2 C_{gs}^2 R_T^2}{g_m^2} \tag{4.40}$$

The optimum input circuit conductance is then given by

$$G_{opt} = \left(G_{in}^2 + \frac{G_{in}}{R_n}\right)^{1/2} \tag{4.41}$$

Equations (4.38) and (4.40) now can be used to compute the minimum noise figure for the measurement conditions as

$$F_{min} = 1 + 2R_n G_{in} + 2(R_n G_{in} + R_n^2 G_{in}^2)^{1/2} \tag{4.42}$$

provided that the input circuit conductance is much smaller than the input conductance of the device.

The noise figure for a given circuit is easily computed from Equation (4.36) and using Equations (4.39) through (4.42). Alternatively, if $F_{min}$ is measured at one frequency, these equations can be used to solve for the spectral density. Once the value of $S_{i0}$ and the other equivalent circuit elements are known, the minimum noise figure and optimum match can be predicted at any other frequency.

Table 4.13 presents appropriate noise parameters evaluated for three $0.5 \times 300$ $\mu$m gate FETs. The devices were biased with $I_d = 5$ mA which is 10% $I_{dss}$, and the calculations are made for a frequency $f_o = 10$ GHz. The computed $F_{min}$ values of 1.60 to 2.30 dB can be compared to a measured $F_{min}$ value of a similar device of 1.9 dB.

The noise model available in the generic SPICE program is for a silicon JFET. The equivalent noise circuit model is shown in Figure 4.34. The thermal noise generated in the drain and source regions of the FET is modeled by the two noise sources, $i_{nr_d}$ and $i_{nr_s}$. The values of the noise sources are given by the following equations:

$$i_{nr} = \sqrt{\frac{4kT}{R}}$$

where, the values of $R$ in $i_{nr_d}$ and $i_{nr_s}$ are $r_d$ and $r_s$, respectively. The shot and flicker noise are modeled by the noise current source $i$ in the drain of the FET and defined by the following equation:

$$i_{nD} = \sqrt{4kTAg_m + \frac{K_f I_D^{a_f}}{f}}$$

**Table 4.13** Calculated Noise Parameters for Three Similar 300 $\mu$m Wide FETs (the devices were biased with a drain current of 5 mA and the calculations are made for a frequency of 10 GHz)

| Noise Model Parameter | FET 1 | FET 2 | FET 3 |
|---|---|---|---|
| $g_m$ (mS) | 15 | 15 | 19 |
| $C_{gs}$ (pF) | .17 | .20 | .20 |
| $R_T$ ($\Omega$) | 5 | 10 | 10 |
| $S_{i0}$ (A²/Hz) | $3 \times 10^{-22}$ | $3 \times 10^{-22}$ | $3 \times 10^{-22}$ |
| $R_n$ ($\Omega$) | 81 | 81 | 50 |
| $G_{in}$ (S) | $5.7 \times 10^{-4}$ | $1.55 \times 10^{-3}$ | $1.55 \times 10^{-3}$ |
| $G_{opt}$ (50 S) | .13 | .23 | .29 |
| $B_{opt}$ (50 S) | $-.53$ | $-.62$ | $-.62$ |
| $F_{min}$ (dB) | 1.60 | 2.30 | 1.93 |

where $g_m$ is the small-signal transconductance of the device, $A$ is a constant whose value is $\frac{2}{3}$ for silicon devices. In GaAs MESFETs the value of $A$ is determined by curve fitting the measured and simulated noise of the device. Generally, the value of $A$ is found to be slightly higher than $\frac{2}{3}$ for GaAs devices. The parameters $K_f$ and $a_f$ determine the flicker noise characteristics of the FETs. Typical values of these parameters for silicon JFETs is $K_f = 10^{-14}$ and $a_f = 1$. Again, these model parameters are evaluated for GaAs MESFETs based on the measured noise figure as a function of frequency. An equivalent circuit as described here is reasonable for GaAs MESFETs down to 0.5 $\mu$m gate length after evaluating the appropriate values of the model parameters. Noise figures of different circuits as estimated using this model are found to be close to the measured values. However, as the noise figure of the circuit approaches 0.5–1.0 dB, there could be a discrepancy of 0.1–0.2 dB in the noise figure as simulated using this model *versus* the measured noise figure.

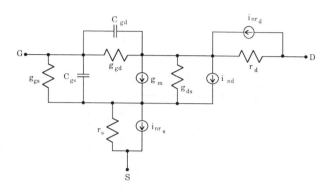

**Fig. 4.34** Noise model for a FET.

## 4.2 DUAL-GATE FET

### 4.2.1 Introduction

Dual-gate FETs were manufactured in the early 1970s [1] as soon as the processing technology for single-gate FETs was sufficiently under control to permit safe deposition of the two-submicron-wide parallel gate strips.

In 1975 Liechti [2] described the main characteristics of this device and its application as gain-controlled amplifier and high-speed modulator. Since then, there has been continuous effort to model, understand, and optimize the device [3–8].

However, the intrinsic behavior of the dual-gate FET could not be satisfactorily understood until the dc bidimensional transfer characteristic of the device was published in 1981 [9]. The reason is that knowledge of the external bias of the dual gate FET (Figure 4.35) does not easily reveal the exact internal bias conditions for each single-gate FET part. The actual potential at node $D1$, which determines the dc operation conditions of both single-gate FETs of the cascode is not directly accessible and can only be controlled via the overall bias voltage $V_{DS}$ and both gate voltages $V_{G1S}$ and $V_{G2S}$ simultaneously, according to a very complicated and not explicitly known relationship.

A clear synoptical solution to this problem is the graphical presentation of the transfer characteristic [9]. This method produces $I_D(V_{G1S}, V_{G2S})$ for every value of $V_{DS}$ and permits us to "see" into the dual-gate FET, to identify internal bias conditions, to model it properly and to interpret correctly the different modes of linear and nonlinear operation of both single-gate FETs of the cascode. The

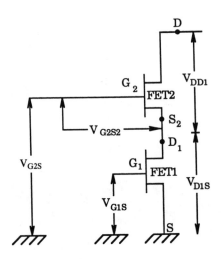

**Fig. 4.35** Voltage distribution within cascode.

method is also useful to explain the operation of circuits using the dual-gate FETs and to anticipate new applications [10].

### 4.2.2 dc Characterization and Basic Device Operation

For dc operation, the dual-gate FET can be considered as a cascode of two single-gate FETs (Figure 4.35). Both FET parts can be driven into saturation using the gate control voltages independently. This interdependence can be illustrated by the dc transfer characteristic (Figure 4.36). The following relations are used for its derivation:

$$V_{DS} = V_{D1S} + V_{DD1} \tag{4.43}$$

$$I_{D1} = I_{D2} = I_D \tag{4.44}$$

$$V_{G2D1} = V_{G2S} - V_{D1S} \tag{4.45}$$

The transfer characteristics can be constructed graphically if the dc output characteristics of the individual single-gate FETs composing the cascode are known. They have to be inversely superimposed as in Figure 4.36 because of (4.43) and (4.44).

It is important to realize that the saturation currents of the two cascode FET need not be equal. This is the case if the two gates have different lengths or the channel resistance between the gates ($R_{12}$) is significantly different from the source

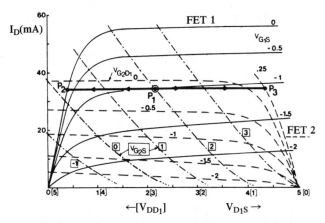

**Fig. 4.36** DC bidimensional transfer characteristic of GaAs dual-gate MESFET. Gate 1: 0.8 $\mu$m; gate 2: 2 $\mu$m; gate width: 200 $\mu$m.

(From C. Tsironis and R. Meierer, "Microwave Wide-Band Model of GaAs Dual Gate MESFETs," *IEEE Trans. Microwave Theory and Techniques,* Vol. MTT-30, No. 3, March 1982. © 1982 IEEE. Reprinted with permission.)

resistance ($R_s$ in Figure 4.38). The control characteristics $I_D(V_{G2S})$ can be derived from (4.45). The diagram of Figure 4.36 contains all information about the internal bias conditions of each individual FET part. For example, setting $V_{G1S} = -1$ V and $V_{G2S} = +2$ V would correspond to point $P_1$ of Figure 4.36, which means a dc current of 34 mA, an internal gate 2 control voltage of $V_{G2D1} = -0.16$ V and channel voltages of $V_{D1S} = 2.1$ V and $V_{DD1} = 2.9$ V over FETs 1 and 2 respectively.

For the graphical construction of the nomogram of Figure 4.36, the dc output characteristics of the partial FETs can be measured directly. A simple electronic circuit consisting of a potentiometer and an operational amplifier (Figure 4.37) permits individual measurement of each internal FET of the dual-gate FET, plotting its characteristics already superimposed, and only draw the $V_{G2S} = $ constant lines using (4.45) [9].

Once the transfer characteristic of the dual-gate FET is known, all possible linear and nonlinear operating conditions of the device can be identified, provided that they do not modify the total drain-to-source bias.

Examples of different basic operations of the dual-gate FET and the approximate location of the corresponding bias point on Figure 4.36 are given in Table 4.14.

Detailed understanding of the device's operation for development and optimization of particular applications obviously requires these data to be available to identify the correct signal processing mechanisms at the various physical locations within the dual-gate FET.

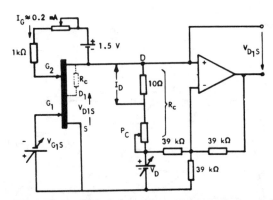

**Fig. 4.37** Measuring setup for dc output characteristics of intrinsic single-gate FET parts of dual-gate FET. The potentiometer $P_t$ is adjusted to make $R_c' = R_c$ where $R_c$ is the channel resistance of the nonsaturated dual-gate FET part. The internal voltage $V_{D1S}$ appears at the output of the operational amplifier and the FET current $I_D$ can be measured across a fixed resistance of 10 $\Omega$.

(From C. Tsironis and R. Meierer, "Microwave Wide-Band Model of GaAs Dual Gate MESFETs," *IEEE Trans. Microwave Theory and Techniques*, Vol. MTT-30, No. 3, March 1982. © 1982 IEEE. Reprinted with permission.)

**Table 4.14** Location and Bias Values Refer to Figure 4.36.
Examples of Basic Operation Modes of Dual-Gate FETs

| Basic Operation | Location | Typical Bias | FET 1 | FET 2 | Ref. |
|---|---|---|---|---|---|
| Linear and Low Noise Amplifier | LNA | $-1V < V_{G2S} < 2V$ $-2V < V_{G1S} < -1V$ | linear | linear | [2] |
| Gain-Controlled Amplifier | GCA | $-2V < V_{G2S} < 0.5V$ $-2.5V < V_{G1S} < -1V$ | linear and nonlinear | linear | [2] |
| Mixer | MIX | $V_{G2S} < -1V$ | nonlinear | nonlinear | [10,11] |
| Self-Oscillating Mixer | SOM | $-0.5V < V_{G2S} < 1V$ $-1V < V_{G1S}$ | nonlinear | linear | [10] |
| Frequency Doubler and Multiplier | FDB | $V_{G2S} < 3V$ $-2.5V < V_{G1S}$ | linear and nonlinear | nonlinear | [12] |
| Image Rejection Mixer[a] | IRM | $V_{G2S} < 3.5V$ $-1.5V < V_{G1S}$ | linear | nonlinear | [24,26] |
| Oscillator | OSC | $-0.5V < V_{G2S} < 2V$ $-1V < V_{G1S} < 0V$ | nonlinear | linear | [13,14] |

[a] An accessible ohmic contact between the two gates can be required for this application.

### 4.2.3 High Frequency Lumped Element Equivalent Circuit

Lumped element equivalent circuits help us to understand the signal processing effects in microwave devices, and thus to optimize both the devices themselves and the associated circuitry for particular applications.

Though accurate lumped element equivalent circuits for single-gate FETs existed since at least 1971 [15], the first attempts to create such models for dual-gate FETs [3–5] were not accurate enough because they were based on the assumption that the two FET parts of the dual-gate FET were equal. This may be true geometrically, if the gate lengths and active layer thickness underneath the two gates are equal, but electrically this is in most cases wrong, as we have seen in Section 4.2.2. The two FET parts are in general biased differently, and therefore the values of the lumped equivalent circuit elements are different.

Assuming that certain knowledge about the internal bias state of the dual-gate FET is available, two approaches can be followed to obtain the values of the lumped elements of an equivalent circuit.

The first lumped equivalent circuit model [6] is shown in Figure 4.39. This is a comprehensive model including all intrinsic lumped elements for the device as well as the parasitic elements which are secondary in nature. Extraction of the element values for such a comprehensive model from measured S-parameters requires a general optimization and S-parameter fitting program. Such a model for dual-gate FETs is valid up to the frequencies where a similar single-gate FET model is valid (i.e., up to approximately 23 GHz). However, as no assumption on the size of the element values is made, simplified analytical expressions for calculating the element values are not possible.

The second approach, described in [8], is based on a simplified model of the dual-gate FET [16] for which closed form expressions can also be used [17] to calculate the bias conditions of the single-gate FETs shown in Figure 4.35. In this method, the graphical step for the construction of the transfer characteristic nomogram described in Section 4.2.2 can be avoided, providing the possibility, with some additional assumptions, of deriving an analytical relationship for the values of the lumped elements of the device model. This method is less precise than the first one and limited to about 10 GHz, but it is more practical to employ.

#### 4.2.3.1 General Method for Modeling the Dual-Gate FET [6,7]

A lumped equivalent circuit of a dual-gate FET is shown in Figure 4.39 based on an illustrative section of the active device area as shown in Figure 4.38.

The two single-gate FET parts of the dual-gate FET (Section 4.2.2) cannot be assumed electrically equal as they are, in general, internally biased differently. Therefore, before proceeding to a realistic model, the internal bias must first be

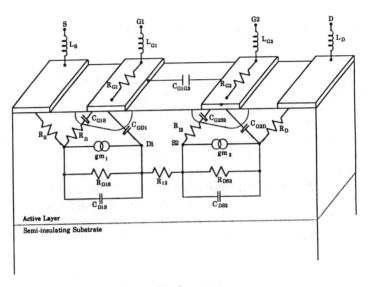

**Fig. 4.38** Dual-gate FET structure with parasitic elements.

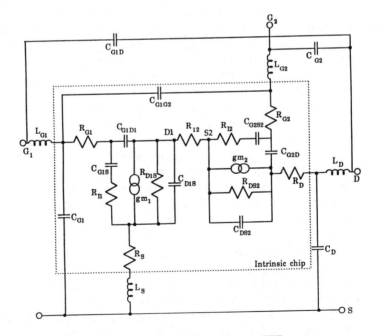

**Fig. 4.39** Lumped element equivalent circuit model for a dual-gate FET.

identified. This can be done by using either *graphical* [9] or *analytical* [17] methods. The graphical method is more efficient for the general modeling procedure, in particular because this method allows separate adjustment of the internal bias conditions of the dual-gate FET to execute a preoptimization by measuring S-parameters of the single-gate FET parts in their actual bias conditions. The preoptimization is based on two-port measurements, and is necessary to determine the starting point for a final three-port optimization. This final optimization is difficult to converge properly if we do not start with an accurate set of values because it involves a large number of independent model elements.

Figure 4.36 illustrates how a dual-gate FET must be biased externally to measure the two-port S-parameters needed for the preoptimization.

If $P_1$ is the external bias ($V_{DS} = 5$ V, $V_{G1S} = -1$ V, $V_{G2S} = 2$ V, $I_D = 34$ mA) with $V_{D1S} = 2.1$ V and $V_{DD1} = 2.9$ V, the S-parameters of FET 2 can be measured if we are able to transform FET 1 into a resistor (i.e., operate in the ohmic region). This can be done at bias point $P_2$. $V_{D1S'}$ is then 0.4 V and $V_{DS'} = V_{DD1} + 0.4$ V = 3.3 V. Also, the gate bias must be transformed to $V_{G1S'} = 0$ V (ohmic region) and $V_{G2S'} = V_{G2D1} + V_{D1S'} = -0.16$ V $+ 0.4$ V = 0.24 V. Consequently, the current $I_D$ remains the same, FET 1 largely represents a resistor, and FET 2 is biased as in the general case (point $P_1$). The two-port S-parameters between $G_1$ and $D$ can now be measured as a first step in the preoptimization, and gate 2 is terminated into a 50 $\Omega$ resistor connected to the third port of the three-port test fixture.

In a second step FET 2 can be transformed into a resistor by biasing it into the ohmic region. This corresponds to point $P_3$ of Figure 4.36. In this case the internal drain to source bias of FET 2 is again 0.4 V, and we obtain $V_{DS''} = V_{D1S} + V_{DD1''} = 2.1$ V $+ 0.4$ V = 2.5 V, $V_{G1S''} = -1$ (as in $P_1$) and $V_{G2S''} = V_{G2D1''} + V_{D1S} = 0.25$ V $+ 2.1$ V = 2.35 V, with $I_D$ remaining constant to 34 mA.

If we set the external bias of the dual-gate FET to these values, the two-port S-parameters of FET 1 can be measured because FET 2 now represents a relatively low ohmic resistor. Again, gate 1 must be terminated into a 50 $\Omega$ load. Figure 4.40 illustrates the double operation executed.

Starting values for the elements of a single-gate FET can be found from S-parameter fitting and optimization by using standard techniques, and the values are used to initialize the optimization routines for cases (b) and (c) of the procedure. This technique involves essentially the model elements of single-gate FETs, and is described in Section 4.1. In the particular case of the device described by the dc characteristics of Figure 4.36, this preoptimization delivers accurate estimates of measured S-parameters for the partial single-gate FETs (Figure 4.41 a,b).

From this point, the complete equivalent circuit of the dual-gate FET can be found by employing direct optimization of three-port S-parameters (Figure 4.40, case (a)). This procedure, although involving a large number of circuit elements to optimize, converges quite rapidly when *random, simplex,* or Powell curve-fitting

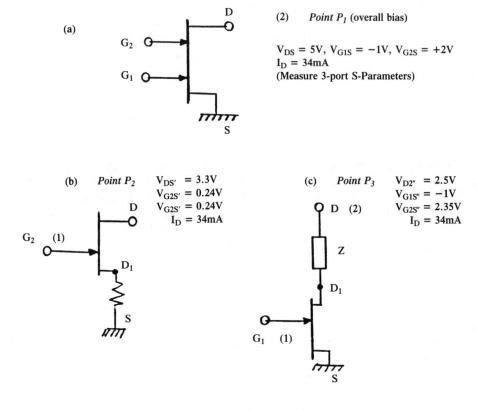

**Fig. 4.40** Pre- and final optimization steps for general modeling of dual-gate FETs.

optimization routines are used. This is because the preoptimization steps in Figure 4.40 (b,c) yield accurate starting values for the three-port optimization. A detailed example for such an optimization procedure is given in [6] and results are reproduced here (Figure 4.42).

All S-parameters except $S_{32}$ (drain-gate 2 feedback) can be very well simulated by this technique up to 11 GHz. The validity of the method can be assumed for higher frequencies as there are no assumptions made, except about the basic choice of a lumped instead of distributed element model. However, as in the case of single-gate FETs, the lumped element equivalent circuit can be sufficiently accurate up to only a maximum frequency of about half the maximum operating frequency of the device. This is even more valid for dual-gate FETs, where additional parasitics, feedback, and mounting fringe elements are present.

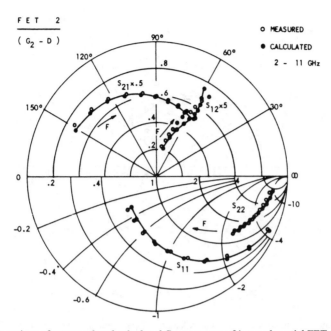

**Fig. 4.41(a)** Comparison of measured and calculated S-parameters of internal partial FET 2. Frequency: 2–11 GHz; Δf = 1 GHz; input port: gate 2; outport port: drain.
(From C. Tsironis and R. Meierer, "Microwave Wide-Band Model of GaAs Dual Gate MESFETs," *IEEE Trans. Microwave Theory and Techniques*, Vol. MTT-30, No. 3, March 1982. © 1982 IEEE. Reprinted with permission.)

### 4.2.3.2 Simplified Method for Modeling Dual-Gate FETs [8]

The simplified equivalent circuit is shown in Figure 4.44. Fewer equivalent circuit elements make it reasonable to attempt either a direct optimization or to develop and use approximate analytic equations to calculate the element values. The reduction of the number of elements is based on two simplifications:

1. The model of the intrinsic single-gate FET contains only six instead of more than nine elements [16] (Figure 4.43).
2. All parasitic elements with the exception of feedback capacitances are neglected.

This, of course, means that the remaining fewer elements become *frequency dependent* and the validity limit of a "lumped" instead of a "distributed" element model is reduced to lower frequencies.

The results of the simplified model, presented in [8], in terms of overall accuracy, when simulated S-parameter values are compared to measured ones,

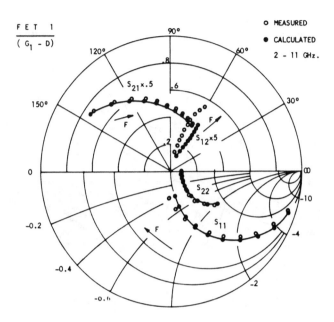

**Fig. 4.41(b)** Comparison of measured and calculated $S$-parameters of partial FET 1. Frequency: 2–11 GHz; $\Delta f = 1$ GHz; input port: gate 1; output port: drain.

are similar to those of the global model [6] (Section 4.2.3.1). This may be explained by the fact that measurements where limited to 11 GHz as the actual device had a maximum operating frequency beyond 20 GHz.

The basic idea behind the development of the simplified model is the possibility of eliminating all series parasitic and feedback elements to achieve a structure easily described by using $Y$ and $Z$ parameters (see Figure 4.44). This, then, permits us to derive the following analytical formulas for the actual values of the model elements.

Assuming $(1 + j\omega C_{G1}R_{11}) \approx 1$ and $(1 + j\omega C_{G2}R_{12}) \approx 1$, then

$$C_{G2} \approx \frac{1}{\omega} \, \mathrm{Im}\left[ \frac{-g_{m2}}{1 + \dfrac{Y_{12}(y_{m1} - j\omega C_{D1})}{Y_{21}(j\omega C_{D1})}} \right]$$

$$g_{02} = \frac{1}{R_{02}} \approx -\omega C_{G2} \, \mathrm{Im}\left[ \frac{Y_{31}}{Y_{21}} \right] - g_{m2}$$

$$C_{D2} \approx -\frac{\mathrm{Im}[Y_{23}]}{\omega} - \frac{C_{G2} g_{02}}{g_{m2} + g_{01} + g_{02}}$$

$$C_{S2} \approx \mathrm{Im}\left[ \frac{-(Y_{23} + j\omega C_{D2})(y_{m1} - j\omega C_{D1})}{\omega Y_{21}} \right]$$

where
$$y_{m1} = g_{m1}e^{-j\omega\tau_1}$$

Also, if

$$C_{G2} \gg g_{m2}\tau_2$$

then

$$\tau_2 \approx -\frac{1}{\omega}\left[\arg\left(\frac{Y_{31}}{Y_{21}}\right) + \frac{\pi}{2}\right]$$

Assuming $(\omega R_{11}C_{G1})^2 \ll 1$:

$$C_{D1} = \frac{-\mathrm{Im}(Y_{12}^{\mathrm{I}})}{\omega}$$

$$C_{S1} = \frac{\mathrm{Im}(Y_{22}^{\mathrm{I}})}{\omega} - C_{D1}$$

$$R_{01} = \frac{1}{\mathrm{Re}(Y_{22}^{\mathrm{I}})}$$

$$g_{m1} = \mathrm{Re}(Y_{21}^{\mathrm{I}})$$

$$C_{G1} = \frac{\mathrm{Im}(Y_{11}^{\mathrm{I}})}{\omega} - C_{D1}$$

$$R_{11} = \frac{\mathrm{Re}(Y_{11}^{\mathrm{I}})}{(\omega C_{G1})^2}$$

$$\tau_1 = \left[-\frac{\mathrm{Im}(Y_{21}^{\mathrm{I}})}{\omega} - g_{m1}R_{11}C_{G1} - C_{D1}\right]/g_{m1}$$

From these formulas and measured S-parameters in a particular noncritical frequency range (in this case, between 4 and 6 GHz), we can easily obtain numerical values of the model elements.

The selection of the frequency range is important as it cannot be too high, so parasitics can still be neglected, but also cannot be too low, so the phases of the reflection factors ($S_{11}$, $S_{22}$, and $S_{33}$) already contain significant information about the input and output capacitances and may be measured with some precision. Such a technique is described in detail for single-gate FET modeling in Section 4.1.

A direct comparison of the element values obtained by using the global model and the simplified one is possible because both approaches use the same measured data (Table 4.15).

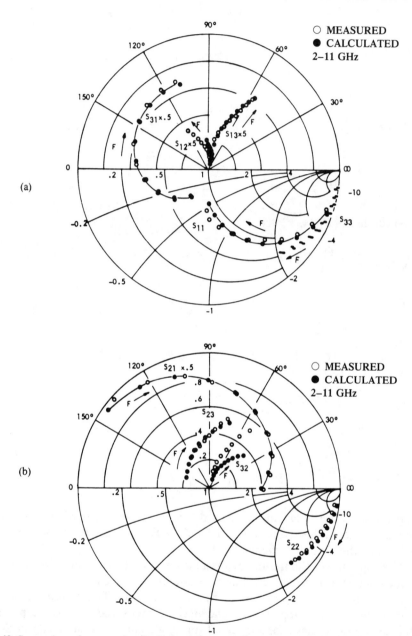

**Fig. 4.42** Comparison of measured and calculated three-port S-parameters of complete dual-gate FET: (a) FET 1; (b) FET 2.

(From C. Tsironis and R. Meierer, "Microwave Wide-Band Model of GaAs Dual Gate MESFETs," *IEEE Trans. Microwave Theory and Techniques*, Vol. MTT-30, No. 3, March 1982. © 1982 IEEE. Reprinted with permission.)

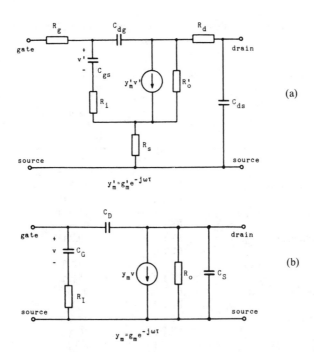

(a)

(b)

**Fig. 4.43** Simplified single-gate FET model.

(From J. R. Scott and R. A. Minasian, "A Simplified Microwave Model of the GaAs Dual-Gate MES-FET," *IEEE Trans. Microwave Theory and Techniques*, Vol. MTT-32, No. 3, March 1984. © 1984 IEEE. Reprinted with permission.)

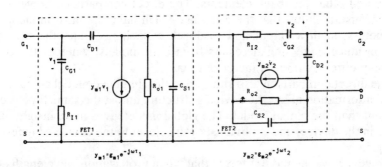

**Fig. 4.44** Simplified dual-gate MESFET equivalent circuit.

**Table 4.15** Direct Comparison Between the Basic Elements of the Global and the Simplified Dual-Gate
FET Equivalent Circuit

| Element | | Global Model | Simplified Model |
|---|---|---|---|
| Feedback Capacitance $(C_{gd})$ | FET 1 | 19.7 fF | 20 fF |
| | FET 2 | 33.2 fF | 17 fF |
| Input Capacitance $\quad(C_{gs})$ | FET 1 | 0.196 pF | 0.181 pF |
| | FET 2 | 0.284 pF | 0.288 pF |
| Output Capacitance $\quad(C_{ds})$ | FET 1 | 20.2 fF | 51 fF |
| | FET 2 | 11 fF | 48 fF |
| Saturation resistance $\quad(R_{ds})$ | FET 1 | 427 Ω | 350 Ω |
| | FET 2 | 854 Ω | 400 Ω |
| Input resistance $\quad\quad(R_i)$ | FET 1 | 9.7 Ω | 17 Ω |
| | FET 2 | 5 Ω | 17 Ω |
| Transconductance $\quad(g_m)$ | FET 1 | 25.1 mS | 24.3 mS |
| | FET 2 | 16.5 mS | 15 mS |
| Time constant $\quad\quad(\tau)$ | FET 1 | 2.3 ps | 0.7 ps |
| | FET 2 | 4.8 ps | 5.6 ps |

The values of the simplified model may be more correctly termed "averaged" over the frequency range as they indirectly include the contribution from parasitic and other feedback elements. The direct comparison, however, is still useful because it shows that the elements which have the most influence on the S-parameters, such as input and feedback capacitances and the transconductances, produce quite similar results from both models, whereas the other elements may differ by up to a factor of four.

The direct comparison also shows that the simplified model contains enough physical information about the device's structure and at the same time is easier to implement than the global model. The global model gives more insight into what happens in the device, and is obviously valid closer to the device's highest operating frequency.

However, we should not forget that when propagation wavelengths through (any) devices become comparable (approximately $\lambda/10$ or more) to the devices' geometry, the lumped element model should be gradually replaced by a distributed element model.

### 4.2.4 Applications of Dual-Gate FETs

The existence of the second control port (gate 2) in the dual-gate FET enables and simplifies the implementation of a number of applications which are not inherent to single-gate FETs. During the 1980s, an important number of applications of the dual-gate FET have been published. Table 4.16 summarizes the most important of these.

**Table 4.16** Some of the Most Important Microwave Applications of Dual-Gate FETs

| Application | Reference |
|---|---|
| Gain Controlled Amplifier | [2] |
| High-Speed Modulator | [2] |
| Gain Controlled Power Amplifier | [19] |
| Mixer | [10,11,24,26] |
| Frequency Discriminator | [20] |
| Frequency Multiplier | [12] |
| Oscillator-Multiplier | [21] |
| Oscillator | [13,14] |
| Switch | [5] |
| Phase Shifter | [22,25] |
| Pulse Regenerator | [23] |

Proper handling of bias and termination of the second control port (gate 2) of the dual-gate FET is the key to successful implementation of most of the applications given in the table. Inadequate termination of gate 2 will, in most cases, cause uncontrollable oscillations. In most of the applications listed, gate 2 serves as a control port, but in one case [22], where *active phase shifters* had been investigated, gate 2 also served as the input port. Handling of the dual-gate FET is more difficult than for the single-gate FET, but a number of functions such as gain control or the combination of two dc and RF decoupled signal paths are inherent to the device, thus making the dual-gate FET a good candidate for MMIC applications.

In general, those circuits have been largely realized in monolithic versions, which fit into standard module or large consumer market applications, such as gain control amplifiers [19], mixers for DBS reception [27], and phase shifters for T/R modules [25].

The monolithic dual-gate FET driver amplifier [28], as shown in Figure 4.45, contains three stages, two with 500 $\mu$m and one of 1800 $\mu$m gate width, and produces 0.5 W output power at 30 dB gain in X-band. The gain can be controlled by over 30 dB and the corresponding insertion phase change is less than 20°.

A monolithic X-band dual-gate FET mixer is shown in Figure 4.46 [28]. Here the dual-gate FET efficiently serves to combine the LO and RF signals without the need of a coupler so that the circuit can be realized entirely on a 10 mil × 10 mil GaAs chip.

Although the individual performance of the dual-gate MESFET mixer within the receiver can only be indirectly estimated, the overall performance of the receiver shows that the mixer has acceptable noise performance and probably some conversion gain as well.

The digital phase shifter [25] shown in Figure 4.47 also exhibits the potential of the dual-gate MESFET as a wideband switching device.

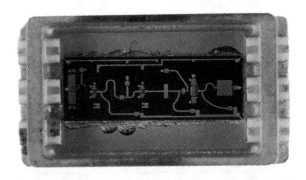

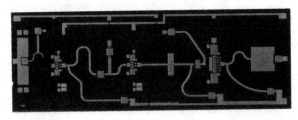

(Courtesy of Texas Instruments.)

**Fig. 4.45**(a) Monolithic X-band dual-gate driver amplifier.

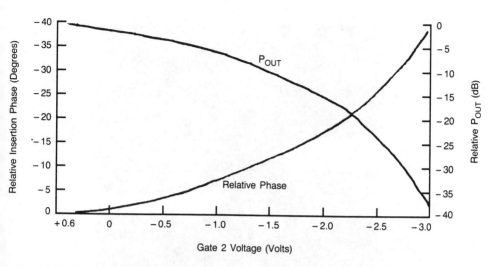

**Fig. 4.45(b)** Typical phase and output power ($T_A$ = 25°C) (packaged device).

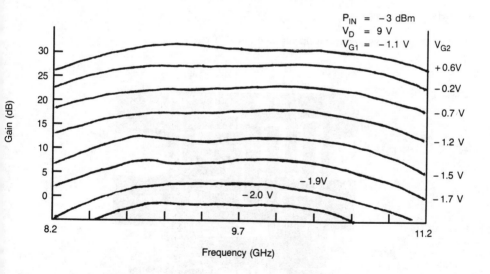

**Fig. 4.45(c)** Typical gain control performance ($T_A$ = 25°C) (packaged device).

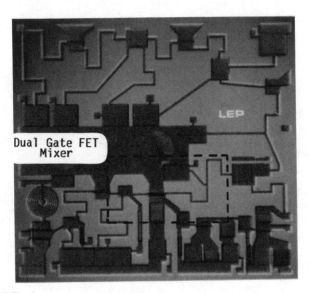

(Courtesy of LEP Philips.)

**Fig. 4.46** One-chip DBS receiver including dual-gate FET mixer.

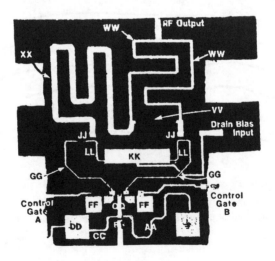

**Fig. 4.47**(a) Active phase-shifter circuit chip.

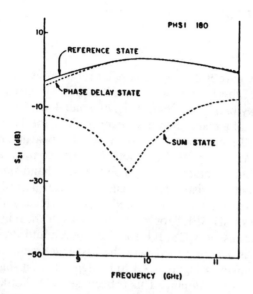

**Fig. 4.47(b)** 180° bit circuit gain *versus* frequency response.

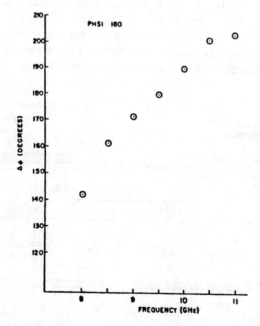

**Fig. 4.47(c)** 180° bit circuit phase shift *versus* frequency response.

## 4.3 SCHOTTKY DIODES

### 4.3.1 Introduction

GaAs Schottky barrier diodes are popular for a variety of applications due to their relative ease of fabrication, low capacitance, and extremely fast switching capabilities. Schottky barrier diodes consist of a metal-semiconductor junction. A cross-sectional view of a planar Schottky barrier diode for use in MMIC is shown in Figure 4.3.1 and the corresponding top view of the structure in Figure 4.3.2. The cross-sectional view depicts the Schottky barrier contact (anode) on the left-hand side and an ohmic contact (cathode) on the right-hand side of the geometry. The Schottky contact consists of the Schottky metal deposited on the active doped layer (in this case, an $n^-$ or $n^+$ doped region), which is $n^-$, the same as the active channel layer of MESFET, or $n^+$, the same as the heavily doped layer used on ohmic contact areas of MESFET such as source and drain. Connection to anode and cathode is accomplished with the first-level metal or airbridge metal. Schottky metal is shown in Figure 4.3.1 to be deposited on slightly etched active layer composed of $n^-$ or $n^+$ doping. This is because the Schottky metal for diodes in MMIC fabrication is deposited at the same process step as for the gates of MESFETs as described in detail in Section 3.4. In the gate process, gate recess is performed before depositing the gate metal, the anode of the Schottky diode goes through the same process step, and hence the recess under the Schottky metal. Associated with this structure are the Schottky-spreading resistance underneath the anode, the junction capacitance, the ohmic contact resistance $R$ of the cathode, and the resistance $R_S$ of the active layer between anode and cathode.

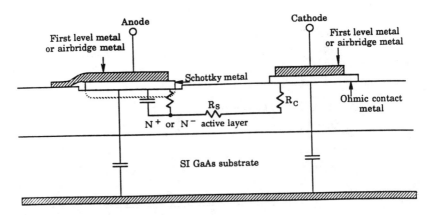

**Fig. 4.3.1** Vertical view of Schottky diode.

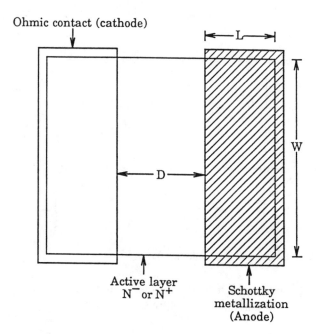

**Fig. 4.3.2** Top view of Schottky diode.

## 4.3.2 Basic Operation

Energy band diagrams for metal and $n$-doped semiconductor are shown in Figure 4.3.3. The term $e\phi_M$ and $e\phi_s$ represent the energy necessary for an electron to escape from the initial energy at the Fermi level (thermal equilibrium state) to the vacuum for the metal or semiconductor, respectively. This quantity is termed as the *work function*. The work functions of metals $e\phi_M$ are on the order of a few electron volts, between 2 and 6 eV, depending on the metal, and less for the semiconductor. If we make contact between semiconductor and metal, charge will flow from the semiconductor to the metal, establishing an equilibrium, which means the Fermi levels on both sides line up as shown in Figure 4.3.3 [13].

In Figure 4.3.4, the energy band diagrams are shown without externally applied voltage (a); forward biased, which means the Schottky metal is positive with respect to the $n$-semiconductor (b); and reverse biased (c). In forward bias, the Schottky barrier is lowered, which allows an increased current flow from the semiconductor to the metal. The current transport in metal-semiconductor barriers is mainly due to majority carriers in contrast to $p$-$n$ junction diodes, where the minority carriers are responsible. For this reason, no junction capacitance is

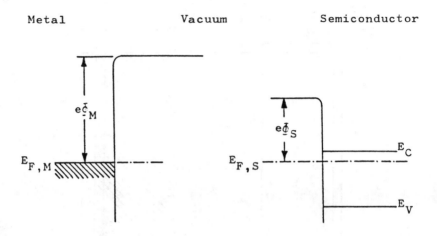

**Fig. 4.3.3** Band diagrams of metal and semiconductor.

present in forward bias, which results in fast switching times of the device. The equation for the diode current is as follows [13]:

$$I = I_S \left[ \exp \left( \frac{V}{V_T} \right) - 1 \right]$$ (4.3.1)

$$I_S = A^* \times S \times T^2 \left[ \exp \frac{-\phi_B(0)}{kT} \right]$$ (4.3.2)

$$\frac{\partial I}{\partial V} = g_D = \frac{I_S}{V_T} \exp \left( \frac{V}{V_T} \right)$$ (4.3.3)

where

$q$ = electronic charge;
$k$ = Boltzmann's constant;
$n$ = ideality factor, usually between 1.02 and 1.2;
$T$ = absolute temperature;
$\phi_B(0)$ = zero bias barrier height, usually close to 0.8 eV;
$A^*$ = modified Richardson's constant;
$S$ = diode area;
$g_D$ = conductance of the diode.

Equation (4.3.1) shows that diode current increases exponentially with the applied bias voltage. The ideality factor $n$ acts as a mathematical parameter to

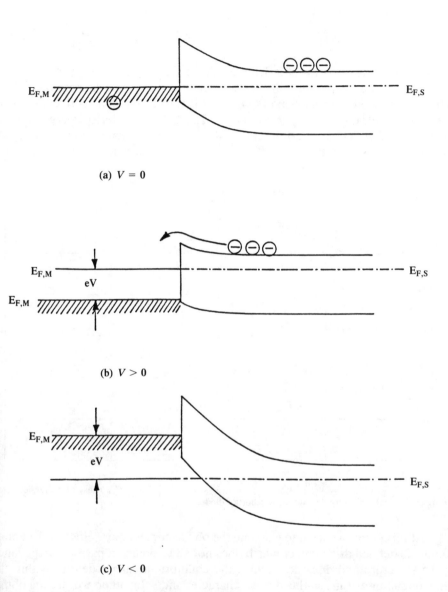

(a) $V = 0$

(b) $V > 0$

(c) $V < 0$

Fig. 4.3.4 Band diagrams for different bias voltages.

correct for various effects, such as damage on the surface of the active layer under Schottky metal due to the metal deposition process, which cause deviation from the ideal exponential characteristic. The following equation also accounts for the series resistance of the diode:

$$I = I_S \left[ \exp\left(\frac{V - IR_S}{nkT}\right) - 1 \right] \tag{4.3.4}$$

Typical *I-V* characteristics of a Schottky diode are shown in Figure 4.3.5, which also includes the temperature-dependent characteristics. The data are taken for a 100 $\mu$m wide diode, where Schottky contact is formed on an $n^+$ active layer with doping concentration of $2$–$4 \times 10^{17}$ cm$^{-3}$. The C-V characteristics are shown in Figure 4.3.6 for diodes with different anode to cathode spacing and various anode widths as represented by $D$ and $L$, respectively in Figure 4.3.2, the width of the diode $w = 100$ $\mu$m.

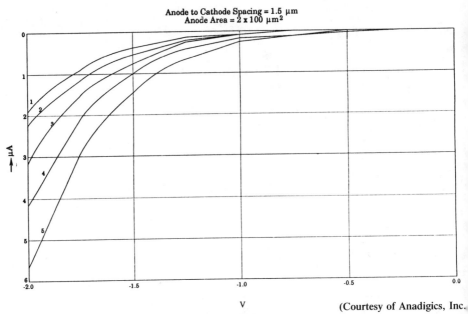

Fig. 4.3.5 Typical *I-V* characteristics of a Schottky diode.

(Courtesy of Anadigics, Inc.)

At this point, we need to examine the ohmic contact more closely. An ohmic contact is defined as a contact which does not add significant parasitic impedance to the structure and does not change the equilibrium carrier density within the semiconductor to affect the device characteristics. In other words, an ohmic contact should have a linear and symmetrical C-V relationship. The ohmic contact is characterized by having no potential barrier.

In practice, the above ideal ohmic contact can only be approximated. A metal-semiconductor contact is approximately ohmic if the semiconductor is very heavily doped. The most common approach is to use metal $n^+$-$n$ or metal $p^+$-$p$ contacts. The energy band gap diagram for metal $n^+$-$n$ structures is shown in

igure 4.3.7. Due to the highly doped layer, a very narrow barrier is formed at the
inction and electrons are able to tunnel from the metal to the semiconductor [13].

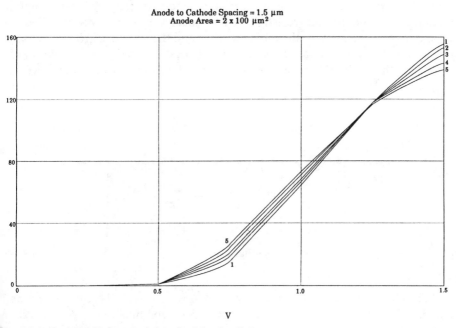

Anode to Cathode Spacing = 1.5 μm
Anode Area = 2 x 100 μm²

Fig. 4.3.6  Typical *I-V* characteristics of a Schottky diode.

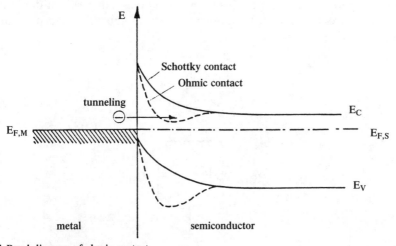

Fig. 4.3.7  Band diagram of ohmic contact.

### 4.3.3 Lumped Element Equivalent Circuit

The lumped equivalent circuit of a Schottky diode is shown in Figure 4.3.8 The diode is modeled as a nonlinear current source $I$ given by (4.3.1) with parallel junction capacitance $C$ and resistance $R$. Connected in series with the diode are the contact resistance $R_C$ of the ohmic contact and the resistance $R_S$ of the $n$-layer between anode and cathode. Parasitic capacitances $C_C$ and $C_A$ are included, which are due to the cathode and anode metal lines-to-ground capacitances, respectively.

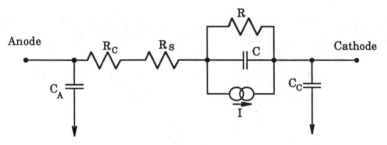

**Fig. 4.3.8** Lumped equivalent circuit of a Schottky diode.

The SPICE diode model is shown in Figure 4.3.9 with the current source $I$ including the nonlinear effect and $r_s$ as the series combination of $R_C$ and $R_S$, as shown in Figure 4.3.1. Parasitic capacitances, $C_C$ and $C_A$, are not included in this basic model, but can be easily incorporated as extrinsic elements in SPICE using subcircuit approach. For ac analysis, the equivalent small signal diode model is shown in Figure 4.3.10 with the nonlinear current source replaced by the diode conductance given in (4.3.3).

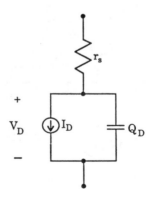

**Fig. 4.3.9** SPICE diode model.

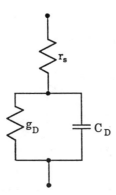

Fig. 4.3.10 Linearized, small-signal diode model.

The *p-n* junction diode model in generic SPICE can be optimized for its model parameters to fit quite accurately to the Schottky diode for dc *I-V* as well as bias-dependent capacitance characteristics. The model in SPICE also provides the breakdown voltage *BV* parameter to define the breakdown current. The bias dependent capacitance is also modeled in SPICE.

The SPICE noise model for a diode is shown in Figure 4.3.11. Two noise sources are present in the equivalent circuit. The diode shot noise due to the current flow is given by the equation:

$$\overline{i_{nD}^2} = 2q \times I_D \times \Delta f \tag{4.3.5}$$

where

$q$ = the electronic charge = $1.6^{-19}$ coulomb;
$I_D$ = diode current;
$\Delta f$ = bandwidth of the external terminating circuit.

In addition, the thermal noise of the diode series resistor is expressed as

$$i_{n_{rs}}^2 = 4kT \frac{1}{R} \Delta f \tag{4.3.6}$$

with $4\,kT = 1.66 \times 10^{-20}$ V-C, and the small signal equivalent resistor:

$$r_d = \frac{1}{g_D} = \frac{kT}{q(I + I_S)} = \frac{V_T}{(I + I_S)} \tag{4.3.7}$$

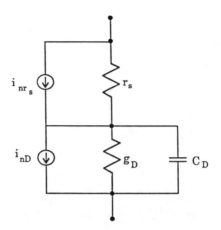

**Fig. 4.3.11** Noise model for a diode.

### 4.3.4 Semidistributed Element Equivalent Circuit

A semidistributed equivalent circuit of the planar Schottky diode in a cross sectional view is shown in Figure 4.3.12. The total series resistance can be mod eled with three lumped resistors. These are the ohmic contact resistance $R_C$, a bulk resistance $R_D$($n$-layer in this case) between the contacts and the so-called Schottky-spreading resistance ($R_1, R_2, \ldots$) primarily due to lateral current flow beneath the anode in the $n$-layer. The ohmic contact resistance $R_C$, as described before, is calculated as follows:

$$R_C = \frac{\rho_C}{WL_T} \tag{4.3.8}$$

where

$W$ = width of the ohmic contact;
$\rho_C$ = specific resistivity of ohmic contact;
$L_T$ = transfer length.

In expression (4.3.8), the current density flow across the interfacial layer i assumed to decrease exponentially with lateral distance from the edge of the ohmic contact. At this point, the current density is greatest. With this approxima tion the characteristic length of the exponential decay is described by $L_T$. The second part of the bulk resistance $R_D$ of the conductive $n$-layer is approximated as follows [10]:

$$R_D = \frac{D}{W} \cdot \rho_s \tag{4.3.9}$$

where

$D$ = separation between anode and cathode;
$\rho_S$ = specific resistivity of the $n$-layer.

Both of these resistances, $R_C$ and $R_D$, are independent of current.

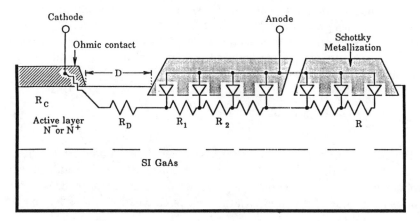

**Fig. 4.3.12** Semidistributed equivalent circuit of a Schottky diode.

The Schottky-spreading resistance beneath the Schottky-barrier junction in the conducting $n$-layer is divided among several resistors $R_1$, $R_2$, . . . $R_n$. This resistance, due to its distributed nature across the nonlinear rectifying junction, becomes current-dependent. With an appropriate number of sections, this nonlinear resistance can be accurately modeled; however, it results in a complicated model. We can show that at very low current densities the maximum Schottky-spreading resistance $R_{max}^{ss}$ is approached due to the negligible voltage drop across the resistor string as compared to the voltage drop across the Schottky-barrier junction, and is given by

$$R_{max}^{ss} = \frac{1}{3} \sum_{i=1}^{n} R_i \tag{4.3.10}$$

or

$$R_{max}^{ss} = \frac{1}{3} \rho_{ss} \cdot \eta_{\square} \tag{4.3.11}$$

where

$\rho_{ss}$ = sheet resistivity of the $n$-layer;
$\eta_\square$ = equivalent number of squares of resistance.

The factor of $\frac{1}{3}$ accounts for the current loss along the lateral path because of nearly uniform injection across the junction as the resistor string increases and the current flow through the junction crowds toward the edge of the anode nearest to the cathode. This is a well known effect of lateral debiasing of the Schottky junction. Thus, the Schottky-spreading resistance monotonically decreases, and eventually has an assymplotically limit of zero. The behavior of $R_{ss}$ is shown in Figure 4.3.13, where $R_{ss}$ is normalized with respect to $R_{max}^{ss}$. The inflection current, denoted by $I_{inf}$ is defined as the forward current at which $R_{ss}$ equals $R_{ss}/2$. This relationship may be expressed as

$$\frac{R_{ss}}{R_{max}^{ss}} = \frac{1}{2} \left\{ 1 - \tanh \left[ \log_{10} \left( \frac{I}{I_{inf}} \right) \right] \right\} \tag{4.3.12}$$

For an accurate extended Schottky-barrier diode model, the series resistance must be divided into the constant component $R_C + R_D$, and the current-dependent $R_{ss}(I)$. This is accomplished either by calculation or measurement of the series resistance of the diode. In the case of calculation, the inflection current is approximated by

$$I_{inf} \approx \frac{\theta}{\eta_\square} \tag{4.3.13}$$

with the empirically determined constant $\theta$ for a given process. Using (4.3.12), $R_{ss}(I)$ can be calculated as a function of $I/I_{inf}$. $R_{ss}$ can also be obtained from the total resistance $R_s$ under forward current as shown in Figure 4.3.14. The low and high current asymptotic limits of $R_s$ identify the region for $R_{ss}(I)$. As shown, the high current limit gives the constant resistance component $R_C + R_D$. Thus, the inflection current is where $R_s(I)$ equals $R_{sinf}$ given in the following equation:

$$R_{sinf} = (R_C + R_D) + 0.5 \times R_{max}^{ss} \tag{4.3.14}$$

The forward voltage $V_f$ across the diode is given as

$$V_f = nV_T \ln \left( \frac{I_f}{I_S} \right) + I_f(R_C + R_D) + I_f R_{ss}(I_f) \tag{4.3.15}$$

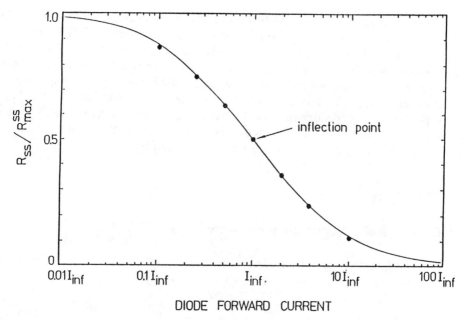

**Fig. 4.3.13** Schottky-spreading resistance *versus* diode forward current.

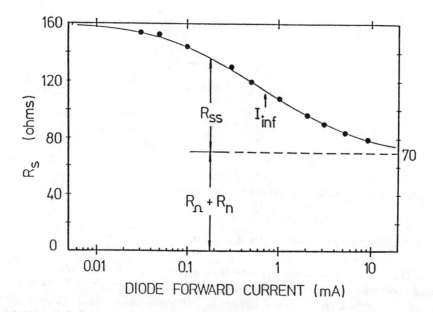

**Fig. 4.3.14** Total diode resistance *versus* diode forward current.

The parameters $n$ and $I_s$ can be determined from a plot of ln ($I$) *versus* $V$ in the region where only the first term on the right-hand side of (4.3.15) dominates. An extended diode model for this equation is shown in Figure 4.3.15. The principal diode models the Schottky-diodes characteristic, except for the voltage drop across $R_{ss}(I)$. The assigned value of the resistor is the maximum Schottky-spreading resistance $R_{max}^{ss}$, which is in parallel with an ideal diode (auxiliary diode). This diode produces a current dependence similar to the one shown in Figure 4.3.5, and is specified by the emission coefficient and saturation current $n_b$ and $I_{sb}$, respectively. Empirical equations for these values are given below:

$$\eta_b = \frac{I_{inf} \times R_{max}^{ss}}{6.435 \ V_T} \tag{4.3.16}$$

or, for $V_T = 0.0259$ V, at room temperature,

$$\eta_b = 6I_{inf} \times R_{max}^{ss} \tag{4.3.17}$$

and

$$I_{sb} = \frac{I_{inf}}{50} \tag{4.3.18}$$

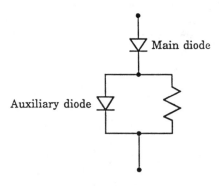

**Fig. 4.3.15** Extended equivalent circuit of a Schottky diode.

## 4.3.5 Application and Layout Considerations

The main applications of Schottky-barrier diodes include rectifiers, microwave switches, and mixer circuits as well as level shifting and ESD protection circuits (described in detail in Chapter 11). In some cases, diodes can be used to realize small capacitors or voltage-controlled capacitors for VCO applications.

Some applications such as level shifting demand optimum performance of the diode in forward bias, whereas others, like VCOs and mixer circuits, require optimum performance in reverse bias. For example, mixer circuits require optimization of such parameters as ideality factor, reverse leakage current, and diode cut-off frequency. Selection of a diode for an application requires evaluation of such parameters as turn-on voltage, forward resistance, ideality factor, reverse breakdown voltage, reverse saturation current, and capacitance variation with voltage.

As shown in Figures 4.3.1 and 4.3.2, the total diode resistance $r_s$ depends on the (i) ohmic contact resistance, (ii) parasitic resistance due to the active layer between anode and cathode, and (iii) the Schottky-spreading resistance because of finite overlap of the Schottky contact. The parasitic resistance due to anode-cathode separation depends on the distance between them, and can be minimized by placing the ohmic contact and Schottky as close to each other as permitted by the processing technology. Devices with minimum separation of 1.5 $\mu$m between anode and cathode can be easily fabricated by use of contact lithography, resulting in a parasitic resistance of about 0.4 $\Omega$ for a 1 mm wide diode with active layer doping concentration of about $1-3 \times 10^{17}/cm^3$ ($n^+$ diode). The capacitance $C_D$, associated with the Schottky diode, depends on the area of the anode and the doping concentration of the active area under it. This capacitance can be minimized by reducing the width of the anode to the limit of the processing technology. An $n^+$ diode with 2 $\mu$m $\times$ 1 $\mu$m anode area on an active layer with doping concentration of about $1-3 \times 10^{17}$ cm$^{-3}$ results in a zero-bias junction capacitance of 2.5–3.0 pF. With a similar structure, with active layer doping concentration of $.7-.9 \times 10^{17}$ cm$^{-3}$ ($n^-$ diode), this junction capacitance can be as low as .8–1 pF. The reverse breakdown voltage for an $n^+$ diode is typically 3–4 V, and for an $n^-$ diode it is as high as 12–15 V. However, the reverse breakdown and capacitance characteristics of the diode also depend on the doping profile of free carriers in the active layer in addition to the peak doping level.

At sufficiently high frequencies, the barrier capacitance dominates the barrier conductance. Thus, a useful figure of merit, the cut-off frequency, is given by

$$\omega_c = 2\pi f_c = \frac{1}{r_s C_D} \tag{4.3.19}$$

This expression neglects the skin effect impedance and bulk spreading resistance of the diode [12]. With a properly optimized diode structure and the doping profile in the active layer, a cut-off frequency of 150 to 250 GHz can be achieved. Superior diode performance can also be achieved by using a single-gate FET with a 0.5 $\mu$m gate length, by connecting source and drain of the FET to operate as cathode and the gate of the FET as anode. By using heavy doping in the channel of such a FET, a cut-off frequency as high as 200 to 300 GHz can be attained.

However, such a diode cannot be used in forward bias because a gate consisting of only 0.5 $\mu$m of metal line will not be able to carry the forward-bias current.

## 4.4 PLANAR LUMPED ELEMENTS

### 4.4.1 Introduction

In any monolithic microwave integrated circuit, a large number of passive components (resistors, inductors, capacitors, transmission line elements, and coupled transmission line components) are necessary. Their properties and their accurate modeling are essential for the valid prediction of the performance of a MMIC. Such components actually outnumber the active devices by about ten to one, and thus deserve a great deal of attention. The characteristics and the constructional and electrical constraints on the passive components must be considered in the first-cut design approach to avoid any unexpected consequences for the final circuit's performance. Fortunately, the process control required for the passive elements is less demanding due to their inherent tolerance to process variations and the theory of operation is less complicated than for nonlinear devices such as FETs and diodes.

A common integrated design philosophy at moderate frequencies tends to replace passive elements by active devices wherever possible because active devices are more easily integrated and require less surface area. In principle this is possible for MMIC designs in some cases. However, in practice, due to the limited intrinsic transistor performance and the parasitics associated with nonlinear devices, which require impedance matching circuits at the input and output port, and the feedback circuitry, which requires such passive elements as resistors, capacitors, inductors, and transmission lines. The circuit designer's goal is to minimize the number of active and passive components to reduce additional parasitics and chip area.

The most important characteristics that need to be considered for passive components in MMICs are given in Table 4.4.1.

**Table 4.4.1**

Range of values;
Tolerance;
Voltage-, current-, and power-handling capabilities;
Linearity;
Stability;
Thermal characteristic;

`able  4.41 cont.

Parasitics: series loss resistance,
            series parasitic inductance,
            stray capacitance to ground and adjacent components.

*Note:* Combinations of the above characteristics can be expressed in terms of the following parameters for passive components:

Quality factor (Q)
Cut-off frequency ($f_c$)
Loss tangent (tan δ)
Resonance frequency ($f_r$)

---

Due to the physical construction of the MMIC's lumped elements, the passive components usually cannot be modeled accurately by a single element in a
simple equivalent circuit. In most cases, the parasitic influences are modeled by
additional circuit elements, for example, shunt capacitances to ground on each
end of the element, or in some cases several distributed sections along its length,
depending on the required accuracy of the model. As a first-order approximation,
the monolithic circuit design philosophy must include *lumped elements* (components with dimensions smaller than about 0.1 λ at the highest operating frequency)
and *distributed elements* (transmission line sections) which can be modeled by
applying the transmission line theory.

The choice of lumped or distributed elements is mainly dictated by the
operating frequency as well as other considerations like chip size and the characteristics mentioned in the Table 4.4.1. Applying these conditions, lumped elements can be realized, and they are suitable from L-band to 20 GHz. Lumped
elements are very attractive to a circuit designer due to their small dimensions.
Above this frequency range truly lumped elements are difficult to realize because
the size of the components generally becomes comparable to a quarter of a wavelength (λ/4) and the parasitics, especially on thin substrates (4–5 mils) with a
ground plane on the back of the wafer, have an immense influence on performance. The required values of the passive components at higher frequencies
fortunately are often small, and distributed elements tend to become convenient
due to the small wavelengths.

Lumped elements as resistors, capacitors, and inductors are extremely useful in microwave circuits and, in some cases, mandatory. Planar inductors are
very useful for matching circuitry, especially at lower frequencies where stub
resonators become physically large. Lumped thin-film capacitors are used for

coupling purposes and bypass applications where large values of capacitance are required. An overview of MMIC passive elements is given in Table 4.4.2.

**Table 4.4.2**

| Type | Value | Q-Factor (10 GHz) | Dielectric Metal | Application |
|------|-------|-------------------|------------------|-------------|
| INDUCTOR | | | | |
| High impedance line (Single-loop, Meander-line, S-line) | 0.01–0.5 nH | 30–60 | Plated gold | Matching |
| Spiral inductor | 0.5–10 nH | 20–40 | Plated gold | Matching, Power supply choke |
| CAPACITOR | | | | |
| Transmission line gap (Broadside-coupled, End-coupled) | 0.001–0.05 pF | | | Coupling, Matching |
| Interdigitated | 0.01–0.5 pF | ~50 | | Coupling, Matching |
| MIM (overlay) | 0.1–100 pF | ~50 | $Si_3N_4$ $SiO_x$ Polyimide | Coupling, Matching, Bypass |
| | | ~25 | $Ta_2O_5$ | |
| RESISTOR | | | | |
| Thin-film | 5 Ω–1 kΩ | | NiCr, TaN | dc biasing, feedback, matching |
| Monolithic | 10 Ω–10 kΩ | | Implanted GaAs | |

## 4.4.2 Planar Inductors

Planar inductors for MMICs can be realized with a plated gold metal layer using either first-level metalization, airbridge metalization, or both. An exception is the spiral inductor, which requires two metal layers, including airbridge, to provide crossover connection to the center of the inductor. A variety of inductor geometries are possible with this scheme.

The simplest planar inductor is the straight line section, also called a high impedance line section, which implements a high impedance (e.g., reactance) by narrowing of a transmission line structure. However, the smallest line width that can be realized in the applied MMIC process limits the characteristic impedance of the line and, together with the length, the value of the obtained inductance. In practice, only small reactances up to 100 $\Omega$ are possible with this geometry, as described in detail in Section 4.7.

The microstrip line section can be easily calculated by the following equations.

$$x_L = z_{0L} \times \sin\left(\frac{2\pi l_L}{\lambda_{gL}}\right)$$

(4.4.1)

for lossless transmission lines, with $l \ll \lambda_g$ and where

$z_{0L}$ = the characteristic impedance of the line is usually on the order of 20 to 120 $\Omega$ for microstrip,
$l_L$ = length of the line,
$\lambda_{gL}$ = wavelength of the line.

Note that, for the calculation of the wavelength $\lambda_{gL}$, the effective dielectric constant for the given line width must be used:

$$\lambda_{gL} = \frac{\lambda_0}{\sqrt{\varepsilon_{\text{eff}}}}$$

(4.4.2)

Parasitic capacitance is associated with the step discontinuity as shown in Figure 4.4.1 and calculated by the following equation:

$$C_L \approx \frac{l_L}{2f \times z_{0L} \times \lambda_{gL}}$$

(4.4.3)

Thus, the length of the line for a given inductance is expressed by

$$l_L = \frac{\lambda_{gL}}{2\pi} \sin^{-1}\left(\frac{\omega L}{z_{0L}}\right)$$

(4.4.4)

Transmission line modeling techniques are described in detail in Section 4.7.

Figure 4.4.2(a) shows a high impedance line. Another often used inductor geometry is the single-loop (Figure 4.4.2(b)). This structure, described by the outer radius $r_o$, inner radius $r_i$, and the angle $\rho$ is already very complicated to model if the mutual inductances of the line segments are to be considered.

In cases where higher reactances are required, the meander-line (Figure 4.4.2(c)) and S-line (Figure 4.4.2(d)) inductor geometries are often employed. These types of structure depend mainly on the mutual coupling between the

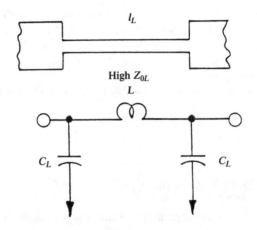

Fig. 4.4.1 Parasitic capacitance associated with step disconuity.

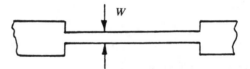

Fig. 4.4.2 (a) High impedance line.

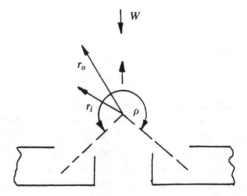

Fig. 4.4.2 (b) Single-loop.

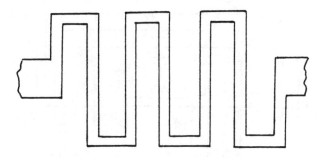

**Fig. 4.4.2** (c) Meander-line.

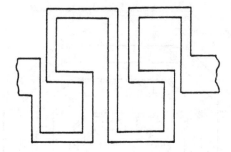

**Fig. 4.4.2** (d) S-line.

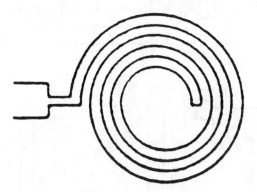

**Fig. 4.4.2** (e) Circular spiral.

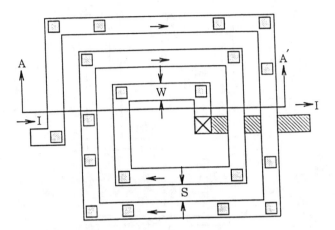

**Fig. 4.4.2** (h) Easy-to-design inductor.

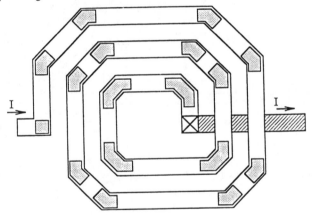

**Fig. 4.4.2** (f) Square inductor.

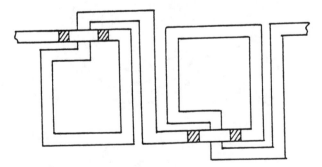

**Fig. 4.4.2** (g) Spiral inductor.

different adjacent line segments and thus achieve a high inductance in a desirable, small area. This layout, however, introduces parasitic capacitances between the lines of the inductor, at the right angle bends, and the capacitance to ground, which reduces both the characteristic impedance and the effective length of the line.

The most popular inductor structures used in MMICs are the high impedance microstrip line and the spiral inductor. Spiral inductors can be laid out in circular (Figure 4.4.2(e)), square (Figure 4.4.2(f)), or octagonal (Figure 4.4.2(g)) shape. All of these geometries strongly depend on the mutual coupling of the individual turns, and thus allow high values of reactances on a small surface area. This type of planar inductor, however, is very difficult to analyze and model due to the complex coupling between adjacent lines. These types of inductors are often measured and the values of the equivalent circuit elements are fit to the measured S-parameters by CAD programs.

Simulators based on electromagnetic field theory can also be employed to model such inductors without the need to fabricate them. The performance of the spiral inductor structures (Figure 4.4.2 structures (e, f, g)) is quite comparable. Thus, the circuit designer often must pick preproven designs from the library and is bound to specific inductance values. Improvement of this situation may be obtained with a novel inductor geometry (Figure 4.4.2(h)), which retains much of the space saving of the spiral inductor but is easier to design and minimizes the effect of the cross-coupling capacitance and performance degradation. Ground planes in MMICs are commonly used, and the substrate is often thinned for heat dissipation and via-hole grounding of components. This, however, increases the effect of the ground plane on the parasitic capacitances of the other passive structures (e.g., resistors and inductors). In addition, fringing capacitance between the closely spaced lines must be taken into account to model the frequency dependence of the line segments.

Another important effect, associated with the closely located ground plane under the substrate, is the mutual inductance between the original coil and the induced mirror coil in the ground plane. This effect reduces the effective inductance of the coil, and thus must be taken into account. The right angle bends of the conductor also introduce discontinuities and further increase the capacitive effects.

Typical quantitative behavior of spiral and single-loop inductors as a function of frequency is shown in Figure 4.4.3.

The resonance frequency of spiral inductors depends on the width of the line and the number of turns. Small inductance spiral inductors (.5–2 nH) have a resonance frequency higher than 20 GHz. Larger inductors (8–10 nH) have a resonance frequency as low as 10–12 GHz. Above resonance frequency, the element behaves capacitively (negative reactance). The resonance frequency of a single-loop is usually high as compared to spiral inductors, as shown in Figure 4.4.3(b).

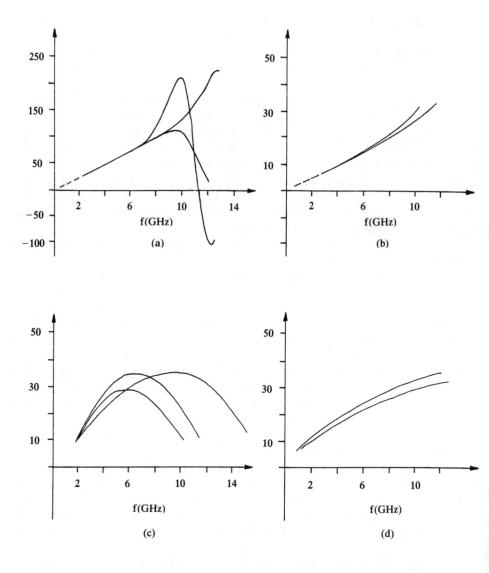

**Fig. 4.4.3** (a) Spiral inductor reactance, (b) single-loop reactance, (c) spiral inductor Q-factor, and (d) single-loop Q-factor *versus* frequency.

Q-factor of spiral and single-loop inductors are shown in Figure 4.4.3(c, d). At lower frequencies, the Q-factor of spiral inductors is comparable to that of a single-loop inductor. However, the Q-factor degrades at higher frequencies for spiral inductors, whereas for single-loop inductors, the Q-factor still increases with increasing frequency.

A cross-sectional view of the spiral inductor is shown in Figure 4.4.4, which depicts the lines with the width $W$, the spacing $S$, and the wafer backside metalization. A simple equivalent circuit of the spiral inductor is shown in Figure 4.4.5. The inductor is modeled by several inductances in series with resistance and associated parasitic capacitance, $C_p$. Most often, three sections of $R$-$L$ combination lumped elements shown in Figure 4.4.5 is sufficient to represent accurately the spiral inductors.

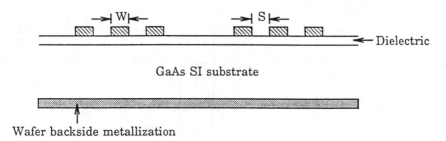

**Fig. 4.4.4** Cross section of a spiral inductor at A–A'.

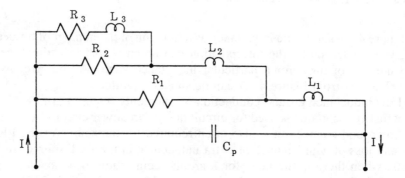

**Fig. 4.4.5** Equivalent circuit for spiral inductor.

Other lumped and semilumped equivalent circuits for spiral inductors are shown in Figures 4.4.6 and 4.4.7. The simplified model in Figure 4.4.6 includes the main element $L$ and the parasitic resistance $R$. $C_c$ represents the capacitive coupling between the spiral inductor lines and $C_g$ represents the capacitance to ground. For large inductors, representing the lumped element model in several sections is useful for accurately modeling the spiral inductor. Figure 4.4.7 presents an elaborate equivalent circuit model, and includes the parasitic resistances $R_c$ and $R_g$. The values of these resistors are in the range of 10–30 k$\Omega$. Such a model is capable of representing the inductors' performance beyond their resonance frequency. Again, such a model can be used to represent the spiral inductor's sections, thus more accurately representing the measured performance.

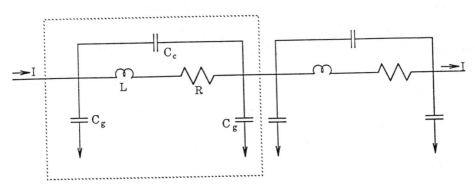

**Fig. 4.4.6** Lumped equivalent circuit for spiral inductors.

Figure 4.4.8 shows the inductance as a function of the number of turns in the spiral. Using such plots, the interpolation can be employed to estimate the required number of turns for a particular inductance value. All other equivalent circuit elements from Figure 4.4.7 can be similarly plotted.

Low inductance ground connection is commonly required in MMICs when the microstrip technique is used for circuit design. In power circuits, a low inductance to ground connection is also often required for the source of the FETs to avoid any loss of gain because of extra inductance in the FET source. In larger circuits, when there is the need for a ground connection to be available in the middle of the layout of the chip, having a via hole available is helpful as opposed to running a long connecting line to the grounding pad at the edge of the chip. As described in Chapter 3, the via hole is etched from the backside of the GaAs substrate, which is typically thinned to 100 $\mu$m, and the hole is filled with plated gold to make connection to the pad metal on the front of the wafer. For a 50 $\mu$m diameter circular or rectangular via hole in a 100 $\mu$m thick substrate, the parasitic resistance is typically 100–200 M$\Omega$ and the inductance is 30–50 pH.

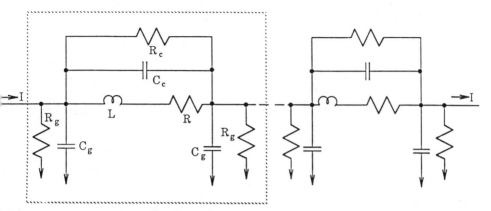

**Fig. 4.4.7** Semilumped equivalent circuit for spiral inductors.

## 4.5 PLANAR CAPACITORS

A variety of planar capacitors are suitable for MMIC as shown in Fig. 4.5.1(a–d). Some of these structures require only a single metalization scheme, such as the (a) broadside coupled and (b) end-coupled microstrip line capacitors and (c) interdigitated capacitor. Overlay capacitors (d), also called MIM (metal-insulator-metal) capacitors use a two-layer metalization scheme with a dielectric film between the electrodes. Table 4.5.1 summarizes the properties of various dielectric films commonly used for MIM capacitors. The first three structures (Figure 4.5.1 (a–c)) mainly depend on dielectric coupling via the substrate. These geometries are suitable for low capacitance, as shown in Table 4.4.2. The main advantage of using these structures is their high reproducibility whereby they can be used for circuit applications in which high precision and small values are required, such as high impedance matching circuits.

The maximum capacitance per unit area for interdigitated capacitors is only 0.3% of that obtained with the overlay type. The capacitive effect in this structure results from fringing between the interleaved fingers and requires the calculation of elliptic integrals to model such capacitors. Parasitic effects in the interdigitated capacitor include series resistance loss in the fingers, dielectric loss, series inductance of the fingers, and shunt capacitance to ground. Interdigitated capacitors have proved to be very useful components in GaAs MMIC due to their relatively high Q and fabrication simplicity.

Different methods have been applied to model interdigitated capacitors. Coupled microstrip line theory can be used to model the individual fingers of the interdigitated capacitor, including the losses and finite metalization thickness. A recent approach in modeling interdigitated capacitors divides the geometry into its

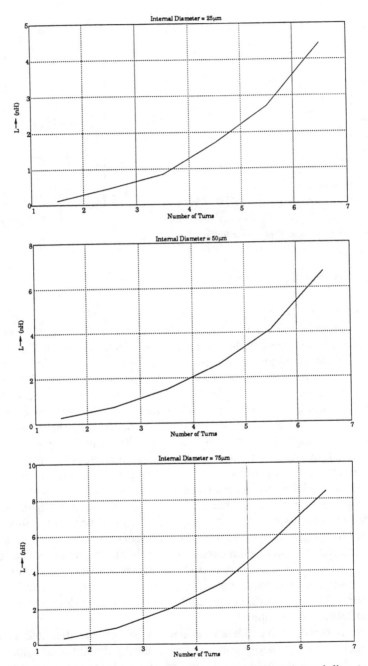

**Fig. 4.4.8** Inductance as a function of the number of turns for different internal diameter structures. (Courtesy of Anadigics, Inc.)

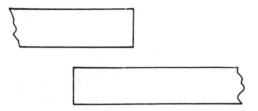

**ig. 4.5.1** (a) Broadside coupled lines.

**Fig. 4.5.1** (b) End-coupled lines.

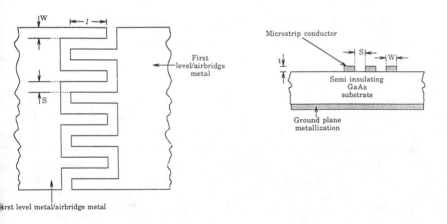

**'ig. 4.5.1** (c) Interdigitated capacitor layout and vertical cross section of an interdigitated capacitor.

basic subcomponents, as shown in Figure 4.5.2: the single microstrip line, coupled microstrip lines, microstrip open end, microstrip unsymmetrical gap, unsymmetrical microstrip 90° bend, and the T-junction. Available computer models of these subcircuits are used to simulate the interdigitated capacitor by using known methods of S-parameter network theory. Series and shunt lumped element equivalent circuit of interdigitated capacitors are shown in Figure 4.5.3.

The overlay capacitor geometry shown in Figure 4.5.4(a,b) employs dielectric films. Capacitance values as high as 10–50 pF can be realized in small areas.

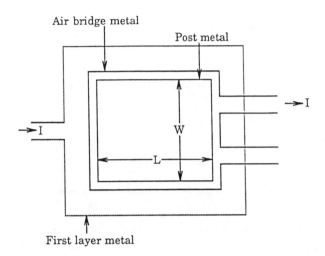

**Fig. 4.5.1** (d) Top view of a MIM capacitor.

**Table 4.5.1** Properties of Some Dielectric Films Used in MIM Capacitors

| Dielectric Film | Capacitance Per Unit Area ($pF/mm^2$) for 2000 Å Thick Film | Q | Thermal Coefficient ($ppm/°C$) | $F_{cv}$ | $F_{cq}$ | Method of Deposition |
|---|---|---|---|---|---|---|
| SiO | 275–325 | medium | 100–500 | low | medium | Evaporated |
| $SiO_2$ | 175–230 | maximum | 50–100 | medium | medium | Evaporated or sputtered |
| $Si_3N_4$ | 300–400 | maximum | 20–40 | high | high | Sputtered or CVD[a] |
| $Ta_2O_5$ | 1000–1200 | moderate | 10–150 | medium | high | Sputtered and anodized |
| $Al_2O_3$ | 350–400 | moderate | 400–600 | high | high | CVD or sputtered |
| Polymide | 30–40[b] | minimum | −400 to −500 | high | medium | Spun and heat treated |

[a] CVD is chemical vapor deposition.
[b] Capacitance/unit area for a 10,000 Å thick film.

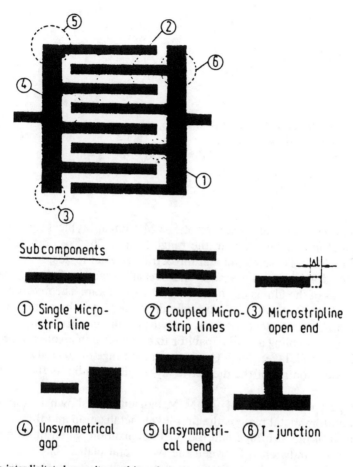

**Subcomponents**

① Single Micro-
   strip line

② Coupled Micro-
   strip lines

③ Microstripline
   open end

④ Unsymmetrical
   gap

⑤ Unsymmetri-
   cal bend

⑥ T-junction

**Fig. 4.5.2** The interdigitated capacitor and its subcomponents.

The dielectric films are suitable for low impedance (power) circuitry, bypass, and blocking applications.

Two useful figures of merit for dielectric films employed in MIM capacitors are the capacitance–breakdown-voltage product:

$$F_{cv} = \left(\frac{C}{A}\right) \cdot V_b$$

$$= \varepsilon_r \cdot \varepsilon_0 \cdot E_b$$
$$= (8\text{--}30) \times 10^3 \text{pF-V/mm}^2$$

(4.5.1)

and the capacitance–dielectric Q-factor product:

$$F_{cq} = \left(\frac{C}{A}\right) Q_d$$

$$= \left(\frac{C}{A}\right) \frac{1}{\tan (\delta_d)} \tag{4.5.2}$$

where

$C/A$ = capacitance per unit area,
$V_b$ = breakdown voltage,
$E_b$ = corresponding breakdown field,
$\varepsilon_r$ = dielectric constant,
$\tan \delta_d$ = dielectric loss tangent.

Breakdown fields on the order of 1–2 MV/cm are typical for good dielectric films. Dielectric constants are in the range from 4–20. Loss tangents can range from $10^{-1}$ to $10^{-3}$. The metallic and dielectric material losses, with the dielectric loss tangent dominating, determine the overall Q-factor of the MIM capacitor. The thickness of the film is on the order of 0.1 to 0.5 $\mu$m. The favored dielectric is silicon nitride ($Si_3N_4$), which is generally deposited after the completion of MESFET fabrication to protect the active devices. Silicon nitride has a relative permitivity of 6–7 enabling a 1 pF capacitor to be realized in an area of 55 $\mu$m square with a 0.2 $\mu$m thick layer. High Q-factors in the range of 50–100 are possible with this dielectric. Control of the dielectric layer thickness limits the tolerance to 5–10%.

The perspective view of the MIM capacitor is shown in Figure 4.5.4(a), where its bottom metal plate is the first-level metalization and the top metal plate is composed of the airbridge and the postmetalization. The thin-film insulator of thickness $d$ is sandwiched between the two metal plates. The area of the MIM capacitor is defined by the overlap area between first-level metal and the postmetal as shown by dimensions $L$ and $W$ in Figure 4.5.4(a) and in the top view of MIM capacitor in Figure 4.5.1(d). The parasitics associated with MIM capacitors are strongly dependent on the geometry of the layout of the capacitor. An important consideration is the connection to the top and bottom plates of the capacitor to avoid any large discontinuities while making such a connection. As shown in Figure 4.5.1(d), there is a large discontinuity in the bottom plate (first-level metalization) connection, whereas such a discontinuity is avoided in the top plate (airbridge) connection by providing multiple connecting lines. An equivalent circuit for the MIM capacitor is shown in Figure 4.5.5(a), where the top and bottom metal plates are modeled as transmission lines $T_1$ and $T_2$ with a MIM capacitor $C$ between them. $G$ represents the loss in the dielectric material of the capacitor, and $C_T$ and $C_B$ are the parasitic capacitance to ground due to the top and bottom metal

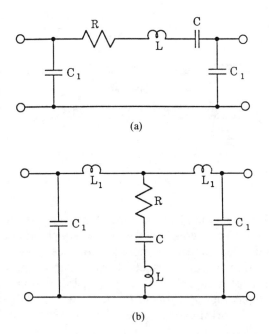

**Fig. 4.5.3** (a) Series equivalent circuit of interdigitated capacitor, and (b) shunt equivalent circuit of interdigitated capacitor.

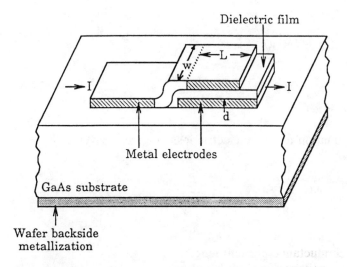

**Fig. 4.5.4** (a) MIM capacitor perspective view on GaAs substrate.

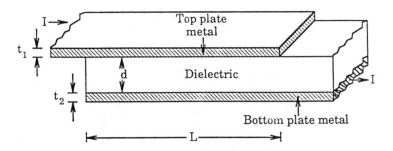

**Fig. 4.5.4** (b) MIM capacitor.

connecting lines. If the geometry of the capacitor's layout is large in any dimension (comparable to $\lambda/10$ of the highest frequency of operation in the circuit), the whole geometry is divided into two or more sections, as shown in Figure 4.5.5(a), for better modeling of the parasitic effects. Figure 4.5.5(b) shows another lumped element model for a MIM capacitor, where the transmission lines $T_1$ and $T_2$ are represented by an equivalent $R_1$, $L_1$ and $R_2$, $L_2$ network. With such a model, the impedance equation for MIM capacitors can be written as

$$Z(\omega) = \frac{Z}{r}\left[\coth\left(\frac{rl}{2}\right) + \frac{rl}{2}\right]$$ (4.5.3)

where

$$r = \sqrt{2(g + j\omega C)(r + j\omega L)}$$

and

$$Z = r + j\omega L$$

The equation for the dielectric loss tangent is given as

$$\tan \delta_d = \frac{g}{\omega C}$$ (4.5.4)

where

$g$ = conductance per unit length,
$C$ = capacitance per unit length.

Hence, with the assumption that $|rl| \ll 1$, which applies for a well designed

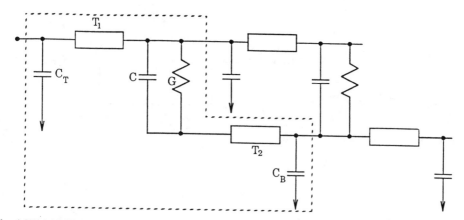

**Fig. 4.5.5** (a) MIM capacitor equivalent circuit.

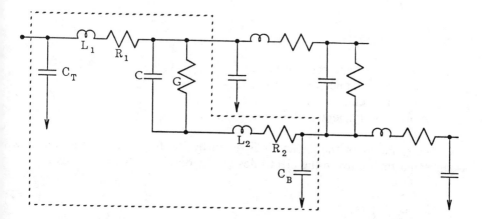

**Fig. 4.5.5** (b) Lumped element equivalent circuit for MIM capacitor.

capacitor, we obtain the following expression:

$$Z(\omega) \approx \frac{1}{j\omega C} + j\frac{2}{3}\omega L + \frac{2}{3}(rl) + \frac{\tan \delta_d}{\omega C} \tag{4.5.5}$$

The equivalent circuit corresponding to (4.5.5) is shown in Figure 4.5.6. We can neglect the inductive term. Note that the total resistance $2rl/3$, is just equal to $2R/3$, where $2R = 2(rl)$ is the sum of the resistance of the top and bottom plates. The factor of $\frac{1}{3}$ is present because the current distribution along the length of the electrodes is not uniform, but has a linear dependence on distance.

The corresponding $R$, $L$, and $C$ elements of such an equivalent circuit can be

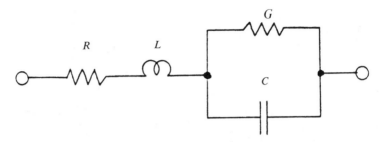

**Fig. 4.5.6** MIM capacitor equivalent circuit.

written as

$$R = \tfrac{2}{3}rl$$
$$C = Cl$$
$$G = \omega C \tan \delta$$
$$L = \tfrac{2}{3}L \cdot l$$

where

$r$ = resistance per unit length,
$C$ = capacitance per unit length,
$L$ = inductance per unit length,
$\tan \delta$ = loss tangent of the dielectric film.

The Q-factor of the capacitor is given by the following expression, if we assume that the losses in the electrodes are predominantely skin losses:

$$Q_c = \frac{3}{2\omega R_s (C/A) l^2} \qquad (4.5.6)$$

where

$R_s$ = surface skin resistivity,
$l$ = electrode length.

If both electrodes are thick compared to the skin depth, the $Q_c$ decreases with frequency. However, in practice, the bottom plate resistance dominates because its metal thickness is less than the skin depth. In this case, the bulk resistance is the dominant factor and $Q_c$ decreases inversely with frequency.

The dielectric Q-factor is given as

$$Q_d = \frac{1}{\tan \delta_d} \qquad (4.5.7)$$

and the total Q-factor is expressed by the following equation:

$$\frac{1}{Q} = \frac{1}{Q_d} + \frac{1}{Q_c} \qquad (4.5.8)$$

The quantitative behavior of interdigitated and overlay capacitors is shown in Figure 4.5.7. Capacitance values of interdigitated capacitors are far below those of MIM capacitors. The variation of two interdigitated capacitors as a function of frequency is shown in Figure 4.5.7(a). The Q-factor of interdigitated capacitors as a function of frequency in Figure 4.5.7(b) is indicated to be twice as high as that of MIM capacitors (Figure 4.5.7(c)).

Spiral inductors and thin-film capacitors may be used to form a resonant circuit. However the Q-factor is very low due to the inductor losses and even degraded by the losses of the MIM capacitor. Higher Q-factors can be achieved by using microstrip resonant elements. The longer lengths of the microstrip line are used to a quarter-wavelength to produce a parallel or series resonance, depending on whether the far end of the stub is short or open circuited. There are three sources of loss: conductor losses, radiation losses, and dielectric losses. Skin losses vary inversely with the substrate thickness and increase as the impedance increases. Distributed elements are recommended, for high-Q narrow-band circuits, provided that the necessary space is available. However, lumped elements are usually the better choice if broadband circuits are to be designed.

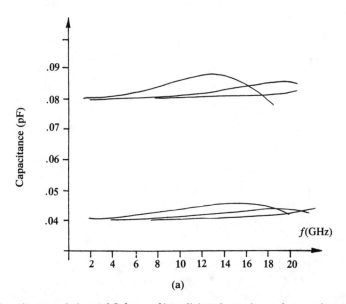

(a)

**Fig. 4.5.7** Capacitance variation and Q-factor of interdigitated capacitor and comparison with Q-factor of MIM capacitor.

**Figure 4.5.7** cont.

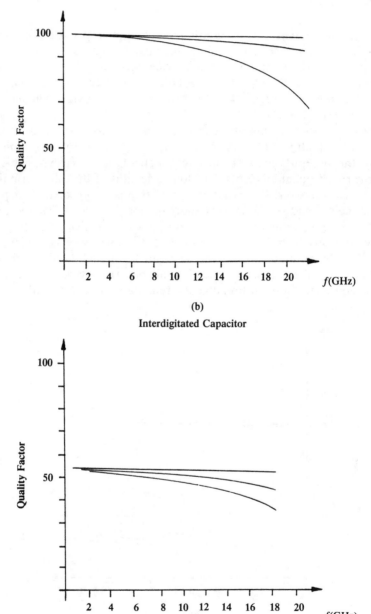

(b)

**Interdigitated Capacitor**

(c)

**MIM Capacitor**

Parasitic capacitance is introduced when the airbridge metal crosses over the first-level metalization line. In a crossover, the airbridge metal is typically used without postmetalization, resulting in an air gap at the crossover point between first-level metal and airbridge metal line, as shown in Figure 4.5.8. The parasitic capacitance in such a structure depends on the area of crossover and the air gap between the two metal lines. The air gap is typically 2–5 $\mu$m, and the parasitic capacitance is given as

$$C_{\text{crossover}} = \frac{\varepsilon_0 \varepsilon_r L \cdot W}{d} \tag{4.5.9}$$

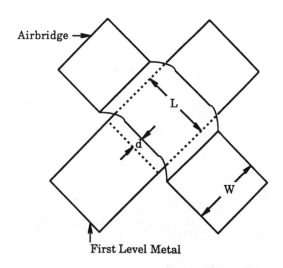

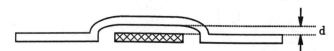

**Fig. 4.5.8** Crossover.

where

$\varepsilon_r$ = relative dielectric constant of the dielectric (air) between the two metal layers at crossover;
$L$ = length of the crossover area;
$W$ = width of the crossover area;
$d$ = vertical gap between the two metal lines.

## 4.6 PLANAR RESISTORS

Planar resistors in MMICs are essential for a variety of applications including load resistors for FETs, termination resistors for hybrid couplers, power dividers and combiners, stabilizing resistors to prevent parasitic oscillations (dampling resistors), feedback resistors for broadband circuits, and biasing circuits.

Planar resistors in MMICs are fabricated in a variety of ways, as shown in Figure 4.6.1, including (a) semiconductor films (implanted resistors), (b) mesa-etched resistors, and (c) deposited metal thin film resistors. The design considerations for planar resistors are sheet resistivity, temperature coefficient, thermal and long-term stability, frequency-dependent performance, and thermal resistance. See Table 4.6.1.

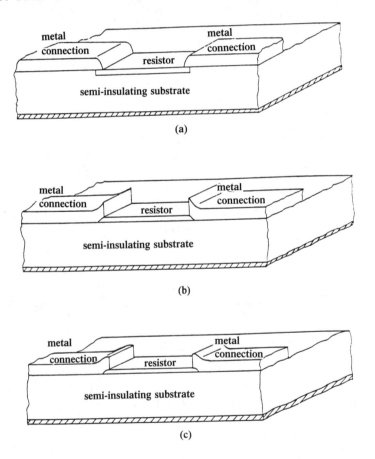

**Fig. 4.6.1** Planar resistors: (a) implanted, (b) mesa-etched, and (c) deposited.

**Table 4.6.1** Properties of Resistive Layers Used for Planar Resistors

| Material | Sheet Resistivity ($\Omega$/square) | Temperature Coefficient (ppm/°C) | Accuracy to Target | Stability | Method of Deposition |
|---|---|---|---|---|---|
| Cr | 10–20 | 3000 | moderate | moderate | Evaporated or sputtered |
| Ta | 30–200 | −100 to 500 | moderate | excellent | Sputtered |
| Ti | 10–100 | 2500 | moderate | moderate | Evaporated or sputtered |
| TaN | 250–300 | −150 to −300 | excellent | moderate | Sputtered |
| NiCr | 40–100 | 200 | excellent | excellent | Evaporated or sputtered |
| Bulk GaAs (Active Layer) | 100–1200 | 3000 | low | excellent | Implanted or epitaxially grown |

For the fabrication of resistors on a semi-insulating substrate, an isolated region of conducting metal or semiconductor film is defined, either by mesa etching, or isolation implant, or implantation of a high-resistive region within the semiconducting substrate. Thus, the material can either be deposited metal film or alternatively the active layer in GaAs, exactly as required for FETs, in conjunction with ohmic contacts. The properties of some resistive films and bulk GaAs resistors for the use in MMIC resistors are shown in Table 4.6.1.

The sheet resistivity of active layer GaAs resistors is typically on the order of 100–1200 $\Omega$ per square. The temperature coefficient of GaAs resistors is about 3000 ppm/°C compared to almost $\frac{1}{10}$ of this value for deposited metal thin-film resistors. This low value, however, is usually achieved after annealing the deposited resistor at high temperatures. Without annealing, the temperature coefficient is still better than for monolithic resistors, but the long-term stability may be poor.

GaAs resistors exhibit a nonlinear behavior at high electric fields when the carriers become velocity saturated. To ensure linearity of the component, the electric field across the resistor must be limited to about 500 V/cm–1 kV/cm as shown by electric field *versus* electron velocity curve in GaAs, in Figure 4.6.2. In a MMIC where the maximum voltage is 5 V, a length of 70–100 $\mu$m will ensure that all the GaAs resistors behave linearly.

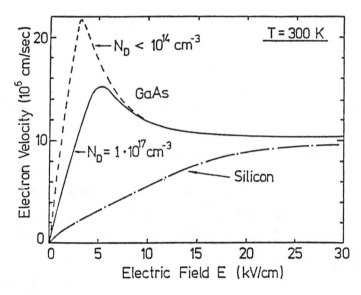

**Fig. 4.6.2** Electric field *versus* electron velocity curve for nonlinear behavior of GaAs resistor.

For deposited metal films, resistivity ranges from 10–300 Ω per square, which covers a different range than the resistivity of GaAs implanted resistors. Typical thin-film resistor materials include nichrome, titanium, tantalum, and tantalum nitride, with nichrome being the most popular. Metal films are usually formed by evaporating a metal layer on the substrate and lifting off the desired pattern by using photolithography.

Due to the electromigration at high current densities in deposited metal films, current densities should not exceed the specified ratings to ensure reliability of the component. Because the deposited layers are very thin, typically 150 nm, the current density must be limited by having an adequate width of the component.

Fabrication tolerances limit the structure to minimum dimensions of about 10 μm. To minimize step discontinuities, we would recommend maintaining approximately the same width for the feeding conductor line as the resistor width, or the use of several connecting lines along the resistor width. Large value of resistors using thin-film metal require long structures to obtain the desired resistor values. The length, however, should be kept far below a quarter-wavelength to ensure a lumped element. The structure otherwise will tend to act as a transmission line, and the VSWR will increase accordingly. Power handling capabilities are important, even in small-signal circuits, because the bias resistors must carry the bias current of the active devices.

The length of the resistive strip for a given width is simply expressed as

$$l = \frac{WR_{dc}}{\rho_s}$$

(4.6.1)

ere

$W$ = the width of the resistor determined by the current requirements through the resistor,
$R_{dc}$ = desired resistance,
$\rho_s$ = sheet resistivity of the resistive layer.

Metal films are usually preferred over monolithic resistors because the latter exhibit a strong temperature dependence, and the previously described nonlinearity. Metal film resistors in general have been found to exhibit less noise than implanted resistors. Thin-film resistors, however, require additional masks and processing steps.

Parasitics of resistors include capacitance to the ground plane and series inductance of the metal strip. A simple equivalent circuit is shown in Figure 4.6.3.

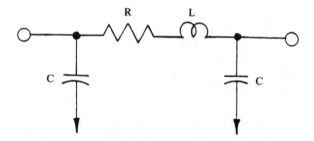

**Fig. 4.6.3** Equivalent circuit for a GaAs resistor.

For use at high frequencies, these parasitics are considered distributed and thus the resistor can be modeled as a transmission line section, but in this case it is a very lossy transmission line section. The analysis leads to an expression of the impedance of the resistor as follows:

$$Z(\omega) = R_{dc}[AB \tanh(A/B)]$$

(4.6.2)

where $R_{dc}$ is the dc resistance of the resistor, and

$$A = \sqrt{1 + j\frac{Z_0}{R_{dc}}\,\theta}$$

$$B = \sqrt{\frac{Z_0}{R_{dc}} \times \frac{1}{j\theta}}$$

$$\theta = \frac{\omega l}{v}$$

(4.6.3)

Here, $Z_0$ and $v$ are the characteristic impedance and phase velocity of the resistive segment, treated as a lossless transmission line.

Considering the thermal power dissipation rating is an important aspect of MMIC design because the dimensions are very small. The heat flux pattern associated with the resistor has been shown to be identical to the electric field pattern. Therefore, the thermal resistance and the capacitance to ground are always directly related, and given by the relationship:

$$C = \frac{\varepsilon_0 \varepsilon_r}{k} \cdot \frac{1}{R_\theta}$$

(4.6.4)

where

$C$ = capacitance to ground,
$\varepsilon_0$ = permittivity of free space,
$\varepsilon_r$ = permittivity of the substrate,
$k$ = thermal conductivity of the substrate,
$R_\theta$ = thermal resistance.

For reliable operation, the resistor temperature must be kept below 200°C. Thus, for a maximum power dissipation of one-third of a watt and an ambient temperature of 100°C, a thermal resistance not greater than 300°C/W is required.

The noise generated by a resistor is modeled in SPICE, as shown in Figure 4.6.4, which includes a noise current generator $i_{nR}$ in parallel with the resistor of value $R$. The noise current is given by the equation:

$$i_{nR} = 4kTR$$

where

$T$ = temperature of the resistor in kelvins,
$k$ = Boltzmann's constant.

As mentioned earlier, the noise generated in implanted resistors formed on

GaAs is much larger than the noise generated by the thin-film resistors. Hence, in low-noise applications, the use of thin-film resistors is recommended.

In certain applications, a single-gate FET can also be used as a resistor. As described in detail in Section 4.1 on the characteristics of a single-gate FET, above knee-voltage the current through the FET saturates and the ac resistance is much larger than a similarly sized bulk resistor can provide. Such a resistor is termed an active resistor, where the FET's gate and source are connected together, resulting in an equivalent circuit with a resistor in parallel with a capacitor, which is essentially the drain-to-source capacitor of the FET when $V_{gs}$ = zero volts. An active resistor is physically smaller than a comparably valued monolithic resistor, and hence provides better performance due to smaller parasitics, such as capacitance-to-ground and transmission line effects.

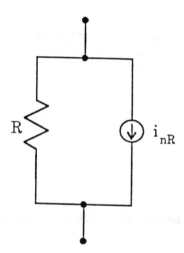

**Fig. 4.6.4** The noise model for a resistor.

## 4.7 TRANSMISSION LINES

### 4.7.1 Introduction

The choice of transmissive media in the design of MMICs is similar to that already successfully implemented in hybrid circuits, such as using quasi-TEM transmission lines rather than non-TEM waveguides. Among the advantages of quasi-TEM lines are their intrinsic wideband behavior, as opposed to the more dispersive nature of non-TEM media, and the easy ground definition that they allow. A disadvantage of the quasi-TEM lines, limiting their use in the millimeter-wave range, is their comparatively large ohmic losses due to the use of strip conductors. Design of a quasi-TEM line can either use both sides of the GaAs substrate (i.e., the front and the back side of the substrate), or lie entirely on the top side. The first case corresponds to the microstrip approach (Figure 4.7.1(a)) and second case is the coplanar or uniplanar [27] approach (coplanar waveguide, Figure 4.7.1(b)). Varieties of the coplanar waveguide are conductor-backed (Figure 4.7.1(c)) and with finite-extent lateral ground planes (Figure 4.7.1(d)), which is a more realistic model for practical coplanar waveguides. The microstrip and coplanar approaches are not simply a different choice of transmission media, but also entail a markedly different circuit design philosophy. Moreover, those two approaches to MMICs are not the only conceivable ones. Indeed, the monolithic approach allows multilayer circuits to be made, in which transmission lines and discrete components are located on the top and bottom sides of the substrate, but also stacked structures separated by dielectric passivation layers are implemented. For example, a microstrip patch array can be printed on the top surface of the substrate, and feed lines coupled to patches are located either on the back side or under a passivating dielectric layer. A multilayer design strategy presents a challenge as it requires a strict control of coupling between different lines and components, which can be achieved in turn only by accurate three-dimensional electromagnetic modeling.

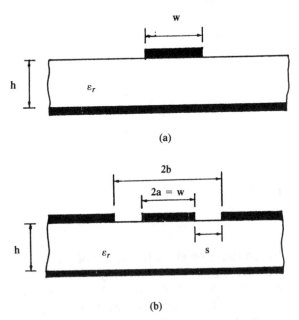

(a)

(b)

**Fig. 4.7.1** Waveguide structures for MMICs: (a) microstrip; (b) coplanar waveguide; (c) conductor-backed coplanar waveguide; (d) coplanar waveguide with finite ground planes.

cont.

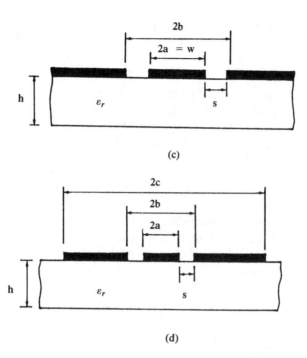

(c)

(d)

Today, the microstrip approach is by far the most popular. Actually, the microstrip on GaAs substrates of thickness ranging from 100 to 400 $\mu$m is practical up to approximately 30 GHz providing a transmission line with reasonably low losses and dispersion. The physically realizable impedance range is wide enough as to cover most practical applications. Series element connection is very easy, whereas parallel connection is more troublesome because of the absence of a top side ground plane. Suitable techniques exist to circumvent this problem, such as the use of top side wrap-around ground planes or via holes to make the bottom plane accessible from the top of the substrate. However, the first solution imposes constraints on circuit layout, and the second increases the IC process complexity. Nonetheless, microstrip circuit design is greatly aided by the possibility of transferring the design experience acquired in hybrid circuits to MMICs, and also exploiting the variety of CAD tools available for microstrip circuit layout, line modeling, and discontinuity modeling. Most CAD tools on the market are not particularly suitable for MMICs, having been developed for hybrid circuits, but the increasing demand for MMIC CAD is rapidly fostering the development of a new generation of tools especially conceived for MMIC design [21].

Although the use of coplanar or uniplanar circuits has been until now rather sparse, and primarily devoted to special applications, the many significant realizations demonstrated during the last few years [17,15,27,30,40] show that the coplanar approach is a design strategy that ultimately can prove more effective than the microstrip approach at least for some classes of circuits. The main advantages offered by coplanar lines can be listed as follows:

- The performances of coplanar waveguides are comparable, and sometimes even better, than those of the microstrip in terms of guided wavelength (Figure 4.7.2), losses, dispersion, and impedance range.
- The coplanar waveguide allows easy series and shunt element connection.
- The coplanar waveguide impedance is almost insensitive to substrate thickness.
- Active elements such as MESFETs can be easily connected as they are intrinsically coplanar.
- The coupling between neighboring lines is reduced owing to the presence of grounded lines between the signal carrying lines.
- On-wafer measurements through coplanar probes are easier and more direct than in microstrip circuits.

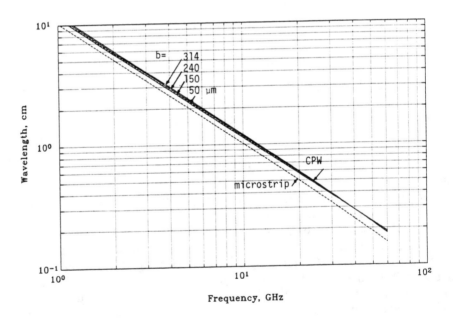

Fig. 4.7.2 Guided wavelength as a function of frequency for 50 $\Omega$ microstrip and coplanar waveguides on GaAs substrate ($\varepsilon_r = 13$). The substrate thickness is $h = 300\ \mu$m.

Moreover, the coplanar approach allows greater flexibility in the use of mixed structures and transitions to slotlines or coupled slotlines, which can be profitably exploited in some applications (e.g., mixers, balancing units). However, coplanar lines also have disadvantages that effectively confine their use to low-power applications:

- Requirement for a thick substrate to avoid spurious coupling with parasitic modes in case of a thin (100 $\mu$m–200 $\mu$m) substrate.
- The power handling capabilities of uniplanar circuits are unsatisfactory, associated with thick wafer requirements, leading to poor heat sinking capabilities.
- To suppress a parasitic slot-like mode, the ground planes have to be connected together by means of air bridges, thereby making circuit realization more complex.
- Far less design experience exists with respect to microstrip circuits design. In particular, CAD tools for coplanar design are confined to line modeling; no CAD-oriented models are available for discontinuities, coupled line parameters and dispersion.

The important parameters for designing transmission lines are the practical impedance ranges, the effect of frequency dispersion, parasitic coupling with other lines, surface waves, or radiation, and the discontinuities. Analytical formulas for microstrip parameters (impedance, effective permittivity, losses) are readily available in the published literature, but analytical models for coplanar lines are not so widely known or available. Therefore, a few basic formulas for coplanar waveguides (impedance, dispersion, losses) are collected for easy reference in Appendix 4A.

### 4.7.2 Microstrip and Coplanar Lines for MMICs

#### 4.7.2.1 Microstrip

Microstrip line models have been the object of investigation since the 1950s. Thus, a variety of analytical approximations exist for the characteristic impedance $Z_0$ and effective permittivity $\varepsilon_{eff}$ of the microstrip, both in the quasistatic (low frequency) approximation [13,39] and to account for frequency dispersion. The most accurate dispersion models available at present for $Z_0$ and $\varepsilon_{eff}$ are probably those in [22,24], respectively. For a more complete discussion of analytical microstrip models, the reader can refer to the books [9,16]. All available analytical models originally refer to zero-thickness lines on homogeneous substrates, and therefore can be grossly inaccurate for narrow strips laid on multilayered substrates, including low $\varepsilon_r$ passivations [21]. Moreover, inaccuracies can also arise for high impedance lines (e.g., 100 $\Omega$–70 $\Omega$ lines which are narrow, approximately 8–30 $\mu$m on a 100 $\mu$m thick substrate) laid on thin substrates, as the influence of the fringing field relative to finite-thickness strips is only approximately accounted in the above-mentioned models.

The behavior of the quasistatic parameters of a microstrip on GaAs substrate is shown in Figure 4.7.3. Typical impedance values are 50 $\Omega$, approximately requiring a $w/h$ ratio ($w$ = strip width, $h$ = substrate thickness) $\approx 0.7$, but also narrower lines of higher impedance (e.g., 70 $\Omega$, $w/h \approx 0.3$) are common, allowing a more compact circuit layout and lower coupling between nearby lines. However, an upper bound arises to the line impedance, not only because of technological constraints on the minimum line width that can be realized, but also because high impedance lines have high ohmic losses, as we will discuss later.

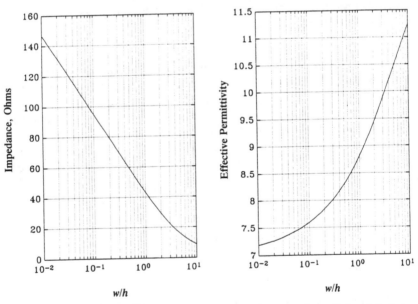

**Fig. 4.7.3** Impedance (left) and effective permittivity (right) for microstrip on GaAs substrate ($\varepsilon_r = 13$).

### 4.7.2.2 Coplanar Waveguides

Ideal coplanar waveguides on an infinitely thick substrate [38] are characterized by $\varepsilon_{\text{eff}} = (\varepsilon_r + 1)/2$, independent of geometry and line impedance. Moreover, the line impedance only depends on the ratio between the slot width and the strip width, or equivalently on the ratio $a/b$, where $a = w/2$, $b = s + a$ (Figure 4.7.4). Notice that $2b$ is the overall lateral extent of the line. The property $\varepsilon_{\text{eff}} = (\varepsilon_r + 1)/2$ also holds for all propagation modes of coupled coplanar waveguides, thereby allowing the design of coupling structures having theoretically infinite directivity. Such a property is partly lost in practical lines on nonideal (finite-thickness) substrates [7,37]. Moreover, in this case, the line impedance also becomes sensitive to the substrate thickness. Figure 4.7.5 shows the behavior of the impedance of coplanar waveguides around 50 $\Omega$ as a function of the shape ratio $a/b$ and of the substrate thickness $h$. As a rule of thumb, the substrate should be at least as thick as the overall lateral extent of the line (i.e., $h > 2b$) to make the influence of $h$ on $Z_0$ almost negligible. For a given $h$, this conversely imposes a limitation on the maximum line dimensions (e.g., for $h = 300\ \mu\text{m}$, we must have $b < 150\ \mu\text{m}$). Although there is no limitation to the impedance range, which only depends on the shape ratio $a/b$, thin substrates require small lines, which, in turn, are affected by heavy ohmic losses.

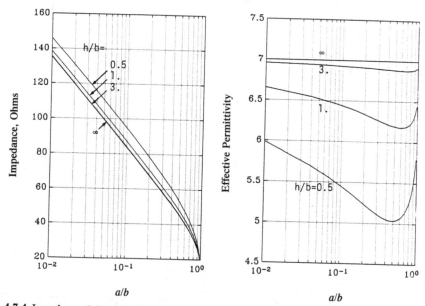

**Fig. 4.7.4** Impedance (left) and effective permittivity (right) for coplanar waveguide on GaAs substrate ($\varepsilon_r = 13$).

Another cause of nonideal behavior is the finite extent $w_{gp} = c - b$ of the lateral ground planes. As a consequence, the line impedance increases and $\varepsilon_{\text{eff}}$ slightly decreases (Figure 4.7.6) [8,37]. In practical circuits, the overall lateral line extent should be kept as small as possible, provided that no spurious coupling

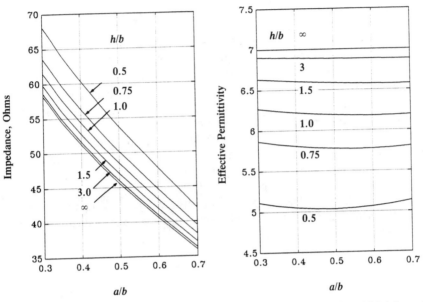

**Fig. 4.7.5** Effect of substrate thickness on impedance (left) and effective permittivity (right) for coplanar waveguide on GaAs substrate ($\varepsilon_r = 13$).

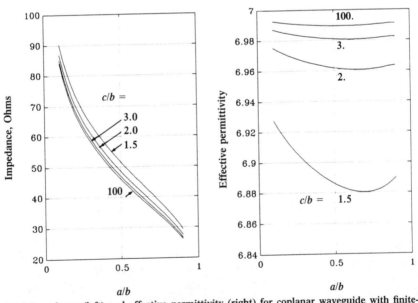

**Fig. 4.7.6** Impedance (left) and effective permittivity (right) for coplanar waveguide with finite-extent lateral ground planes on GaAs substrate ($\varepsilon_r = 13$). The substrate thickness is $h = 300$ $\mu$m, and the lateral ground-plane spacing is $2b = 200$ $\mu$m.

arises between neighboring lines and the impedance level of the line is not seriously affected. A reasonable compromise, actually displayed in most practical coplanar circuits, is to have $c \approx 3b$, at least.

A last variety of coplanar waveguide is the so-called conductor-backed coplanar waveguide [6,37]. There seemingly are many reasons to advocate the use of conductor backing in uniplanar circuits. GaAs substrates have poor mechanical properties; in particular, thin substrates (e.g., $h < 100~\mu m$) are very brittle. Thus, coplanar circuits should be made on rather thick substrates, which, however, have poor thermal properties (high thermal resistance seen by active devices). Conductor backing allows thinner substrates to be used, and connecting the back of the circuit to a suitable heat sink.

Unfortunately, backside conductor also has some disadvantages. Apart from problems connected to spurious coupling with parasitic modes, which will be discussed later, backside conductor lowers the impedance level of the line and makes it again dependent on $h$ (See Figure 4.7.7). In other words, backside conductor leads to a mixed coplanar-microstrip structure. If the aim is to obtain coplanar rather than microstrip behavior, the substrate ought to be approximately as thick as in the case where no backing is present (i.e., we should have $h > 2b$). Now, if we consider the actual substrate thickness needed to allow reasonable power dissipation for even medium-power devices, we find that very low $h$ is needed (e.g., $h \approx 30$–$60~\mu m$). If such a low thickness can be locally obtained through substrate thinning and connection to a heat sink, making a whole conductor-backed coplanar circuit on such a thin substrate does not seem to be a good policy because the requirement $h > 2b$ leads to extremely small lines with unacceptable ohmic losses. As a conclusion, backside conductor does not seem to yield an entirely acceptable solution for high-power circuits. In low-power application, to which the coplanar approach seems to be confined at present, the usefulness of backside conductor should be discussed independently of thermal considerations. Additional comments on the effect of backside conductor on spurious coupling will be presented below.

Finally, for a discussion of the effect of slight line asymmetries and of a metallic cover, the reader can refer to [4,8], respectively.

### 4.7.2.3 Impedance Limits

Limitations to the useful impedance range arise from two sources. First, strip and slot widths cannot be smaller than a certain value, due to processing limitations as well as fringing effects. Second, widths cannot be larger than a suitably small fraction of the guided wavelength. Otherwise, the structure would behave as a planar resonator rather than as a transmission line. For example, we can take as a minimum strip and slot width the value of $5~\mu m$ and as a maximum the values of $\lambda_g/8$. Actually, $1~\mu m$ line widths can now be obtained with comparative ease, but such small lines are irregular and have unacceptably high ohmic losses.

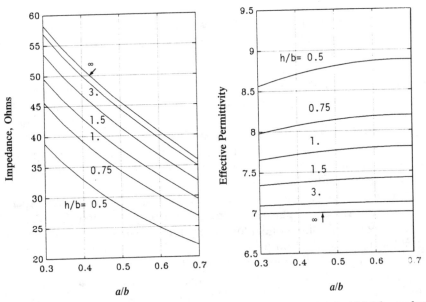

**Fig. 4.7.7** Effect of substrate thickness on impedance (left) and effective permittivity (right) for conductor-backed coplanar waveguide on GaAs substrate ($\varepsilon_r = 13$).

The impedance range of the microstrip is simply obtained, at each frequency, by taking the minimum strip width (maximum impedance) and the maximum strip width (minimum impedance). The result is shown in Figure 4.7.8 for infinitely thin lines on GaAs substrate. It can be noticed that the impedance range also depends on substrate thickness. For coplanar waveguides, the impedance limits are obtained by combining maximum strip width with minimum slot width (lower limit) or minimum strip width with maximum slot width (upper limit). The result is shown in Figure 4.7.9, again for infinitely thin lines on an infinitely thick GaAs substrate.

The useful impedance ranges of microstrips and coplanar waveguides are comparable, although the coplanar waveguide is slightly superior in this respect. In both cases, the range decreases with increasing frequency. For a practical coplanar waveguide on finite-thickness substrate, the impedance range shifts slightly toward higher values. However, line designers must also be aware of the presence of losses, which are heavier on high impedance microstrips and on high and low impedance coplanar waveguides.

### 4.7.2.4 Frequency Dispersion

Frequency dispersion of the line parameters, above all $\varepsilon_{\text{eff}}$, should play a

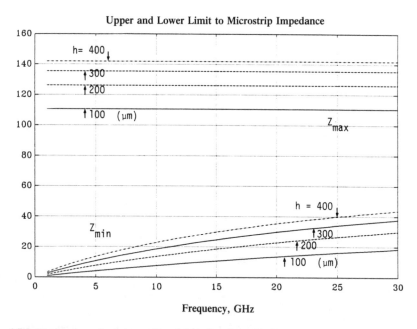

**Fig. 4.7.8** Maximum and minimum impedance for microstrip lines on GaAs substrate ($\varepsilon_r = 13$), for several substrate thicknesses. The maximum impedance $Z_{max}$ corresponds to $w = 5\ \mu$m, the minimum impedance $Z_{min}$ to $w = \lambda_g/8$.

minor role in MMICs due to the small values taken by $h/\lambda_0$ ratio in the frequency band of interest. Whereas microstrip dispersion has been extensively studied, dispersion in coplanar waveguides is less well known. However, enough data exist to allow an approximation of dispersion for coplanar waveguides as well. An analytical approximation to the frequency dependence of $\varepsilon_{eff}$ for a coplanar waveguide has been presented [14] and is reported in Appendix 4A. Moreover, numerical data on conductor-backed coplanar waveguides and coplanar waveguides with finite-extent ground planes can be found in [31,12], respectively. Therefore, the following general conclusions can be drawn:

- For both microstrip and coplanar lines, dispersion amounts to an increase in $\varepsilon_{eff}$ from the low frequency value to the asymptote of $\varepsilon_r$. The qualitative behavior of $\varepsilon_{eff}(f)$ is the same for both structures.
- Line dispersiveness is affected by substrate thickness in approximately the same way for both microstrips and coplanar lines: the thinner the substrate, the less dispersive the line. However, in coplanar waveguides an important role is played by the line dimensions $a$ and $b$: narrow lines are less dispersive than wide ones.
- Microstrips and coplanar waveguides having the same dimensions have

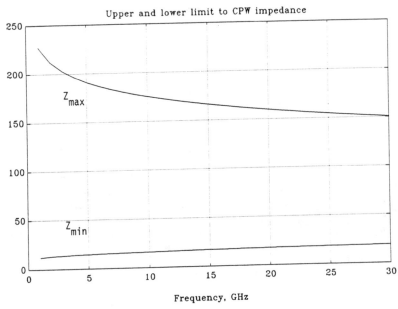

**Fig. 4.7.9** Maximum and minimum impedance for coplanar waveguides on an infinitely thick GaAs substrate ($\varepsilon_r = 13$). The maximum impedance $Z_{max}$ corresponds to $w = 5$ $\mu$m, $s = (b - a) = \lambda_g/8$; the minimum impedance $Z_{min}$ to $w = \lambda_g/8$; $s = (b - a) = 5$ $\mu$m.

   comparable dispersion, but conductor-backed coplanar waveguides and those with finite ground planes are less dispersive than conventional, ideal coplanar waveguides.

- The effect of dispersion on $Z_0$ is more controversial because no unique definition for $Z_0(f)$ exists. In the so-called power-current definition [22], the impedance slightly increases with increasing frequency for microstrip lines, and may be expected to behave the same way for coplanar waveguides.

An example of the frequency behavior of $\varepsilon_{eff}$ for 50 $\Omega$ lines on 300 $\mu$m GaAs substrate is shown in Figure 4.7.10. Microstrip dispersion is modeled as in [24], and for the coplanar waveguide the analytical expression proposed in [14] is used. Although neither the microstrip nor the coplanar waveguide are particularly dispersive over the frequency range where MMICs operate, frequency dispersion ought to be taken into account when designing narrow band components.

### 4.7.2.5 Parasitic Coupling and Radiation

Parasitic coupling on uniform lines can occur either because the quasi-TEM

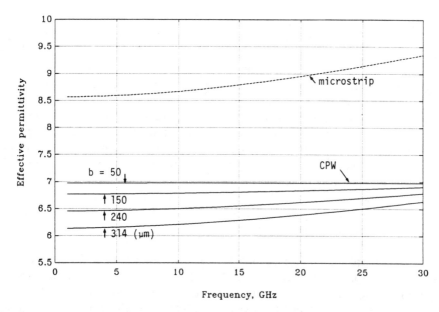

**Fig. 4.7.10** Effective permittivity dispersion for 50 Ω microstrip (above) and coplanar waveguide (below) on a 300 μm thick GaAs substrate ($\varepsilon_r = 13$).

field of the line couples with other quasi-TEM fields (line-to-line coupling) or because coupling occurs with surface waves or free-space radiation. Coupling between quasi-TEM modes and other guided or radiated waves is significant only in the presence of phase velocity synchronism. If this condition occurs, circuit operation deteriorates severely due to power conversion to spurious modes or radiation.

A different sort of coupling can occur in the presence of line discontinuities. In such cases, higher-order line modes are excited and the related power can easily convert into surface waves or free-space radiation. Such effects modify the circuit behavior of line discontinuities such as impedance steps or open circuits, and can be approximately modeled by concentrated radiation conductances. We shall try to collect some ideas on spurious coupling mechanisms so as to give design criteria on this fairly complex matter.

### Line-to-Line Coupling

While closed waveguides offer an extremely high degree of isolation between lines and circuit components, the open nature of planar circuits and, in particular, the small size of MMIC's lead to a comparatively high degree of unwanted coupling between elements. A possible design philosophy [21] could consist of exploiting rather than avoiding coupling. However, although this approach is certainly possible for certain kinds of concentrated planar elements (e.g., spiral inductors), it seems to fail as a general rule. In fact, not only is it still to be demonstrated that a MMIC can be made to work in the presence of fairly strong interelement coupling, but also under such conditions design ought to be based on an electromagnetic CAD tool able to perform a global three-dimensional analysis and optimization of the circuit, which are still far from practical use given their current status. Therefore, today's design is often based on keeping elements far enough from each other to render coupling negligible. Unfortunately, design criteria are often based on quasistatic coupling estimates, therefore neglecting the possible occurrence of surface-wave excitation and discontinuity radiation. Although such phenomena usually play a minor role, their effect is felt in high-isolation devices, like microstrip couplers, the directivity of which can be significantly worse than theoretical predictions (e.g., not more than 20 dB at 10 GHz [16]).

As an example of a coupling estimate, let us consider a situation wherein two parallel matched lines are located at a distance $D$ between line centers (see the inset in Figure 4.7.11). Such a line pair can be considered as a matched four-port coupler; the parasitic influence between lines can be expressed in terms of the near-end coupling (maximum coupling between ports 2 and 1 occurring at the coupler centerband frequency) and far-end coupling (coupling between ports 2 and 3, commonly referred to as coupler directivity).

The near-end coupling is mainly influenced by lateral field confinement. In this respect, coplanar waveguides are superior to microstrip because a grounded plane lies between two parallel lines. In Figure 4.7.11(a), this is true only for $D/h > 1.66$ (approximately) because, for this $D/h$ ratio, and keeping the strip and slot widths of the lines constant, the width of the ground plane lying between the two coupled guides shrinks to zero. For lower $D/h$ values, the coupling given in Figure 4.7.11(a) refers to coupled coplanar waveguides without intermediate ground plane, a structure commonly referred to as a coplanar coupler. For this configuration, the coupling grows actually larger than for microstrips of equal width.

Far-end coupling is mainly due to the phase velocity mismatch between the even and odd modes of the coupled line pair; it increases with increasing line length, increasing frequency, and decreasing line spacing. Fairly extensive measurements of near- and far-end coupling between 50 $\Omega$ microstrip lines on GaAs substrate can be found in [3]. In Figure 4.7.11(b) far-end coupling measurements from [3] have been collected for several line lengths and spacings. Clearly, from set A, coupling values of up to 10 dB can be found in long lines (several wavelengths) for $D/h$ values greater than one, but also (sets C and D) going under 30 dB coupling is difficult for very high $D/h$ ratios, in contrast with estimates derived from static criteria. The reader can refer to [3] for a more detailed discussion on

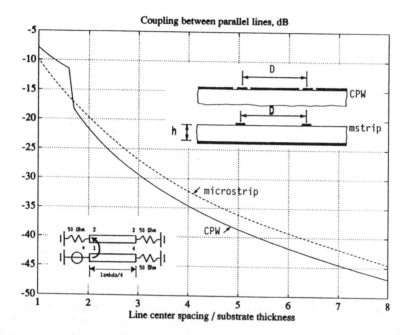

**Fig. 4.7.11(a)** Theoretical near-end coupling in dB (i.e., coupling between port 1 and 2) for pairs of end-matched parallel microstrips and coplanar waveguides on GaAs ($\varepsilon_r = 13$) substrate, as a function of line center spacing $D$. Both microstrips and coplanar waveguides are 50 $\Omega$ lines when isolated ($D \to \infty$). For microstrips, the substrate thickness is $h = 300$ $\mu$m; coplanar waveguides are on an infinitely thick substrate.

the effect of screen lines. As a rule of thumb, the distance between the centers of short parallel lines should be at least $D/h > 2$ to grant a theoretical minimum 20 dB near-end isolation.

As a further example of a practical situation where coupling arises between a microstrip line and a nearby wrap-around ground plane, let us consider the configuration shown in Figure 4.7.12 [42]. If we call $w_0$ the strip width needed to obtain an isolated 50 $\Omega$ line, the figure shows the strip narrowing needed to compensate for the added capacitive effect of the ground plane, thereby keeping constant the line impedance at 50 $\Omega$. Note, however, that in this way conductor losses increase with respect to the isolated line. Therefore it is safe to set $s > h$ in layout design. A more conservative approach, depending on the availability of space in the MMIC, requires $s \geq 3h$.

*Coupling with Spurious Modes in Coplanar Waveguides*

Coupling with spurious modes is an important issue in coplanar design. Actually, coplanar waveguides also support a variety of quasi-TEM or non-TEM

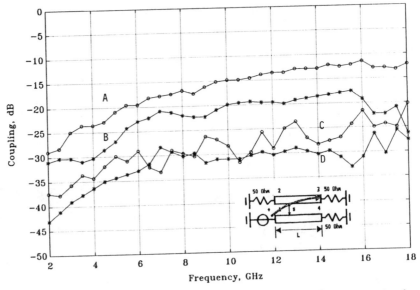

**Fig. 4.7.11(b)** Measured [3] far-end coupling (i.e., coupling between port 1 and 3) between pairs of parallel 50 $\Omega$ microstrip lines on a 300 $\mu$m thick GaAs substrate as a function of frequency, varying the coupled section length $L$ and the strip separation $s$: (A) $s = 500\ \mu$m, $L = 6.1$ mm; (B) $s = 500\ \mu$m, $L = 2.0$ mm; (C) $s = 2\ \mu$m, $L = 6.1$ mm; (D) $s = 4\ \mu$m, $L = 2.0$ mm.

(slot-like) modes which must be suppressed as much as possible. Namely,

- All coplanar lines support a parasitic slot-like mode which is odd with respect to the central conductor. Such a coupled-slot wave is a true quasi-TEM mode when the lateral ground planes have finite extent [8] and can be excited both by discontinuities and by synchronous coupling. To suppress this mode, the lateral ground planes must be connected at short intervals (less than $\lambda_g/4$) by means of airbridges.
- Coplanar lines, on a wafer with backside metalization, with finite-extent lateral ground planes also support a microstrip-like mode in which all strips have the same potential [20]. Although synchronous coupling with this mode is impossible due to the large difference of effective permittivities, the microstrip-like mode can be excited at discontinuities (typically, at short circuits). Although mode conversion at short circuits is not dramatic from a quantitative point of view [20], suppressing this spurious mode is practically impossible. We would need to connect the upper and lower ground planes through via-holes and wrap-arounds, which would be precisely what we should avoid doing by the use of coplanar waveguides. Thus, this sort of parasitic coupling may be a further reason why conductor backing in practice will be less appealing than expected.

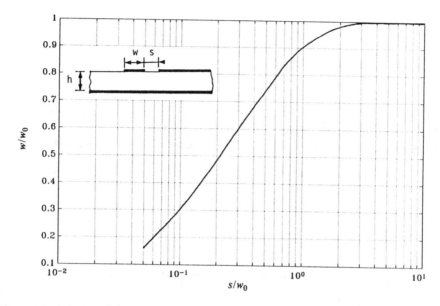

**Fig. 4.7.12** Influence of a parallel lateral ground plane on the impedance $Z$ of a microstrip line on GaAs substrate $\varepsilon_r = 12.7$ [42]. $Z(w_0, s \rightarrow \infty) = 50\ \Omega$; $w/w_0$ is the strip width reduction needed to compensate for the effect of the ground plane placed at a normalized distance $s/w_0$ so as to keep constant the line impedance ($Z(w,s) = 50\ \Omega$). The result is fitted on the two cases $w_0 = 109\ \mu m$, $h = 150\ \mu m$ and $w_0 = 146\ \mu m$, $h = 200\ \mu m$ [42].

### Coupling to Surface Waves

Surface waves are TE or TM modes supported by the grounded dielectric layer whereon strips lie. Moreover, some structures also allow for the propagation of quasi-TEM waves (e.g., the coplanar waveguide with backside metalized plane). The structures of interest are the grounded dielectric waveguide, the partially filled parallel plate waveguide and dielectric-filled parallel-plate waveguide (Figure 4.7.13). The first and second structures support a fundamental TM mode with zero cut-off frequency, whereas the fundamental mode of the last structure is TEM. (Surface wave coupling is discussed in [16, Chapter 15], mainly for microstrips; in [32] a discussion can be found on conductor-backed coplanar waveguides.) As a general remark, synchronous coupling with surface waves is unlikely to occur in the frequency band where MMICs operate; however, surface waves can be excited at discontinuities.

Taking into account that for all waveguiding structures shown in Figure 4.7.14 the effective permittivity of the fundamental surface mode is given by the

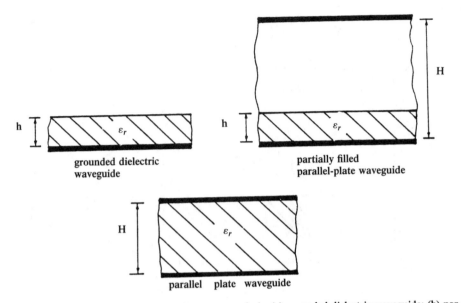

**Fig. 4.7.13** Waveguide structures for surface wave analysis: (a) grounded dielectric waveguide; (b) partially filled parallel-plate waveguide; (c) parallel-plate waveguide.

lowest-order solution of the following dispersion relation:

$$\frac{1}{\varepsilon_r}\sqrt{\frac{\varepsilon_r - \varepsilon_{\text{eff}}}{\varepsilon_{\text{eff}} - 1}} = \frac{\tanh\left(2c_0 f(H - h)\sqrt{\varepsilon_{\text{eff}} - 1}\right)}{\tan\left(2c_0 fh\sqrt{\varepsilon_r - \varepsilon_{\text{eff}}}\right)} \tag{4.7.1}$$

where $c_0$ is the free-space velocity of light. The following remarks can be made:

- For microstrip lines, coupling can occur with the $TM_0$ mode mainly at discontinuities. Synchronous coupling can only occur at very high frequencies and with lines of very high impedance (see Figure 4.7.14). Such coupling is suppressed for lines on GaAs substrates having arbitrary impedance if the operating frequency $f$ is such as $f < 30.6/h$, where $f$ is expressed in GHz and $h$ in mm. If the microstrip is covered, the dispersion properties of the surface wave slightly change at low frequencies (Figure 4.7.14), but the same remarks apply.
- Coplanar waveguides on finite-thickness substrates can support the same kind of surface wave as microstrips, only the surface wave propagates along the *back* of the line. The same remarks apply as for the microstrip concerning the maximum operating frequency, but notice that synchronous coupling can actually occur at a lower frequency than for microstrip, owing to the lower $\varepsilon_{\text{eff}}$ of coplanar lines.
- Finally, quasi-TEM modes in conductor-backed coplanar waveguides can-

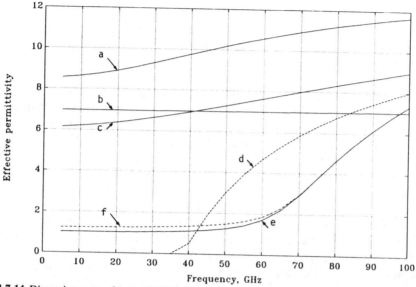

**Fig. 4.7.14** Dispersion curves for quasi-TEM modes and surface waves on 300 $\mu$m thick GaAs substrate: (a) 50 $\Omega$ microstrip line; (b) 50 $\Omega$ coplanar waveguide with or without lower ground plane, $b$ = 50 $\mu$m; (c) 50 $\Omega$ coplanar waveguide (no lower ground plane), $b$ = 314 $\mu$m; (d) $TM_1$ mode in parallel-plate waveguide, $H$ = 300 $\mu$m; (e) $TM_0$ mode in grounded dielectric waveguide, $h$ = 300 $\mu$m; (f) $TM_0$ mode in partially filled parallel-plate waveguide, $H$ = 1.5 $\mu$m, $h$ = 300 $\mu$m.

not possibly couple with TM surface waves, but synchronous coupling is possible at very high frequency with the first TM mode of the parallel-plate waveguide formed by the lower and upper ground planes. Again, a very conservative estimate on the maximum operating frequency can be $f <$ 30.6/$h$. Unfortunately, the parallel-plate parasitic waveguide made by the upper and lower ground planes does support a TEM fundamental mode with $\varepsilon_{eff} = \varepsilon_r$ [32]. As this mode has higher $\varepsilon_{eff}$ than the quasi-TEM coplanar mode, it can be directly excited at discontinuities or launchers by the quasi-TEM coplanar mode along a direction oblique with respect to the line axis. This sort of spurious excitation has been experimentally demonstrated in [32].

As a conclusion, synchronous coupling with surface waves can usually be avoided in practical MMICs. The same remark does not apply to surface wave excitation from discontinuities, which can also be enhanced if the circuit is shielded [21]. The conductor-backed coplanar waveguide is possibly more critical in this regard than conventional coplanar waveguides.

### 4.7.2.6 Losses

Apart from being a concern in many classes of circuits (e.g., filters), losses impose limitations on the useful impedance range and prevent line dimensions from being reduced under a certain value so as to increase the upper operating frequency of the circuit. Line losses basically fall into three classes: conductor losses, due to resistive dissipation within metalization; dielectric losses, due to finite substrate conductivity (usually modeled through a loss tangent, $\tan \delta$); and radiation losses. Radiation mechanisms have already been discussed. Here, radiation losses will be considered in the context of the Q-factor of resonators. The signal attenuation associated with the above-mentioned physical causes will be referred to as $\alpha_c$, $\alpha_d$, and $\alpha_r$, respectively. Although semiconductor substrates are usually affected by comparatively high dielectric losses, in practical lines on GaAs semi-insulating layers we almost always have $\alpha_c \gg \alpha_d$.

The accurate evaluation of conductor losses in strip conductors is intrinsically difficult as an electromagnetic problem. Moreover, the actual parameters of thin metallic films can be remarkably different from those of bulk materials due to the presence of surface irregularities, metal cavities originated from sputtering, and the effect of multilayered metalization where there are both good and poor conductors used as contact layers. To minimize conductor losses, strip thickness should be suitably larger than the skin depth in the frequency range of interest. Assuming $t = 3 \times$ skin depth, we obtain for the minimum strip thickness the values shown in Figure 4.7.15. The value of $t = 3$ $\mu$m seems to be a reasonable estimate for any kind of conductor at frequencies above 3 GHz. However, bulk conductivity values are used in the figure. We can expect that the effective conductivity of thin layers will be lower, leading to higher skin depth and the need for thicker stips. For reference, the skin depths of many conductors used in metalization as conducting or adhesion layers are shown in Figure 4.7.16 as a function of frequency.

Models for microstrip conductor losses can be found in [2,29] (see also [16, Chapter 7]). Such models are based on the so-called incremental inductance rule and assume fully developed skin effect (i.e., metalization must be thick with respect to the skin depth). Concerning coplanar waveguides, conductor losses were evaluated in [28] for lines in air (see Appendix 4A for a more complete discussion). Such formulas can be readily extended to the case where the line lies on a dielectric substrate. Results are in good agreement with those recently computed in [19].

Despite the widespread belief that coplanar lines are lossier than microstrip [36], a comparison based on equal impedance lines reveals that this is not always the case [19]. As a general rule, microstrip shows increased conductor losses with increasing impedance due to the grater resistance of narrow strips, whereas the loss behavior of coplanar waveguides is nonmonotonic. High losses occur both for low impedance values (corresponding to narrow slots and high edge coupling between strip and lateral ground planes) and for high impedances (corresponding to narrow strips). Therefore, an optimum value exists for the line impedance, usually around 50 $\Omega$. However, an additional degreee of freedom exists in coplanar waveguides, such that the overall line dimension $2b$ can be made larger,

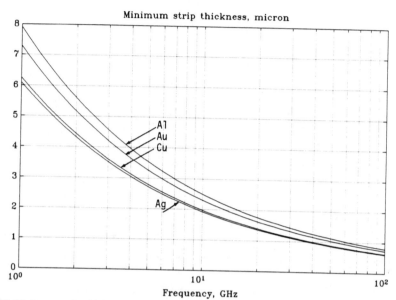

**Fig. 4.7.15** Minimum strip thickness (taken as $t_{min} = 3\delta$, $\delta$ skin depth) as a function of frequency for several conductors.

thereby achieving lower losses (but also a more dispersive structure). Therefore, losses of medium impedance lines are substantially similar for both microstrips and coplanar waveguides whereas the latter are superior for high impedances, microstrip for low impedance values. These remarks are supported by the results shown in Figure 4.7.17, where conductor losses and overall losses are compared at 20 GHz for microstrip and coplanar waveguide of the same impedance, as a function of $Z_0$. Dielectric losses are evaluated as in [2] for microstrip; for the coplanar waveguide, an expression derived from [16, Chapter 13]) as been used (see Appendix 4A). Dielectric losses are almost constant as a function of the line impedance.

Owing to the combined effect of ohmic and radiation losses, the Q-factor of both microstrip and coplanar resonators on GaAs cannot exceed values of approximately 100. In both cases, an optimum impedance level exists for which minimum losses are achieved. For microstrip lines, this is due to the fact that radiation losses follow a trend which is opposite to conductor losses with respect to the parameter $w/h$. In coplanar lines, a well defined optimum impedance exists only if radiation losses are subdominant because the radiation Q-factor $Q_r$ has the opposite behavior with respect to the ohmic Q-factor $Q_o$ (including dielectric and conductor losses). Namely, $Q_r$ is minimum approximately when $Q_o$ is maximum and *vice versa*. For a discussion of microstrip resonators and their quality factor, the reader can refer to [1,2]; for coplanar resonators, see the papers [5,10,11,40].

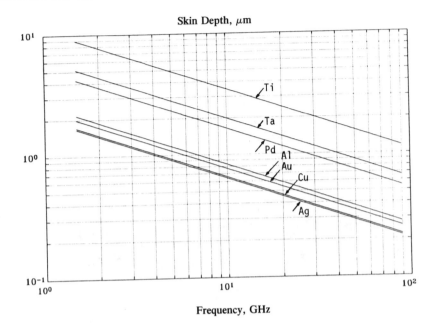

Skin Depth, $\mu m$

Frequency, GHz

**Fig. 4.7.16** Skin depth as a function of frequency for several conductors.

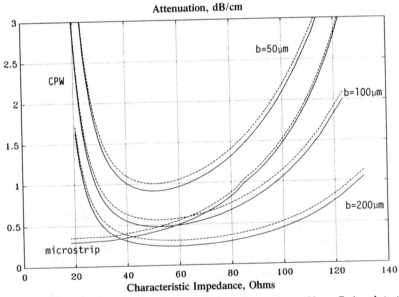

Attenuation, dB/cm

Characteristic Impedance, Ohms

**Fig. 4.7.17** Attenuation in dB/cm for microstrip and coplanar waveguide on 100 $\mu m$ GaAs substrate ($\varepsilon_r = 13$) at $f = 20$ GHz as a function of line impedance. Continuous line: conductor loss $\alpha_c$; dashed line: overall attenuation (conductor + dielectric loss, $\alpha_c + \alpha_d$). A 3 $\mu m$ thick copper metalization is assumed, and the loss angle of the substrate is taken as tan $\delta = 0.002$.

An example of Q-factor for a GaAs short-circuit coplanar waveguide resonator is shown in Figure 4.7.18 (from [5]). We suggest that performance estimates of coplanar waveguide resonators based on simplified radiation models such as those used [10,11] should be given only a qualitative value as the radiation mechanism is typically complex and, for practical substrates, involves above all surface waves, the propagation of which can, in turn, be heavily influenced by the environment (e.g., by truncated dielectric layers or metallic housing). Conclusions drawn from a full-wave model [19] reveal that there should be no substantial differences in radiation losses between open-circuit discontinuities in microstrip circuits and short-circuit discontinuities in coplanar circuits.

## 4.7.3 Line Discontinuities

Microstrip discontinuities in hybrid MICs have been the object of wide interest since the late 1960s. Nonetheless, we cannot say that their influence on the behavior of the circuit has been completely and satisfactorily assessed. In MMICs the frequency and thickness rescaling should suggest that the effect of

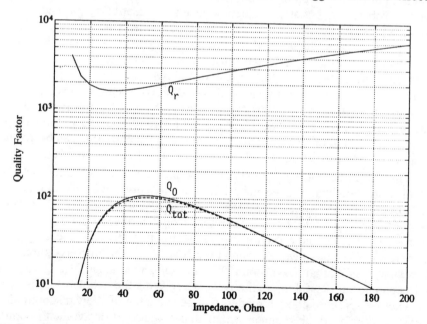

**Fig. 4.7.18** Quality factor of $\lambda/2$ short-circuit coplanar waveguide resonator on 500 $\mu$m thick GaAs substrate ($\varepsilon_r = 13$) at $f = 10$ GHz as a function of the line impedance. The line thickness is $t = 3$ $\mu$m and the ground-plane spacing is $2b = 500$ $\mu$m. $Q_r$ is the *radiation* Q-factor, $Q_0$ is the *ohmic* Q-factor, and $Q_{tot}$ is the *overall* Q-factor.

discontinuities would be not greater than in hybrid MICs, but rather somewhat smaller. However, spurious radiation may become of importance.

Unfortunately, no reliable model yet exists for microstrip discontinuities on GaAs substrates. Early models were based on lumped-parameter equivalent circuits, the elements of which could be derived from quasistatic capacitance and inductance estimates. (See [9,16,26] for a comprehensive review of quasistatic parameter models.) Examples of the most frequently encountered discontinuities, together with the related lumped-parameter equivalent circuits, are shown in Figure 4.7.19. Discontinuities are hardly avoidable in a microstrip design based MMIC circuit, either because they are a consequence of circuit layout (bends, T-junctions, crossings) or because they are exploited as circuit elements within specific components (e.g., gaps as coupling sections in microstrip filters or impedance steps in matching sections). Lumped models are fast and easily included in CAD programs; however, they are usually inaccurate beyond a certain frequency, which depends on the kind of discontinuity considered, as discussed later.

A more accurate and wideband modeling strategy was introduced in 1972 by means of the so-called closed waveguide microstrip model [41]. This model allows conventional mode-matching techniques to be approximately applied to the microstrip problem, and yields frequency-dependent scattering parameters of the discontinuities. As an alternative, from the closed-waveguide approach, lumped-parameter models can be derived with frequency-dependent elements. Typically, these elements are nearly constant up to a resonance frequency, which usually corresponds to the onset of higher-order modes in part of the structure. Resonances of this kind, which may actually be avoided in circuit design, cannot be estimated from the simple quasistatic model, which is, in turn, inaccurate for frequencies close to resonance. Because the range of quasistatic model validity depends on the kind of discontinuity considered, no general rule can be given. For a more detailed discussion, the reader may refer to [16, Chapter 10].

The closed-circuit waveguide approach has been the most accurate characterization method that can be introduced, without too many difficulties, into a CAD circuit analysis tool; its accuracy has also recently been tested on MMICs. However, the approach fails in accurately modeling discontinuities such as the impedance step [25], and does not apply at all to other structures (e.g., gaps). So, a significant effort has been expended recently to obtain full-wave three-dimensional analyses [25] that also account for radiation phenomena [23]. Although such models can be successful in qualitatively interpreting the behavior of discontinuities, whether accurate enough models will be available to allow discontinuities to be exploited as a circuit element rather than minimized or avoided is doubtful. Note that the cumulative effect of layout discontinuities (e.g., line bends) and interelement coupling can lead to unexpected circuit behaviors [21], which are difficult to understand even with the most sophisticated analysis tools available today.

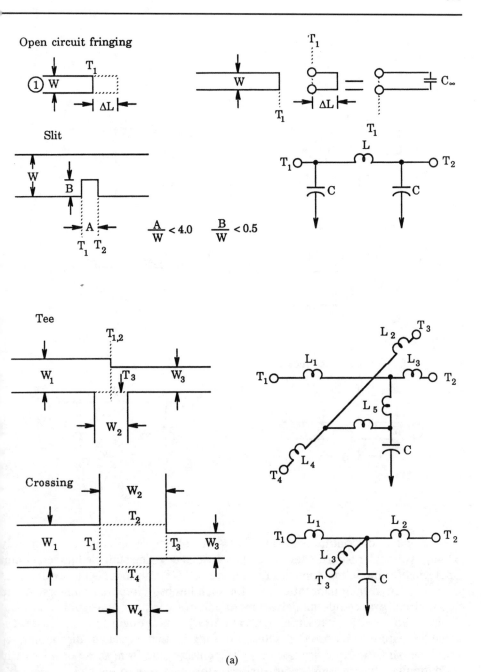

(a)

**Fig. 4.7.19** Microstrip discontinuities and their lumped equivalent circuits.

cont.

cont.

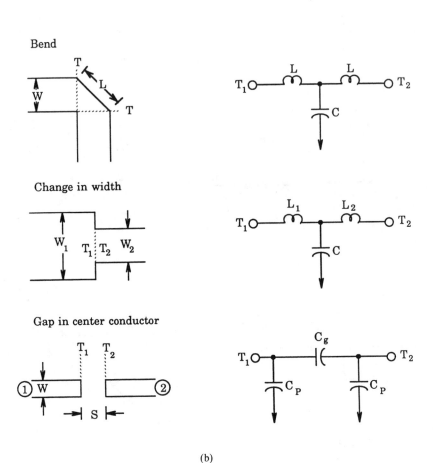

(b)

Coplanar discontinuities have been the object of interest only recently [33,40]. Although no particular difficulties arise in devising suitable lumped circuit topologies for modeling coplanar discontinuities [33], no models are available at present, even in the quasistatic case, for such lumped circuit parameters. As no closed-circuit waveguide model seems to hold for coplanar waveguides, any coplanar discontinuity analysis, either static or dynamic, ought to be three-dimensional in nature. Measured results [33] suggest that coplanar discontinuities introduce parasitics that are not very different in value from those of microstrip discontinuities having similar dimensions. However, due to the lack of available models, the only way to assess the effect of discontinuities on the behavior of uniplanar MMICs seems at present to be the experimental approach.

## APPENDIX 4A

## 4A ANALYTICAL FORMULAS FOR COPLANAR LINES

### Quasistatic Parameters

Quasistatic expressions for the line parameters have been derived through approximate conformal mapping techniques, and are accurate if the substrate thickness is not small with respect to the line width $2b$. Although the exact asymptotic limit is obtained for $h \to 0$, confine the use of these expressions to $h > b/2$. In all of the following formulas, $K(k)$ is the complete elliptic integral of the first kind, of argument $k$, while $k' = \sqrt{1 - k^2}$. The ratio $K(k)/K(k')$ can be accurately approximated as follows:

$$\frac{K(k)}{K(k')} \approx \frac{1}{\pi} \log \left(2 \frac{1 + \sqrt{k}}{1 - \sqrt{k}}\right), \quad 0.5 \le k^2 < 1 \tag{4A.1}$$

$$\frac{K(k')}{K(k)} \approx \frac{1}{\pi} \log \left(2 \frac{1 + \sqrt{k'}}{1 - \sqrt{k'}}\right), \quad 0 < k^2 \le 0.5 \tag{4A.2}$$

We have the following for the characteristic impedance ($Z_0$) and effective permittivity ($\varepsilon_{\text{eff}}$).

- Coplanar waveguide on infinitely thick substrate [38]:

$$Z_0 = \frac{30\pi}{\sqrt{\varepsilon_{\text{eff}}}} \frac{K(k')}{K(k)} \tag{4A.3}$$

$$\varepsilon_{\text{eff}} = \frac{\varepsilon_r + 1}{2} \tag{4A.4}$$

$$k = a/b \tag{4A.5}$$

- Coplanar waveguide with finite-thickness substrate [7,16,37]:

$$Z_0 = \frac{30\pi}{\sqrt{\varepsilon_{\text{eff}}}} \frac{K(k')}{K(k)} \tag{4A.6}$$

$$\varepsilon_{\text{eff}} = 1 + \frac{\varepsilon_r - 1}{2} \frac{K(k')}{K(k)} \frac{K(k_1)}{K(k_1')} \tag{4A.7}$$

where

$$k = a/b \tag{4A.8}$$

$$k_1 = \frac{\sinh\,(\pi a/2h)}{\sinh\,(\pi b/2h)} \tag{4A.9}$$

- Conductor-backed coplanar waveguide [6]:

$$Z_0 = \frac{60\pi}{\sqrt{\varepsilon_{\text{eff}}}} \frac{1}{\dfrac{K(k)}{K(k')} + \dfrac{K(k_2)}{K(k_2')}} \tag{4A.10}$$

$$\varepsilon_{\text{eff}} = \frac{\dfrac{K(k)}{K(k')} + \varepsilon_r\dfrac{K(k_2)}{K(k_2')}}{\dfrac{K(k)}{K(k')} + \dfrac{K(k_2)}{K(k_2')}} \tag{4A.11}$$

where

$$k = a/b \tag{4A.12}$$

$$k_1 = \frac{\tanh\,(\pi a/2h)}{\tanh\,(\pi b/2h)} \tag{4A.13}$$

- Coplanar waveguide with finite-thickness substrate and finite-extent ground planes [8,16,37]:

$$Z_0 = \frac{30\pi}{\sqrt{\varepsilon_{\text{eff}}}} \frac{K(k_3')}{K(k_3)} \tag{4A.14}$$

$$\varepsilon_{\text{eff}} = 1 + \frac{\varepsilon_r - 1}{2} \frac{K(k_3')}{K(k_3)} \frac{K(k_4)}{K(k_4')} \tag{4A.15}$$

where

$$k_3 = \frac{a}{b} \sqrt{\frac{1 - (b/c)^2}{1 - (a/c)^2}} \tag{4A.16}$$

$$k_4 = \frac{\sinh\,(\pi a/2h)}{\sinh\,(\pi b/2h)} \sqrt{\frac{1 - [\sinh\,(\pi b/2h)/\sinh\,(\pi c/2h)]^2}{1 - [\sinh\,(\pi a/2h)/\sinh\,(\pi c/2h)]^2}} \tag{4A.17}$$

*Coplanar Waveguide Dispersion*

An analytical expression of the frequency-dependent behavior of the effective permittivity for a coplanar waveguide with finite substrate is given in [14]. The range of validity is such that it covers most practical substrates. The formula is based on the well known Schneider expression for microstrip dispersion, with parameters fitted on numerical results. We have

$$\sqrt{\varepsilon_{\text{eff}}(f)} = \sqrt{\varepsilon_{\text{eff}}(0)} + \frac{\sqrt{\varepsilon_r} - \sqrt{\varepsilon_{\text{eff}}(0)}}{1 + A(f/f_{TE})^{-1.8}} \qquad (4A.18)$$

where

$$A = \exp\{[0.54 - 0.64 \log(2a/h) + 0.015(\log(2a/h))^2] \log(2a/(b-a))$$
$$+ 0.43 - 0.86 \log(2a/h) + 0.540(\log(2a/h))^2\} \qquad (4A.19)$$

$$f_{TE} = c_0/(4h\sqrt{\varepsilon_r - 1}) \qquad (4A.20)$$

## Coplanar Waveguide Losses

Although the recent history of coplanar waveguides in hybrid circuits originates from a paper by Wen [38], the coplanar waveguide is actually much older, and traces of its use as a balanced printed antenna feed line can be found in the literature of the 1950s. In particular, a remarkable paper by Owyang and Wu [28] published in 1958 analyzes what is now called a coplanar waveguide in air, giving both the electrical parameters and the conductor losses. Owing perhaps to its title, which is rather misleading from the point of view of today's terminology, this paper was almost ignored in the later literature. In particular, although the extension of coplanar waveguides to dielectric substrates is trivial, the result for conductor losses was ignored for a long time. In 1979 a new expression based on the incremental inductance rule was proposed in [9], which, however, is more complex but less accurate than Owyang and Wu's. Additionally, in the original paper [28] there is a misprint in the final formula for the attenuation, which turns out to be wrong by a factor of two. This mistake was corrected in [5], and also the reference book by Hoffman [16] is aware of [28] and gives a corrected version of the formula which, unfortunately, is again wrong by a factor ten. The formula as it is presented here was tested againt the numerical results published by Jackson [19] and the agreement found was good. Owyang and Wu's corrected expression is shown here together with the expression for dielectric losses. The attenuation is expressed in dB per unit length:

$$\alpha_c = \frac{8.68 R_s(f) \sqrt{\varepsilon_{\text{eff}}}}{480\pi K(k)K(k')(1 - k_2)}$$

$$\times \left\{ \frac{1}{a} \left[ \pi + \log\left(\frac{8\pi a(1-k)}{t(1+k)}\right) \right] + \frac{1}{b} \left[ \pi + \log\left(\frac{8\pi b(1-k)}{t(1+k)}\right) \right] \right\} \qquad (4A.21)$$

$$\alpha_d = 27.83 \frac{\tan\delta}{\lambda_0} \frac{\varepsilon_r}{2\sqrt{\varepsilon_{\text{eff}}}} \frac{K(k_1)}{K(k_1')} \frac{K(k')}{K(k)} \qquad (4A.22)$$

# REFERENCES

## *Section 4.1*

1.  D. D. Heston and R. E. Lehmann, "X-Band Monolithic Variable Gain Series Feedback LNA," *IEEE Microwave and Millimeter Wave Monolithic Circuits Symp. Dig.*, 1988, pp. 79–81.

2.  R. B. Culbertson and D. C. Zimmerman, "A 3-Watt X-Band Monolithic Variable Gain Amplifier," *IEEE Microwave and Millimeter Wave Monolithic Circuits Symp. Dig.*, 1988, pp. 121–124.

3.  A. N. Riddle and R. J. Trew, "A New Measurement System for Oscillator Noise Characterization," *IEEE MTT-S Int. Microwave Symp. Dig.*, Vol. 1, 1987, pp. 509–512.

4.  T. S. Howard and A. M. Pavio, "A Distributed Monolithic 2–18 GHz Dual-Gate FET Mixer," *IEEE Microwave and Millimeter Wave Monolithic Circuits Symp. Dig.*, 1987, pp. 27–30.

5.  M. J. Schindler, and A. M. Morris, "DC–40 GHz and 20–40 GHz MMIC SPDT Switches," *IEEE Microwave and Millimeter Wave Monolithic Circuits Symp. Dig.*, 1987, pp. 85–88.

6.  R. J. Bayruns, A. D. M. Chen, J. Gilbert, R. Goyal, and J. Wang, "A GaAs MESFET 7 Gb/s Dynamic Decision Circuit I.C.," *IEEE Microwave and Millimeter Wave Monolithic Circuits Symp. Dig.*, 1988, pp. 27–30.

7.  C. A. Mead, "Schottky Barrier Gate Field Effect Transistor," *Proc. IEEE,* Vol. 54, February 1966, pp. 307–308.

8.  J. Michael Golio, "Ultimate Scaling Limits for High-Frequency GaAs MESFETs," *IEEE Trans. Electron Devices,* Vol. ED-35, July 1988, pp. 839–848.

9.  M. Weiss and D. Pavlidis, "Power Optimization of GaAs Implanted FET's Based on Large-Signal Modeling," *IEEE Trans. Microwave Theory and Tech.,* Vol. MTT-355, February 1987, pp. 175–188.

10. Y. Tajima, B. Wrona, and K. Mishima, "GaAs FET Large-Signal Model and Its Applications to Circuit Designers," *IEEE Trans. Electron Devices,* Vol. ED-28, February 1981, pp. 171–175.

11. J. M. Golio, P. A. Blakey, and R. O. Grondin, "A General CAD Tool for Large-Signal GaAs MESFET Circuit Design," *IEEE MTT-S Int. Microwave Symp. Dig.*, 1985, pp. 417–420.

12. J. M. Golio, J. R. Hauser, and P. A. Blakey, "A Large-Signal GaAs MESFET Model Implemented on SPICE," *IEEE Circuits and Devices,* Vol. 1, September 1985, pp. 21–30.

13. N. Scheinberg, R. Bayruns, and R. Goyal, "A Low-Frequency GaAs MESFET Circuit Model," *IEEE J. Solid State Circuits,* Vol. 23, April 1988, pp. 605–608.

14. H. Statz, P. Newman, I. W. Smith, R. A. Pucel, and H. A. Haus, "GaAs FET Device and Circuit Simulation in SPICE," *IEEE Trans. Electron Devices,* Vol. ED-34, February 1987, pp. 160–169.

15. W. R. Curtice and M. Ettenberg, "N-FET, a New Software Tool for Large-Signal GaAs FET Circuit Design," *RCA Review,* Vol. 46, September 1985, pp. 321–339.

16. M. A. Khatibzadeh, R. J. Trew, and I. J. Bahl, "Large-Signal Modeling of GaAs Power FET Amplifiers," *IEEE MTT-S Int. Microwave Symp.,* 1987, pp. 107–110.

17. H. A. Willing, C. Rauscher, and P. deSantis, "A Technique for Predicting Large-Signal Performance of a GaAs MESFET," *IEEE Trans. Microwave Theory and Tech.,* Vol. MTT-26, December 1978, pp. 1017–1023.

18. P. H. Ladbrooke and S. R. Blight, "Low-Field Low-Frequency Dispersion of Transconductance in GaAs MESFET's with Implications for Other Rate-Dependent Anomalies," *IEEE Trans. Electron Devices,* Vol. ED-35, March 1988, pp. 257–266.

19. P. Canfield, J. Medinger, and L. Forbes, "Buried-Channel GaAs MESFET's with Frequency-Independent Output Conductance," *IEEE Electron Device Letters,* Vol. EDL-8, March 1987, pp. 88–89.

20. M. A. Smith and T. S. Howard, "RF Nonlinear Device Characterization Yields Improved Modeling Accuracy," *IEEE MTT-S Int. Microwave Symp. Dig.*, 1986, pp. 381–384.

21. S. A. Maas, *Microwave Mixers,* Artech House, Dedham, MA, 1984, pp. 24–26.

22. H. Abe and Y. Aono, "11-GHz GaAs Power MESFET Load-Pull Measurements Utilizing a New Method of Determining Tuner Y Parameters," *IEEE Trans. Microwave Theory and Tech.,* Vol. MTT-27, May 1979, pp. 394–399.

23. J. M. Cusack, S. M. Perlow, and B. S. Perlman, "Automatic Load Contour Mapping for Microwave Power Transistors," *IEEE Trans. Microwave Theory and Tech.,* Vol. MTT-22, December 1974, pp. 1146–1152.

24. D. Poulin, "Load-Pull Measurements Help You Meet Your Match," *Microwaves,* Vol. 11, November 1980, pp. 61–65.

25. R. A. Soares, "Novel Large Signal S-Parameter Measurement Technique Aids GaAs Power Amplifier Design," *Proc. 7th European Microwave Conf.,* September 1977, pp. 113–117.

26. Y. Mitsui, M. Nakatani, and S. Mitsui, "Design of GaAs MESFET Oscillator Using Large-Signal S-Parameters," *IEEE Trans. Microwave Theory and Tech.,* Vol. MTT-25, December 1977, pp. 981–984.

27. R. L. Vaitkus, "Uncertainty in the Values of GaAs MESFET Equivalent Circuit Elements Extracted from Measured Two-Port Scattering Parameters," *Proc. IEEE-Cornell Conference on High-Speed Semiconductor Devices and Circuits,* 1983, pp. 301–308.

28. R. L. Vaitkus, "Realistic Small-Signal Parameter Extraction Methods for Microwave Transistors," *GaAs Microwave Monolithic Integrated Circuits and Radar Cross Section Workshop Dig.,* Phoenix IEEE Waves and Devices Group, 1988.

29. R. A. Minasian, "Simplified GaAs MESFET Model to 10 GHz," *Electron Letters,* Vol. 13, September 1977, pp. 549–551.

30. FET Model Extraction Using Fetfitter 1.2, Cascade Microtech, Inc., Beaverton, OR, 1988.

31. S. N. Prasad, J. B. Beyer, and I-S Chang, "Power-Bandwidth Considerations in the Design of MESFET Distributed Amplifiers," *IEEE Trans. Microwave Theory and Tech.,* Vol. MTT-36, July 1988, pp. 1117–1123.

32. J. M. Golio, R. J. Trew, and G. N. Maracas, "A Modeling Technique for Characterizing Ion-Implanted Material Using C-V and DLTS Data," *Solid-State Electronics,* Vol. 27, April 1984, pp. 367–373.

33. A. Madjar and F. J. Rosenbaum, "A Large-Signal Model for the GaAs MESFET," *IEEE Trans. Microwave Theory and Tech.,* Vol. MTT-29, August 1981, pp. 781–788.

34. K. Doganis et al., "General Optimization and Extraction of IC Device Model Parameters," *IEEE Trans. Electron Device,* Vol. ED-30, September 1983.

35. K. Doganis, "SUXES Computer Program," Stanford Distribution Center, Office of Technology Licensing, Stanford University, Stanford, CA, February 1983.

36. K. Lehovec and R. Zuleeg, "Voltage-Current Characteristics of GaAs JFETs in the Hot Electron Range," *Solid-State Electronics,* Vol. 13, August 1981, pp. 1415–1426.

37. C. D. Hartgring, "An Accurate JFET/MESFET Model for Circuit Analysis," *Solid-State Electronics,* Vol. 25, 1982, pp. 233–240.

38. T. Ducourant, "3GHz, 150 mV, 4 bit GaAs Analogous to Digital converter," *Proc. GaAs IC Symp.,* 1986, pp. 209–212.

39. W. R. Curtice, "A MESFET Model for Use in the Design of GaAs Integrated Circuits," *IEE Trans. Microwave Theory Tech., Vol. MTT-28, 1980, pp. 448–456.*

40. C. Camacho-Penalosa, "Modeling Frequency Dependence of Output Impedance of a Microwave MESFET at Low Frequencies," *Electron. Letters,* June 1985, pp. 527–529.

41. J. M. Golio and R. J. Trew, "Profile Studies of Ion-Implanted MESFETS," *IEEE Microwave and Millimeter Wave Monolithic Circuits Symp. Dig.,* 1983, pp. 22–26.

42. J. R. Hauser, "Characteristics of Junction Field Effect Devices with Small Channel Length-to-Width Ratios," *Solid-State Electronics,* Vol. 10, 1967, pp. 577–587.

43. H. Fukui, "Design of Microwave GaAs MESFET's for Broadband Low-Noise Amplifiers," *IEEE Trans. Microwave Theory and Tech.*, Vol. MTT-27, July 1979, pp. 1031–1034.

44. M. A. Khatibzadeh, R. J. Trew, N. A. Masnari, and J. M. Golio, "Optimum Profiles for Low-Noise Ion-Implanted GaAs MESFETs," *IEEE Int. Electron Device Meeting, Tech. Dig.*, Washington, DC, December 1983, pp. 605–608.

45. R. A. Pucel, H. Statz, and H. A. Haus, "Signal and Noise Properties of Gallium Arsenide Microwave Field-Effect Transistors," *Advances in Electronics and Electron Physics*, Vol. 38, 1975, pp. 195–265.

46. A. F. Podell, "A Functional GaAs FET Noise Model," *IEEE Trans. Electron Devices*, Vol. ED-28, May 1981, pp. 511–517.

## Section 4.2

1. J. A. Turner *et al.*, "Dual Gate Gallium Arsenide Microwave Field-Effect Transistor," *Electronics Letters*, Vol. 7, November 1971, pp. 661–662.

2. C. A. Liechti, "Performance of Dual-Gate MESFETs as Gain-Controlled Low-Noise Amplifiers and High-Speed Modulators," *IEEE Trans. Microwave Theory Tech.*, Vol. MTT-23, June 1975, pp. 461–469.

3. S. Asai, F. Murai, and H. Kodera, "GaAs Dual-Gate Schottky Barrier FETs for Microwave Frequencies," *IEEE Trans. Electron Devices*, Vol. ED-22, October 1975, pp. 897–904.

4. T. Furutsuka, M. Ogawa, and N. Kawamura, "GaAs Dual Gate MESFETs," *IEEE Trans. Electron Devices*, Vol. ED-25, June 1978, pp. 580–586.

5. J. L. Vorhaus, W. Fabian, P. B. Ng, and Y. Tajima, "Dual Gate GaAs FET Switches," *IEEE Trans. Electron Devices*, Vol. ED-28, 1981, pp. 204–211.

6. C. Tsironis and R. Meierer, "Microwave Wideband Model of GaAs Dual Gate MESFETs," *IEEE Trans. Microwave Theory Tech.*, Vol. MTT-30, 1982, pp. 243–251.

7. R. S. Tucker *et al.*, "S-Parameter Model of the Dual Gate GaAs MESFET," *Electronics Letters*, Vol. 19, January 1983, pp. 39–40.

8. J. R. Scott and R. A. Minasian, "A Simplified Microwave Model of the GaAs Dual-Gate MESFET," *IEEE Trans. Microwave Theory Tech.*, Vol. MTT-32, March 1984, pp. 243–247.

9. C. Tsironis and R. Meierer, "DC Characteristics Aid Dual-Gate FET Analysis," *Microwaves*, July 1981, pp. 71–73.

10. C. Tsironis, R. Meierer, and R. Stahlmann, "Dual Gate MESFET Mixers," *IEEE Trans. Microwave Theory Tech.*, Vol. MTT-32, March 1984, pp 247–255.

11. S. C. Cripps, O. Nielsen, D. Parker, and J. A. Turner, "An Experimental Evaluation of X-Band Mixers Using Dual-Gate GaAs MESFETs," *Proc. 7th Eur. Microwave Conf.*, Copenhagen, 1977, pp. 101–104.

12. P. T. Chen, C. Li, and P. H. Wang, "Performance of a Dual Gate GaAs FET as a Frequency Multiplier at Ku-band," *IEEE Trans. Microwave Theory Tech.*, vol. MTT-27, May 1977, pp. 411–415.

13. C. Tsironis and P. Lesartre, "X and Ku Band Dual Gate MESFET Oscillators Stabilized Using Dielectric Resonators," *Proc. 11th Eur. Microwave Conf.*, Amsterdam, 1981, pp. 469–474.

14. J. S. Joshi and R. S. Pengelly, "Ultra Low Chirp GaAs Dual Gate FET Microwave Oscillators," *IEEE MTT-S Symp. Digest* Washington, 1980, pp. 379–382.

15. P. Wolf, "Microwave Properties of Schottky-Barrier Field Effect Transistors," *IBM J. Res. Develop.*, Vol. 14, 1970, pp. 125–141.

16. R. A. Minasian, "Simplified GaAs MESFET Model to 10 GHz," *Electronics Letters*, Vol. 13, September 1977, pp. 549–551.

17. R. A. Minasian, "Modelling DC Characteristics of Dual Gate MESFETs," *IEEE Proc. Solid-State Electronics*, Vol. 130, 1983, pp. 182–186.

18. G. Vendelin and M. Omori, "Circuit Model for the GaAs MESFET Valid to 10 GHz," *Electronic Letters,* Vol. 11, 1975, pp. 60–61.

19. B. Kim, H. Q. Tserng, and P. Saunier, "GaAs Dual-Gate Power FET for Operation up to K-Band," *IEEE Int. Solid-State Circuit Conf.,* February 1983, pp. 200–201.

20. R. S. Pengelly, "Broad and Narrow-band Frequency Discriminators Using Dual Gate GaAs FETs," *Proc. 9th Eur. Microwave Conf.,* Brighton, 1979, pp. 326–330.

21. A. S. Chu and P. T. Chen, "An Oscillator up to K-band Using Dual-Gate GaAs MESFET," *IEEE MTT-S Symp. Digest.,* Washington, 1980, pp. 379–386.

22. C. Tsironis, P. Harrop, and M. Bostelmann, "Active Phase Shifters at X-Band Using GaAs MESFETs," *Proc. ISSCC,* New York, 1981, pp. 140–141.

23. W. Filencky, F. Ponse, and H. Beneking, "Applications of Dual-Gate GaAs MESFETs for Fast Pulse Shape Regeneration Systems," *Electronic Letters,* Vol. 6, March 1980.

24. S. C. Cripps, O. Nielsen, and J. Cockrill, "An X-Band Dual Gate MESFET Image Rejection Mixer," *IEEE MTT-S Symp. Digest,* Ottawa, 1978, pp. 300–302.

25. J. Vorhaus *et al.*, "Monolithic Dual Gate MESFET Digital Phase Shifter," *IEEE Trans. Microwave Theory Tech.,* Vol. MTT-30, 1982, pp. 982–992.

26. C. Tsironis, "BRFET: A Band Rejection FET Amplifier and Mixer Applications," *IEEE MTT-S Symp. Digest,* Dallas, 1982, pp. 271–273.

27. C. Kermarrec *et al.*, "Monolithic Circuits for 12 GHz DBS Reception," *Proc. IEEE MMIC Symp.,* 1982, pp. 5–10.

28. Texas Instruments, Technical Information and Data Sheet on TGA8024 Monolithic X-Band Dual Gate Driver Amplifier.

## *Section 4.3*

1. R. A. Pucel, "Design Considerations for Monolithic Microwave Circuits," *IEEE Trans. Microwave Theory and Tech.,* Vol. MTT-29, No. 6, June 1981, pp. 513–534.

2. K. Wilson, "Other Circuit Elements for MMICs," *GEC J. Research,* Vol. 4 No. 2, 1986, pp. 126–133.

3. —— *Technology and Design Considerations of Monolithic Microwave Integrated Circuits.*

4. P. R. Shepherd, "Analysis of Square-Spiral Inductors for Use in MMICs," *IEEE Trans. Microwave Theory and Tech.,* Vol. MTT-34, No. 4, April 1986, pp. 467–472.

5. R. Garg and I. J. Bahl, "Characteristics of Coupled Microstiplines," *IEEE Trans. Microwave Theory and Tech.,* Vol. MTT-27, No. 7, July 1979, pp. 700–705.

6. H. M. Greenhouse, "Design of Planar Rectangular Microelectronic Inductors," *IEEE Trans. Parts, Hybrids and Packaging* Vol. PHP-10, June 1974, pp. 101–109.

7. R. Esfandiari, D. W. Maki, and M. Siracusa, "Design of Interdigitated Capacitors and their Application to Gallium Arsenide Monolithic Filters," *IEEE Trans. Microwave Theory and Tech.,* Vol. MTT-31, No. 1, January 1983, pp. 57–64.

8. J. P. Mondale, "An Experimental Verification of a Simple Distributed Model of MIM Capacitor for MMIC Applications," *IEEE Trans. Microwave Theory and Tech.,* Vol. MTT-35, No. 4, April 1987, pp. 403–408.

9. E. Pettenpaul, H. Kapuska, A. Weinberger, H. Mampe, J. Luginsland, and I. Wolff, "CAD Models of Lumped Elements on GaAs up to 18 GHz," *IEEE Trans. Microwave Theory and Tech.,* Vol. MTT-36, No. 2, February 1988, pp. 294–304.

10. D. B. Estreich, "A Simulation Model for Schottky Diodes in GaAs Integrated Circuits," *IEEE Trans. Computer-Aided Design of Integrated Circuits and Systems,* Vol. CAD-2, No. 4, April 1983, pp. 106–111.

11. M. V. Schneider, "Electrical Characteristics of Metal-Semiconductor Junctions," *IEEE Trans. Microwave Theory and Tech.,* Vol. MTT-28, No. 11, November 1980, pp. 1169–1173.

12.  K. S. Champlin and G. Eisenstein, "Cut-off Frequency of Submillimeter Schottky-Barrier Diodes," *IEEE Trans. Microwave Theory and Tech.*, Vol. MTT-26, No. 1, January 1978, pp. 31–34.

13.  S. M. Sze, *Physics and Semiconductor Devices*, 2nd Ed., Wiley and Sons, New York, 1981.

14.  T. W. Crowe and R. J. Mattauch, "Conversion Loss in GaAs Schottky-Barrier Mixer Diodes," *IEEE Trans. Microwave Theory and Tech.*, Vol. MTT-34, No. 7, July 1986, pp. 753–760.

15.  K. W. Low, D. K. Schroder, R. C. Clarke, A. Rohatgi, and G. W. Eldridge, "Low Leakage Current GaAs Diodes," *IEEE Trans. Electron Devices*, Vol. ED-28, No. 7, July 1981, pp. 796–800.

16.  L. E. Dickens, "Spreading Resistance as a Function of Frequency," *IEEE Trans. Microwave Theory and Tech.*, Vol. MTT-15, No. 2, February 1967, pp. 101–109.

17.  M. McColl, "Conversion Loss Limitations on Schottky-Barrier Mixers," *IEEE Trans. Microwave Theory and Tech.*, Vol. MTT-25, No. 1, January 1977, pp. 54–59.

18.  S. Y. Liao, "Microwave Circuit Analysis and Amplifier . . . "

19.  R. A. Soares, ed., *GaAs MESFET Circuit Design*, Artech House Norwood, MA, 1988, p. 479.

20.  S. Mao, S. Dones, and G. D. Vendelin, "Millimeter-Wave Integrated Circuits," *IEEE Trans. Microwave Theory and Tech.*, Vol. MTT-16, No. 7, July 1968, pp. 455–461.

21.  M. Caulton and H. Sobol, "Microwave Integrated Circuit Technology—A Surgey," *IEEE J. Solid-State Circuits* Vol. SC-5, No. 6, December 1970, pp. 292–303.

22.  E. W. Mehal and R. W. Wacker, "GaAs Integrated Microwave Circuits," *IEEE Trans. Microwave Theory and Tech.*, Vol. MTT-16, No. 7, July 1968, pp. 451–454.

23.  T. C. Edwards, *Foundations for Microwave Circuit Design*, John Wiley and Sons, New York, 1981.

## *Section 4.7 & Appendix 4.A*

1.  E. Belohoubek and E. Denlinger, "Loss Considerations for Microstrip Resonators," *IEEE Trans. Microwave Theory Tech.*, Vol. MTT-23, No. 6, June 1975, pp. 522–526.

2.  E. J. Denlinger, "Losses of microstrip lines," *IEEE Trans. Microwave Theory Tech.* Vol. MTT-28, No. 6, June 1980, pp. 513–522.

3.  H. J. Finlay, J. A. Jenkins, R. S. Pengelly, and J. Cockrill, "Accurate Coupling Predictions and Assessments in MMIC Networks," *IEEE GaAs Int. Circ. Symp. Tech. Dig.*, 1983, pp. 16–19.

4.  V. Fouad Hanna and D. Thebault, "Theoretical and experimental investigation of asymmetric coplanar waveguides," *IEEE Trans. Microwave Theory Tech.*, Vol. MTT-32, No. 12, December 1984, pp. 1649–1651.

5.  G. Ghione, C. Naldi, and R. Zich, "Q-factor evaluation for coplanar resonators," *Alta Frequenza*, Vol. LII, No. 3, June 1983, pp. 191–193.

6.  G. Ghione and C. Naldi, "Parameters of coplanar waveguides with lower ground planes", *Electronics Letters*, Vol. 19, No. 18, Sept. 1983, pp. 734–735.

7.  G. Ghione and C. Naldi, "Analytical formulas for coplanar lines in hybrid and monolithic MICs," *Electronics Letters*, Vol. 20, No. 4, February 1984, pp. 179–181.

8.  G. Ghione and C. Naldi, "Coplanar Waveguides for MMIC Applications: Effect of Upper Shielding, Conductor Backing, Finite-Extent Ground Planes, and Line-to-Line Coupling," *IEEE Trans. Microwave Theory Tech.*, Vol. MTT-35, No. 3, March 1987, pp. 260–267.

9.  K. C. Gupta, R. Garg, and I. J. Bahl, *Microstrip Lines and Slotlines*, Artech House, Norwood, MA, 1979.

10.  A. Gopinath, "Losses in coplanar waveguides," *IEEE Trans. Microwave Theory Tech.*, Vol. MTT-30, No. 12, July 1982, pp. 1101–1104.

11.  A. Gopinath, "A comparison of coplanar waveguide and microstrip for GaAs MMIC's," *IEEE MTT-S Digest*, 1979, pp. 109–111.

12. B. J. Janiczak, "Analysis of coplanar waveguide with finite ground planes", *AEU,* Vol. 38, No. 5, May 1984, pp. 341–342.

13. E. Hammerstad, F. Melhus, O. Jensen, F. Bekkadal, *Simulation of Microwave Components,* Elab Report No STF44 F80127, December 1980.

14. G. Hasnain, A. Dienes, J. R. Whinnery, "Dispersion of Picosecond Pulses in Coplanar Transmission Lines," *IEEE Trans. Microwave Theory Tech.,* Vol. MTT-34, No. 6, June 1986, pp. 738–741.

15. T. Hirota, Y. Tarusawa, H. Ogawa, "Uniplanar MMIC Hybrids—A proposed new MMIC structure," *IEEE Trans. Microwave Theory Tech.,* Vol. MTT-33, No. 6, June 1987, pp. 576–581.

16. R. Hoffmann, *Integrierte Mikrowellenschaltungen,* Springer-Verlag, Berlin, 1983; English translation: (H. Howe, ed.) *Handbook of Microwave Integrated Circuits,* Artech House, Norwood, MA, 1987.

17. M. Houdart, "Coplanar lines: application to broadband microwave circuits," *Proc. 6th Eur. Microwave Conf.,* 1976, pp. 49–53.

18. R. W. Jackson, D. M. Pozar, "Full-wave analysis of microstrip open-end and gap discontinuities," *IEEE Trans. Microwave Theory Tech.,* Vol. MTT-33, No. 10, October 1985, pp. 1036–1042.

19. R. W. Jackson, "Considerations in the Use of Coplanar Waveguide for Millimeter-Wave Integrated Circuits," *IEEE Trans. Microwave Theory Tech.,* Vol. MTT-34, No. 12, December 1986, pp. 1450–1456.

20. R. W. Jackson, "Mode conversion due to discontinuities in modified grounded coplanar waveguide," *IEEE MTT-S Digest,* 1988, pp. 203–206.

21. R. H. Jansen, R. G. Arnold, and I. G. Eddison, "A comprehensive CAD approach to the design of MMIC's up to MM-wave frequencies," *IEEE Trans. Microwave Theory Tech.,* Vol. MTT-36, No. 2, February 1988, pp. 208–219.

22. R. H. Jansen and M. Kirschning, "Arguments and an Accurate Model for the Power-Current Formulation of Microstrip Characteristic Impedance," *AEU,* Vol. 37, No. 3–4, 1983, pp. 108–112.

23. P. B. Katehi and N. G. Alexopulos, "Frequency-dependent characteristics of microstrip discontinuities in millimeter-wave integrated circuits," *IEEE Trans. Microwave Theory Tech.,* Vol. MTT-33, No. 10, October 1985, pp. 1029–1035.

24. M. Kirschning and R. H. Jansen, "Accurate model for effective dielectric constant of microstrip with validity up to millimeter-wave frequencies", *Electronics Letters,* Vol. 18, 1982, p. 272.

25. N. H. Koster and R. H. Jansen, "The microstrip discontinuity: a Revised Description," *IEEE Trans. Microwave Theory Tech.,* Vol. MTT-34, No. 2, January 1986, pp. 213–223.

26. R. Mehran, *Grundelemente des Rechnergestutzten Entwurfs von Mikrostreifenleitungs-Schaltungen,* H. Wolff Verlag, Aachen, West Germany.

27. M. Muraguchi, T. Hirota, A. Minakawa, K. Ohwada, and T. Sugeta, "Uniplanar MMIC's and their applications", *IEEE Trans. Microwave Theory Tech.,* Vol. MTT-36, No. 12, December 1988, pp. 1896–1901.

28. G.H. Owyang and T. T. Wu, "The approximate parameters of slot lines and Their Complement," *IRE Trans. Antennas and Propagation,* Vol. AP-6, No. 1, January 1958, p. 49.

29. R. A. Pucel, D. J. Massé, and C. P. Hartwig, "Losses in microstrip," *IEEE Trans. Microwave Theory Tech.,* Vol. MTT-16, No. 6, June 1968, pp. 342–350.

30. M. Riaziat, S. Bandy, and G. Zdasluk, "Coplanar Waveguides for MMICs," *Microwave Journal,* Vol. 30, No. 6, June 1987, pp. 125–131.

31. Y. C. Shih and T. Itoh, "Analysis of conductor-backed coplanar waveguide", *Electronics Letters,* Vol. 18, No. 12, June 1982, pp. 538–540.

32. H. Shigesawa, M. Tsuji, and A. A. Oliner, "Conductor-backed slot line and coplanar waveguide: dangers and full-wave analysis," *IEEE MTT-S Digest*, 1988, pp. 199–202.

33. R. N. Simons and G. E. Ponchak, "Modeling of some coplanar waveguide discontinuities", *IEEE Trans. Microwave Theory Tech.*, Vol. MTT-36, No. 12, December 1988, pp. 1796–1803.

34. H. Sobol, "Radiation conductance of open-circuit microstrip", *IEEE Trans.*, Vol. MTT-19, No. 11, November 1971, pp. 885–887.

35. H. Sobol, "Applications of integrated circuit technology to microwave frequencies", *Proc. IEEE*, Vol. 59, No. 8, August 1971, pp. 1200–1211.

36. B. Spielman, "Dissipation loss effects in isolated and coupled transmission lines," *IEEE Trans. Microwave Theory Tech.*, Vol. MTT-25, No. 8, August 1977, pp. 648–655.

37. C. Veyres and V. Fouad Hanna, "Extension of the application of conformal mapping techniques to coplanar lines with finite dimensions", *Int. J. Electronics*, Vol. 48, No. 1, 1980, pp. 47–56.

38. C. P. Wen, "Coplanar waveguide: a surface strip transmission line suitable for nonreciprocal gyromagnetic device applications," *IEEE Trans. Microwave Theory Tech.*, Vol. MTT-17, No. 12, December 1969, pp. 1087–1090.

39. H. A. Wheeler, "Transmission Line Properties of a Strip on a Dielectric Sheet on a Plane," *IEEE Trans. Microwave Theory Tech.*, Vol. MTT-25, No. 8, August 1977, pp. 631–646.

40. D. F. Williams and S. E. Schwarz, "Design and performance of coplanar waveguide bandpass filters," *IEEE Trans. Microwave Theory Tech.*, Vol. MTT-31, No. 7, July 1983, pp. 558–566.

41. J. Wolff, G. Kompa, and R. Mehran, "Calculation method for microstrip discontinuities and T-junctions," *Electronics Letters*, Vol. 8, pp. 177–179, April 1972.

42. E. Yamashita, K. R. Li, E. Kaneko, and Y. Suzuki, "Characterization method and simple design formulas of MCS lines proposed for MMICs," *IEEE MTT-S Digest*, 1987, pp. 685–688.

## ACKNOWLEDGEMENTS

The editor wishes to thank the following for their contributions to this chapter: Dr. Christos Tsironis of Focus Microwaves, Inc., Point-Claire, Quebec for contributing Section 4.2; Prof. Giovanni Ghione of the Department of Electronics, Politecnico de Milano, Italy for contributing Section 4.7; Dr. Madhu Gupta of Hughes Research Center, Torrance, California for critically reviewing Section 4.1.

# Chapter 5
# MMIC Design Considerations and Amplifier Design

*M. Kumar, R. Goyal and T. H. Chen*

## 5.1 INTRODUCTION

Solid-state devices are replacing vacuum tube (klystrons or traveling wave tubes, for example) for amplification of microwave signals at low and medium power levels. Power levels of several watts in X and K bands and up to a few hundred watts in L and S bands have been achieved by solid-state amplifiers. Solid-state devices require low voltage for operation and are compact and light weight. Weight and size impose severe limitations on the choice of components and systems in military and space applications. Low cost circuits are required for commercial applications. Unlike hybrid MICs, MMICs are not generally tunable. However, some on-chip tuning capability can be provided to reduce the overall development cost. The availability of some standard MMICs as building blocks allows the system designer to demonstrate the system performance using these well-characterized components. A standard cell library approach reduces the overall development cost.

## 5.2 DESIGN CONSIDERATIONS FOR MMICs

There are some general considerations for designing MMICs, which include amplifiers, mixers, switches, and phase shifters. The following constraints are imposed by technology, frequency of operation, and device properties:

1. Chip size.
2. Thermal design and wafer thickness.
3. Low-inductance grounds and low-capacitance crossovers.

4. Suitable choice of the propagation mode.
5. Other design considerations:
    (a) Consistency of the design with manufacturing capabilities.
    (b) On-chip *versus* off-chip biasing.
    (c) On-chip *versus* off-chip bypass capacitors.
    (d) Tolerance to the process variation.
    (e) Accounting for packaging and bondwire effects.

### 5.2.1 Chip Size

The chip size is one of the most important considerations because the cost and yield of the MMICs are related directly to it. Low cost is the key factor for affordable MMICs for the large volume applications (e.g., phased array radars). Higher density circuits can be designed to reduce the chip size. However, the circuit yield on a wafer depends on the number of components in a given area and the product of the processing yields associated with individual fabrication steps. Because active devices require the most processing steps, the overall yield is determined by the number of active components on a circuit and the device technology. The other potential problem is undesirable coupling, such as back gating and side gating as well as microwave coupling, within the circuit elements. The coupling effect determines the density of individual components that can be packed within a circuit to keep the overall size small. Use of lumped and active components (e.g., FETs for matching) reduces the overall size of the circuit. However, use of the FETs to simulate resistors or capacitors in matching circuits introduces high dc power dissipation and noise, which may not be desirable in circuits such as low-noise and high efficiency amplifiers.

### 5.2.2 Thermal Design and Wafer Thickness

The choice of thickness of the substrate is based on the following consider-ations: wafer handling, thermal resistance, propagation loss and quality factor ($Q$) of the circuit element, minimum line width requirement, higher-order mode propa-gation, and ease of manufacture and tolerance of transmission line impedance *versus* wafer thickness. Power dissipating devices, such as a FET, require the thinnest substrates to achieve low thermal resistance, particularly for the power applications. However, the loss and impedance of the microstrip line is inversely proportional to the substrate thickness, whereas the inductance and quality factor of the spiral inductor are proportional to the wafer thickness. Therefore, the choice of the substrate thickness is a compromise among many factors, such as the thermal impedance (power dissipation in the circuit, which is the most impor-tant one for power applications), the operating frequency, the line width tolerance for high impedance lines, *et cetera*. In practice, 0.1 mm thick substrate is widely used for power amplifier circuits and circuits requiring substrate via holes, and thickness in the range up to 0.4 mm is suitable for small-signal and low-noise circuits. In power amplifier circuits, the thermal impedance can be reduced by selectively thinning the GaAs substrate under the active area. This technology is now becoming available.

### 5.2.3 Low-Inductance Grounds and Crossovers

The microstrip mode of propagation is preferred to reduce the overall size of the circuit, because no ground plane is required on the topside of the substrate. The plated through substrate via-hole technique provides ground connections on the chip wherever it is needed. A circuit may require a number of ground connections, and this can be accomplished by a number of via-hole connections. There are two techniques of providing low inductance to ground connection. In the substrate plated through via-hole technique, the via holes are etched and plated through the hole to provide the connection to the top metalization. Then, the backside is plated to provide a common ground to the circuit elements. For MMICs, the inductance of a 40 × 40 $\mu$m via hole through a 0.1 mm substrate has an inductance of approximately 30 pH, which is equivalent to 1.8 $\Omega$ at 10 GHz. This should be taken into account in the circuit simulation.

The other technique is the *wrap-around ground,* which requires a topside metalization pattern near the periphery of the chip that can be connected to the backside ground plane, as well as connections to the various places where circuit elements must be connected to the ground plane. This requires additional crossovers and more surface area. For small-signal circuits, where thicker substrates are used and the via-hole is not easily etched through the thick substrate, the wrap-around ground may be preferable. For power amplifiers, where low inductance grounds are essential because source lead inductance reduces the gain of the amplifier, a multiple via-hole approach is preferable.

Interconnects are often required in MMICs; for example, to connect several source pads of the FET and crossover of two transmission lines. An airbridge crossover provides the interconnect between various elements without wirebonding. The thickness and the width of an airbridge crossover design should meet the current handling criteria. This provides a low capacitance crossover capability. The crossover capacitance depends on the crossover and the gap between the airbridge and the first-level metal.

### 5.2.4 Propagation Modes

Four basic modes of propagation are available to a MMIC designer:

1. Microstrip, which requires a backside ground plane.
2. Slotline, which is the inverse of the microstrip mode.
3. *Coplanar waveguide* (CPW), which consists of a center conductor separated by a slot from the two adjacent ground planes.
4. Coplanar stripline.

Among the four modes, the slotline is not TEM-like and is not used for MMICs. Microstrip and coplanar waveguide modes have been used widely for MMICs. However, CPW has the disadvantage of having the ground plane on the top surface; hence, the circuit size becomes large. However, it provides an easy connection of a shunt element to the ground plane. CPW and coplanar stripline modes require thick substrates, which is a disadvantage for power amplifiers where low thermal resistance is required. A higher impedance range can be achieved by CPW and coplanar strip line modes than the microstrip. The impedance of the microstrip depends on the ratio of the line width to the substrate thickness. The highest impedance achievable for a microstrip requires a narrow line. For loss and size considerations, the impedance range achievable with microstrip is 10–100 Ω for a 0.1 mm thick substrate.

With these considerations, microstrip is used widely for most circuits, whereas CPW is used only in cases where the heat dissipation needs of a thick substrate is not a problem, such as in low-noise amplifiers and small-signal control components, or when low-inductance ground connections are not available.

### 5.2.5 Other Design Considerations

The design's ease of manufacture is very important. The accurate modeling of the active and passive elements and matching circuits is a critical step in designing a MMIC. Chapter 4 examines this aspect in detail. Besides active elements such as FETs and diodes, the other elements available to the designer are lumped elements: MIM and interdigital capacitors; thin film and GaAs resistors; spiral inductors, *et cetera*. The manufacture of a MMIC design depends on the sensitivity of the performance on the variations of these circuit elements. The design should contain specifications of a reasonable parameter range for the circuit elements to allow for process variations. Although distributed element occupy more space than lumped elements, they have lower loss and offer the best manufacturing control, being very process-tolerant. A successful design technique is to select matching circuits that are more tolerant to the process variations. Since ratios are less sensitive to process variations than absolute values, a good design should depend upon the ratios of resistance, capacitance, inductances, or other parameters rather than their absolute values.

The main considerations for on-chip *versus* off-chip biasing network and bypass capacitors depend on the desired values of the capacitors and inductors.

MIM capacitors of 10–20 pF values and spiral inductors up to 10–12 nH are typically put on-chip for these purposes. However, it is important that as the spiral inductors values are increased their resonance frequency goes down. We must make sure that the resonance frequency of the spiral inductor is out of the frequency band of interest. As the MIM capacitor values are increased, the total area is increased as well as the yield of the capacitors decrease.

While the design of the MMICs and the size of the chip are finalized, it is important to decide on a package. This will determine the length of the bond wires at the input and output as well as grounding if no plated through substrate via-hole technique is used for grounding. During the final phase of circuit design, the inductance due to bond wires must be taken into account. The effect of bond wire inductance can be reduced by using multiple bond wires for one connector, particularly for grounding purposes. Depending on the thickness of the chip and the height of the inside package, the top cover of the package could significantly affect the performance of the overall MMIC. It is also important to take into account these effects before the design is finalized.

## 5.3 BIASING TECHNIQUES

An important part of the amplifier design is the proper selection of a biasing network. Considerable effort normally is spent in designing an amplifier for a given performance (e.g., gain, bandwidth, noise, *et cetera*). However, if a proper biasing network is not chosen, the amplifier may be unnecessarily sensitive to device characteristics resulting from process variations. The purpose of a good dc bias design is to select the proper quiescent point and hold it constant over the variations in the device characteristics and temperature. The design considerations for biasing circuits are efficiency, noise, oscillation suppression, single-source power supply, and RF chocking.

There are several ways of biasing a GaAs FET. Some of the basic practical dc biasing networks are given in Table 5.1. The biasing circuits, *a, b, e,* and *h* require two power supplies of opposite polarity; circuits *c* and *d* require only a single positive or negative power supply; and circuits *f* and *g* require two power supplies of the same polarity. The circuits *a, b,* and *e* have sources connected to ground with minimum possible source inductance (achieved by the plated through substrate via-hole technique) and therefore provide the maximum gain. When two power supplies are used, the gate (negative) voltage is applied before the drain voltage (positive), to prevent transient burnout of the GaAs FET. One technique to apply both voltages at the same time is to provide a long RC time constant network in the drain supply and a short RC time constant network in the gate supply. The bias circuits *c* and *d* use a source resistor and are widely used for small and medium power applications. The source resistor provides automatic

**Table 5.1** Basic practical dc biasing networks

| Bias Circuit | Bias Voltage | Amplifier Characteristics | Comments |
|---|---|---|---|

**Table 5.1a**

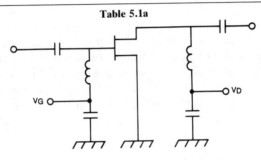

| (a) | $V_d = 3V$ <br> $V_g = -1V$ | Low noise <br> High gain <br> High power <br> High efficiency | Biasing network is part of matching <br> Insensitive to bias current |

**Table 5.1b**

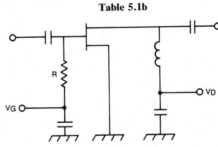

| (b) | $V_d = 3V$ <br> $V_g = -1V$ | Moderately noisy <br> High gain <br> High power <br> High efficiency | Biasing network is part of matching <br> Insensitive to bias current <br> High value of $R$ provides higher <br> isolation between gate and <br> power supply |

**Table 5.1** cont.

| Bias Circuit | Bias Voltage | Amplifier Characteristics | Comments |
|---|---|---|---|

**Table 5.1c**

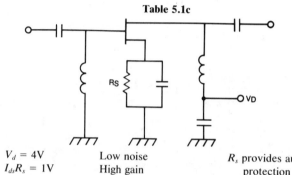

(c)

$V_d = 4V$
$I_{ds}R_s = 1V$

Low noise
High gain
Medium power
Low efficiency
Gain adjustment
 by varying Rs

$R_s$ provides automatic transient
 protection sensitive to bias
 current

**Table 5.1d**

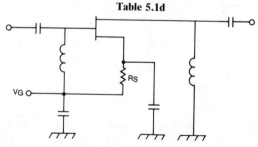

(d)

$V_g = -4V$
$I_dR_s = 1V$

Low noise
High gain
Medium power
Low efficiency

$R_s$ provides automatic transient
 protection sensitive to bias
 current

**Table 5.1** cont.

| Bias Circuit | Bias Voltage | Amplifier Characteristics | Comments |
|---|---|---|---|

**Table 5.1e**

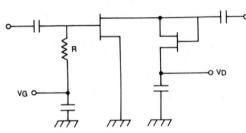

| (e) | $V_d = 5V$ <br> $V_g = -1V$ | Low noise <br> High gain <br> Medium power <br> Low efficiency | Broadband at lower frequencies sensitive to bias voltage |

**Table 5.1f**

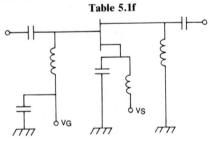

| (f) | $V_g = -7V$ <br> $V_s = -5V$ | Low noise <br> High gain <br> High power <br> High efficiency | Negative supplies only |

**Table 5.1** cont.

| Bias Circuit | Bias Voltage | Amplifier Characteristics | Comments |
|---|---|---|---|

**Table 5.1g**

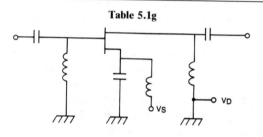

(g) | $V_d = 7$ <br> $V_s = 2$ | Low noise <br> High gain <br> High power <br> High efficiency | Positive supplies only

**Table 5.1h**

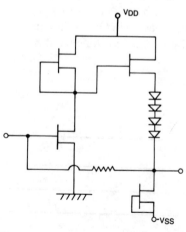

(h) | $V_{dd} = 5V$ <br> $V_{ss} = 5V$ | Moderately noisy <br> Low gain <br> Low efficiency <br> Medium power | Directly coupled <br> Ultrabroadband

transient protection. The gate is reversely biased with respect to the source by the series resistor $R_s$. The circuits $c$ and $d$ provide low efficiency because of the additional power dissipation in the source resistors. The negative feedback resistor $R_s$ decreases the effect of variations of $I_d$ with respect to temperature and $I_{dss}$.

The circuit $e$ uses an active bias network for a common source FET. This technique provides broadband performance in a lower frequency region and is sensitive to the bias voltages. The circuit $f$ uses all negative power supplies, whereas the circuit $g$ uses all positive power supplies. In circuit $f$, the difference between $V_g$ and $V_s$ is the gate-to-source bias voltage, which in this case is $-2$ V and the drain-to-source bias is the same value and in opposite polarity to the $V_s$. In circuit $g$, the drain-to-source bias is the difference between $V_d$ and $V_s$ and the gate-to-source bias is of the same value and opposite polarity to the $V_s$. The circuit $h$ provides input and output terminals at ground potential and, hence, can be directly coupled. The level shift diodes are used to shift the dc voltage level. The dc bias is stable because of the feedback resistor $R_f$.

The selection of the dc quiescent point in a GaAs FET depends on the particular application, such as low-noise, small-signal, power, or efficiency. Typical I-V characteristics of a GaAs FET are shown in Figure 5.1. For low-noise and low-power amplifier, the FET can be biased at about $0.15\ I_{dss}$ and $V_{ds} = 3\text{–}4$ V, (quiescent point A). A higher power gain and still lower noise figure, though not as low as in quiescent point A, can be achieved by biasing at low $V_{ds}$ (3–4 V) and high $I_{ds} = 0.8$ to $0.9\ I_{dss}$ (quiescent point B). The higher output power is achieved by selecting the quiescent point C with $I_{ds} = 0.5$ to $0.6\ I_{dss}$ and $V_{ds} = 8\text{–}10$V. For obtaining higher efficiency or to operate in GaAs FET in class AB or B mode, the drain to source current must be decreased (quiescent point D).

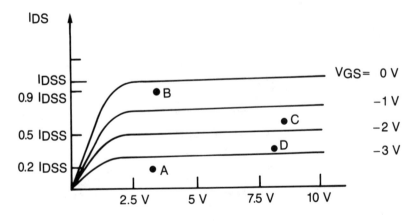

**Fig. 5.1** Typical GaAs FET I-V characteristic and recommended quiescent points for various applications.

## 5.4 MICROWAVE AMPLIFIER DESIGN

### 5.4.1 Design Considerations

Before discussing design considerations for microwave amplifiers, it is appropriate to describe different types of microwave amplifiers in terms of applications, functional features, frequency bands, matching techniques, and circuit topology. Once the classifications and the characteristics of the microwave amplifiers have been clarified, we will proceed to discuss design considerations. Microwave amplifiers can be categorized according to their applications and functional features:

1. General purpose amplifiers or the gain blocks.
2. Low-noise amplifiers.
3. Power amplifiers.
4. High-linearity amplifiers.
5. High-efficiency amplifiers.
6. Variable gain and power amplifiers.
7. Logarithmic amplifiers.

The main goal of a gain block is to obtain sufficient power gain, whereas the rest of the amplifiers just mentioned are designed for a specific application besides power gain. The application of a special purpose amplifier is self-explanatory. For example, a low-noise amplifier is designed specifically for a low-noise performance, a power amplifier is designed specifically for maximum output power. However, in a practical application the amplifier may require more than one specific purpose. For instance, a variable gain amplifier may also need low-noise or high-linearity performance and a logarithmic amplifier may require not only the logarithmic relationship between the output voltage and the input power but also lower noise figures and good efficiency.

The microwave amplifiers can be classified into four categories in terms of their required frequency bandwidths:

1. Narrowband amplifier (<20% bandwidth).
2. Broadband amplifier (>20% to <2 octaves bandwidth).
3. Ultra-wideband amplifier (multioctave bandwidth).
4. Direct coupled amplifier.

For example, a spot frequency amplifier belongs to the narrowband category, a 6–18 GHz amplifier belongs to the broadband category, a 2–20 GHz amplifier belongs to the ultra-wideband category, and a dc to 2 GHz amplifier belongs to the direct coupled category. In addition to these two classifications, microwave amplifiers also can be classified according to their matching technique or circuit topology into the following categories:

1. Reflective (reactive) match amplifier.
2. Lossy (resistive-reactive) match amplifier.
3. Feedback amplifier.
4. Active match amplifier.
5. Distributed (traveling wave) amplifier.
6. Balanced amplifier.
7. Push-pull amplifier.

The reflective match amplifier uses only reactive elements in the matching circuits. The lossy match amplifier uses both resistive and reactive elements in the matching circuit [1]. The resistive elements can be introduced in the matching circuit as either series or shunt elements. Figure 5.2 shows typical resistive-reactive matching circuits. The resistive elements in the lossy match amplifier normally serve two purposes: stabilizing the devices and equalizing the gain over wideband. The negative feedback in a feedback amplifier can be either parallel or in series or both [2]. The most commonly used parallel feedback uses a resistor, an inductor, and a blocking capacitor in series as the feedback path from the output to the input. Common lead series feedback often has been used in conjunction with the parallel feedback or alone. Figure 5.3 illustrates a feedback amplifier with both parallel and series feedback. Negative feedback has the advantage of increasing the stability factor, improving input and output return losses, and flattening gain over widebands. However, these advantages are obtained at the expense of the maximum gain available.

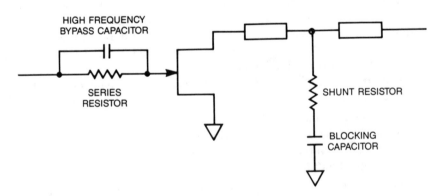

**Fig. 5.2** Schematic of resistive-reactive matching circuit.

The active match amplifier [3], shown in Figure 5.4, uses common-gate and common-drain FETs to improve the input and output matches, respectively. A common-drain FET is rarely used in the microwave circuit because good output

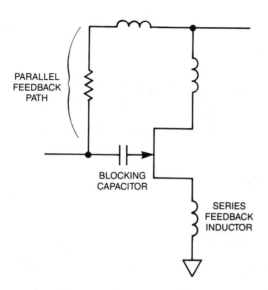

**Fig. 5.3** Schematic of feedback amplifier with parallel and in series feedbacks.

match is much easier to obtain from other types of matching techniques than good input match. Normally, the active match is applicable only for frequency below 10 GHz or so, due to the FET's parasitic degradation. A distributed amplifier [4] has the unique capability of adding the transconductance of several FETs without paralleling their input or output capacitances. This is accomplished by linking the input or output capacitors of the devices with series inductors to form an artificial low-pass transmission line in the input or output circuit. By terminating the two transmission lines with resistive loads at their idle ports, the unwanted signals are absorbed by the resistive loads while the desired signals are added in phase at the output of the amplifier. This results in an increased gain-bandwidth product with flat gain and low VSWRs. Figure 5.5 is a simplified schematic of the distributed amplifier.

Figure 5.6 shows the arrangement of a balanced amplifier [5] using two single-ended reflective match amplifiers of the same performance between the output and the input ports of two quadrature 3 dB couplers, such as Lange couplers. Single-ended amplifiers usually are mismatched for flat gain, low-noise figure, and good stability. The reflections from the amplifier are absorbed in the isolated ports. The use of a pair of balanced units ensures that the input and output of each amplifier is terminated by an impedance close to 50 Ω. This normally guarantees good stability. The push-pull amplifier [6], shown in Figure 5.7, is used in the power amplification to obtain low distortion or high efficiency. For the current MESFET technology, the basic circuit of the push-pull amplifier consists

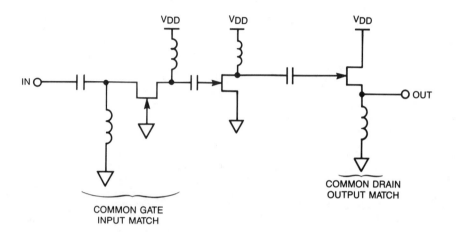

**Fig. 5.4** Schematic of an amplifier using a common-gate and common-drain FET for matching circuits.

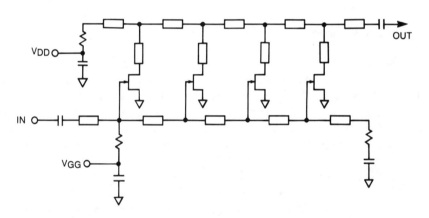

**Fig. 5.5** Schematic of a distributed amplifier.

of a pair of FETs driven 180° out of phase; a 180° power divider; and a 180° power combiner. The 180° power combiner is a 180° hybrid coupler, whereas the 180° power divider can be either a 180° hybrid coupler or an active paraprase (180°) splitter. Due to the losses of the hybrids and the additional power consumption of the active paraprase splitter, the efficiency achieved is still far below (≈50%) the theoretical value (78% in class B mode). In order to completely eliminate the 180° power divider and combiner, a pair of complimentary devices will be required. The complimentary GaAs-AlGaAs *heterojunction bipolar transistor* (HBT) [7], currently under development, is a very attractive candidate to be used in the push-pull amplifier.

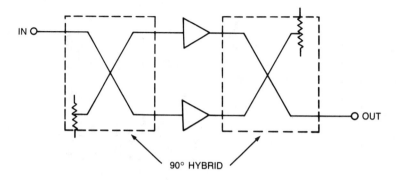

**Fig. 5.6** Schematic of a balanced amplifier.

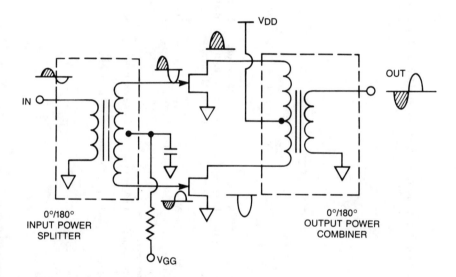

**Fig. 5.7** Schematic of a push-pull amplifier.

In the practical design of an amplifier, a combination of matching techniques is adopted to improve the performance achieved by using a single technique. For example, a combination of common lead inductor series feedback and reflective matching techniques can be used to simultaneously obtain low noise, low input VSWR, and reasonable gain, which is not possible using either technique alone. A qualitative comparison of various circuit topologies [8] is summarized in Table 5.2.

Besides the general design considerations of MMICs described in Section 5.2, some specific considerations in the amplifier design need to be addressed. The

**Table 5.2 A Qualitative Comparison of Circuit Topologies**

| | RMA | LMA | FBA | AMA | TWA | BA | PPA |
|---|---|---|---|---|---|---|---|
| Maximum achievable bandwidth | Octave | Multioctave | Multioctave dc coupled | Multioctave | Multioctave | 2-Octave | Multioctave |
| Chip size | Moderate | Moderate | Small | Small–moderate | Moderate | Large | Large |
| Noise figure | Low | Moderate–high | Low–moderate | Moderate–high | Moderate | Low–moderate | Moderate |
| Fabrication tolerance | Tight | Moderate | Loose | Loose–Moderate | Very loose | Loose | Tight |
| Cascadability | Poor | Moderate | Moderate–good | Good | Good | Very good | Moderate |
| I/O VSWRs | Poor | Moderate | Moderate–good | Very good | Good | Very good | Moderate |
| FET size-gain | Small | Small–moderate | Moderate–large | Moderate–large | Large | Moderate | Moderate–large |
| Efficiency | Moderate–high | Moderate–high | Low–moderate | Low–moderate | Low | Moderate–high | High |
| Stability | Poor | Good | Good | Moderate | Good | Good | Moderate |
| Power density | Moderate–high | Moderate | Low–moderate | Low–moderate | Low | High | Very high |

design considerations of amplifiers are aimed at three primary goals: performance, reliability, and cost. These three goals are not completely independent but somewhat related to one another. Separate considerations will be discussed for each goal, but we should keep in mind the need for tradeoff and compromise among them.

To minimize the cost of a single amplifier means not only maximizing the ratio of the amplifier's yield to its area but also minimizing the total cost of design, manufacture and testing. Increased circuit density reduces the amplifier's area, however, it normally decreases the yield, too, as has been discussed in Section 5.2. Using different matching techniques or circuit topologies also reduces the amplifier's area. By choosing the proper matching technique, we can simultaneously achieve small size and higher yield. The manufacturing cost is determined by the designer as well as the processing personnel. The designer can eliminate unnecessary processing steps by selecting proper circuit elements, thus reducing the manufacturing cost and increasing the yield. The designer can greatly affect the testing cost by incorporating automatic test schemes that provide access to internal modes and some relationship between dc measurement and RF performance.

Although the reliability of the materials is a main concern, the designer has the obligation to anticipate most of these problems (e.g., degradation due to high temperature, variation due to changing temperature, and electromigration). Consideration of the proper heat dissipation in the design can greatly reduce temperature degradation problem. The incorporation of a temperature compensation circuit can minimize the performance variation due to changing temperature. Also electromigration can be avoided by careful consideration of the current carrying capability of interconnect lines.

Many characteristics in the performance of an amplifier need to be considered. The most important ones are the following:

1. Frequency bandwidth.
2. Power gain and gain flatness over frequency band.
3. Input and output return losses.
4. Stability.
5. Noise figure.
6. Output power and efficiency.
7. Intermodulation performance.
8. Dynamic range.

Although the specifications may not include all of them, it is good practice to take these considerations into account while emphasizing the required concerns. The power gain, stability, noise figure, saturated power output, 1 dB compression point, and intermodulation performance will be briefly described.

### 5.4.1.1 Amplifier Power Gain

Four different definitions of the power gain of an amplifier are used frequently in the design of microwave amplifiers: the power gain, $G_p$ (also the operating power gain); the available power gain, $G_a$; the insertion power gain, $G_i$; and the transducer power gain $G_t$. These are defined as follows:

$$G_p = \frac{\text{Power delivered to the load}}{\text{Power input to the amplifier}} = G_p\,(Z_l)$$

$$G_a = \frac{\text{Power available from the amplifier}}{\text{Power available from the source}} = G_a\,(Z_s)$$

$$G_i = \frac{\text{Power available from the amplifier}}{\text{Power delivered to the load directly from the source of the amplifier were not present}} = G_i\,(Z_l)$$

$$G_t = \frac{\text{Power delivered to the load}}{\text{Power available from the source}} = G_t\,(Z_s,\,Z_l)$$

Among these four definitions of the power gain, the transducer power gain has been found to be the most useful one.

### 5.4.1.2 Amplifier Stability

The stability of an amplifier is the figure of merit against oscillating. It can be determined from the S-parameters of the active device, the input and output matching circuits, and the terminations. A two-port network is said to be unconditionally stable at a given frequency if it is stable against oscillation for all passive source and load impedances. If a two-port network is not unconditionally stable, it is potentially unstable and is said to be conditionally stable. In the second case, some passive source and load impedances can cause the two-port network to oscillate.

In terms of reflection coefficients and S-parameters of the two-port network the conditions for unconditional stability at a given frequency are

$$|\Gamma_s| < 1,\ |\Gamma_L| < 1,\ |\Gamma_{in}| < 1,\ \text{and}\ |\Gamma_{out}| < 1$$

where

$$\Gamma_{in} = S_{11} + \frac{S_{12}S_{21}\Gamma_L}{1 - S_{22}\Gamma_L}$$

$$\Gamma_{out} = S_{22} + \frac{S_{12}S_{21}\Gamma_s}{1 - S_{11}\Gamma_s}$$

For $|\Gamma_s| < 1$ and $|\Gamma_L < 1|$, the necessary and sufficient conditions for unconditional stability are

$$K = \frac{1 - |S_{11}|^2 - |S_{22}|^2 + |D|^2}{2|S_{12}S_{21}|} > 1$$

$$|D| = |S_{11}S_{22} - S_{12}S_{21}| < 1$$

In the conditionally stable case ($K < 1$), the loci of $\Gamma_s$ and $\Gamma_L$ in the Smith chart, where values of $\Gamma_s$ and $\Gamma_L$ produce $|\Gamma_{out}| = 1$ and $|\Gamma_{in}| = 1$, are called the *input and output stability circles*, respectively. The radii ($r$) and centers ($c$) of the stability circles in the $\Gamma_s$ and $\Gamma_L$ planes, respectively, are given as follows. Input stability circle:

$$r_s = \left| \frac{S_{12}S_{21}}{|S_{11}|^2 - |D|^2} \right|$$

$$c_s = \frac{(S_{11} - DS_{22}^*)^*}{|S_{11}|^2 - |D|^2}$$

Output stability circle:

$$r_L = \left| \frac{S_{12}S_{21}}{|S_{22}|^2 - |D|^2} \right|$$

$$c_L = \frac{(S_{22} - DS_{11}^*)^*}{|S_{22}|^2 - |D|^2}$$

If $|S_{11}| < 1$, the inner area of the Smith chart, which is outside of the output stability circle, represents the stable region. On the other hand, if $|S_{11}| > 0$, the intersection of the inner areas of the Smith chart and the output stability circle represents the stable region. In Figure 5.8 the shaded areas represent the stable regions in the $\Gamma_L$ plane. A similar illustration of the $\Gamma_s$ plane is shown in Figure 5.9. A potentially unstable FET can be made unconditionally stable by either adding series or parallel resistors to the device or by applying negative feedback. However, proper input and output terminations normally are used in narrowband amplifier design to ensure stability within and outside the frequency band of interest.

The maximum transducer power gain of an unconditionally stable two-port network can be obtained under simultaneous conjugate match condition. In this situation, the values of the transducer power gain, the operating power gain, and the available power gain all are the same and are given by

$$\text{MTG} = \frac{|S_{21}|}{|S_{12}|} (K - \sqrt{K^2 - 1})$$

If the two-port network is potentially stable ($K < 1$), the maximum stable gain is defined as the value of MTG when $K = 1$; namely:

$$\text{MSG} = \frac{|S_{21}|}{|S_{12}|}$$

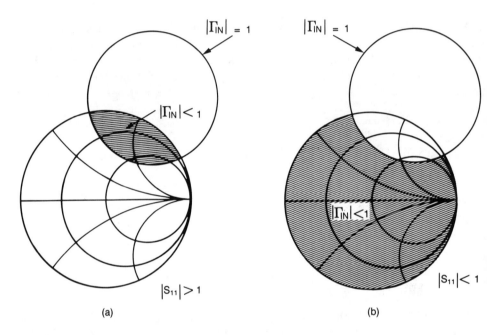

(a)    (b)

**Fig. 5.8** Output stability circle in $\Gamma_L$ plane.

### 5.4.1.3 Amplifier Noise

Thermal noise is always generated randomly by the thermal agitation of electrons in resistors and ohmic resistances of semiconductor devices. In addition active devices with current flowing through them exhibit shot noise. Shot noise arises from random fluctuations of charge carriers in a dc current. In a GaAs FET this shot noise is generated in the channel and also induces noise at the gate. The induced noise at the gate is produced by the channel noise voltage. A figure of merit that represents the noise performance of an amplifier is the noise figure, $F$. The noise figure of any linear, noisy two-part network is defined as the ratio of the total available noise power at the output of the two-port network to the noise power at the output that would be available if the two-port network were noise

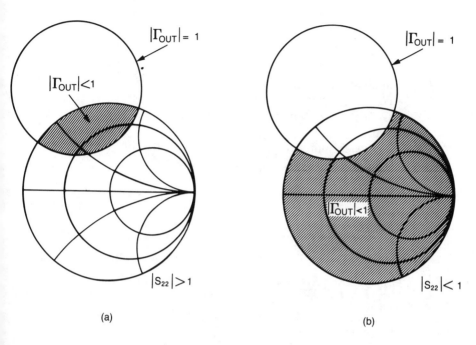

**Fig. 5.9** Input stability circle in $\Gamma_s$ plane.

free. Alternatively, $F$ can be defined as the ratio of the available signal-to-noise power ratio at the input to the available signal-to-noise power ratio at the output. That is, $F$ represents the amount of signal-to-noise power degradation after the signal has passed through a noisy two-port network.

The noise figure for a cascaded amplifier having $n$ stages with gain values, $G_1, G_2, \ldots, G_n$, and noise figure values $F_1, F_2, \ldots, F_n$ is

$$F = F_1 + \frac{F_2 - 1}{G_1} + \frac{F_3 - 1}{G_1 G_2} + \cdots + \frac{F_n - 1}{G_1 G_2 \ldots G_n - 1}$$

The noise measure $F_m$ is defined as the noise figure of an amplifier having infinite identical stages cascaded in series with gain value $G$ and noise figure value $F$; namely:

$$F_m = \frac{F - 1/G}{1 - 1/G}$$

The noise figure of a linear noisy two-port network is given by

$$F = F_{\min} + \frac{R_n}{G_s} |Y_s - Y_o|^2$$

where $F_{\min}$ = minimum noise figure

$Y_s = G_s + jB_s$ = source admittance

$Y_o = G_o + jB_o$ = optimum source admittance that gives minimum noise figure

$R_n$ = equivalent noise resistance that represents how sensitive the noise figure is with respect to the source admittance

$F$ can also be expressed in terms of reflection coefficients $\Gamma_s$ and $\Gamma_o$:

$$F = F_{\min} + \frac{(4R_n|Z_o|)|\Gamma_s - \Gamma_o|^2}{(1 - |\Gamma_s|^2)|1 + \Gamma_o|^2}$$

where $\Gamma_s = (1 - Y_s)/(1 + Y_s)$

$\Gamma_o = (1 - Y_o)/(1 + Y_o)$

### 5.4.1.4 Amplifier Nonlinear Behavior

An amplifier is nearly linear only when the output power is much lower than its maximum power capability. This is because the FET, which is the heart of an amplifier, is a nonlinear element. Only under the small-signal operating condition can the FET be treated as a nearly linear element. Many types of nonlinear phenomena have been defined; however, the ones of greatest concern are harmonic generation, gain compression and power saturation, intermodulation, and AM-PM conversion. These nonlinear phenomena introduce distortion in the amplified signal.

When signal passes through an amplifier (or a nonlinear system, in general), harmonics of the excitation frequency or frequencies are generated in the output terminal. However, the nonlinear behavior of the FETs is not the only cause of the harmonic generation in amplifiers. It can be even more severe, in the case of waveform distortion due to voltage clipping caused by the limited power supply. However, harmonic generation is not a serious problem because the harmonics can be rejected by filters except in multioctave amplifiers.

As input power increases, the output power of the amplifier does not increase at a rate proportional to the increase of the input power; that is, the power gain is decreased from its small-signal value. The output power at which the gain has compressed by 1 dB below the small-signal gain is called the *1 dB gain compression point*, $P - 1$ dB. Normally the power gain will drop rapidly for output powers above $P - 1$ dB and then reach 0 dB gradually at a saturated output power. Saturation occurs in amplifiers (or any linear circuit) because of the limited power supply.

When two or more sinusoidal signals at different frequencies enter an amplifier, the output of the amplifier contains additional frequency components, which are linear combinations of the excitation frequencies. These mixing frequencies are called *intermodulation* (IM) *products*. For example if two signals at frequencies $F_1$ and $F_2$ are applied to an amplifier, the generated IM products will be at frequencies in $mF_1 \pm nF_2$ and $mF_2 \pm nF_1$, where $m + n$ is called the *order of IM product*. Even-order IM products usually are far removed in frequency from the signals that generate them and often of little concern. Third-order IM products, $2F_1 - F_2$ and $2F_2 - F_1$, which are the strongest odd-order products, are closest in frequency to the fundamental signal and often cannot be removed by filters, so they usually dominate the *IM distortion* (IMD) at moderate saturation levels. The $n$th-order intermodulation intercept point $Pn$th is the extrapolated point where the power in the $n$th-order IM product and fundamental signal are equal when the amplifier is assumed to be linear. Figure 5.10 shows the input-output power curves for fundamental, second-order and third-order IM products. The slopes at low power levels are 1 dB/dB, 2 dB/dB, and 3 dB/dB, respectively. Similarly, an $n$th-order IM product varies by $n$ dB/dB with input power level.

A phase distortion called *AM-PM conversion* occurs when an *amplitude modulated* (AM) signal is applied to a nonlinear amplifier causing a phase shift. The phase shift is a function of the instantaneous amplitude of the signal and results in a phase noise in the signal. The AM-PM conversion is defined as the phase change in the output signal for a 1 dB increment of the input power. This type of distortion can be a serious problem in a system in which the signal's phase is important.

### 5.4.2 Procedure for General Design of an Amplifier

Before going into the design examples, we shall describe the general procedure for designing microwave amplifiers:

1. *Selection of process:* Based on the application of the amplifier, the process required to fabricate it can be a low-noise, low-power, or high-power process. From the required frequency bandwidth of operation, the process can be 1 $\mu$m, 0.5 $\mu$m, or even 0.25 $\mu$m. From the cost point of view, the process can be either ion-implanted or epitaxial-based process. As a rule of thumb, for designing an amplifier, 1 $\mu$m process can be used up to 12–15 GHz, 0.5 $\mu$m process up to $\approx$30 GHz, and a 0.25 $\mu$m process up to $\approx$50 GHz. Normally, ion-implanted processing is more cost effective because of its better uniformity. However, epitaxial-based process has better performance in terms of low noise and high power.

2. *Selection of circuit topology:* The most suitable circuit topology is selected to achieve the required performance, such as bandwidth, input-output, VSWR, noise figure, output power, gain, and IP3 or IMD.

3. *Selection of device size and number of stages:* The number of the stages for the required overall gain is determined by estimating the power gain per stage within the frequency band with the given process and the circuit topology. The size of FETs is based on the required output power, dc power limitation, and realization of the matching elements at the operating frequency.

4. *Selection of quiescent point:* The selection of quiescent point is based on whether it requires a low-noise, high-linearity, high-gain, high-power performance, or any combination of these characteristics. The biasing network can be either part of the matching circuit or an additional circuit.

5. *Matching circuit design:* Normally, in designing the matching circuit (for small signal and power amplifiers), we start from the output matching circuit and proceed backward, except for the low-noise amplifier where the input is matched to its optimal source impedance for minimum noise figure then proceeds toward the output. To choose the most suitable matching circuit, the simple lumped circuit elements such as inductors, capacitors, and resistors can be used at the beginning to reduce the time required in the selection process. Once the circuit configuration has been determined, distributed circuit components such as microstrip line and equivalent circuit of the actual lumped elements such as MIM capacitor, interdigital capacitor, and spiral inductor can be used to replace the simple lumped elements. Final optimization for the overall amplifier is always required to obtain the optimal value of each element for the best performance.

6. *Layout of the entire amplifier:* During the layout process, if any circuit element or interconnection cannot be physically realized, adjustment on the element values is necessary, and the entire amplifier also may need reoptimization.

### 5.4.3 Design Examples

#### 5.4.3.1 6–18 GHz Lossy Match Amplifier

As an example, the specifications of the amplifier are:

Frequency: 6–18 GHz
Gain: $\geq$ 10 dB
Gain Ripple: $\leq$ 2.0 dB
Input/Output:

Return loss: $\geq$ 10 dB
Output power: $\geq$ 18 dBm
(1 dB compression point)

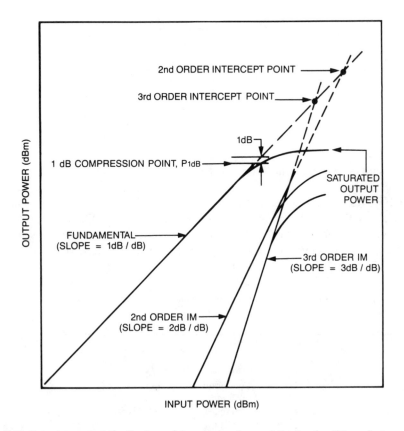

**Fig. 5.10** Output power of the fundamental, second-order, and third-order IM products *versus* input power.

The first step of the procedure is to select a proper process. Normally, a 1 $\mu$m (gate length) process can produce MESFET having $f_t$ of 12–15 GHz. Obviously, this type of process cannot provide sufficient gain at the high end of the required frequency band. Therefore, a process with shorter gate length is desired in this case. For example, the half-micron foundry power process, which has $f_t$ of 22 GHz and provides a maximum available gain of about 7.5 dB at 18 GHz, should be able to meet the requirement. The lumped-element equivalent circuit and its associated circuit element values of a typical FET are shown in Figure 5.11 and Table 5.3, respectively. The second step is to choose the optimal operating point of the

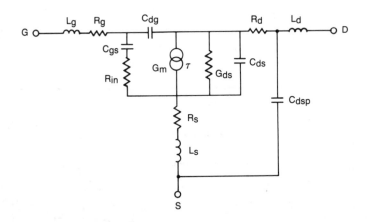

**Fig. 5.11** Lumped element equivalent circuit of a FET.

selected FET to meet the required performance. The gate-to-drain breakdown voltage of this FET is 12 V and the pinch-off voltage is $-3$ V. Therefore, the drain-to-source voltage cannot exceed 9 V. To get as much output power as possible from each unit gate width of FET and provide sufficient gain, the bias point is chosen as $V_{ds} = 7$ V and $I_{ds} = 60\%$ $I_{dss}$.

**Table 5.3** Equivalent Circuit Element Values ($V_{ds} = 7$ V, $I_{ds} = 60\%$ $I_{dss}$)

| | | |
|---|---|---|
| $G_m$ | (mS/mm) | 177.0 |
| $G_{ds}$ | (mS/mm) | 14.0 |
| $C_{gs}$ | (pF/mm) | 1.235 |
| $C_{dg}$ | (pF/mm) | 0.1454 |
| $C_{ds}$ | (pF/mm) | 0.1128 |
| $R_{in}$ | ($\Omega$-mm) | 0.8654 |
| $R_s$ | ($\Omega$-mm) | 0.373 |
| $R_d$ | ($\Omega$-mm) | 0.400 |
| $\tau_d$ | (ps) | 3.1273 |
| $R_g$ | ($\Omega$/mm) | 132.0 |
| $L_g$ | (pH/mm) | 301.6 |
| $L_d$ | (pH/mm) | 121.1 |
| $L_s$ | (pH/mm) | 82.28 |
| $C_{dsp}$ | (pF/mm) | 0.1036 |
| $L_{via}$ | (pH) | 36.0 |

The maximum available gain at 18 GHz is about 7 dB. Hence, providing a 10-dB gain over the frequency band requires two stages. Due to the inherent power gain roll off about 6 dB/octave of FET, the matching circuits have to taper the gain in the opposite way to compensate the gain roll off of the FET. The gain taperings are accomplished by the input and interstage matching circuits by using lossy match technique to provide good input VSWR. To obtain a larger output power density, the output matching circuit is implemented with a reactive match, which avoids additional loss from the resistive element. A flat transfer gain of the output matching circuit is essential for good output VSWR. This means that the output of the FET in the second stage is conjugationally matched to 50 Ω for maximum gain. Figure 5.12 shows the frequency responses of the input, interstage, and output matching circuits as well as two FETs.

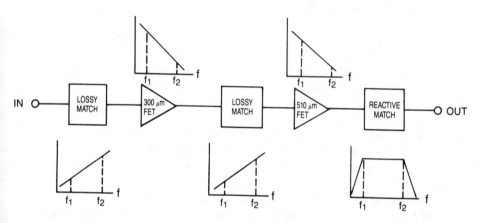

**Fig. 5.12** Frequency response of the input, interstage, and output matching circuits of a 6–18 GHz amplifier.

Determination of the FET's size is based on the output power, power gain, and easy realization requirements of the matching circuits. The maximum output power density at 1 dB compression point of the FET for the given bias point is about 400 mW/mm, when matched for maximum output power. Therefore, an FET having gate width of 300 $\mu$m for the first stage and an FET having 510 $\mu$m for the second stage should be sufficient to meet the power and gain requirements when matched for maximum gain.

The resistors in the input and interstage matching circuits are used to flatten the gain over the frequency band and to stabilize the FETs. They can be in either series or parallel configuration. Normally, the parallel resistor makes matching over a broadband easier than the series resistor does. Therefore, the parallel resistors are utilized in the input and interstage matching circuits.

After the preliminary design, the values of each circuit element should be close to its optimal value. Then, the entire circuit needs to be optimized for the best overall performance. Before the optimizing of the entire circuit, the idealized lumped circuit elements, such as capacitors and inductors also are replaced by the equivalent circuits used to realize them, such as MIM capacitor, interdigital capacitor, spiral inductor, or distributed elements. The final completed circuit is illustrated in Figure 5.13. The predicted small-signal performance of the final circuit is shown in Figure 5.14. Table 5.4 shows the large signal model of the FET. The predicted output power of the amplifier at 1 dB compression point as a function of frequency is shown in Figure 5.15.

### 5.4.3.2  6–18 GHz Feedback Amplifier

For comparison, in this section, we describe another example of broadband amplifier design. Besides the lossy match amplifier, the feedback amplifier frequently has been used to obtain more than one octave band width in a single-ended amplifier. In the following design, the FETs and the quiescent point of the FETs are held the same as those in the previous section. The design goal is to meet the specifications in the previous section whenever it is possible. The considerations in selecting the process and determining the quiescent point and the FET sizes are similar to those in the lossy match amplifier. However, for comparison, here we use the same FETs and quiescent point as those of the lossy match amplifier.

In realizing the matching circuits, we first apply the negative feedback to the FET itself to equalize the gain, stabilize the device, and bring the input and output impedances to 50 Ω, as close as possible while maintaining sufficient gain. The series feedback inductor normally is utilized to improve input VSWR, but it degrades the power gain at the high end of the frequency band. Therefore, only the parallel feedback is used in this design. Table 5.5 shows the stability factor, the maximum stable or available gain, and the impedances for simultaneously conjugate matching, if the device is unconditionally stable with or without the negative feedback, respectively, for a 300 μm FET. Similar information for a 510 μm FET is shown in Table 5.6. As shown in these tables, the devices have become unconditionally stable at all frequencies and their maximum stable gains also have been equalized over the frequency band.

Figure 5.16 shows the frequency responses of the two FETs and those required for the input, interstage, and output matching circuits. By applying parallel feedback, the amplifier design has been simplified to the simultaneous impedance transformations; that is, transforming the input of the first stage and the output impedance of the second stage to 50 Ω and transforming the input impedance of the second stage to the conjugate of the output impedance of the first

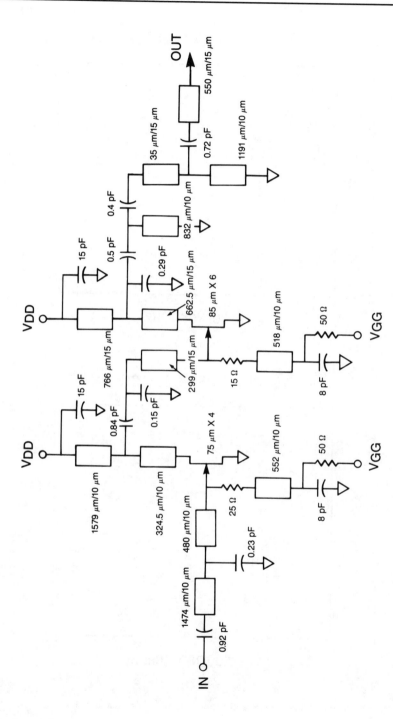

**Fig. 5.13** Schematic of a 6–18 GHz amplifier.

## 6−18 GHz LOSSY MATCH AMP.

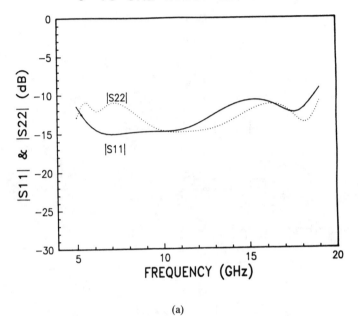

(a)

## 6−18 GHz LOSSY MATCH AMP.

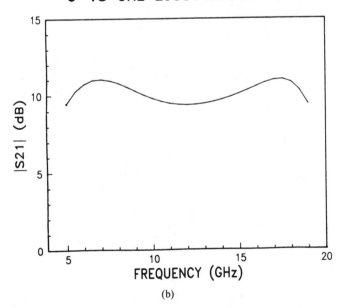

(b)

**Fig. 5.14** Predicted small-signal performance of the 6–18 GHz lossy match amplifier.

## 6–18 GHz LOSSY MATCH AMP.

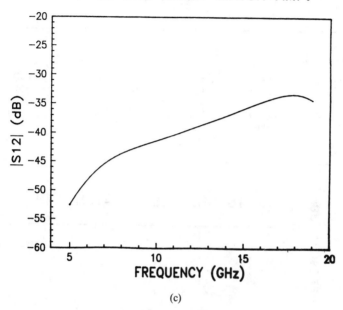

(c)

**Table 5.4 Large-Signal Model of a 0.5 μm FET**

1. The gate-controlled drain current:

$$I_{ds} = (A_0 + A_1V_1 + A_2V_1^2 + A_3V_1^3)(1 + \lambda V_0)\tanh[\alpha V_0(t)]$$

where $V_1 = V_{in}(t - \tau)[1 + \beta V_{dcm} - \beta V_0(t)]$
$\tau = \tau_d V_0(t)/V_{dcm}$

2. The drain-gate avalanche current:

$$I_{dg} = \begin{cases} [V_{dg}(t) - V_b(t)]/R_{avl} & V_{dg} < V_b(t) \\ 0 & V_{dg} < V_b(t) \end{cases}$$

where $V_b(t) = V_{bk} + R_{bk}I_{ds}$

3. The forward-biased gate current:

$$I_{gs} = \begin{cases} [V_{in}(t) - V_{bi}]/G_{fw} & V_{in}(t) > V_{bi} \\ 0 & V_{in}(t) < V_{bi} \end{cases}$$

where $V_{bi}$ = built-in voltage

*cont'd*

**Table 5.4** cont.

Values per mm gate width:

| $A_0$(mA) | $A_1$(mA/V) | $A_2$(mA/V$^2$) | $A_3$(mA/V$^3$) | $\alpha$ | $\beta$ | $\lambda$ | $V_{dcm}$(V) | $G_{fw}$(mS) |
|---|---|---|---|---|---|---|---|---|
| 319.57 | 224.36 | 42.42 | 1.025 | 1.0 | 0.04 | 0. | 5.0 | 700.0 |

| $V_{bk}$(V) | $R_{bk}$($\Omega$) | $R_{avl}$($\Omega$) | $R_s$($\Omega$) | $V_{bi}$(V) |
|---|---|---|---|---|
| 11 | 22.9 | 110.0 | 0.373 | 0.75 |

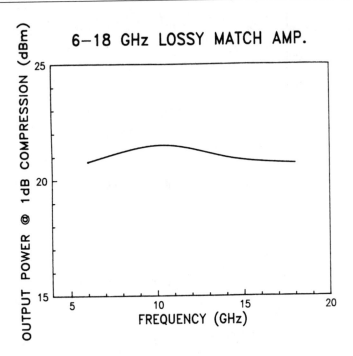

**Fig. 5.15** P-1 dB *versus* frequency of the 6–18 GHz lossy match amplifier.

stage. Of course, final optimization is still needed after the preliminary design in order to obtain the optimal overall performance. Realization of the idealized circuit elements by the actual lumped elements (i.e., MIM capacitors and spiral inductor and distributed elements, such as transmission line) also have been carried out before the final optimization. The optimized circuit is shown in Figure 5.17. Figures 5.18–5.20 illustrate the predicted small-signal performances of the

**Table 5.5** Stability Factor, Maximum Available-Stable Gain for 300 $\mu$m FET

| Freq GHz | K | Sign* BI | Maximum Gain Available dB | Stable dB | Conjugate Match $\gamma$ Source Mag | Ang | Load Mag | Ang |
|---|---|---|---|---|---|---|---|---|
| 1 | .071 | + | | 22.40 | | | | |
| 2 | .142 | + | | 19.40 | | | | |
| 3 | .212 | + | | 17.65 | | | | |
| 4 | .283 | + | | 16.43 | | | | |
| 5 | .354 | + | | 15.49 | | | | |
| 6 | .425 | + | | 14.73 | | | | |
| 7 | .496 | + | | 14.10 | | | | |
| 8 | .567 | + | | 13.56 | | | | |
| 9 | .639 | + | | 13.10 | | | | |
| 10 | .711 | + | | 12.70 | | | | |
| 11 | .783 | + | | 12.35 | | | | |
| 12 | .856 | + | | 12.04 | | | | |
| 13 | .929 | + | | 11.77 | | | | |
| 14 | 1.003 | + | 11.22 | | .980 | 137.6 | .963 | 87.4 |
| 15 | 1.077 | + | 9.62 | | .900 | 140.3 | .824 | 88.3 |
| 16 | 1.152 | + | 8.76 | | .867 | 142.8 | .774 | 89.2 |
| 17 | 1.228 | + | 8.07 | | .846 | 145.1 | .744 | 90.2 |
| 18 | 1.305 | + | 7.49 | | .830 | 147.1 | .724 | 91.3 |
| 19 | 1.383 | + | 6.98 | | .819 | 149.0 | .711 | 92.4 |
| 20 | 1.461 | + | 6.52 | | .810 | 150.7 | .703 | 93.6 |
| 21 | 1.541 | + | 6.11 | | .802 | 152.3 | .698 | 94.8 |
| 22 | 1.621 | + | 5.73 | | .797 | 153.8 | .695 | 96.0 |
| 23 | 1.702 | + | 5.38 | | .793 | 155.2 | .694 | 97.2 |
| 24 | 1.783 | + | 5.06 | | .789 | 156.4 | .695 | 98.4 |
| 25 | 1.863 | + | 4.75 | | .787 | 157.6 | .697 | 99.7 |

**Table 5.5 Cont'd**

|  |  |  |  |  | Conjugate Match γ |  |  |  |
|  |  | | | | | | | |
|  |  | | Maximum Gain | | Source | | Load | |
|  |  | | | | | | | |
| Freq GHz | K | Sign* B1 | Available dB | Stable dB | Mag | Ang | Mag | Ang |
|---|---|---|---|---|---|---|---|---|
| With shunt negative feedback: | | | | | | | | |
| 1 | 1.605 | + | 2.13 | | .149 | 75.7 | .160 | 149.2 |
| 2 | 1.628 | + | 2.02 | | .148 | 78.7 | .144 | 163.0 |
| 3 | 1.625 | + | 2.08 | | .178 | 83.9 | .142 | 168.0 |
| 4 | 1.616 | + | 2.20 | | .217 | 88.7 | .142 | 170.3 |
| 5 | 1.602 | + | 2.35 | | .259 | 93.0 | .143 | 171.6 |
| 6 | 1.585 | + | 2.54 | | .303 | 96.8 | .146 | 172.4 |
| 7 | 1.565 | + | 2.77 | | .348 | 100.4 | .150 | 172.8 |
| 8 | 1.543 | + | 3.04 | | .394 | 103.8 | .155 | 173.1 |
| 9 | 1.519 | + | 3.33 | | .441 | 107.1 | .162 | 173.5 |
| 10 | 1.493 | + | 3.66 | | .488 | 110.4 | .171 | 174.1 |
| 11 | 1.465 | + | 4.02 | | .535 | 113.8 | .184 | 175.0 |
| 12 | 1.436 | + | 4.40 | | .581 | 117.3 | .199 | 176.4 |
| 13 | 1.407 | + | 4.80 | | .627 | 120.9 | .219 | 178.4 |
| 14 | 1.377 | + | 5.21 | | .672 | 124.5 | .244 | −178.9 |
| 15 | 1.347 | + | 5.62 | | .714 | 128.3 | .276 | −175.5 |
| 16 | 1.316 | + | 6.00 | | .754 | 132.1 | .313 | −171.4 |
| 17 | 1.285 | + | 6.33 | | .789 | 135.9 | .356 | −166.9 |
| 18 | 1.256 | + | 6.60 | | .819 | 139.6 | .402 | −162.1 |
| 19 | 1.232 | + | 6.75 | | .844 | 143.2 | .448 | −157.3 |
| 20 | 1.217 | + | 6.76 | | .861 | 146.5 | .485 | −152.6 |
| 21 | 1.217 | + | 6.59 | | .872 | 149.6 | .509 | −148.1 |
| 22 | 1.237 | + | 6.23 | | .875 | 152.3 | .515 | −144.1 |
| 23 | 1.280 | + | 5.70 | | .873 | 154.7 | .506 | −140.4 |
| 24 | 1.348 | + | 5.05 | | .866 | 156.9 | .487 | −137.1 |
| 25 | 1.439 | + | 4.32 | | .858 | 158.7 | .462 | −134.1 |

* For B1 positive the amplifier is unconditionally stable and for B1 negative it is potentially unstable.

**Table 5.6** Stability Factor, Maximum Available-Stable Gain for 510 $\mu$m FET

| | | | Maximum Gain | | Conjugate Match $\gamma$ | | | |
| | | | | | Source | | Load | |
| Freq GHz | K | Sign B1 | Available dB | Stable dB | Mag | Ang | Mag | Ang |
|---|---|---|---|---|---|---|---|---|
| 1 | .084 | + | | 22.40 | | | | |
| 2 | .168 | + | | 19.41 | | | | |
| 3 | .252 | + | | 17.67 | | | | |
| 4 | .337 | + | | 16.46 | | | | |
| 5 | .422 | + | | 15.53 | | | | |
| 6 | .508 | + | | 14.79 | | | | |
| 7 | .595 | + | | 14.19 | | | | |
| 8 | .683 | + | | 13.68 | | | | |
| 9 | .772 | + | | 13.25 | | | | |
| 10 | .863 | + | | 12.89 | | | | |
| 11 | .955 | + | | 12.57 | | | | |
| 12 | 1.050 | + | 10.94 | | .938 | 149.9 | .859 | 112.1 |
| 13 | 1.146 | + | 9.75 | | .903 | 152.3 | .786 | 113.4 |
| 14 | 1.245 | + | 8.89 | | .883 | 154.3 | .750 | 114.8 |
| 15 | 1.346 | + | 8.18 | | .869 | 156.1 | .729 | 116.0 |
| 16 | 1.449 | + | 7.57 | | .860 | 157.8 | .717 | 117.3 |
| 17 | 1.553 | + | 7.04 | | .853 | 159.2 | .711 | 118.5 |
| 18 | 1.659 | + | 6.56 | | .849 | 160.5 | .709 | 119.7 |
| 19 | 1.764 | + | 6.13 | | .845 | 161.7 | .710 | 120.9 |
| 20 | 1.868 | + | 5.73 | | .843 | 162.8 | .713 | 122.1 |
| 21 | 1.968 | + | 5.37 | | .842 | 163.8 | .717 | 123.2 |
| 22 | 2.061 | + | 5.04 | | .841 | 164.7 | .723 | 124.4 |
| 23 | 2.144 | + | 4.73 | | .841 | 165.6 | .730 | 125.5 |
| 24 | 2.214 | + | 4.44 | | .841 | 166.3 | .737 | 126.6 |
| 25 | 2.267 | + | 4.17 | | .842 | 167.1 | .745 | 127.7 |

*cont'd*

**Table 5.6** cont.

| | | | Maximum Gain | | Conjugate Match γ | | | |
| | | | | | Source | | Load | |
| Freq GHz | K | Sign B1 | Available dB | Stable dB | Mag | Ang | Mag | Ang |
|---|---|---|---|---|---|---|---|---|
| With shunt negative feedback: | | | | | | | | |
| 1 | 1.303 | + | 8.24 | | .237 | 63.9 | .269 | 162.0 |
| 2 | 1.306 | + | 8.20 | | .314 | 81.8 | .280 | 162.8 |
| 3 | 1.304 | + | 8.21 | | .407 | 95.6 | .303 | 161.1 |
| 4 | 1.300 | + | 8.23 | | .494 | 106.3 | .332 | 160.1 |
| 5 | 1.293 | + | 8.26 | | .571 | 114.8 | .367 | 160.2 |
| 6 | 1.285 | + | 8.29 | | .635 | 122.0 | .405 | 161.4 |
| 7 | 1.274 | + | 8.33 | | .690 | 128.0 | .445 | 163.6 |
| 8 | 1.261 | + | 8.37 | | .735 | 133.1 | .485 | 166.6 |
| 9 | 1.247 | + | 8.40 | | .772 | 137.6 | .526 | 170.4 |
| 10 | 1.232 | + | 8.43 | | .803 | 141.4 | .565 | 174.7 |
| 11 | 1.217 | + | 8.44 | | .829 | 144.8 | .601 | 179.6 |
| 12 | 1.204 | + | 8.41 | | .849 | 147.8 | .632 | −175.0 |
| 13 | 1.195 | + | 8.34 | | .865 | 150.4 | .658 | −169.3 |
| 14 | 1.193 | + | 8.20 | | .877 | 152.7 | .677 | −163.3 |
| 15 | 1.200 | + | 7.98 | | .884 | 154.8 | .687 | −157.0 |
| 16 | 1.220 | + | 7.66 | | .888 | 156.7 | .688 | −150.5 |
| 17 | 1.254 | + | 7.24 | | .888 | 158.4 | .681 | −143.8 |
| 18 | 1.305 | + | 6.73 | | .886 | 159.9 | .666 | −137.0 |
| 19 | 1.376 | + | 6.13 | | .883 | 161.3 | .645 | −130.2 |
| 20 | 1.467 | + | 5.47 | | .879 | 162.5 | .620 | −123.3 |
| 21 | 1.582 | + | 4.76 | | .874 | 163.7 | .592 | −116.4 |
| 22 | 1.721 | + | 4.01 | | .869 | 164.7 | .562 | −109.5 |
| 23 | 1.885 | + | 3.24 | | .865 | 165.7 | .530 | −102.6 |
| 24 | 2.076 | + | 2.46 | | .860 | 166.7 | .498 | −95.8 |
| 25 | 2.293 | + | 1.67 | | .857 | 167.6 | .466 | −88.9 |

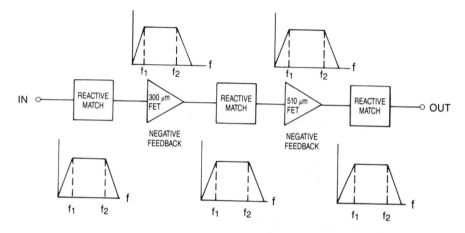

**Fig. 5.16** Frequency response of the input, interstage, and output matching circuits.

final circuit design. The predicted output power at 1 dB compression point as a function of frequency is also shown in Figure 5.21.

### 5.4.3.3 8.5–10.5 GHz 2 W Power Amplifier

To illustrate the design procedure of power amplifiers, we are going to describe an example of designing a 2 W power amplifier having the following specifications:

Frequency band: 8.5–10.5 GHz
Output power at 1 dB compression point: 2 W
Power gain at 1 dB compression point: ≥ 10 dB
Gain ripple: ≤ 1.5 dB
Input return loss: ≥ 10 dB

The considerations for selecting a proper process for the power amplifier are power density, power added efficiency, and maximum available-stable gain of the FET over the frequency band of interest. The 0.5 $\mu$m process used in the previous two examples provide approximately 400 mW/mm of power density, 25% power-added efficiency, and 8 dB maximum available gain at 10.5 GHz. Therefore, this process fulfills the requirements for the power amplifier. The determination of the quiescent point is based on the breakdown voltage, class A operation, maximum output power, and high-power added efficiency as described in the section on the lossy match amplifier. The optimal quiescent point is $V_{ds} = 7$ V and $I_{ds} = 60\%$ of $I_{dss}$.

The next step is to choose the best circuit topology for the power amplifier.

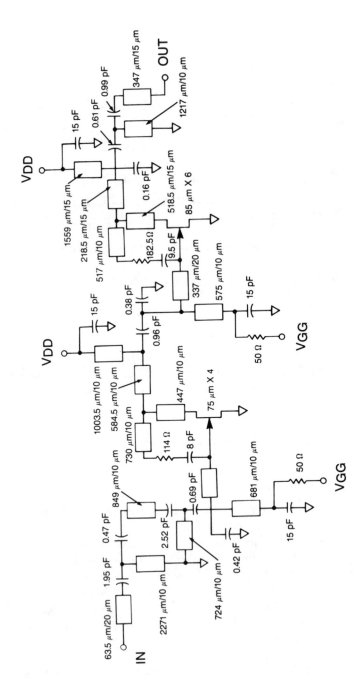

**Fig. 5.17** Schematic of the optimized 6–18 GHz feedback amplifier.

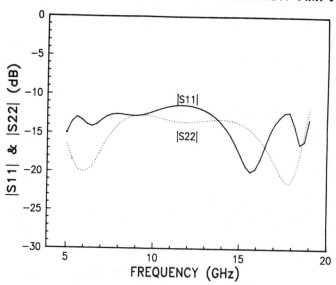

**Fig. 5.18** Small-signal performance of the 6–18 GHz feedback amplifier ($S_{11}$ and $S_{22}$).

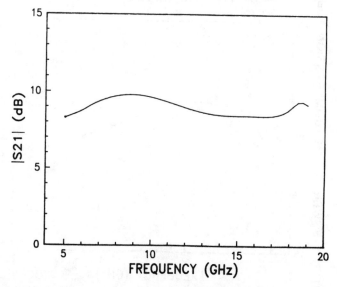

**Fig. 5.19** Small-signal performance of the 6–18 GHz feedback amplifier ($S_{21}$).

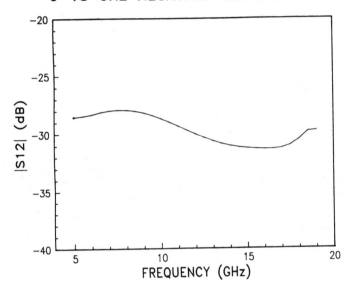

**Fig. 5.20** Small-signal performance of the 6–18 GHz feedback amplifier ($S_{12}$).

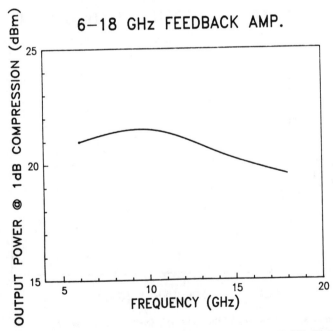

**Fig. 5.21** Output power at 1 dB compression point *versus* frequency of the G-18 GHz feedback amplifier.

Two stages of FETs are required to obtain 10 dB of power gain over the frequency range from 8.5 GHz to 10.5 GHz. The frequency response of the matching circuits are shown in Figure 5.22. The positive gain slope of the interstage and output matching circuits are required to compensate for the 6 dB/octave gain fall off of the FETs. A flat frequency response of the input matching circuit is desired to give good input VSWR.

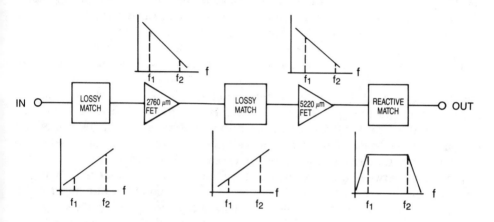

**Fig. 5.22** Frequency response of the matching circuit of a 2 W amplifier.

The procedure of designing the matching cicuits for the power amplifier is given in the following steps:

1. Determine FET size to provide sufficient output power.
2. Obtain the optimal load impedance of the last stage for the maximum output power from the load-pull measurement or from the load-pull simulation using the large-signal model [9].
3. Design the output matching circuit as an impedance transformer to transfer 50 Ω into the optimal load impedance of the output stage.
4. Design the interstage matching circuit with a positive gain slope to flatten the overall power gain of the amplifier.
5. Design the input matching circuit to provide a conjugate match between the input impedance and 50 Ω so that good input return loss can be obtained.
6. Optimize the input and interstage matching circuits for the best performance in terms of input VSWR, power gain, and gain ripple.
7. Perform analysis using large-signal model to obtain output power, power-added efficiency, and power gain at 1 dB compression point.
8. Repeat any previous steps necessary if the large-signal performance requirements are not satisfied.

To deliver 2 W (33 dBm) of output power to the load, assuming that 1 dB of loss in the output matching circuit is due to mismatch and ohmic loss and 0.4 W/mm of power density, requires about 6.4 mm of gate width for the FET of the output stage. The last stage can provide about 5 dB of power gain at 1 dB compression point. Therefore, the input stage is required to produce 29.5 dBm of output power, if we assume a 0.5 dB ohmic loss in the interstage match. Normally, the output power at 1 dB compression point of the FET is about 1–1.5 dB lower if the FET is matched for maximum gain instead of maximum output power. So, if the input stage is matched for the gain and the factor described earlier is taken into account, 3.2 mm of gate width is required to deliver 29.5 dBm of power.

The design of the FET's geometry is based on the considerations for the thermal dissipation minimizing the gate resistance as well as the discontinuity between the gate bus and the feed and uniform feed on each gate finger. The thermal resistance of the device is determined by the substrate thickness, the gate finger width, and the gate-to-gate spacing. However, the substrate thickness, which is 100 $\mu$m for the 0.5 $\mu$m process, normally is not a variable for the circuit designer. The short gate finger width provides low gate resistance and low thermal resistance but requires a larger number of gate fingers, which means that the discontinuity between the gate bus and the feed is greater and feeding each gate finger uniformly is more difficult. Therefore, the selection of the gate finger width and the number of gate fingers is compromised among these factors. The gate-to-gate spacing is determined mainly by consideration of the thermal resistance. Hence, we chose a 115 $\mu$m single-gate finger width and 55 $\mu$m of gate-to-gate spacing. Thus, 28 gate fingers are required for the 2.76 mm FET and 56 gate fingers for the 5.52 mm FET.

By using a load-pull simulator [9] and the large signal model shown in Table 5.4 for the 0.5 $\mu$m process, the load-pull simulation results of the 2.76 mm FET are shown in Figure 5.23. The impedance loci for the maximum gain is plotted in the same figure for comparison. As can be seen from Figure 5.23, the maximum gain impedance loci is located within the −1 dB power contours.

The output matching circuit is used to transform the 50 $\Omega$ into the optimal load impedance of the 5.52 mm FET. The interstage matching circuit is designed to flatten the gain over the required frequency band. Therefore, it provides the best match between the output of the input FET and the input of the output FET at the high end of the frequency band, and it is gradually mismatched at the low end of the band. Finally, the input match transforms the input impedance of the input FET to 50 $\Omega$.

After the preliminary design of the matching circuits, the small-signal performance of the amplifier is close to the design goals. Then the entire circuit, with its fixed output match, is optimized to improve the input VSWR and the gain ripple. The optimized circuit is shown in Figure 5.24 and its large-signal performance is analyzed using a harmonic balance simulator (described in detail in Chapter 8) and the large-signal model. The output power at 1 dB compression point as a function of frequency is shown in Figure 5.25.

frequency
range

ghz

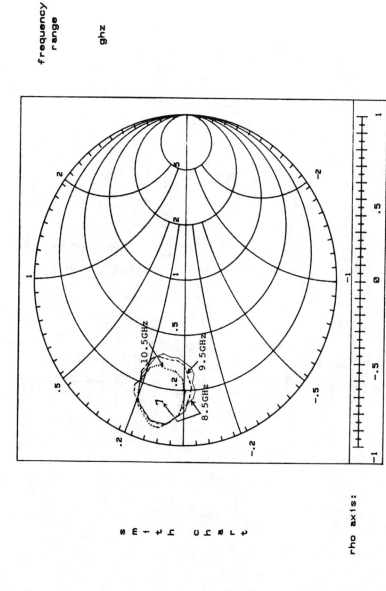

rho axis:

swing ratio

**Fig. 5.23** Load-pull contours of a 2.76 mm FET.

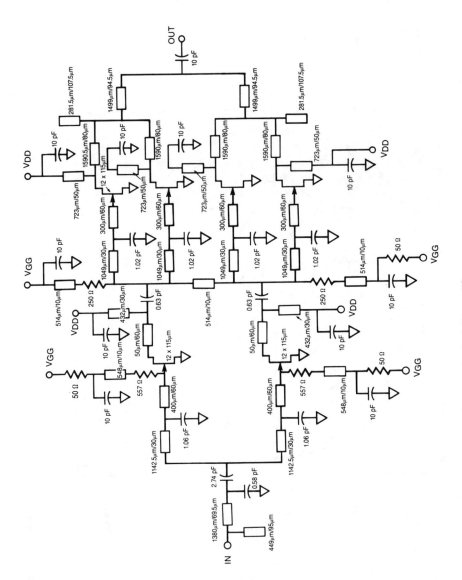

**Fig. 5.24** Schematic of an optimized circuit of a 2 W amplifier.

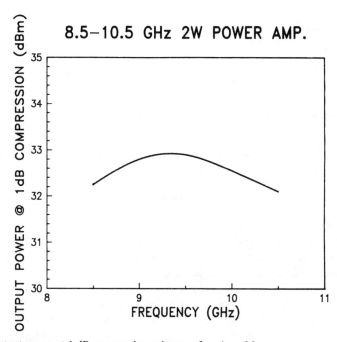

**Fig. 5.25** Output power at 1 dB compression point as a function of frequency.

### 5.4.4 2–18 GHz Distributed Amplifier

A schematic diagram of an *n*-cell FET distributed amplifier is shown in Figure 5.26. In this circuit, the input and output impedance of the FET are combined with microstrip lines to form artificial lossy transmission lines of different characteristic impedances and propagation constants. These lines are coupled with the transconductances of the FETs. The gate and drain lines essentially are loaded constant-*k*[10] lines. Series microstrip lines are connected to the FET drains to compensate for the difference of the phase delays in these lines due to their different propagation constants.

Many other variations are also possible: use of series gate or drain capacitors as well as a tapered drain line to increase output power and power added efficiency; use of dual-gate instead of single-gate FETs to increase power gain; use of "constant *R* network" [10] as opposed to the "constant *k* network" to increase the gain band width product; use of varying gate width to improve the input and output VSWRs' and so forth. When drain-line losses are small compared with gate-line losses, the small-signal gain expression for an *n*-cell amplifier can be given approximately as follows [11]:

$$G = \frac{g_m^2 n^2 Z_o^2}{4} \left(1 - \frac{\alpha_g l_g n}{2} + \frac{\alpha_g^2 l_g^2 n^2}{6}\right)^2$$

for values of $\alpha_g l_g n \leq 1$

where $n$ = number of FETS

$\quad\quad g_m$ = transconductance per FET

$\quad\quad \alpha_g = \frac{1}{2} \frac{r_g w^2 C_{gs}}{l_g}$  $Z_o$ = gate line attenuator per unit length

$$Z_o = \sqrt{L_g \Big/ \left(C_g + \frac{C_{gs}}{l_g}\right)} = \sqrt{L_d \Big/ \left(C_g + \frac{C_{ds}}{l_d}\right)}$$

$\quad\quad$ = length of gate (drain) transition line per unit cell

$\quad L_g(L_d)$ = per unit length inductance of the gate (drain) lines

$\quad C_g(C_d)$ = per unit length capacitance of the gate (drain) lines

$\quad\quad r_g$ = gate resistance per unit length

Thus, we observe that, for a given FET, as the number of cells $n$ is increased, the gain increases to the upper limit for the gate-line attentuation, $\alpha_g l_g n \leq 1$; clearly, in the presence of attentuation, there is an optimum number of cells $n$ that maximizes gain at the high end of the frequency band. In Figure 5.27, gain for a given FET is plotted as a function of the number of cells. The gain of the amplifier increases with additional cells until an optimum number is reached. Any number above the optimum value actually degrades the gain, starting at the high end of the frequency band. Plots similar to Figure 5.27 are illustrated in Figures 5.28 and 5.29 for a given $n$, where gains are plotted as functions of the gate width of the FET and the drain-to-source current, respectively, which show optimum values as well.

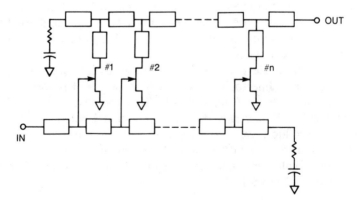

**Fig. 5.26** Schematic diagram of an $n$-cell FET distributed amplifier.

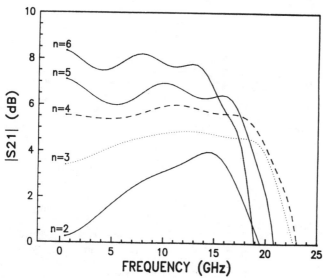

Fig. 5.27 Gain *versus* frequency of the distributed amplifier with various number of stages.

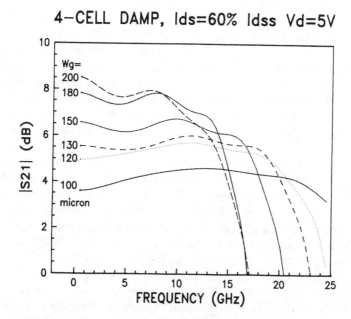

Fig. 5.28 Gain *versus* frequency of the distributed amplifier with various gate widths of FET.

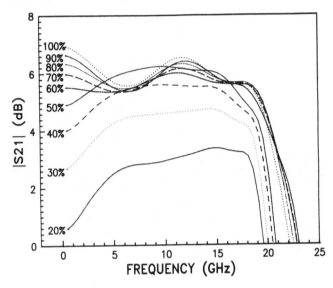

## 4x130um DAMP, Ids=20% − 100%, 10% Idss

**Fig. 5.29** Gain *versus* frequency of the distributed amplifier with various biased drain-to-source currents.

Determination of the number of cells $n$, the gate width of the FET, and the drain-to-source current are based on the gain, band width requirements, output power, and noise figure. Both large number of cells and low bias current provide a better noise figure and wider bandwidth but lower gain and output power. In the present design of a 2–18 GHz gain block, the frequency band is predetermined whereas gain and output power are as high as possible. Therefore, the number of cells $n$, the gate width of the FET, and the bias current are chosen to maximize gain and output power while maintaining the required bandwidth and keeping the circuit as simple as possible.

As a rule of thumb, the upper limit of the frequency band of a distributed amplifier with gain of 5–6 dB can be close to the $f_t$ of the FET to be used in the circuit. Therefore, the process whose FET has $f_t$ of about 20 GHz should be adequate for the present design. It also provides higher output power. Figure 5.27 clearly shows that four cells give the maximum gain with reasonable gain flatness over the 2–18 GHz frequency range. As described in the previous sections, this process provides maximum power when the FET is biased at $V_{ds} = 7$ V and $I_{ds} = 60\% I_{dss}$. From Figure 5.28 we can see that under this bias condition, the amplifier provides about 6 dB gain over the frequency range from 1 to 19 GHz.

The gate width of FETs is maximized in order to produce as much output power as possible while maintaining sufficient gain over the required frequency band. The gate width of FETs was chosen at 135 $\mu$m based on the simulated results illustrated in Figure 5.29. The terminating resistance at the idle ports of the gate and drain lines are chosen to be equal to their corresponding characteristic impedance. Their values should be close to 50 $\Omega$ to provide good input and output VSWRs.

The schematic of the circuit optimized for higher gain, smaller gain ripple, and good VSWRs is shown in Figure 5.30. The predicted small-signal performance in the 1–20 GHz frequency band is shown in Figure 5.31. A gain of 6 dB $\pm$ 0.5 dB was obtained in the 2–18 GHz band. The input and output return losses are better than 10 dB over the same band. To observe the power performance, large-signal analysis was done using a harmonic balance circuit simulator. The predicted output power at 1 dB compression point as a function of frequency in the 2–18 GHz band is shown in Figure 5.32. The output power is 200 mW in the 2–18 GHz band.

### 5.4.5 2–6 GHz Feedback Gain Module

Design of a 2–6 GHz MMIC dual-stage feedback amplifier, where each stage is fed back independently to achieve high performance and excellent isolation in the reverse direction, is presented in this section. The greater feedback of the first stage establishes the input match by trading gain and noise for better input VSWR. The second stage is fed back less than the first stage and accounts for a larger percentage of the gain than does the first stage. The output VSWR is attributed to the second stage active device characteristics and the second-stage feedback.

The active devices of the MMIC amplifier are 600 $\mu$m MESFETs with 0.5 $\mu$m gate lengths. The devices are ion implanted with plated T-gates and airbridged source interconnects. Thermal resistance for each device from the active channel to the backside of the substrate is 80° C/W and under typical bias conditions ($I_{dd} \cdot V_{dd}$ = 0.5 W) and the rise in channel temperature is 40°C. The MMIC was fabricated using a selective ion-implantation process, where the FET active layers and resistors are realized by implanting directly into an undoped LEC substrate. The wafers are thinned to 5 mils, and the backside of the substrate is metalized with a gold-based metal suitable for eutectic die attach.

The MMIC design incorporates all of the bias circuitry on-chip and eliminates much of the additional assembly and bonding associated with off-chip elements (Figure 5.33). The drain supply voltage is RF bypassed by a shunt capacitor and isolated from the amplifier circuit by a series spiral choke inductor. Via hole grounds are not needed for this circuit since the bypass capacitors and source resistors are connected to ground using bond wires. The lack of via holes increases total process yield and reduces the cost per circuit.

The nominal bias point is defined as $V_{dd}$ = +5 V, $I_{ds}$ of stage one = 45 mA

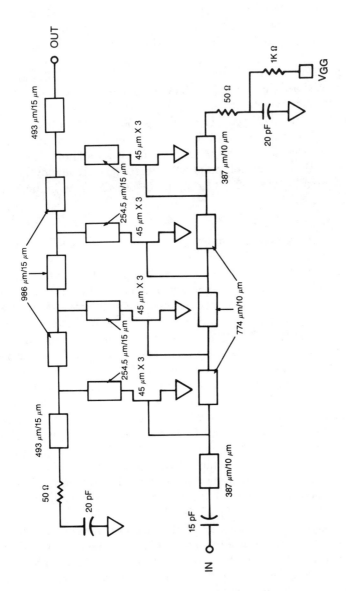

**Fig. 5.30** Schematic of the optimized distributed amplifier.

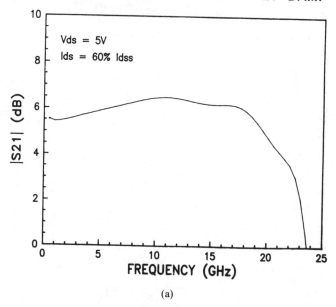

(a)

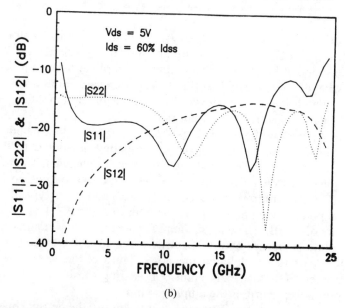

(b)

**Fig. 5.31** Predicted small-signal performance of the distributed amplifier.

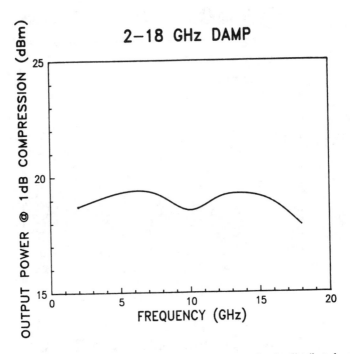

**Fig. 5.32** Output power at 1 dB compression point *versus* frequency for the distributed amplifier.

and $I_{ds}$ of stage two = 75 mA, for a total $I_{dd}$ = 120 mA. The RF performance for gain, VSWR, power out, and noise figure are shown in the Figures 5.34 through 5.37. The gain shape of Figure 5.34 shows that the amplifier has an up slope across the frequency band, a nominal gain of 12 dB, and total gain variation of less than 1 dB. A $\leq$ 2:1 match is easily achieved at the input of the amplifier, whereas a $\leq$ 1.2:1 VSWR is typical for the output port, as shown in Figure 5.35.

The output power of this amplifier is shown in Figure 5.36. The output power curve has a positive slope with respect to frequency attributed to the greater amount of RF output power being fed back at lower frequencies. The total output power variation is within 1 dB from 2–6 GHz. Saturated output power is also plotted and a 2–3 dB variation between the P − 1 dB and $P_{sat}$ is typical. The third-order intercept point is a minimum of 10 dB higher than the P − 1 dB point and has a nominal value of 31 dBm. The noise figure *versus* frequency is plotted in Figure 5.37. The maximum occurs at 2 GHz, where the total feedback power is the greatest. The major factor defining the noise figure is the first-stage feedback resistance and the first-stage bias point. Optimizing these two constraints improves the total noise figure of the amplifier chip.

This circuit topology also allows the user to choose the optimal bias point for a specific application. Source inductance, bias current ($I_{dd}$), and drain voltage ($V_{ds}$

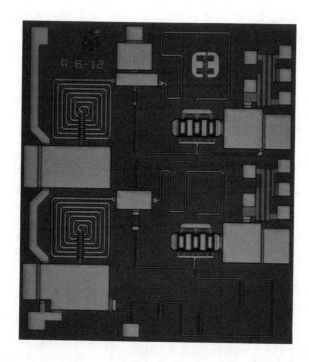

Fig. 5.33 Dual-stage 2–6 GHz amplifier (courtesy of Harris Microwave Semiconductor).

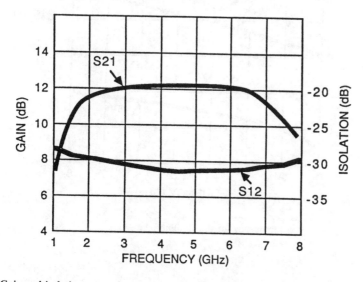

Fig. 5.34 Gain and isolation *versus* frequency for the amplifier.

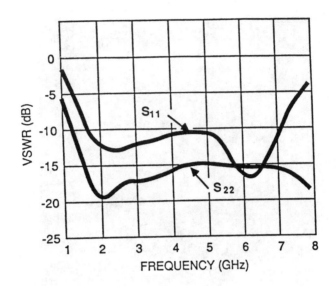

**Fig. 5.35** Input-output VSWR *versus* frequency for the amplifier.

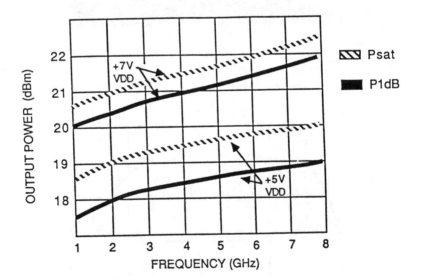

**Fig. 5.36** Output power *versus* frequency for the amplifier.

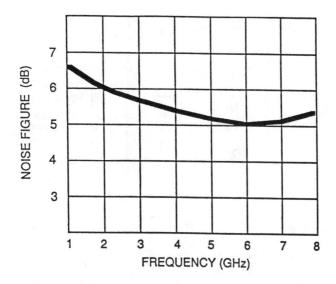

**Fig. 5.37** Noise figure *versus* frequency for the amplifier.

all can be used to vary the RF performance of this MMIC amplifier. Gain, gain flatness, output power, *et cetera* can be tailored for an application by changing the quiescent bias condition of the circuit. Gain and gain flatness are dependent upon the bias currents of the two stages, whereas output power is strongly dependent upon the supply voltage and the bias current of the output stage. Figures 5.38 and 5.39 define the variations in amplifier performance parameters as functions of the bias point.

Gain is insensitive to the supply voltage $V_{dd}$ and a change of 0.5 dB is typical as the voltage is changed from +5 V to +9 V. The bias current, on the other hand, has a strong correlation with $C_{gs}$ and $g_m$ of the active FETs. Changing the bias current from 25 to 50% $I_{dss}$ can increase $C_{gs}$ and $g_m$ by as much as 27 and 45%, respectively. The resulting change in gain is appreciable and can be to the designer's advantage.

The bias current determines the overall gain magnitude of the amplifier. The total gain flatness is dependent upon the bias point of the FETS. With changing $I_{ds}$, the intrinsic capacitances of the FET change affecting primarily the gain magnitude and gain flatness. A gain tradeoff exists between the magnitude and total variation. The bias point must be determined according to the required specifications.

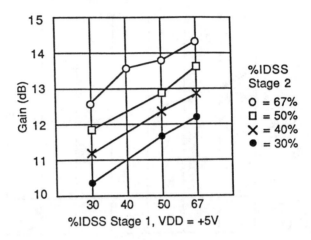

**Fig. 5.38** Minimum gain level *versus* dc bias for the amplifier.

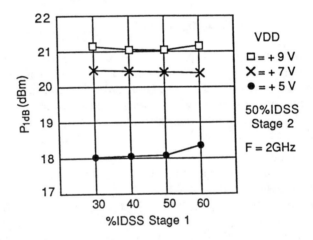

**Fig. 5.39** Output power *versus* dc bias for the amplifier.

The output power of the amplifier is directly dependent upon the second stage of the amplifier. The first stage is capable of providing enough linear output power to drive the second stage into compression and saturation. Figure 5.39 shows the dependence of the output power on both the first-stage and second-stage bias conditions. Moderate output power of 17 dBm is achievable even with both stages operating at 30% $I_{dss}$, and +7 V $V_{dd}$. The gain and power performance of the two chips when cascaded is given in Figures 5.40 and 5.41. As was expected, the gain is seen to be 29 dB with an associated gain variation of ± 1.0 dB for the 2–6 GHz frequency band. Input and output VSWRs are both below 2 : 1,

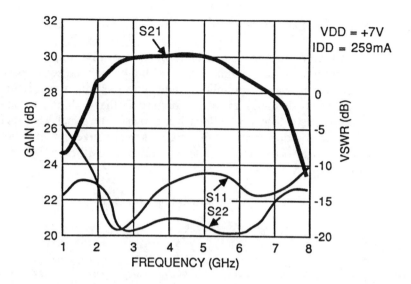

**Fig. 5.40** Module performance *versus* frequency for two amplifiers cascaded on a single carrier.

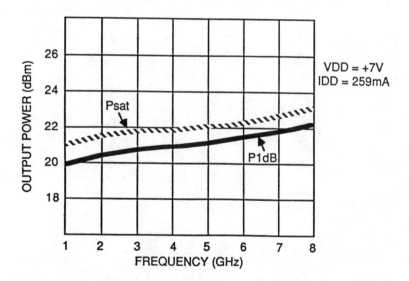

**Fig. 5.41** Module output power *versus* frequency for two amplifiers cascaded on a single carrier.

and isolation greater than 50 dB. Average supply current for the two-chip module was 260 mA with $V_{dd}$ equal to 7 V to obtain higher output power of 22 dBm.

### 5.4.6 Low-Noise Amplifier Design

The circuit requirement to be considered is a low-noise amplifier operating over a 3–6 GHz frequency range. This circuit is for a broadband sensor system in which accurate measurement of amplitude and phase are the key performance aims. This application demands an amplifier circuit that exhibits a flat, well-controlled gain response; good phase linearity; and good terminal VSWRs over a broadband width. In addition, because the circuit is the receiver preamplifier, low-noise performance is also required. Table 5.7 summarizes the key specification parameters for the LNA circuit.

**Table 5.7 Key Parameters for the LNA Circuit**

| Parameter | Symbol | Minimum | Maximum | Units | Comment |
|---|---|---|---|---|---|
| Frequency | $F$ | 3.0 | 6.0 | GHz | |
| Gain | $G$ | 15.0 | | dB | |
| Noise figure | $NF$ | | 3.7 | dB | |
| Return loss | $S_{11}$ | | −9.9 | dB | |
| | $S_{22}$ | | −9.9 | dB | |
| *Linearity:* | | | | | |
| Amplitude phase | $\Delta G_p$ | | 0.5 | dB/rms | Measured at powers |
| | $\Delta\phi$ | | 2.5 | Deg/rms | up to 1 dB |
| | | | | | comparison |
| dc power consump-tion | $P_{dc}$ | | 500 | mW | |
| Test temperature | $T_{amb}$ | 25 | 25 | °C | |
| Operating temperature range | $T_{op}$ | −54 | +75 | °C | Operational survival |

*Note:* All chips to be RF on wafer probeable.

### 5.4.6.1 Initial Design Decisions

Three generic types of amplifier circuit could be considered to fulfill these specifications; namely, reactively matched [12], shunt feedback [13], or traveling wave [14]. Each option should be examined by designing a prime element amplifier and analyzing the predicted performance against the specifications. These amplifier designs are set up on microwave circuit simulator files using S-parameter blocks for the appropriate standard foundry FET cells. The results of these analyses are presented in Table 5.8, from which the choice of preferred circuit topology can be readily made. The review clearly shows that the traveling wave amplifier approach should be discontinued due to its low gain and high noise.

**Table 5.8** Predicted Performance of Parameters

| Parameter | Reactively matched amplifier | Shunt feedback amplifier | Traveling wave amplifier |
|---|---|---|---|
| Bandwidth | <Octave | >Octave | >Decade |
| Gain | High | Moderate | Low |
| Noise figure | Meets specifications | Marginal | Fails specifications |
| Gain flatness | Fails specifications | Meets specifications | Meets specifications |
| Return loss | Tradeoff with NF to specifications | Tradeoff with NF to specifications | Meets specifications |
| dc power drain | Meets specifications | Meets specifications | Very poor |
| Stability | Poor | Good | Good |
| Sensitivity | Moderate (tuned circuits dominate performance) | Low (resistor in feedback dominates performance) | Low |

In the case of the reactively matched circuit topology, a design would be sensitive to passive component and FET parameter variation. Further, the inherent multisection input, interstage, and output matching networks make gain flatness very difficult to achieve and also consume large areas of GaAs. The shunt feedback design, in contrast, often provides optimum performance tradeoffs for small- and medium-signal level amplifiers of 1–2 octave bandwidths. The particular merits offered by the feedback circuit are excellent gain flatness and amplifier stability coupled with good input and output match, all within a relatively small GaAs area. With these advantages, the initial choice of circuit should be the shunt feedback topology because this offers the best solution to the original specifications.

The process with 0.5 $\mu$m gate length having 20 GHz operation limit meets the circuit objectives. Briefly, this process is a 10 mask level IC process based on ion-implanted material, which uses 0.5 $\mu$m gate FETs as the active devices and has GaAs via holes for minimum length ground returns.

The design of a shunt feedback amplifier starts by considering the single-stage circuit of Figure 5.42. The first step is to select the FET gate periphery and the basic prime element values for the passive components. In this case, the FET gate width was selected to maximize $g_m$ for gain and $I_{dss}$ for output power against the need to keep dc power consumption below the specification value. The FET characterization data base showed that the optimum compromise for the 3–6 GHz frequency range would be a 600 $\mu$m gate width FET.

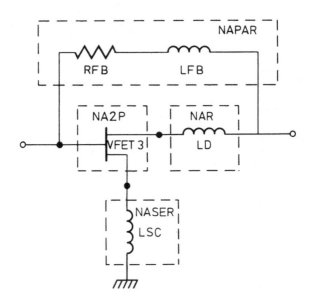

**Fig. 5.42** Single-stage shunt feedback amplifier circuit.

The initial passive element values can be determined using the design equations of Niclas et al. [15] as follows:

1. From the condition of perfect match at the lowest operating frequency:

$$R_{fb} = g_m Z_0^2 = 125 \ \Omega$$

where
$Z_0 = 50 \ \Omega$ (system characteristic impedance)
$g_m = 50$ mS (FET transconductance at low noise bias)

2. From the condition that the inductor $L_d$ should resonate with the FET output capacitance at the upper operating frequency:

$$L_d = X_c/W = 1.3 \text{ nH}$$

3. Assume that initially there is no common lead inductance:

$$L_{sc} = 0$$

With these initial values the circuit is optimized for target specification.

Cascading two of the single-stage circuit blocks quickly realizes the two-stage amplifier. This two-stage circuit is then optimized against the target specification to complete the initial design study phase. From the results of the single- and two-stage optimizations, we conclude that the design would just satisfy the gain and noise figure requirements.

Having established that the basic design is satisfactory, the next step is to ensure that the design will still meet the target specifications when its element values are varied through the known process tolerance range. In other words, the circuit is analyzed to determine its sensitivity to process variations. This analysis sensitivity is achieved using the TUNE mode within TOUCHSTONE or for SUPER COMPACT by using the ANA VAR facility. In the circuit of Figure 5.42, analysis shows that the most sensitive component is the feedback resistor. This is because any FET variations are inherently desensitized by the negative feedback and inductors are accurately realized on an MMIC.

The design task now starts to move toward a MMIC floor plan with the addition of the parasitics associated with the prime circuit elements and the incorporation of bias networks. The parasitic models and element values used are from within the design data base as a function of the prime element size and value. Subsequent optimizations change the prime element values, and the correct parasitic component values are included for circuit analysis.

Figure 5.43 shows a conventional bias insertion network comprising RF choke, bypass capacitor, and damping resistor. The purpose of the resistor is to

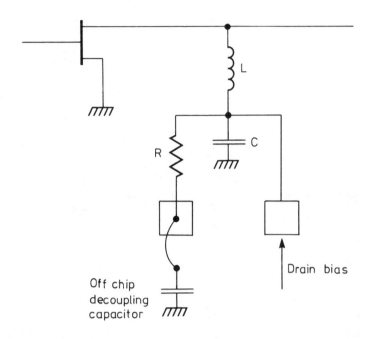

**Fig. 5.43** MMIC bias insertion network.

dampen any low frequency instabilities within the FET drain bias network. Selection of the bias component values is largely a compromise between the highest value desired and the GaAs chip area consumed. In this case, nominal values of 6 nH and 10 pF on-chip are selected. Additional bypass capacitance is normally provided off-chip in the final IC package environment. The circuit is reoptimized using the full parasitic model for all the passive components. The schematic and the performance of the two-stage amplifiers are shown in Figures 5.44 and 5.45, respectively.

The shunt feedback topology cannot be used for the input stage, because the additional noise generated by the feedback resistor would further degrade the cascaded noise figure. If a nonfeedback FET stage is used, a compromise between low noise and input match has to be made. An elegant solution to this compromise is the deliberate introduction of a series feedback inductance in the FET source [16]. By careful selection of the inductance value, the FET's physical input resistance can be increased to equal the input noise resistance. Hence, an input matching network can be synthesized to achieve both optimum gain and low-noise operation.

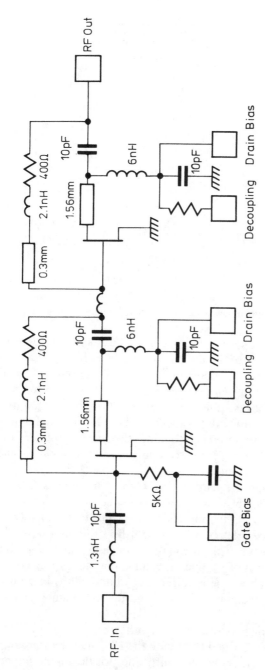

**Fig. 5.44** Optimized prime element two-stage amplifier.

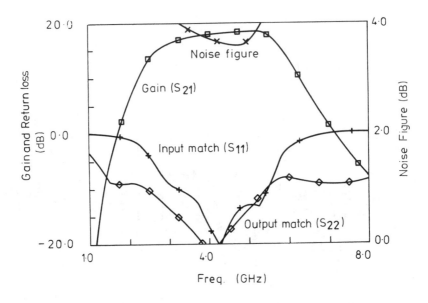

**Fig. 5.45** Predicted performance of two-stage shunt feedback amplifier with parasitic component models.

Several (often conflicting) factors must be taken into consideration when determining the required FET gate periphery, including noise figure and associated gain, stability, dc power consumption, input and output impedances, sensitivity to mismatch, and output power capability. Reference to an available FET data base shows that a 600 $\mu$m gate width FET is the optimum choice for the design frequency band. In order to reduce the FET's minimum noise figure, a small unit gate width is required [17]; again, references to foundry data shows that a 75 $\mu$m unit gate width is the preferred choice requiring eight parallel gate stripes to give the final FET structure.

Having determined the required FET size, the next task is to examine the impedance control available using inductive source feedback. A mapping technique using the ANA VAR function of SUPER COMPACT enables comparison of $S_{11}^*$ and $\Gamma_{opt}$ (the source reflection coefficient for minimum noise figure) as functions of source inductance. This mapping is shown in Figure 5.46. It is seen that as $L_s$ increases, the real part of $S_{11}^*$ increases and approaches $\Gamma_{opt}$. At $L_s = 0.6$ nH the values of $S_{11}^*$ and $\Gamma_{opt}$ cover the required LNA bandwidth. In order to realize and control such a small inductance value a via grounding technology is required in the MMIC process.

An input matching network is now synthesized using the SYNTHESIS function of SUPER COMPACT and written to the FET mapping circuit file previously described. Optimizing the prime element values for the minimum input return loss and noise figure gives the response shown in Figure 5.47. A good input match and

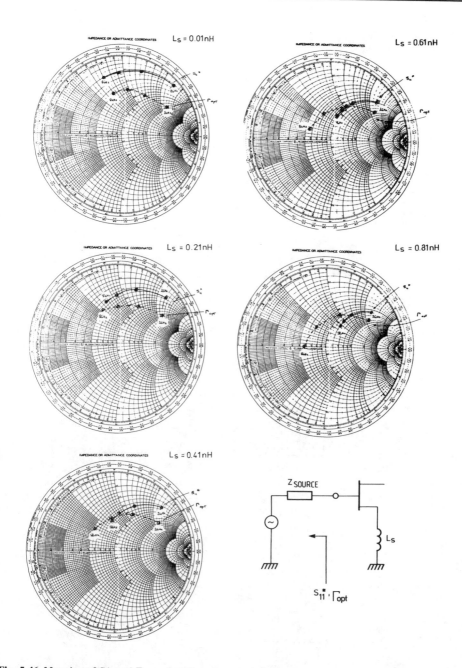

**Fig. 5.46** Mapping of $S_{11}^*$ and $\Gamma_{opt}$ as function of source inductance.

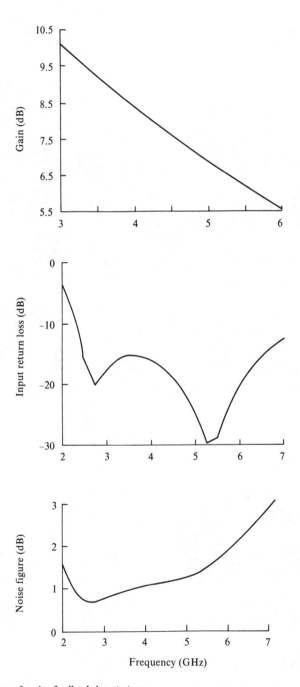

**Fig. 5.47** Response of series feedback input stage.

noise figure is obtained, but the inherent gain slope of the device is still present. In reactively matched amplifiers, the gain slope is conventionally compensated for by a controlled degree of mismatch between stages; however, with source inductance, present FET input and output become interactive making this technique difficult to realize. The gain slope, therefore, is compensated for in this design by adjusting the feedback resistor and inductor values in the original two-stage amplifier. Figure 5.48 shows the result of doubling the feedback inductor and halving the feedback resistor values in the two-stage feedback amplifier design. Over the 3–6 GHz band an 8 dB gain slope is achieved, which more than compensates for the first-stage gain slope.

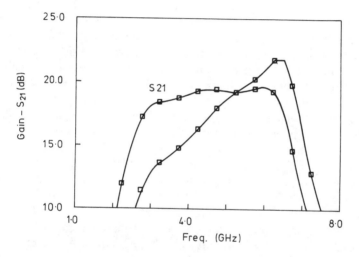

**Fig. 5.48** Gain slope compensation capability of shunt feedback amplifier.

The amplifier design now continues by cascading the series feedback and two-stage shunt feedback circuits into a single three-stage amplifier. The complete three-stage amplifier design moves toward a MMIC realization by optimizing the prime element values, adding the full parasitic models of the components and reoptimizing. During the various optimization iterations different targets are specified, according to which amplifier stage is being optimized. For the first stage, the optimization targets are input match and noise figure, for the second stage, maximum flat gain and minimum noise figure are targeted; and for the third stage, the targets are maximum flat gain and output match. Throughout the optimization a careful check is kept on the stability of each stage by calculating the K-factor and plotting stability circles. The TUNE mode of TOUCHSTONE, whereby component values can be adjusted manually, is of considerable use during this optimization. Indeed, this manual mode often can be more effective than the optimization routines in obtaining a satisfactory circuit response.

With the three stage amplifier configuration, shown in Figures 5.49 and 5.50, a satisfactory electrical performance is obtained and the design appears far more attractive for MMIC realization. In Figure 5.50, each element has been replaced by its equivalent circuit model, taking into account all the parasitics associated with an element. The circuit shown in the figure will be used in analyzing the three-stage LNA circuit shown in Figure 5.54.

### 5.4.6.2 Sensitivity Analysis

The best method of examining the sensitivity of a MMIC design is to vary each component type over its process limits, with each element varied in the same direction. This technique reflects the action of any semiconductor process. For example, because all capacitors essentially are fabricated at the same process stage, any variation from the process mean (to a first-order approximation) must affect all the circuit capacitors in the same way. Figures 5.51 and 5.52 illustrate the effects of typical component variations on the three-stage amplifier circuit. The results of the sensitivity analyses on both gain and noise figure are shown for variations in resistor, capacitor, and inductor values. From these results it is clear that the noise figure has a low sensitivity to such variations, a fact that vindicates the original choice of a feedback topology for the MMIC realization. As might be expected, the gain response has its greatest sensitivity to resistor variations. This effect is caused almost entirely by the feedback resistors in the second and third gain stages. Indeed, these feedback resistors are responsible for the gain slope (see Figure 5.48). Examining the important input match requirement during the sensitivity analysis highlights a sensitivity with inductance. As might be expected the source inductance in the series feedback part of the input stage is the major influence. However, the very accurate control of source inductance provided by the through GaAs via hole means that this is not a design hazard.

These sensitivity analyses on the circuit's passive components show that, even for the worst tolerance limits, it is possible to achieve reasonable yields within the prime specification values. In practice, the foundry process to be used exhibits much smaller spreads than the tolerance limits analyzed. The conclusion therefore is that, for the passive components at least, the design is robust enough for MMIC realization. However, the analysis also shows that the feedback resistor control is important in ensuring flat gain.

Turning to the active components in the design, the FETs also have to be examined for sensitivity to process-induced variation. How this is done in the CAD depends very much on the way FET variability is presented within the foundry process data base. There are basically two ways of describing FET variations in a circuit tolerance analysis: S-parameter tolerancing and tolerancing the

# Three Stage L.N.A. Circuit Diagram

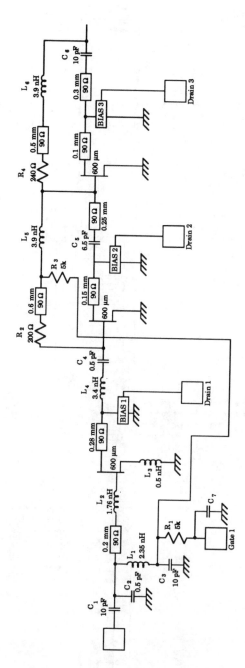

**Fig. 5.49** Full circuit diagram of final three-stage LNA design.

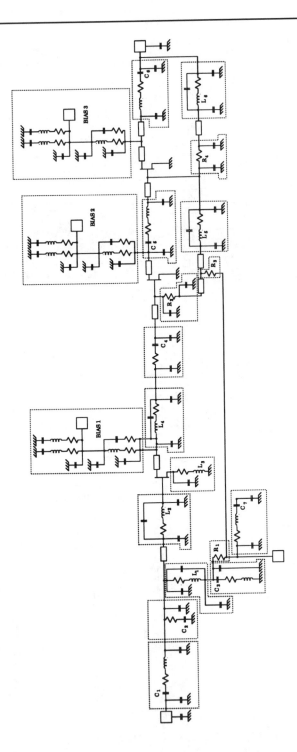

**Fig. 5.50** Full circuit design of final three-stage LNA design showing the equivalent circuit models of each element.

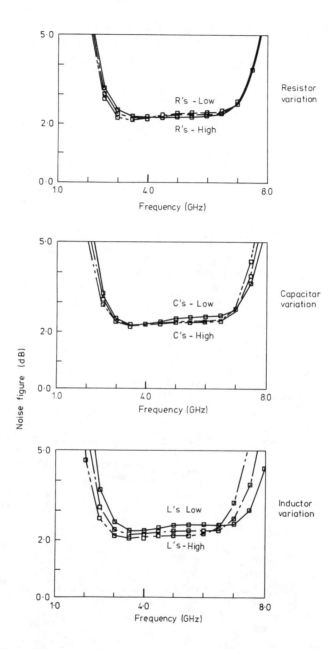

**Fig. 5.51** Sensitivity analysis of noise figure.

454

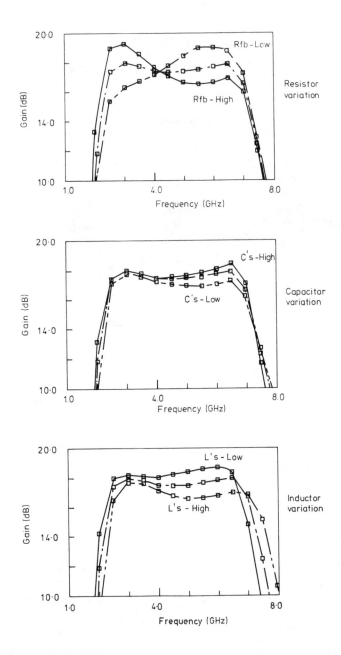

**Fig. 5.52** Sensitivity analysis of gain.

element values of a FET equivalent circuit. Although S-parameter tolerancing is convenient because the spreads are obtained directly from the S-parameter measurements, it is difficult to correlate the resulting data with the physical parameter variations within the device. Equally, it is difficult to determine what S-parameter sets really represent the worst case. For these reasons, the method preferred in the foundry process data base is to tolerance the FET equivalent circuit model elements. Since the FET model elements (e.g., $gm$ and $C_{gs}$) are related directly to physical device parameters (e.g., gate length, doping profile, gate recess) a direct correlation can be made between process variations and resulting circuit performance.

For the three-stage amplifier the FET sensitivity analysis is performed by writing the FET equivalent circuit into the amplifier design file and then tolerancing the FET model elements over the foundry process specification limits. Shown in Figure 5.53 are the effects of tolerancing $gm$ on amplifier performance. Here it can be seen that, for low $gm$ values, the circuit gain could drop below the minimum specification value. This problem can be easily corrected by reoptimization resulting in only minor adjustment to some circuit element values.

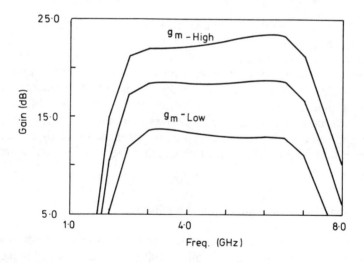

**Fig. 5.53 Effect of intrinsic FET *gm* variation on amplifier gain.**

Completing the sensitivity analysis on all the key FET parameters ($C_{gs}$, $g_m$, $R_s$, $R_d$, *et cetera*) shows that the circuit design can be optimized for insensitivity to the known foundry process variations. This combined with the sensitivity analyses on the passive components reveals that a MMIC circuit design capable of meeting the original objectives can be achieved. Thus, the design cycle is complete to the final circuit design stage of the CAD procedure and the MMIC floor plan layout can begin. The amplifier circuit diagram is shown in Figure 5.54, having grown in complexity considerably from the initial circuit of Figure 5.44. Comparison of these circuit diagrams clearly underlines the need and impact of accurate simulation and modeling techniques in handling what has become a complex design problem.

### 5.4.7 Circuit Performance

A photograph of a completed chip is shown in Figure 5.55. The measured noise figure of 4 dB appears to be disappointing when compared to the 3.7 dB target. However, the difficulties of measuring noise figures on wafers leads to uncertainties of < 0.5 dB. This measurement uncertainty is caused by difficulties in calibrating the system containing the losses of probes, connectors, and coax lines before the low noise measurement equipment. Subsequent noise evaluations on these chips using conventional low-noise jig techniques revealed noise figures of 3.5 dB.

However, actually in the vast majority of cases, any chip that passed the noise figure specification also passed the other performance specifications and *vice versa*. From these results we can see that processing a mask set dedicated to this circuit would produce more than 900 chips with over 400 meeting specifications from a single 3-inch diameter GaAs wafer. Figure 5.56 compares the simulated results of gain, noise figure, and input match with the mean of the yielded circuits as measured on the microwave wafer prober. The very close correlation between design values and measured values is excellent proof that the original foundry design data, the process repeatability, and the design route all were of a high quality.

### 5.4.8 Combining Techniques for Power Amplifiers

The output power of a monolithic power amplifier is limited by heat dissipation through the substrate, which normally is 100 $\mu$m thick. Some of these limitations were described in detail in Section 5.2. To obtain higher power, use of a large gate periphery device is required, but this results in poor thermal impedance and low input and output impedance. To match a low impedance over a wideband of the order of < 1 $\Omega$ (for a 4–5 W device) to 50 $\Omega$, requires a large matching circuit and results in poor input and output return losses. The techniques to accomplish this goal can be combined either on-chip or off-chip.

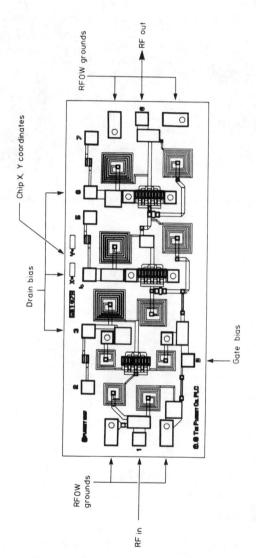

**Fig. 5.54** Physical layout of low noise amplifier MMIC.

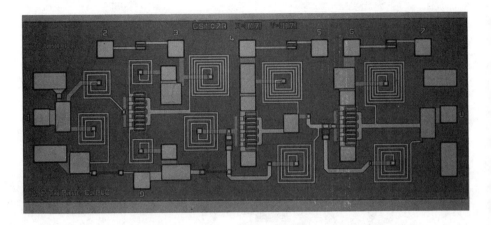

**Fig. 5.55** Photograph of completed low noise amplifier MMIC (Courtesy of Plessey).

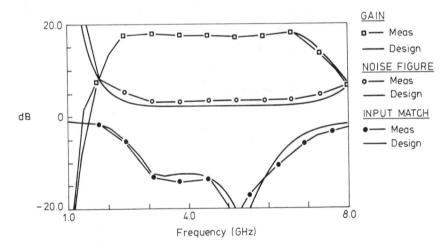

**Fig. 5.56** Comparison of LNA designed and measured gain, noise figure, and input match.

### 5.4.8.1 On-Chip Combining

On-chip combining is accomplished by clustering the devices or a number of device cells onto the equivalent of a single chip. Matching to 50 $\Omega$ is provided by considering the input impedance of the combined cells or FETs. This is illustrated in Figure 5.57. However, because the input impedance of the combined FET (consisting of many cells or FETs) is small and the number of cells that can be clustered is limited by the heat dissipation, the total gate periphery that can be used in this technique is limited to approximately 5 mm at 10 GHz.

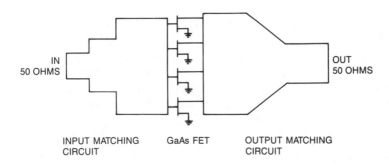

INPUT MATCHING      GaAs FET      OUTPUT MATCHING
CIRCUIT                                  CIRCUIT

**Fig. 5.57** Cluster matching technique.

The other method to avoid the problems mentioned earlier is to match the individual cells or FETs whose impedance is higher (e.g., 2–3 $\Omega$ for a 1 mm FET) to a higher impedance, such as 25, and then provide a matching circuit to the combined impedance of the prematched cells or FETs. The example of such an approach is illustrated in Figure 5.58. This technique is still limited to moderate power levels. The limitations come from the overall chip size, yield, and the thermal impedance. The power levels achieved by this technique is currently limited to 4–5 W at 10 GHz. The chip size is limited to about 10–12 mm$^2$ for a reasonable yield of approximately 50%, if it is to remain affordable.

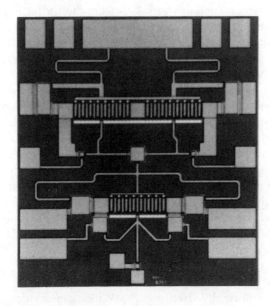

**Fig. 5.58** Photo of a X-band 2 W amplifier (courtesy of MSC).

Still another technique to combine the FET amplifiers on the same chip is through a balanced approach. At higher frequencies, $> 10$ GHz, the $\lambda/4 \simeq 2$ mm, and the normal chip size that can be fabricated with reasonable yield is approximately 10 mm$^2$. Thus, a Lange coupler at the input and output can be accommodated on the same chip with a 10% increase in the chip size. However, the loss of a Lange coupler on a 100 $\mu$m thick substrate is higher compared to a higher substrate thickness such as 200–300 $\mu$m.

### 5.4.8.2 Off-Chip Combining

On-chip combining is limited by the number of devices that can be combined and effectively matched in a small area. In off-chip combining, several small monolithic chips, having moderate power levels, can be combined through techniques used in hybrid amplifiers to obtain higher power levels. For example, several 2–4 W monolithic chips can be combined to obtain 10–20 W output power. Several techniques for this are used in hybrid amplifiers. One of the criteria in selecting the combining technique is the combining efficiency of the combiner. The off-chip combiner circuits are fabricated on a thick ceramic substrate, such as alumina, to reduce the loss in the combiner and hence to increase the combining efficiency. The desirable characteristics of a combining network are loss, minimum phase and amplitude imbalance, low input VSWR, and even distribution of dissipated heat. The main disadvantage with use of hybrid power dividers and combiners is that the gain and power losses from their insertion losses reduce the overall power-added efficiency. Several MMIC amplifiers can be combined either with an $n$-way combiner (radial or planar) or in a binary structure.

A schematic of an $n$-way planar power divider-combiner structure is shown in Figure 5.59. The divider-combiner has the advantage of low loss, moderate bandwidth, and good amplitude and phase balance. The combining efficiency of such a power divider-combiner is about 90%. The inherent redundancy in the $n$-way combiner makes it possible to obtain a graceful degradation characteristic. In an $n$-way amplifier combiner with $F$ failed amplifiers, the power output relative to maximum output power is given by $P_0/P_{max} = (N - F)/N^2$.

Figure 5.60 is a photograph of a 5.5 GHz 14 W amplifier consisting of four 4 W amplifiers with power-added efficiency of 25% and 6 dB gain combined using a four-way power divider and a four-way power combiner. The power divider and combiner are fabricated on a 0.63 mm thick alumina substrate. The divider and combiner have a loss of 0.2 dB each. The total output power of 14 W obtained with 22.8% power-added efficiency.

In a binary structure the power is combined using two-way dividers and combiners. A schematic of such a structure is shown in Figure 5.61. The two-way dividers-combiners normally used are the Wilkinson divider-combiner and the

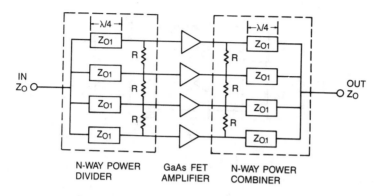

**Fig. 5.59** *n*-way planar power divider-combiner.

**Fig. 5.60** Photograph of a 5.5 GHz 14 W amplifier (courtesy of Inder Bahl, ITT).

Lange coupler. The Lange coupler usually is preferred, because it provides good isolation and low VSWR over a broad frequency range. In Figure 5.62 the combining efficiency of a binary structure is plotted as a function of numbers of devices combined and the loss in each two-way divider-combiner. The combining efficiency reduces as the number of devices increases, becoming impractical beyond eight devices due to higher losses and significantly lower combining efficiency.

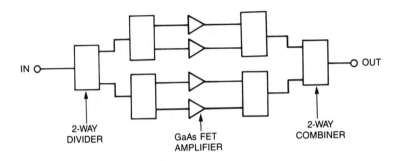

**Fig. 5.61** Binary combining structure.

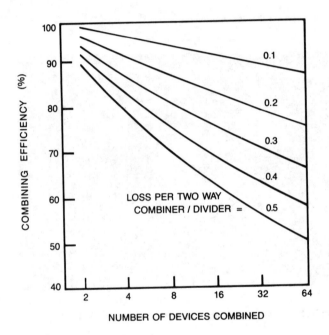

**Fig. 5.62** Combining efficiency *versus* number of combined devices for a binary combiner.

## 5.5 ON-CHIP TUNING

MMICs offer the advantage of a high level of integration for reduced cost, size, and weight and improved reliability. However, currently the yield of such circuits is low. Circuits are still at the development level and have not reached production. Because of the long time required to redesign a MMIC, it is essential that some ability to tune the circuit is incorporated to achieve a quick turn around time and obtain the desired results in the first design iteration. Even though a large variety of microwave circuit analysis, synthesis, and circuit optimization software tools are available today, the models for highly process-dependent circuit elements (e.g., FETs, spiral inductors, MIM capacitors) are not very accurate. To obtain more accurate and predictable equivalent circuit models, the characterization of a large number of components from a large number of wafers is required. Certain tuning capabilities can be introduced on the circuit, which allows a designer to tweak a circuit to achieve the desired results in the first design iteration. The following two techniques provide such a capability in the circuit.

## 5.5.1 Tuning Technique Using an Addition of Elements

A technique commonly used in internally matched FETs and hybrid amplifiers is to design the circuit elements to 80–90% of the desired value then bond additional components to it to increase the element value to 100–120%. Thus, a tuning capability of up to ± 20% in the element value is provided. The additional circuit elements are fabricated on the same chip but not connected.

The power amplifier is a suitable candidate for such an approach, since achieving maximum power and efficiency is the key requirement. Normally, the characteristics of a power FET are sensitive to the process variation resulting in the variation in output power and efficiency. Figure 5.63 shows an output matching circuit of a 4 W amplifier operating over a 5–6 GHz range [18]. The tuning pads are added to provide additional path lengths for open stubs or capacitive loading to a transmission line. Additional MIM capacitors having common ground also are provided. These capacitors and tuning pads can be added to the circuit by bonding a gold wire or ribbon to obtain the optimized performances.

The other method of tuning an amplifier is to optimize the performance by varying the operating point of the FET. In an amplifier design shown in Figure 5.64 operating from a single supply, the gate bias can be varied by changing the resistance value in the biasing network. Here, several resistors of various values are fabricated but not connected. We can connect various combinations of these values using bond wires and an optimized performance can be achieved.

## 5.5.2 Airbridge Removal Technique

Building an airbridge is the single most important MMIC fabrication step, although it is not available in MIC technology, and it can be used as a means of providing on-chip tuning capability. In this technique, instead of adding additional components to increase the element value (e.g., from 80 to 120%), the tuning elements are already added to the circuit by airbridge connections. By disconnecting the airbridge connections, various element values can be obtained. Examples

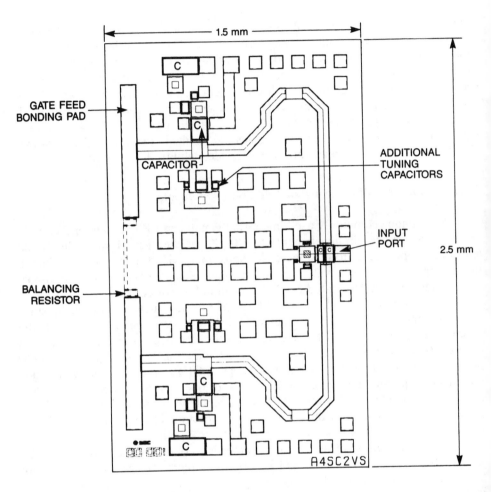

**Fig. 5.63** Output matching circuit of a 4 W amplifier (Courtesy of MSC).

of a MIM capacitor, spiral inductor, and a thin film capacitor are shown in Figure 5.65. In case of an MIM capacitor, four capacitors $A$, $B$, $C$, and $D$ are connected in parallel with an airbridge. By disconnecting the airbridges, various values of capacitors can be obtained. Similarly, two parallel resistors are connected by an airbridge, and the spiral inductor can be tapped into point $A$ or $B$ to achieve different values of the inductance.

In both techniques just described, the circuit can be tuned to improve the performance. A detailed analysis can take into account the variation in element values due to process variations and uncertainties in the modeling techniques. The statistical distribution of each element value should also be included.

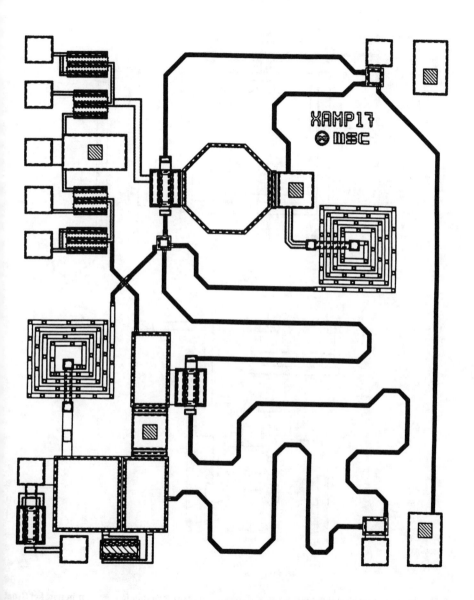

Fig. 5.64 Layout of an X-band amplifier illustrating the tuning capability of the gate bias (Courtesy of MSC).

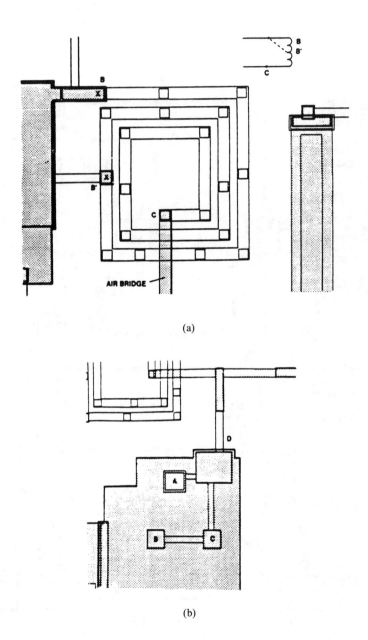

(a)

(b)

**Fig. 5.65** Layouts of (a) spiral inductor, (b) MIM capacitor, and (c) Two thin film resistor in parallel (from [19], pp. 135, 138).

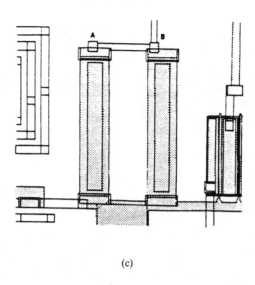

(c)

**Fig. 5.65** cont.

Once the circuit has been tuned for the desired performance, a question remains, How do we reduce the cost of the final circuit without going through a redesign and long processing cycle? The tuning of each component by these methods are impractical and not cost effective. To reduce the overall cost and have a product in a short time, the following technique can be used.

The airbridge, or the final metal, is the last processing step before the completion of the frontside processing of a MMIC wafer. Several wafers can be processed up to the step before deposition of the final metal and only a few wafers, preferably one in three or four lots of wafers (each lot consisting of six to eight wafers) can be processed completely. The circuits are tested, tuned, and a typical optimum circuit design is obtained. Then, an airbridge mask can be modified to incorporate the right element values by connecting the required components by airbridges. Thus, a final product can be achieved without repeating a complete design iteration.

## ACKNOWLEDGEMENTS

The editor wishes to thank the following for their contributions to this chapter: Dr. J. Tenedorio of Harris Microwave for providing design information for the completion of Section 5.4.5; Dr. I. G. Eddison of Plessey Research Caswell, Ltd., England for providing design details to help the completion of Section 5.4.6.

# REFERENCES

1. K. B. Niclas, "On Design and Performance of Lossy Match GaAs MESFET Amplifiers," *IEEE Trans. Microwave Theory Tech.*, Vol. MTT-30, November 1982, pp. 1900–1907.

2. C. E. Weitzei and D. Scheitlin, "Single-Stage GaAs Monolithic Feedback Amplifiers," *IEEE Trans. Microwave Theory Tech.*, Vol. MTT-33, November 1985, pp. 1244–1249.

3. W. C. Petersen, "A Monolithic GaAs 0.1 to 10 GHz Amplifier," *IEEE MTT-S Int. Microwave Symp. Dig*, June 1981, pp. 354–355.

4. James B. Beyer et al., "MESFET Distributed Amplifier Design Guidelines," *IEEE Trans. Microwave Theory Tech.*, Vol. MTT-32, March 1984, pp. 268–275.

5. T. Dao et al., "A Low Phase Noise MMIC Hybrid 3.0W Amplifier at X-Band," *IEEE Int. Microwave Symp. MTT-S Dig*, 1986, pp. 459–462.

6. L. C. Witkowski et al., "An X-Band 4.5 Watt Push-Pull Power Amplifier," *GaAs IC Symp.*, 1985, pp. 117–120.

7. David A. Sunderland and P. Daniel Dapkus, "Optimizing N-P-N and P-N-P Heterojunction Bipolar Transistors for Speed," *IEEE Trans. Electron Devices*, Vol. ED-34, February 1987, pp. 367–377.

8. Karl Niclas, "Multi-Octave Performance of Single-Ended Microwave Solid-State Amplifiers," *IEEE Trans. Microwave Theory Tech.*, Vol. MTT-32, August 1984, pp. 896–908.

9. Walter R. Curtice and M. Ettenberg, "A Non-Linear GaAs FET Model for Use in the Design of Output Circuits for Power Amplifiers," *IEEE Trans. Microwave Theory Tech.*, Vol. MTT-33, December 1985, pp. 1383–1384.

10. Eric M. Chase and Wayne Kennan, "A Power Distributed Amplifier Using Constant-$R$ Networks," *IEEE MTT-S, Int. Microwave Symp. Dig.*, 1986, pp. 811–815.

11. Yalcin Ayasli et al., "A Monolithic GaAs 1-13 GHz Traveling-Wave Amplifier," *IEEE Trans. Microwave Theory Tech.*, Vol. MTT-30, July 1982, pp. 976–981.

12. M. W. Green et al., "GaAs MMIC Yield Evaluation," *Proc. 17th European Microwave Conf.*, Rome, 1987.

13. P. Rigby, J. R. Suffolk and R. S. Pengelly, "Broadband Monolithic Low Noise Feedback Amplifiers," *IEEE MTT-S Int. Microwave Symp., Dig.*, 1983.

14. Y. Ayasli et al., "A Monolithic 1-13 GHz Traveling Wave Amplifier," *IEEE Trans. Microwave Theory Tech.*, Vol. MTT-30, July 1982.

15. K. B. Niclas et al., "The Matched Feedback Amplifier; Ultrawideband Microwave Amplification with GaAs MESFETs," *IEEE Trans. Microwave Theory Tech.*, Vol. MTT-28, April 1980.

16. R. E. Lehman and D. D. Heston, "X-Band Monolithic Series Feedback LNA," *IEEE Trans. Microwave Theory Tech.*, Vol. MTT-33, December 1985.

17. H. Fukui, "Design of Microwave GaAs MESFETs for Broadband Low Noise Amplifiers," *IEEE Trans. Microwave Theory Tech.*, Vol. MTT-27, July 1979

18. T. Bambridge and B. Malloy, "MMIC Matching Technology for Power and Efficiency," *J. Monolithic Technology*, April 1988.

19. R. Goyal and S. S. Bharj, "MMIC: On-Chip Tunability," *Microwave J.*, April 1987.

# Chapter 6
# MMIC Design: Nonlinear and Control Circuits

**Y. Tajima, W. S. Titus, A. M. Pavio, S. Miller,**
**R. Kaul, J. Cordero and R. Goyal**

## 6.1 MIXER CIRCUIT DESIGN

### 6.1.1 Introduction

Mixer circuits convert signal frequencies from the baseband to the transmitting frequency or from the received frequency to the baseband. Oftentimes the *intermediate frequency* (IF) is used between these frequencies and the conversion is done in several steps. This conversion is achieved by the nonlinear characteristics of diodes and transistors. Devices are especially designed for mixing operations by maximizing the nonlinear characteristics.

When a radio frequency (RF) signal is injected into a nonlinear device, it creates harmonics. If two signals are injected into the device simultaneously, two signals are mixed and various spurious signals generated. Usually only one of these mixed signals is the desired output, and the rest must be suppressed. Different suppression mechanisms classify the types of mixers, such as image rejection mixers, balanced mixers and double balanced mixers. System requirements determine the type of mixers to be used.

Thus, at the onset of the mixer design, the device and mixer types must be selected. Diodes can be excellent nonlinear devices but may not be readily available in the monolithic integrated circuit environment. FETs can be used as nonlinear devices under certain bias conditions. Dual-gate FETs can provide additional isolation between the ports.

From the standpoint of circuit simulation and design of nonlinear circuits, such as mixers, there is no design software available that can be used to optimize the mixer circuit. A nonlinear simulator is still very inadequate, and the computer time to perform any optimization would be prohibitively long. Microwave monolithic mixers developed to date have been a product of engineers' ingenuity using limited computer tools for the design.

In Section 6.1.4, design and performance of a MMIC image rejection mixer is described. By adopting a novel distributed structure using dual-gate FETs, image rejection performance was achieved without using a broadband balun, which is a difficult structure to fabricate on a planar structure.

## 6.1.2 LINEARIZATION

Neither of the nonlinear programs (SPICE or harmonic balance) has yet been fully utilized in microwave mixer design due to the computing requirements during the optimization. Instead, the linearized approach has shown a powerful alternative in producing actual designs. In the design of monolithic image rejection mixer described in Section 6.1.4, this linearized approach was used successfully.

In the linearized approach, nonlinear devices are analyzed separately, either analytically or by numerical method using Harmonic Balance or SPICE. Examples of both cases are shown in Section 6.1.3, for diodes, single-gate FETs, and dual-gate FETs. In the analysis, the parameter most critical to the signal conversion is the conversion transconductance; all other parameters are assumed to be linear. Derivation of conversion transconductance is not computer intensive and manageable within the standard computing capacity.

Design of a mixer will reduce to linear circuit design using the standard linear-frequency domain circuit analysis programs. It still is cumbersome because we have to deal with different frequencies at different ports. However, computing time to perform intensive optimization necessary for the design can be reasonably short. A design example is demonstrated in Section 6.1.4.

### 6.1.3 Device Models

#### 6.1.3.1 Single-Gate FETs

Single-gate FETs can be used as mixing diodes by injecting both signal and local oscillators into the device. Nonlinear performance is obtained from either the saturation characteristics of drain current or the cut-off of transconductance. A successful mixer design requires that the optimum mixing mode and bias point be chosen, allowing for the design of an appropriate matching circuit.

Various FET $I/V$ fitting equations are published, and some important mixer parameters can be obtained from these equations. The analysis is useful in understanding the mixing mechanism of the device, selecting the mixing mode and bias conditions, and performing the design based on the linearization approximation, which is the only useful means at this moment of designing and optimizing the circuit to some performance goals. Figure 6.1 shows the two modes of mixer operation, where $U$ and $S$ stand for the unsaturated and saturated regions of the FET. The $U$ region utilizes the Schottky gate characteristics; however, the noise figure ($NF$) is high due to the large current existent with this mode. In the $S$ region, nonlinear characteristics of $g_m$ due to the pinch-off is the source of the mixing. Generally this is the preferred mode for better $NF$.

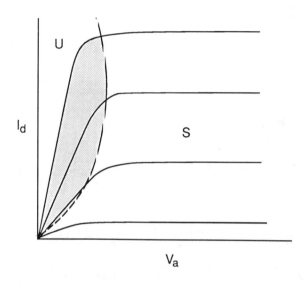

**Fig. 6.1** Two mixing modes of single-gate FETs.

Existence of a local oscillator power drive is prerequisite to the mixer operation. The local power determines the dc current level as well as the mixing mechanism. In the case of single-gate FETs, this local power typically is fed into the gate terminal, modulating the drain current. The large variation of drain current is considered to be the dominant contribution to the nonlinear operation of the FETs. We will examine how the mixing takes place when the drain current variation is excited by the local oscillator and RF signals at the gate terminal.

Let us assume that two signals, $V_1$ and $V_2$, are fed to the gate terminal as follows:

$$V_g = V_0 + V_1 \cos \omega_1 t + V_2 \cos \omega_2 t \tag{6.1}$$

$V_1$ and $V_2$ have different frequencies $\omega_1$, $\omega_2$. $V_1$, representing the local power has a large amplitude compared to $V_2$, which represents the RF signal. $V_0$ is the bias point.

Now, drain current $I_{ds}(V_g)$ will be modulated by the gate voltage drive in Equation (6.1). $I_{ds}$ can be expanded into a Fourier series, using the normalized gate voltage $x$, as follows:

$$I_{ds} = I_{ds}(x) = I_{ds0} + \sum_{m=0}^{\infty} (a_m \cos mx + b_m \sin mx) \tag{6.2}$$

where

$$x = \pi(V_g - V_0)/(V_p - V_0)$$
$$a_m = \frac{1}{\pi} \int_{-\pi}^{\pi} I_{ds}(x) \cos mx \, dx$$
$$b_m = \frac{1}{\pi} \int_{-\pi}^{\pi} I_{ds}(x) \sin mx \, dx$$
$$V_p = \text{pinch-off voltage}$$
$$I_{ds0} = \text{dc current}$$

By applying Equation (6.1) into (6.2) we can obtain the RF components ($\omega_2$) of the drain current:

$$I_{ds}(\omega_2) = 2 \sum_{m=1}^{\infty} b_m J_0 (m\alpha) J_1(m\alpha\delta) \cos \omega_2 t \tag{6.3}$$

and the IF component ($\omega_1 - \omega_2$) of the drain current:

$$I_{ds}(\omega_1 - \omega_2) = 2 \sum_{m=1}^{\infty} a_m J_1 (m\alpha\delta) J_1(m\alpha) \cos(\omega_1 - \omega_2)t \tag{6.4}$$

where

$$\alpha = V_1\pi/(V_p - V_0)$$
$$\alpha\delta = V_2\pi/(V_p - V_0)$$

and $J_n$ ($n = 0, 1, \ldots$) are the Bessel functions.

If we assume that the $V_2$ (RF signal) is small compared to $V_1$ ($\delta \ll 1$), we can derive transconductance $g_m$ from the ratio of amplitudes of drain current $I_{ds}(\omega_2)$ and RF signal $V_2$ as

$$g_m = \frac{|I_{ds}(\omega_2)|}{V_2} = \frac{\pi}{(V_p - V_0)} \cdot \sum_{m=1}^{\infty} m b_m J_0(m\alpha) \tag{6.5}$$

We can also derive conversion transconductance $g_c$ as the amplitude ratio of drain current at IF ($\omega_1 - \omega_2$) and RF signal at $\omega_2$

$$g_c = \frac{I_{ds}(\omega_1 - \omega_2)}{V_2} = \frac{\pi}{V_p - V_0} \cdot \sum_{m=1}^{\infty} m a_m J_1(m\alpha) \tag{6.6}$$

Equations (6.5) and (6.6) present transconductance and conversion transconductance as functions of bias voltage as well as the local drive level. In order to calculate these values, we must know $I_{ds}$ as a function of gate voltage. We can use any of several models that simulate the FET IV curves. One of these models, developed by H. Statz et al., is described here. In this model, the source-drain current for the intrinsic FET is given by

$$I_{ds}(V_{gs}, V_{ds}) = \frac{\beta(V_{gs} - V_T)^2}{1 + b(V_{gs} - V_T)} (1 + \lambda V_{ds}) \tanh(\alpha V_{ds}) \text{ for } V_{gs} > V_T \tag{6.7}$$

$$I_{ds}(V_{gs}, V_{ds}) = 0 \text{ for } V_{gs} < V_T$$

where

$I_{ds}$ = drain current
$V_{ds}$ = drain voltage
$V_{gs}$ = gate voltage
$V_T$ = threshold voltage

and where the four parameters $\beta$, b, $\alpha$, $\lambda$, and the pinch-off or threshold voltage are determined by least-squares fits to experimental data. An example of the closeness of the fit is illustrated by Figure 6.2.

Equations (6.5), (6.6), and (6.7) demonstrate an analytical solution for the mixer problem using a single-gate FET. To complete the FET modeling, the remaining frequency-dependent elements of the FET equivalent circuit need to be specified. An efficient first-order modeling can be made by linearizing the remaining elements of the FET equivalent circuit and evaluating them at the quiescent point of the LO swing. Given this assumption, the mixer analysis problem can be reduced to the analysis of the simplified FET mixer circuit shown in Figure 6.3.

Elements in the simplified FET mixer circuit are specified by fitting the model to S-parameter data measured at the optimum bias point. All of these elements are considered constant with frequencies (RF, LO, IF), and conversion transconductance $g_c$ from (6.6) is used to calculate the conversion characteristics of the FET mixer.

Note that the operating frequency at the gate port is RF, and at the drain port the operating frequency is IF. Transconductance $g_m$ from (6.6) is used to calculate LO to IF isolation.

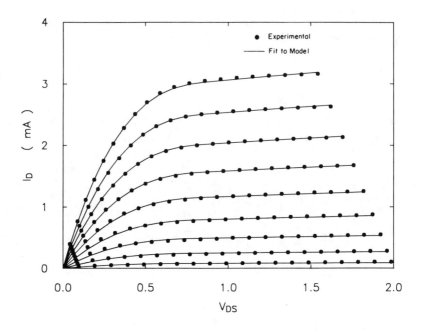

**Fig. 6.2** Closeness of the fit of experimental points to the model of equation 6.7.

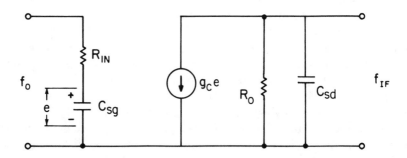

**Fig. 6.3** Simplified FET mixer circuit.

### 6.1.3.2 Dual-Gate FETs

The dual-gate FET has a great potential for mixer applications because of its three separate terminals and the inherently good isolation between LO and RF signals, when the RF is applied to the first gate, the LO to the second gate, and the IF extracted from the drain. The device can be modeled as a cascode arrangement of single-gate FETs as shown in Figure 6.4. Unfortunately, the convenience of the second gate is offset by the increased complexity in determining the best mode of mixer operation and the optimum bias point. Figure 6.5 shows the four modes of mixing operation, where $U$ and $S$ stand for the unsaturated and saturated regions of the individual FETs in the cascode model.

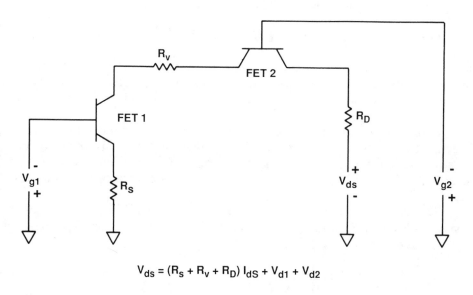

$$V_{ds} = (R_s + R_v + R_D) I_{dS} + V_{d1} + V_{d2}$$

**Fig. 6.4** Cascode representation of a dual-gate FET.

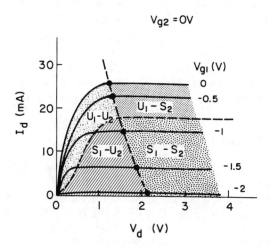

**Fig. 6.5** Four regimes of operation of a dual-gate FET.

The total source-drain voltage for the dual-gate FET model is as follows:

$$V_{ds} = V_{d1} + V_{d2} + I_{ds}(R_S + R_V + R_D) \tag{6.8}$$

and the current is continuous so that:

$$I_1(V_{g1}, V_{d1}) = I_2(V_{g2} - V_{d1} - I_{ds}(R_S + R_V), V_{d2}) \tag{6.9}$$

because the effective source-gate voltage is $V_{g2} - V_{d1} - I_{ds}(R_S + R_V)$. Equations (6.8) and (6.9) are highly nonlinear when $I_1$ and $I_2$ are described by equations like Equation (6.7), so they must be solved by a numerical iterative root-finding algorithm.

Complication in Equation (6.8) and (6.9) also makes it difficult to obtain the generalized analytical expression for transconductance and conversion conductance as a function of LO drive level, as was shown in the previous section for the single-gate FET case. Here, we present a numerical approach to obtain these values.

The transconductance, $g_m$, is defined as the partial derivative of $I_{ds}$ with respect to $V_{g1}$ with $V_{g1}$ with $V_{ds}$ and $V_{g2}$ held constant, while $V_{d1}$ and $V_{d2}$ adjust so that the current remains continuous:

$$g_m = \frac{\partial I_1/\partial V_{g1} + \partial I_2/\partial V_{d2} + \partial I_2/\partial V_{g2}}{\partial I_1/\partial V_{d1} + \partial I_2/\partial V_{d2} + \partial I_2/\partial V_{g2}} \tag{6.10}$$

The conversion transconductance is calculated by applying a LO voltage to the second gate, so that the second gate voltage can be written as follows:

$$V_{g2} = V_{g20} + V_{LO} \cos(\omega t) \tag{6.11}$$

where the time origin has been chosen to make the waveforms symmetric about $t = 0$. In this case the transconductance can be expanded as follows:

$$g_m(t) = \sum_{k=0}^{\infty} g_n \cos(k\omega t) \tag{6.12}$$

where $g_m(t)$ is obtained from Equations (6.10) and (6.11).

The collocation method is used to solve for $g_k$, which is to say, we truncate Equation (6.12) after $k = N$ terms and solve the $N + 1$ linear equations:

$$g(k\pi/N) = \sum_{n=0}^{N} g_n \cos(nk\pi/N), \; k = 0, N \tag{6.13}$$

The $g_0$ term gives an effective transconductance, and the $g_1$ term gives the conversion transconductance that can be used to predict isolation and conversion gain properties of dual-gate FET mixers. Figure 6.6 shows the first three Fourier coefficients of the transconductance plotted *versus* the LO amplitude for a given gate 1, gate 2, and drain voltage. Repeated analysis at various bias points will determine the optimum bias point for mixer operation.

A similar mixer model to that shown in Figure 6.3 is used for dual-gate FETs by employing the conversion transconductance (called $g_1$ earlier) that is calculated from the nonlinear program. A similar model can be applied to the second gate and drain with another transconductance to calculate LO to IF isolation.

### 6.1.3.3 Diodes

Schottky diodes have been the primary mixing devices in the microwave technology because of their low noise figure. Diodes operate under the local drive in a passive mode, which is much quieter than FETs under the active condition with high bias current. Signal conversion loss is minimal due to its high nonlinearity.

Diodes have two terminal configurations, however, which make it difficult to isolate the signal from local and IF powers. Various configurations, such as diode stars and rings, can be used to create a neutral point for the IF terminals. Various baluns and quadrature couplers are used to isolate RF and local signals. These associated circuits and configurations are the critical part of a diode mixer design, determining the performance and application.

A good mixing diode requires minimum parasitic resistances. To achieve the minimum noise figure performance, the active layer is usually grown on the highly doped contact layer, as shown in Figure 6.7(a), which is not compatible with FET structure (Figure 6.7(b)). Monolithic integration of diode mixers with FET amplifiers must accomodate some degradation of performance because of the lack of highly doped layers. Even so, diode mixers can present a better noise figure than FET mixers built on the same material.

Diode mixers can be built on separate (mixer) material from the FET circuit. Such circuits have been reported for various frequency bands.

## 6.1.4 Distributed Mixer Design

In this section we discuss the monolithic mixer circuit designs where the device modeling described above is applied. The first circuit is an image rejection mixer using dual-gate FETs as the mixing device in a distributed configuration. The novel FET arrangement and selective phasing provided by the circuitry give

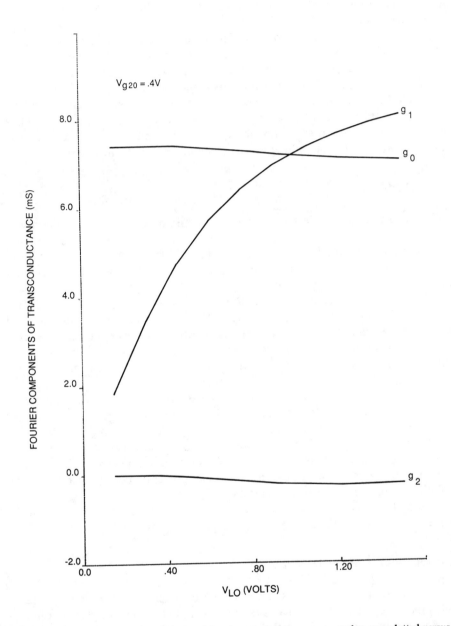

**Fig. 6.6** The first three Fourier coefficients of the transconductance, $g_0$, $g_1$, and $g_2$, are plotted *versus* the LO amplitude for $V_{g1} = 0.0$ V and $V_{g20} = 0.4$ V.

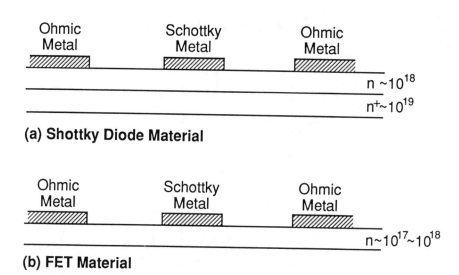

**(a) Shottky Diode Material**

**(b) FET Material**

**Fig. 6.7** (a) Doping concentrations in active layers for Schottky diode used for mixer applications; (b) FET active layer doping.

this mixer an image frequency signal separation capability. The second circuit described is a single-ended type of mixer that can provide broadband conversion characteristics (2–26 GHz). The circuit is completed with an IF filter and IF amplifier. Segregation of the image frequency, however, must be provided by other means.

### 6.1.4.1 Monolithic Image Rejection Mixer

A broadband FET image rejection mixer can be designed by using a distributed arrangement of *dual-gate FETs* to provide broadband image rejection. As shown in the circuit schematic of Figure 6.8, the RF and IF lines form a traveling wave structure while the LO is fed in phase through a four-way power divider. The novel arrangement and selective phasing of these lines gives this mixer an image separation or rejection capability from approximately 6–21 GHz RF.

This mixer operates under the principle of phase cancellation. In operation, the signal ($\omega_{rf} = \omega_{lo} + \omega_{if}$) travels on the gate transmission line and couples to the first gate terminals of dual-gate FETs with a phase delay of $\theta$ ($\omega_{rf}$) between each FET. The signal is mixed with the local signal, which is fed to the second gates of all the FETs in phase. As a result, IF signals, generated at each FET, have the same delay $\theta$ as the RF signals on the signal line. These IF signals are similarly incorporated into the drain distributed line, which presents the phase delay of $\theta$

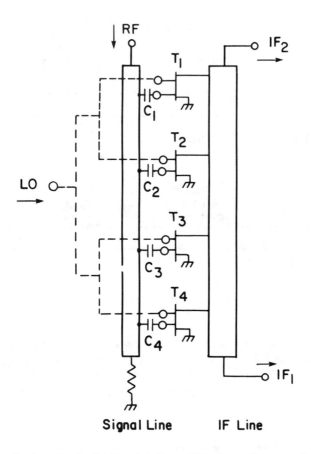

**Fig. 6.8** Circuit schematic of a distributed dual-gate FET mixer with image rejection operation.

($\omega_{if}$) between the FETs at IF frequency. Thus, the IF signal will propagate on the drain line in one direction, but propagation to the other direction will be cancelled.

When the signal in the image frequency band ($\omega_{im} = \omega_{lo} - \omega_{if}$) is mixed with the local power at FETs, the phase relation of IF signals from the FETs will be opposite to the phase shift presented by the signal frequency. The phase relation of these signals is listed in Table 6.1 for clarification. The propation of the image IF signal on the drain line will be opposite to that of the IF generated by the RF signal. Image frequency separation thus is accomplished.

To accomplish the total separation or rejection of image signal requires not only the phase relation but a strict amplitude control of IF signals from each FET. This was done by tailoring the RF excitation of each FET by the series capacitors to the gate terminals. A capacitor in series with a gate will reduce the RF signal excitation. Capacitor size was designed to achieve the maximum IF bandwidth by Chebyschev response.

**Table 6.1**

| FET | Signal in USB | | Image from USB | | If from Image | |
|---|---|---|---|---|---|---|
| | Frequency | Phase Shift | Frequency | Phase Shift | Frequency | Shift |
| 1 | $\omega_{LO} + \omega_{IF}$ | 0 | $\omega_{LO} - \omega_{IF}$ | 0 | $\omega_{IF}$ | 0 |
| 2 | $\omega_{LO} + \omega_{IF}$ | 0 | $\omega_{LO} - \omega_{IF}$ | 0 | $\omega_{IF}$ | $-0$ |
| 3 | $\omega_{LO} + \omega_{IF}$ | 20 | $\omega_{LO} - \omega_{IF}$ | 20 | $\omega_{IF}$ | $-20$ |
| 4 | $\omega_{LO} + \omega_{IF}$ | 30 | $\omega_{LO} - \omega_{IF}$ | 30 | $\omega_{IF}$ | $-30$ |

Figure 6.9 shows a *Monolithic Image Rejection Mixer* (MIRM), type A, mixer chip with two IF ports. The *upper sideband* (USB) converted signal ($\omega_{rf} - \omega_{lo}$) is directed to port $IF_1$ and the *lower sideband* (LSB) converted signal ($\omega_{lo} - \omega_{rf}$) is directed to port $IF_2$. This mixer uses a fixed frequency narrow band LO circuit at 12 GHz to convert a 14–20 GHz RF (USB) and a 6–10 GHz RF (LSB) to a broadband 2–8 GHz IF. A relative image separation of approximately 20 dB was predicted for this design. The design goal was to maximize the image separation while preserving the flattest possible response for the USB signal out of port $IF_1$, sacrificing some conversion gain and bandwidth as a result.

Figure 6.10 shows a composite graph of the measured performance existing the $IF_1$ port of one MIRMA mixer at a low current bias of $V_{g1}$ (gate 1 bias) = $-1.3$ V, $V_{g2}$ (gate 2 bias) = $-1.0$ V, and $V_d$ (drain bias) = 1 V. Figure 6.11 shows the response exiting the $IF_2$ port for the same mixer at the same bias point as in Figure 6.10, 10–20 dB image rejection up to 6 GHz is observed.

The RF to IF isolation shown in Figures 6.9 and 6.10 is 10–25 dB. The LO to RF isolation for this mixer is better than 25 dB. The dc and RF were taken of the dual-gate test FETs on these wafers and were used to specify our dual-gate FET model parameters, as discussed in previous sections. Using this model, the dual-gate FET nonlinear analysis program calculated $g_m$ curves as a function of gate 1 and gate 2 bias as shown in Figure 6.12. At the various gate 1 and gate 2 biases labeled 1 to 7 in Figure 6.12, the conversion transconductance was calculated as a function of $V_{LO}$, the voltage of an applied LO signal to the second gate. These results are tabulated in Table 6.2.

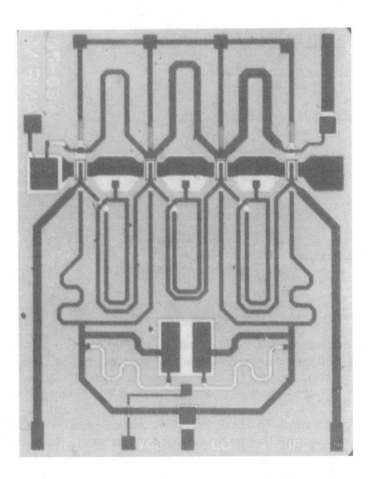

**Fig. 6.9** A 14–20 GHz RF (monolithic image rejection mixer (MIRMA), type A) chip.

To check the validity of the mixer simulation, a simulation of the measurements at low drain current was made. The dc modeling results indicate that the conversion transconductance at bias points 2 and 3 is approximately 6–8 mS. Using $g_1 = 6$ mS in our simplified DGFET model, excellent agreement between measurement and theory was obtained. Figure 6.10 shows the measured and simulated response at the $IF_1$ port. Figure 6.11 shows the measured and simulated response at the $IF_2$ port. These results indicate good agreement between the measured results and our simulation.

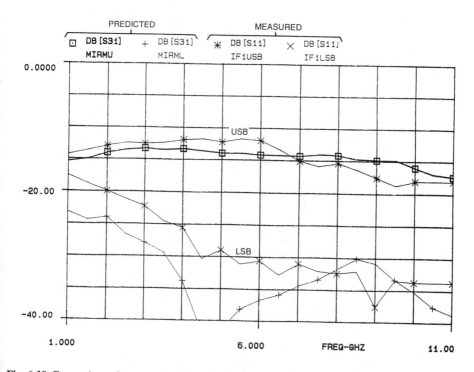

**Fig. 6.10** Comparison of measured and simulated MIRMA USB and LSB conversion gain to IF₁ port.

## 6.1.4.2 *Monolithic Single Ended Broadband Mixer* [2]

A broadband single-ended mixer can be designed by using a distributed topology similar to the image rejection mixer (Figure 6.8). In this case, the local power is injected into the FETs with the same phase delay as the RF signal by using the traveling wave configuration (Figure 6.13). As a result, IF signals from all of the FETs will have the constant phase. All of the IF signals will be combined to the output port by using a matching circuit. Note that IF signals from both RF and image signals will be summed at the same port. The nature of the traveling wave configuration can provide an extreme broadband performance for the RF and LO signals.

The 3 × 5 mm chip shown in Figure 6.14 is completed with a distributed dual-gate FET mixer, an elliptic IF filter, and a single-stage IF amplifier. This

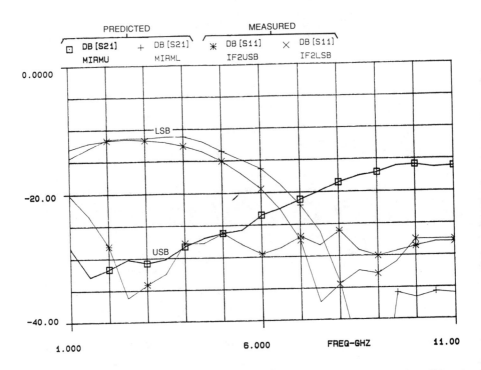

**Fig. 6.11** Comparison of measured and simulated MIRMA USB and LSB conversion gain to IF₂ port.

circuit down-converts 2 to 26 GHz RF signals to a 400 MHz to 2.5 GHz IF band with 7 dB of conversion gain by using a 12 dBm LO drive. Table 6.3 summarizes the measured results of this circuit.

The topology of this mixer, shown in Figure 6.13, has two broadband RF and LO artificial transmission line inputs incorporating four dual-gate FETs, the combined drains of which form the IF output. The RF and LO lines were optimized to give approximately 20 dB return loss and equal voltage at each gate for the RF and LO bands. Capacitors in series with each device capacitance were used to help equalize the RF and LO signal distributed to each FET.

Although the RF, LO and IF bands (RF = 2–20 GHz, LO = 4–22 GHz, and IF = 1–3 GHz) were selected to avoid overlapping, a large LO to IF isolation is required especially near the low end of the LO band to avoid saturating the following linear IF circuitry. Improved mixer conversion gain will also result if the LO signal is shorted at the IF port of the mixer. A one-section elliptical filter was designed to provide better than 30 dB rejection for 4–7 GHz and approximately 3 dB loss at 2 GHz.

To recover the loss from this filter and to improve the rejection properties

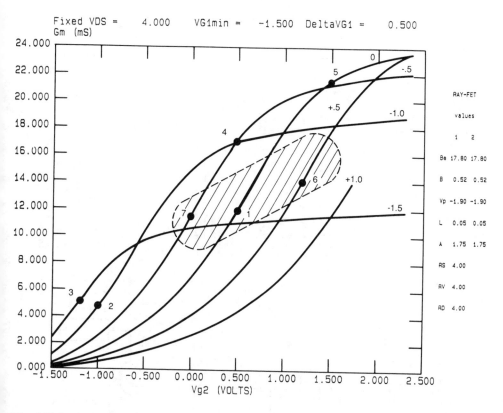

**Fig. 6.12** Dual-gate FET $g_m$ (mS) curves.

above 6 GHz, a one-stage common source IF amplifier with feedback was designed to give 8–10 dB gain and better than 2 : 1 input and output VSWRs for 1–3 GHZ.

Figure 6.15 shows the predicted RF, LO, and IF port matching responses of the complete mixer chip (including mixer, filter, and IF amplifier) from dc to 22 GHz. The results show a good IF port return loss better than 10 dB for 1–3 GHz as well as excellent RF and LO matches for their respective bands. Figure 6.16 shows the RF to IF, LO to IF, and LO to IF isolation predicted response of the complete mixer circuit. The combined effects of the mixer, filter, and amplifier predict better than 30 dB LO to IF isolation for 4–22 GHz and better than 25 dB LO to RF isolation from dc to 22 GHz.

Figure 6.17 shows the predicted conversion gain response of the complete mixer chip. The RF and LO circuit responses were loaded into this linear circuit file by multiplying the RF by $1 \times 10^{-4}$ and adding that number to 2 GHz. Figure

**Table 6.2** A 10 dBm LO Signal Applied to a Matched 50 $\Omega$ Port at Gate 2

| Bias Pt. | $V_{g1}$ | $V_{g2}$ | $g_1$ (Conv. Trans.) |
|----------|----------|----------|----------------------|
| 1 | 0.0 | +0.5 | 9.5 mS |
| 2 | −1.0 | −1.0 | 8.0 mS |
| 3 | −1.5 | −1.3 | 5.9 mS |
| 4 | −0.5 | +0.5 | 7.8 mS |
| 5 | +0.0 | +1.5 | 6.0 mS |
| 6 | +0.5 | +1.2 | 9.5 mS |
| 7 | −0.5 | 0.0 | 9.4 mS |

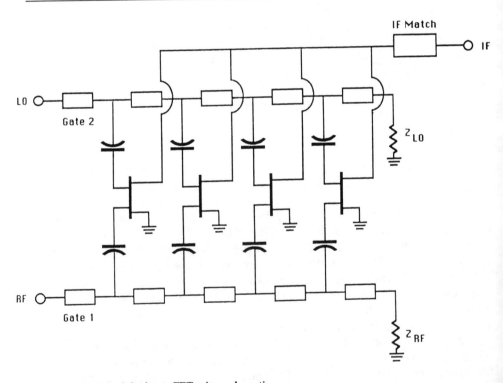

**Fig. 6.13** Distributed dual-gate FET mixer schematic.

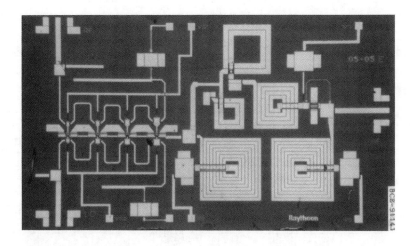

**Fig. 6.14** 2–26 GHz MMIC frequency converter with mixer, filter, and IF amplifier.

**Table 6.3** 2–26 GHz MMIC Frequency Converter

| | |
|---|---|
| RF bandwidth | 2–26 GHz |
| IF bandwidth | 400 MHz to 2.5 GHz |
| NF | 11 dB DSB (20 GHz) |
| Gain | 7 dB (3–8 dB over IF band) |
| Isolation LO to RF | 35 dB     (1–12 GHz) |
| | 20–25 dB   (12–26 GHz) |
| LO to IF | 25 dB     (4–26 GHz) |
| RF to IF | 25 dB     (4–26 GHz) |
| VSWR (LO, RF) | <2.16     (1–26 GHz) |
| LO power | 12 dBm   (2.5–26.5 GHz) |
| Measurement Temperature | 33°C |

6.18 shows the measured results of conversion gain *versus* RF and RF to LO and LO to RF isolation for an IF of 500 MHz.

The analysis shows 10 dB conversion gain dropping by 8 dB at 20 GHz for a 16 dBm applied LO. This gradual decrease in conversion gain at high frequencies can be diminished by improving the differential phase match between the dual-gate FETs on the RF and LO lines. Additional flattening can be expected as the LO drive saturates the first few FETs in the LO transmission line and provides

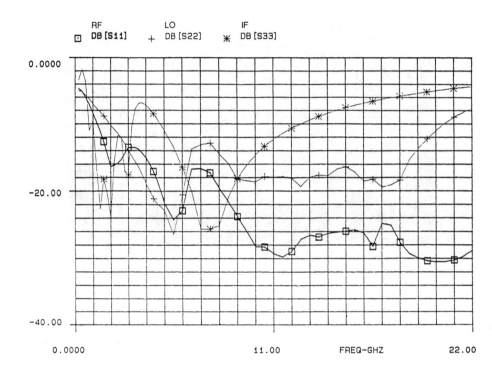

**Fig. 6.15** Predicted RF, LO, and IF port return loss results for the cascaded mixer, filter, and amplifier chip.

increased drive to the later FETs in the line.

The 2–26 GHz frequency converter was fabricated in $2 \times 10^{17}$ cm$^{-3}$ VPE-doped GaAs material with 0.7 mm E-beam defined gate lengths. Measured results confirmed the validity of the design approach used to predict the performance shown above.

The measured conversion gain of this chip is shown in Figure 6.18, plotted as a function of RF from 2–26 GHz at a 500 MHz IF, indicating a nominal 7 dB conversion gain. Similar conversion gain performance was measured for IFs up to 2.5 GHz. Figure 6.19 shows the measured conversion gain of this chip as a function of IF for a 16 GHz RF signal, showing the 500 MHz to 2.5 GHz IF bandwidth. Also shown in Figure 6.19 are the RF to IF isolation *versus* RF and the LO to IF isolation *versus* LO frequency, indicating 25 dB of isolation for both for 4–26 GHz.

The LO to RF isolation is 35 dB for 2–12 GHz and 20 dB up to 26 GHz. Return losses at RF and LO ports are better than 10 dB over most of the 2–26 GHz bandwidth. These results were very close to the predicted values shown in Figures 6.15, 6.16, and 6.17.

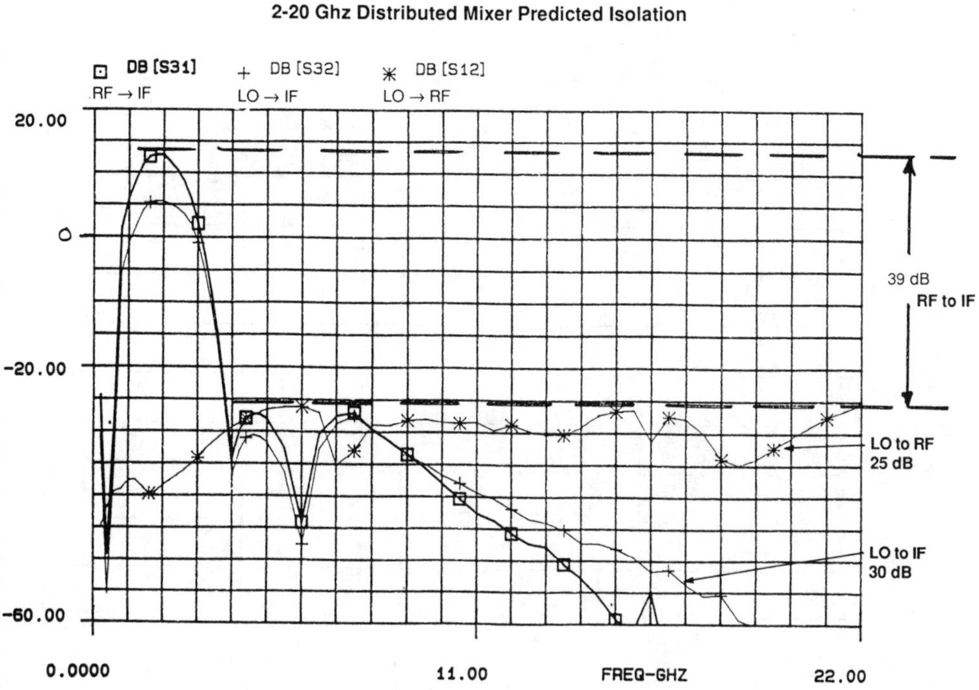

**Fig. 6.16** Predicted RF to IF, LO to IF, and LO to RF isolation for the cascaded mixer, filter, and amplifier chip.

## 6.2 PHASE SHIFTER DESIGN

### 6.2.1 Introduction

A number of different methods are available for shifting phase. The switched line, reflection, and loaded line phase shifters all have potential for octave bandwidth but require distributed line lengths and therefore a relatively large MMIC substrate area (Figure 6.20). A fourth type of phase shifter that provides broadband performance, presented by Garver [1], consists of a switched combination of high-pass and low-pass filters. A low-pass filter made up of series inductors and shunt capacitors provides phase delay to signals passing through it. A high-pass filter composed of series capacitors and shunt inductors provides phase advance. By adding two SPDT switches to these filters as shown in Figure 6.21, a phase shifter with broadband performance exceeding an octave can be realized. This type of phase shifter has the added benefit of not needing distributed elements, and it can be made very compact for MMIC realization. The actual size and

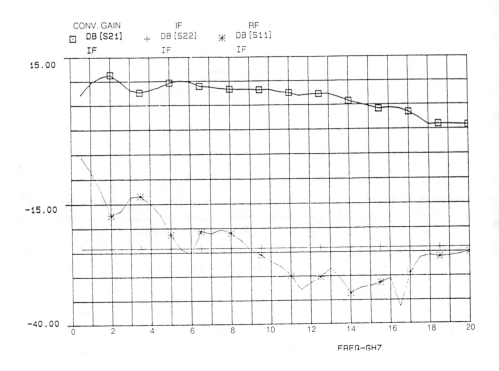

**Fig. 6.17** Predicted conversion gain and RF return loss as a function of RF ($\omega_{RF} \times 10^{-4} + 2$ GHz) and IF return loss at 2 GHz for the cascaded mixer, filter, and amplifier chip.

bandwidth of the phase shifter depends on the size and bandwidth of the two $1 \times 2$ switches required for each phase bit.

Currently, MESFETs are the most common switching elements utilized for GaAs MMIC applications. Compared to pin diodes, the total capacitance shunting the high impedance (pinched off) state is large. This capacitance must either be resonated or somehow included in the design of impedance matching sections. For both cases, the bandwidth will be severely limited. If the input and output switches could be eliminated completely, we would expect to realize the intrinsic bandwidth of the high-pass and low-pass sections alone. Additionally, significant substrate area would be saved, as switches usually need more space than filter sections. The following discussion describes one such approach to achieving a broadband phase shifter in monolithic form.

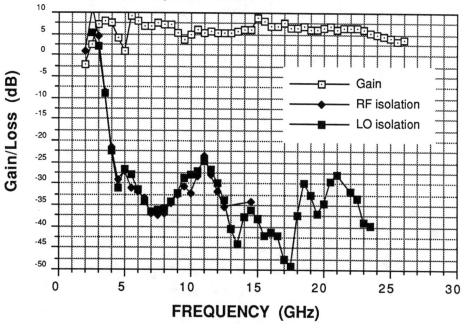

## 2-26 GHz MMIC Frequency Converter

**Conversion Gain / RF, LO to IF Isolation**

**Fig. 6.18** Measured conversion gain *versus* RF and RF to IF and LO to IF isolation of the MMIC frequency converter chip for a 500 MHz IF.

### 6.2.2 Design Approach

The design approach used for this particular phase shifter is unique, as the MESFETs used for switching between the two filters are in fact part of the filters. The low resistance of the FET in the "on" state (when gate voltage is zero) is used to short out unwanted elements and can be used as a series-connected capacitor and resistor in the "off" state (when gate voltage is between pinch-off and break-down) [2].

Consider the circuit shown in Figure 6.22. In this circuit, there are six discrete FETs, connected into two sets of T-configurations. The series elements of one T and the shunt element of the other T are controlled from a single gate

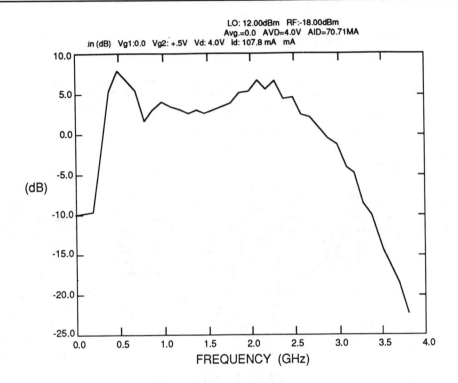

**Fig. 6.19** Measured conversion gain versus IF of the MMIC frequency converter chip for a fixed 16 GHz RF.

control voltage. When control voltage $V_1$ is zero and $V_2$ is beyond the pinch-off voltage $V_p$ of the switch FETs, the circuit of Figure 6.22 can be reduced to the form shown in Figure 6.23, if the "off" state FET impedance is represented as a capacitor and the "on" state impedance as a resistor. If the resistors $r_1$ and $r_2$ are small compared to the impedances in parallel with them, the circuit can be further simplified to the form shown in Figure 6.24, which is a five section low-pass filter.

In the opposite switch state (that is, when $V_2$ is equal to zero and $V_1$ is more negative than the FET pinch-off voltage), the circuit of Figure 6.22 reduces to that shown in Figure 6.25. If $r_3$ and $r_4$ again are small in value, the circuit can be simplified to the form shown in Figure 6.26. If capacitors $C_4$ and $C_5$ have a sufficiently large impedance compared to $L_2$ and $L_3$, the circuit behaves like a five section high-pass filter. The parameter assumptions for $r_1$ through $r_4$, $C_4$ and $C_5$ can be met through the use of controlled optimization of the circuit.

In summary, the bridge circuit shown in Figure 6.22 allows us to realize a high-pass, low-pass phase shifter without external switches. The FET capacitance no longer is an undesired parasitic and, in fact, has been absorbed into the filter structure. Hence, maximum bandwidth can be realized and the circuit can be very compact.

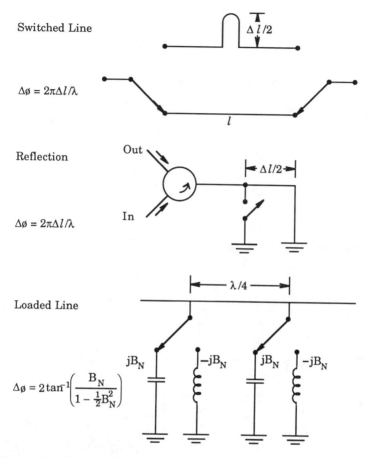

**Fig. 6.20** Three types of phase shifters.

This approach also can be applied to a single T-configuration. The circuit shown in Figure 6.27 has a total of five FETs with two series and one shunt FET controlled from a single gate control voltage. The remaining single series and single shunt FET also are controlled from a common gate control voltage. When control voltage $V_1$ is zero and $V_2$ is between the pinch-off voltage and breakdown voltage of the switch FETs, the circuit of Figure 6.20 can be reduced to the form shown in Figure 6.28. If $r_1$, $r_2$, and $r_4$ are small, the circuit can be further reduced to that shown in Figure 6.29. If $C_3$ has a sufficiently large impedance, the circuit behaves as a three section low-pass filter.

For the opposite switch state, when $V_1$ is beyond pinch-off and $V_2$ is zero, the circuit in Figure 6.27 reduces to that in Figure 6.30. Again, if $r_3$ and $r_5$ are made small by design, the circuit is reduced further to that shown in Figure 6.31. This circuit behaves as a three section high-pass filter if the impedance of $C_5$ is sufficiently large.

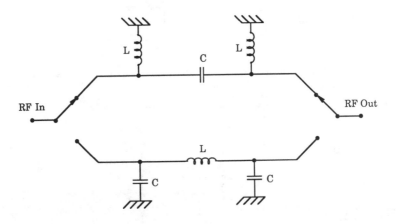

**Fig. 6.21** A typical low-pass high-pass phase shifter.

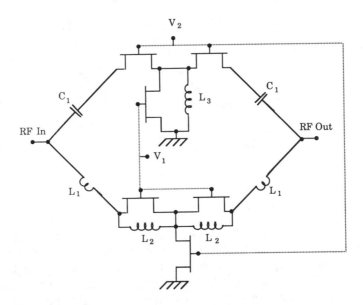

**Fig. 6.22** A single bit phase shifter.

The circuit shown in Figure 6.20 can also be used as a high-pass, low-pass phase shifter with a lesser number of sections and therefore less transmission phase change with frequency than the five-section circuit as previously described. This will allow easier matching of the high-pass and low-pass phase slopes for the broadband phase shifters requiring less total phase shifts in the 45° to 90° range.

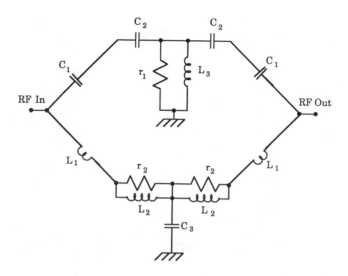

**Fig. 6.23** Equivalent circuit of a single bit phase shifter, when $V_1 = 0$ and $|V_2| > |V_p|$.

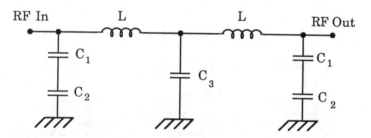

**Fig. 6.24** Simplified equivalent circuit of a single bit phase shifter, when $V_1 = 0$ and $|V_2| > |V_p|$.

### 6.2.3 Design Example

The following design example is used to demonstrate the validity of the design approach. The specifications for the design generally were very loose due to the new circuit concepts. A three-bit phase shifter was decided upon, as this is the minimum number of bits needed for a practical phase shifter. The specifications are listed in Table 6.4 and show the design goals for the phase shifter. Parameters such as switching speed, power handling, and logic interface were considered secondary and were not specifically optimized.

The critical element of this design is the MESFET. Because the FET is used not only for switching but for realization of precise capacitor values in the filter

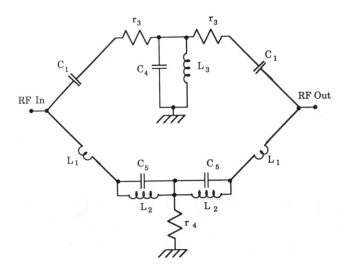

**Fig. 6.25** Equivalent circuit of a single bit phase shifter diagram, when $V_2 = 0$ and $|V_1| > |V_p|$.

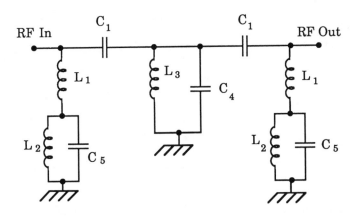

**Fig. 6.26** Simplified circuit of a single bit phase shifter, when $V_2 = 0$ and $|V_1| > |V_p|$.

structures, an accurate model is crucial. The model used in this design is shown in Figure 6.32. When the FET is in the "on" state, it can be modeled as a single resistance designated $R_{on}$. This resistance is the sum of the contact resistances and the channel resistances. For the opposite state, the gate-to-source and gate-to-drain model is a resistor and capacitor in series. The capacitance $C_g$ represents approximately half of the gate capacitance with the channel fully depleted. The resistor $r_g$ represents the charging resistance of $C_g$. The source to drain capacitance $C_{sd}$ represents the fringing capacitance between the source and drain electrodes. The term $R_d$ represents the RF losses associated with $C_{sd}$.

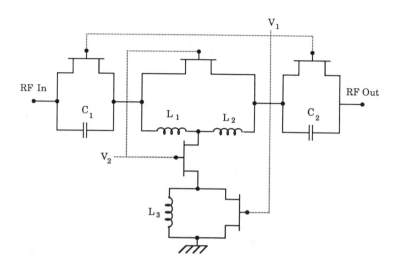

**Fig. 6.27** High-pass low-pass T-configured phase shifter.

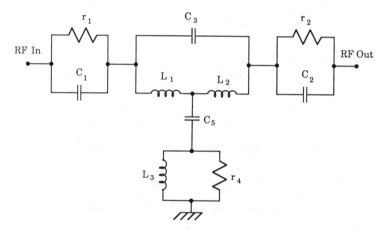

**Fig. 6.28** Equivalent circuit of a high-pass low-pass T-configured phase shifter, when $V_1 = 0$ and $|V_2| > |V_p|$.

The values of these various model elements are related to the substrate doping levels, pinch-off voltage, and physical dimensions [2]. The gate length for this design was chosen to be 1 $\mu$m for optical processing. A shorter gate would reduce the channel resistance and gate capacitance. The pinch-off voltage was chosen to be −4 V with a single recess process.

Note that all elements of the model can be expressed in terms of the gate width, $W$. Using this as an optimized variable in the linear (frequency domain)

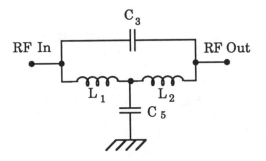

**Fig. 6.29** Simplified circuit of a high-pass low-pass T-configured phase shifter, when $V_1 = 0$ and $|V_2| > |V_p|$.

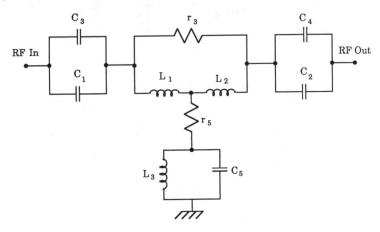

**Fig. 6.30** Equivalent circuit of a high-pass low-pass T-configured phase shifter, when $V_2 = 0$ and $|V_1| > |V_p|$.

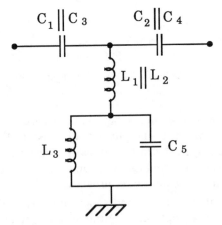

**Fig. 6.31** Simplified circuit of a high-pass low-pass T-configured phase shifter, when $V_2 = 0$ and $|V_1| > |V_p|$.

**Table 6.4** Design Specifications

| Parameter | Goal |
|---|---|
| Type | Digital 3 bit |
| Frequency range | 2–8 GHz |
| Insertion loss | Minimum |
| Amplitude tracking | ± 1 dB |
| Phase error | 10° rms |
| VSWR | < 2.0 : 1 |

circuit simulation programs (e.g., Super Compact, Touchstone), the FET model can be effectively scaled up or down in size automatically. These programs have very powerful optimization routines, but for best efficiency they must have starting values that are reasonable. In the simplest forms the topologies given in Figures 6.22 and 6.27 are lumped-element high-pass or low-pass filters. Therefore, an ideal filter with the proper phase shift (180°, 90°, or 45°) at band center, in this case 5 GHz, was used as a starting point. Many references are available on derivation of element values for this filter [3]. The most efficient way of deriving the final element values required in Figures 6.22 and 6.27 is to use a random optimization routine with a phase comparison subroutine. This allows the error function to consider not only insertion loss and the reflection coefficient but amplitude balance and the phase shift between two states.

As with any design, once a suitable performance criteria has been met with the computer model, a full sensitivity analysis must be performed. Any elements that show a high sensitivity to normal process variations must have that sensitivity reduced if possible. Adjusting the error function weighting or optimization goals and searching for a new operating point may reduce element sensitivity. This particular design has low sensitivities for all elements due to its basically passive nature. Typically, a 10% variation in element value can be tolerated and still meet the tracking requirements listed in Table 6.4. During the optimization phase, the physical realizability of the elements must be considered, including process limitations. Capacitance per unit area, transmission line widths and spacings, FET periphery limitations, *etcetera* must be incorporated into the model.

Capacitor size is limited by the dielectric thickness and the plate area. As the plate length or width approaches an appreciable portion of a wavelength, a simple lumped model no longer can be used. This condition is best avoided by keeping

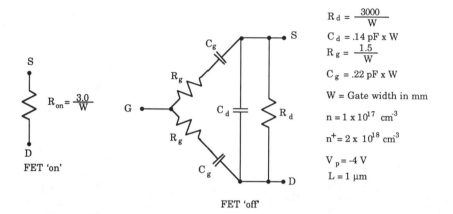

$$R_d = \frac{3000}{W}$$

$$C_d = .14 \text{ pF} \times W$$

$$R_g = \frac{1.5}{W}$$

$$C_g = .22 \text{ pF} \times W$$

W = Gate width in mm

$n = 1 \times 10^{17}$ cm$^{-3}$

$n^+ = 2 \times 10^{18}$ cm$^{-3}$

$V_p = -4$ V

$L = 1$ μm

$$R_{on} = \frac{3.0}{W}$$

FET 'on'

FET 'off'

**Fig. 6.32 Planar switch MESFET model.**

lengths or widths less than 10° long at the highest operating frequency. If this is not possible, a model of the distributed effect of the plates should be used. Minimization of plate area has a generally positive effect on circuit yield as well.

Minimum transmission line widths are limited by the process used and possibly the conductor losses. Line widths on the order of 10 μm are not overly lossy and are easily processed. Narrower lines can be fabricated but will have increased loss that may not be acceptable. If a larger inductance is required, a spiral inductor can be used.

Spacing between adjacent RF transmission lines should be kept at a minimum of three line widths to avoid coupling. Higher impedance lines will need to be spaced further. Parallel RF line sections that are short in wavelength can be spaced closer. If there is any question whether there is coupling, it should be modeled. Spacing between RF and dc lines can be closer, typically one-half that required between RF lines.

The MESFET layout must consider the maximum gate finger width based upon the operating frequency range. Gate control will degrade as the finger approaches one-quarter wavelength in width. The number of fingers also is limited by the total physical width of the FET, which is also required to be less than λ/4. For lower frequency designs, these FET parameters are not a problem but will be a factor for designs above 10 GHz. The FET also must be isolated from the gate bias lines. Isolation of the gate structure from the bias circuit is best achieved for broadband designs by use of series resistors located in the gate bias line. A value of 1 kΩ to 2 kΩ located closely to the FET will provide sufficient isolation. The resistance accuracy is not important, so an open-gate resistor can be used. This has the added advantage of eliminating a process step that would be required for thin film resistors.

Due to the frequency range of this design, a spiral inductor was used to achieve the required inductance values. Because no reliable programs were available for analysis of a multiturn spiral, the inductors were limited to $1\frac{3}{4}$ turns. It is composed of only a pair of coupled lines, so it could be readily analyzed with available programs. A 10 $\mu$m line width and 15 $\mu$m line spacing was chosen for this design. The spiral layout and model are shown in Figure 6.33. Note the negative line lengths for the 90° corner model. These are predicted in [4] and verified by measurement. To avoid coupling across the inner section of the inductor, and therefore voiding the simple model we have established, the spacing should be at least one substrate thickness across (4 mils in this case). The upper limitation on size occurs when the inductor becomes self-resonant. For this design, self-resonance occurs near the upper band edge frequency, when the size reaches about 450 $\mu$m$^2$. In any case an inductor of this size should not be used, as it consumes considerable substrate area. Two smaller inductors in series can be used in place of one large one.

The final schematics of the three individual bits are shown in Figures 6.34 through 6.36. The total periphery of each FET, the capacitor values, and the inductor side lengths (refer to Figure 6.33) are listed. The layout of the complete three-bit phase shifter is shown in Figure 6.37. A final optimization was performed to determine the length and impedance of the interbit connecting lines required for best performance.

The layout was arranged to avoid unwanted coupling between adjacent RF lines. A more compact layout may have been possible, but coupling effects would then need to be accounted for. The only parasitic effects added to the model at this point were the shunt capacitance to ground from the bottom plates of the capacitors and the FET source-drain fingers. They had a small effect on this design but higher frequency designs would show a larger effect. Discontinuity effects due to bends in the RF lines were avoided by using two 45° bends in place of one 90° bend. No compensation is required for a 45° bend in a high impedance line and, for this particular design, the 50 $\Omega$ lines were not compensated either. The effect of mitering the 45° bends is small for the frequency range of this design. After including the layout parasitics in the model, it was reoptimized. This should be a retweaking rather than a redesign of the circuit. If a parasitic is expected to have a large effect, as with a millimeter wave circuit design, the effect should be estimated at the outset and included in the model. This is especially true for a three-bit phase shifter. This particular circuit has eight states to consider during final optimization. If the performance of the design deviates greatly when the layout parasitics are included, the number of variables involved will make it difficult, if not impossible, to achieve best performance. Reoptimization was not difficult in this case, as parasitic effects were minimal and the layout was done to assure they remained so.

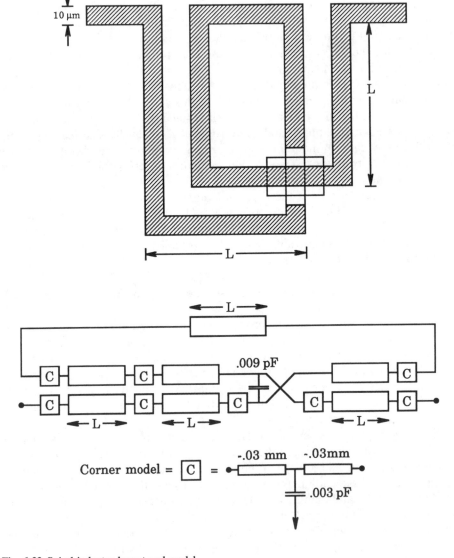

**Fig. 6.33** Spiral inductor layout and model.

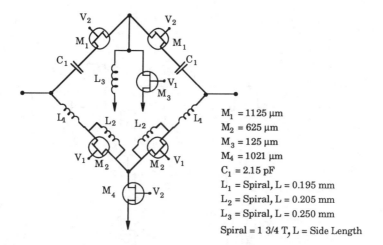

**Fig. 6.34** 180° bit schematic.

$M_1 = 1125 \, \mu m$
$M_2 = 625 \, \mu m$
$M_3 = 125 \, \mu m$
$M_4 = 1021 \, \mu m$
$C_1 = 2.15 \, pF$
$L_1 = Spiral, L = 0.195 \, mm$
$L_2 = Spiral, L = 0.205 \, mm$
$L_3 = Spiral, L = 0.250 \, mm$
Spiral = 1 3/4 T, L = Side Length

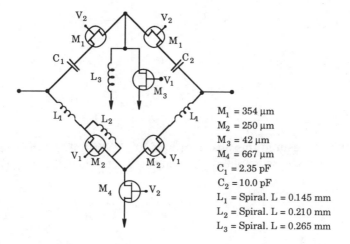

**Fig. 6.35** 90° bit schematic.

$M_1 = 354 \, \mu m$
$M_2 = 250 \, \mu m$
$M_3 = 42 \, \mu m$
$M_4 = 667 \, \mu m$
$C_1 = 2.35 \, pF$
$C_2 = 10.0 \, pF$
$L_1 = Spiral. \, L = 0.145 \, mm$
$L_2 = Spiral. \, L = 0.210 \, mm$
$L_3 = Spiral. \, L = 0.265 \, mm$

The predicted phase performance and the measured values are shown in Figure 6.38. As we can see, the measurement matches the theory very well over the design bandwidth. Plots of insertion loss and return loss, both predicted and measured, are shown in Figures 6.39 through 6.42. The specifications generally

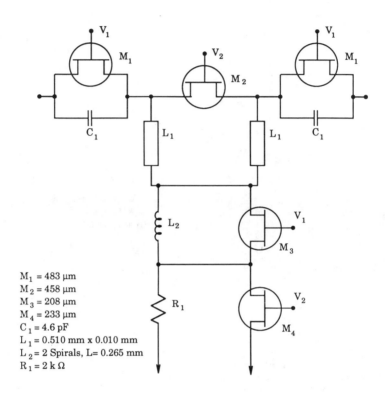

M₁ = 483 µm
M₂ = 458 µm
M₃ = 208 µm
M₄ = 233 µm
C₁ = 4.6 pF
L₁ = 0.510 mm x 0.010 mm
L₂ = 2 Spirals, L= 0.265 mm
R₁ = 2 kΩ

**Fig. 6.36** 45° bit schematic.

were met, with a measured rms phase error of 10°, VSWR of less than 2 : 1 across the band, and a worst case amplitude tracking between states of ± 1.7 dB versus a 1 dB goal.

### 6.2.4 Conclusion

The feasibility of designing a broadband digital phase shifter in MMIC form has been verified. In fact, this phase shifter probably could not be built other than in MMIC form due to the required phase and amplitude requirements. The phase shifter has shown not only the feasibility of the circuit concept but also that with the available design tools (Super Compact, Touchstone) and careful modeling, simulation, and layout, predicted performance can very accurately match the measured result.

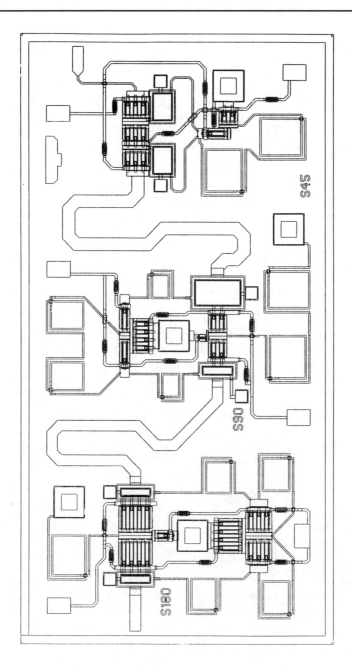

**Fig. 6.37** 3-bit phase shifter layout.

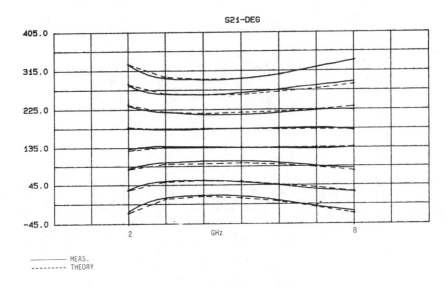

**Fig. 6.38** Measured *versus* predicted phase performance.

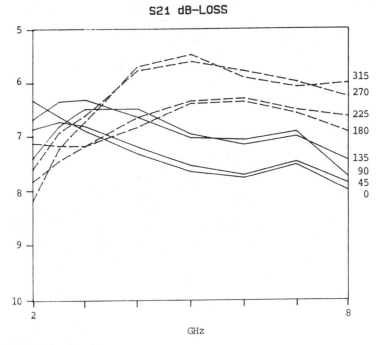

**Fig. 6.39** Predicted insertion loss.

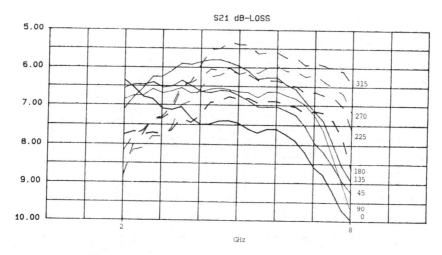

**Fig. 6.40** Measured insertion loss.

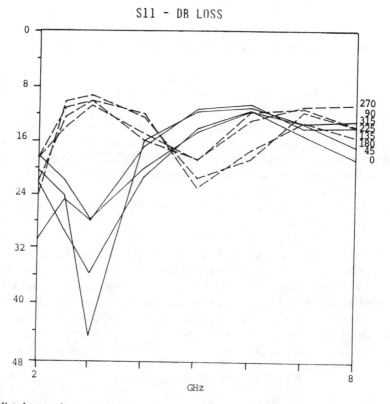

**Fig. 6.41** Predicted return loss.

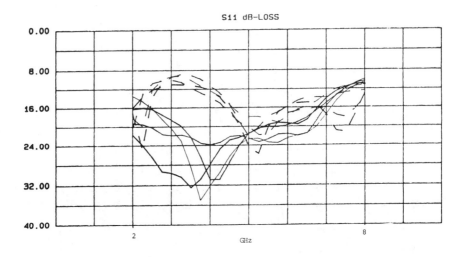

**Fig. 6.42** Measured return loss.

## 6.3 DOUBLE AND SINGLE BALANCED MIXER DESIGN

Traditionally, the mixer has been the most difficult element to design and analyze in modern microwave receiver systems. The vast majority of these systems employ passive diode mixers as the frequency converting element, but active FET implementations are beginning to appear. These mixers typically employ large transmission line baluns used in three-dimensional structures, although completely planar 2–18 GHz double balanced mixers have been demonstrated [1–3]. Conventional mixer designs such as these are not feasible for monolithic implementation because their passive elements require excessive GaAs area. Hence, a completely new design concept must be used in the development of GaAs monolithic microwave mixers.

The mixer, which can consist of any device capable of exhibiting nonlinear performance, is essentially a multiplier. That is, if at least two signals are present, their product will be produced at the output of the mixer. From basic trigonometry, we know that the product of two sinusoids produces a sum and a difference frequency. Either of these frequencies can be selected with the IF filter. Unfortunately, no physical nonlinear device is a perfect multiplier; thus, they contribute noise and produce a vast number of spurious frequency components. If the small-signal case is assumed and modulation is ignored, the mixer output spectra can be defined as follows:

$$\omega_d = n\omega_p \pm \omega_s \qquad (6.14)$$

The desired component usually is the difference frequency ($|\omega_p + \omega_s|$ or $|f_p - f_s|$) but sometimes the sum frequency ($f_s + f_p$) is desired, when building an up-converter or a product related to a harmonic of the LO, as in a subharmonically pumped mixer.

A mixer can also be analyzed as a switch that is commutated at a frequency equal to the pump frequency $\omega_p$. This is a good first-order approximation of the mixing process for a diode (or an FET), because it is driven from the low resistance state (forward bias) to the high resistance state (reverse bias) by a high-level LO signal.

Recently, the growing interest in GaAs monolithic circuits again began to heighten interest in active MESFET mixers. This is indeed fortunate, because properly designed FET mixers offer some distinct advantages over their passive counterparts. This is especially true in the case of the dual-gate FET mixer, whose additional port allows for some inherent LO to RF isolation and at times it can replace single-balanced passive approaches. The possibility of conversion gain rather than loss also is an advantage, as the added gain may eliminate the need for excess amplification, thus reducing system complexity.

There are several drawbacks to designing active mixers. Unlike a diode mixer, where the design engineer can make excellent first-order performance approximations with linear analysis and where the diode always mixes reasonably well almost independent of the circuit, these conditions are unfortunately lacking in active mixer designs. Simulating performance, especially with a dual-gate device, requires some form of nonlinear analysis tool to obtain any circuit information other than small-signal impedance. An analysis of the noise performance is even more difficult.

The dominant nonlinearity of the FET is its transconductance [4–6], which typically, especially with JFETs, is a square-law function. Hence, it makes a very efficient multiplier with reasonably low spurious products. The small-signal circuit [7] shown in Figure 6.43 denotes the principal elements of the FET that must be considered in a model. The parasitic resistances $R_g$, $R_d$, and $R_s$ are small compared to $R_{ds}$ and can be considered constant; but, they are important in determining the noise performance of the circuit. The mixing products produced by parametric pumping of the capacitances $C_{gs}$, $C_{dg}$, and $C_{ds}$ typically are small and add only second-order effects to the total circuit performance. Time-averaged values of these capacitances can be used in circuit simulation with good results.

This leaves the FET transconductance, $g_m$, whose primary nonlinearity is an extremely strong nonlinear dependance as a function of gate bias. The greatest percentage change in transconductance occurs near the pinch-off voltage, with the most linear change with respect to gate voltage occurring in the center of the bias range. As the FET is biased toward $I_{dss}$, the transconductance function again becomes nonlinear. It is in these most nonlinear regions that the FET is most

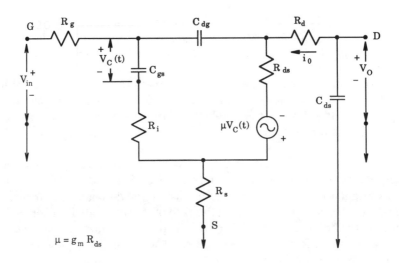

**Fig. 6.43** Small-signal GaAs FET equivalent circuit with voltage source representation.

efficient as a mixer.

Single-ended as well as balanced mixers also can be designed using dual-gate FETs [8–10]. Dual-gate devices offer several advantages over conventional devices such as ease of LO injection, improved isolation, and added gain. However, they are considerably less stable, hence added care must be used when designing such circuits.

The operation of a dual-gate FET can be understood easily, if the FET is considered as a cascade connected FET pair. Using this concept for the FET, the drain characteristics for the pair can be approximated by combining the characteristics of each intrinsic FET. A convenient representation of the drain characteristics is shown in Figure 6.44. With this representation, the operating point for FET 1 can be found as a function of gate 1 and 2 bias as well as its drain-to-source voltage.

The operating point can vary significantly, depending on how the FET is biased. Typically, gate 1 is used for signal injection with gate 2 biased ($V_{gs2} \cong 0$) for FET operation in the low-noise mode (shaded area in Figure 6.44). Gate 2 is also used for LO injection. Applying the LO at gate 2 is in effect drain pumping the first FET; hence, FET 1 is the primary mixing element. The operation is reversed if a sufficiently high bias voltage ($V_{gs2} > 2$) is applied to gate 2. With these bias conditions, FET 1 acts as an RF preamplifier, and FET 2 becomes the primary mixing element.

Dual-gate FET mixers can be analyzed in a limited way with the aid of a nonlinear simulator such as LIBRA, Microwave HARMONICA, or SPICE. Conversion performance obtainable with dual-gate mixers is comparable to that ob-

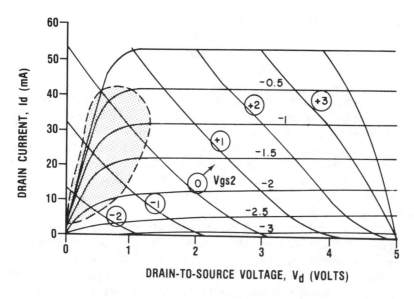

**Fig. 6.44** Dual-gate FET I-V characteristics as a function of $V_{gs1}$ and $V_{gs2}$.

tained with conventional devices [7,11], with the exception of slightly degraded noise figure and possibly more gain.

Unfortunately, the best mixer conversion loss is obtained for large values of LO voltage. With this in mind and knowing the FET parasitic element values that can be obtained from small-signal S-parameter measurements, an estimate of the required LO power can be made. From conventional circuit theory, the power dissipated in the gate circuit is

$$P_{LO} = 0.5(\omega_p \tilde{C}_p V_p)^2 R_{in} \qquad (6.15)$$

where $R_{in}$ is the input resistance, $\tilde{C}_p$ is the time averaged value of $C_{gs}$, and $V_p$ is the pinch-off voltage. When the total gate periphery of the FET is small, the optimum LO power will be modest (3–6 dBm), but as the FET becomes larger, the term $C_p$ in Equation (6.15) begins to dominate. It is not uncommon for a large FET to require 20 dBm of LO drive power. The amount of LO power required for maximum conversion loss at a particular FET size also can be reduced by selecting or designing the FET for the lowest possible value of pinch-off voltage. Because, for a given gate periphery, the values of $R_{in}$ and $C_p$ change little compared to the change in the pinch-off voltage obtained when changing the FET doping profile, a dramatic improvement in LO efficiency is obtained when using low-noise FETs with pinch-off voltages in the 1.5 V range as compared to power FETs that exhibit pinch-off voltages in the 4–5 V range.

The design problem for any type of mixer can be divided into two main areas: the nonlinear element and the balun. If a monolithic implementation is desired, the most practical choices for nonlinear elements are planar Schottky diodes, single-gate FETs, and dual-gate FETs.

In the monolithic realm, the balun problem is further constrained by chip area and backside processing requirements. If a 2–18 GHz passive mixer balun were used, it would be approximately 2 cm in length, which is an order of magnitude too large for a monolithic circuit realization. Thus, active baluns or lumped-element transformers are the only viable options. The problem is further complicated in that it is desirable to employ baluns that approximate low-frequency equivalent center tapped transformers. A center tapped balun is a convenient way to extract the IF frequency from a conventional double-balanced diode mixer. However, a center tapped transmission line balun cannot be fabricated.

A new balun topology, which can be readily implemented using monolithic technology, eliminates these problems and provides a virtual center tap. Because the balun uses common gate and common source circuit techniques, an ideal 180° phase shift occurs for the signals present between the upper and lower halves of the circuit (Figure 6.45). Typical broadband amplitude and phase performance, for a balun designed with no center tap and resistive terminations at the reverse end of the drain transmission line, are shown in Figure 6.46a and 6.46b. The balun, which is shown in Figure 6.46c, measures 1.87 × 1.73 mm. As we can see, the balun exhibits excellent balance through the design band of 2–18 GHz. The performance of a center tapped balun, designed for the same frequency range is slightly degraded due to the removal of the drain terminations.

If two such baluns are used in conjunction with a diode or FET ring to form a double balanced mixer, the IF signal appearing at the ring terminals propagates (in phase) down both arms of the balun and can be summed at a common node, thus forming a virtual center tap. This center tap can be used for IF extraction or grounded to complete the IF return path. Because active baluns are not reciprocal, a combining or dividing structure will be needed on the RF port depending on whether the mixer is used as a up- or down-converter (Figure 6.47).

The frequency limitations of the RF and LO ports are determined by the distributed amplifier-like sections, which can be designed to operate over extremely large bandwidths. The IF frequency response can be designed to exhibit broadband performance as well. This mixer concept also can be extended to include double ring mixer topologies. Double ring approaches have the added advantage of allowing the IF frequency response to overlap the RF and LO frequency bands, thus making IF extraction even easier.

By using this technology, both single and double ring designs were fabricated. The conversion loss characteristics, at an IF frequency of 4 GHz, of a typical single-ring down-converter is shown in Figure 6.48. As we can see in the illustration, the performance is comparable to conventional diode designs requiring similar amounts of pump power (12 dBm). With the LO drive shown, the mixer exhibited a 1 dB compression point (referred to the input) of 4 dBm (Figure 6.49), LO-to-RF isolation of 20 dB (average), LO-to-IF isolation of 30 dB (average), and RF-to-IF isolation greater than 20 dB (Figure 6.50).

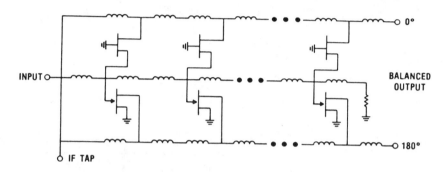

**Fig. 6.45** Lumped element equivalent circuit of center tapped balun.

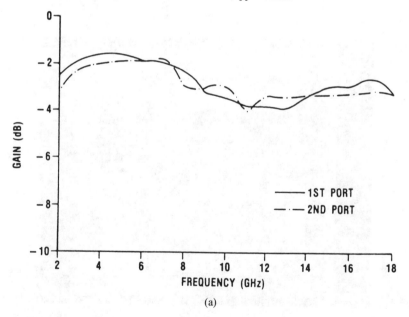

**Fig. 6.46** Monolithic (dividing) balun: (a) frequency response and amplitude balance; (b) differential phase performance; (c) chip photograph.

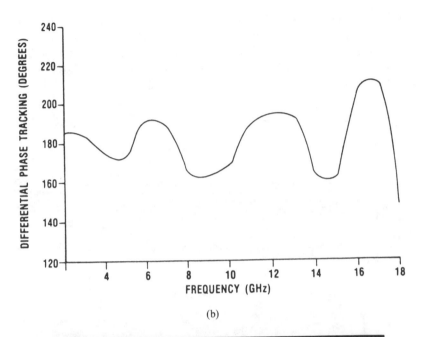

(b)

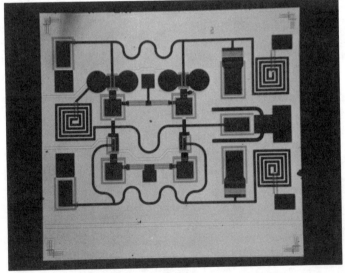

(c)

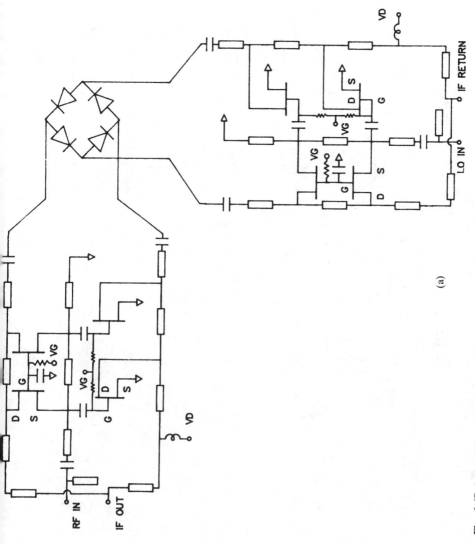

**Fig. 6.47** Active balun diode mixers: (a) down-converter mixer circuit diagram; (b) up-converter mixer circuit diagram.

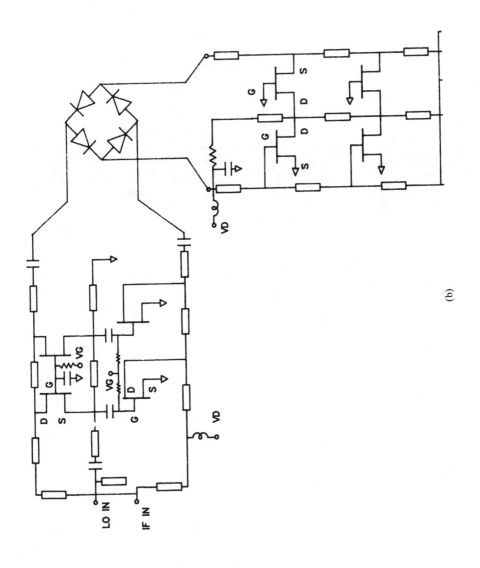

(b)

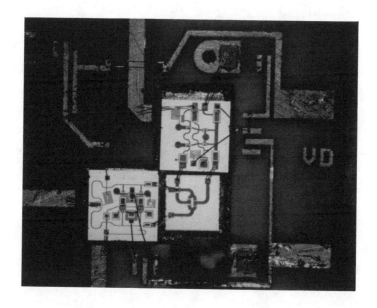

(a)

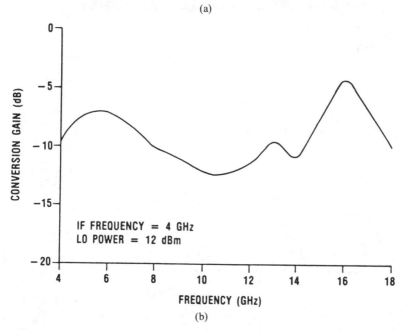

(b)

**Fig. 6.48** Monolithic double balanced diode mixer: (a) MMIC photograph; (b) conversion loss performance of down-converter mixer.

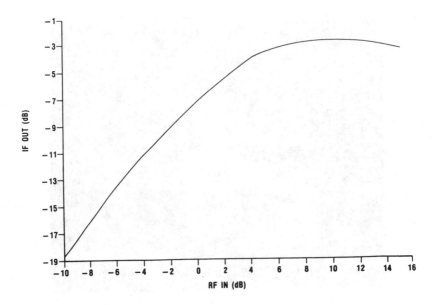

**Fig. 6.49** RF compression characteristics of double balanced monolithic diode mixer.

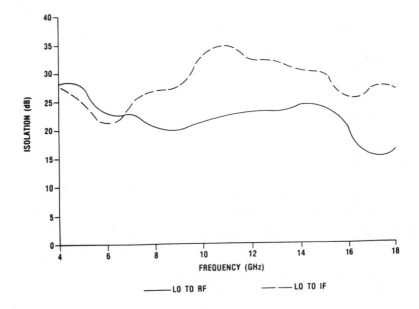

**Fig. 6.50** Isolation characteristics of double balanced monolithic diode mixer.

The conversion performance as a function of frequency, which was measured at an IF frequency of 500 mHz for the double-double balanced design is shown in Figure 6.51. Although the mixers employ diodes as the nonlinear elements, the conversion loss (gain) of the double ring design is somewhat greater than a conventional structure because of the gain associated with the baluns. The isolation characteristics, which are comparable to hybrid designs, are shown in Figures 6.52. The compression characteristics of this configuration was also measured with conventional levels of LO drive (13 dBm) and is shown in Figure 6.53. As we can see, the mixer exhibited a 1 dB compression point (referred to the input) of 8 dBm, which is comparable to the best hybrid designs.

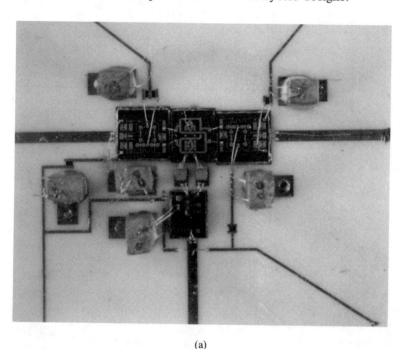

(a)

**Fig. 6.51** Monolithic double-double balanced mixer: (a) chip photograph; (b) conversion loss performance.

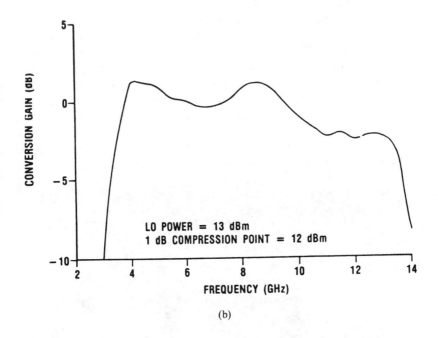

(b)

Figure 6.51 cont.

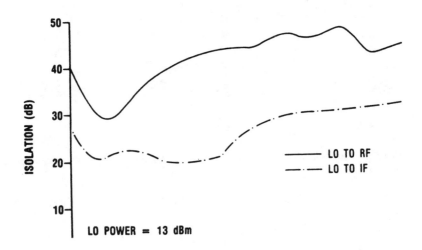

Fig. 6.52 Isolation performance of monolithic double-double balanced mixer.

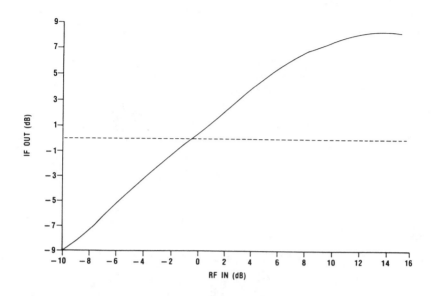

**Fig. 6.53** RF compression characteristics of double-double balanced mixer.

A slightly different topology, which can be readily implemented using mono-lithic technology, employs active baluns in conjunction with a unique distributed dual-gate FET mixer structure. The proposed circuit employs a single balun, which can be either active or the passive lumped element type (transformer, differential line, *etcetera*), and distributed dual-gate FET mixer sections. Trans-mission line models for the balun and mixer are shown in Figure 6.54. The number of distributed sections employed depends on the bandwidth, conversion gain, and impedance matching requirements. In this design only two sections were required to achieve adequate distributed performance; however, greater conversion gain probably could have been obtained if more sections were employed at the expense of chip complexity.

As we can see in the circuit diagram, one mixer section employs a common source topology and the other uses a common gate topology, which are inherently antiphasal. Thus, an ideal 180° phase shift occurs for the signals present between the upper and lower halves of the circuit, eliminating the requirement for a second balun. Both the LO and RF voltages, which are present at the FET drains of each mixer section, are also out of phase by 180°, while the IF voltages are in phase. By summing the output of both mixer sections, an independent IF port is obtained and the RF and LO signals are cancelled. Thus, the mixer structure is completely double balanced.

The frequency limitations of the mixing portion of the circuit is determined

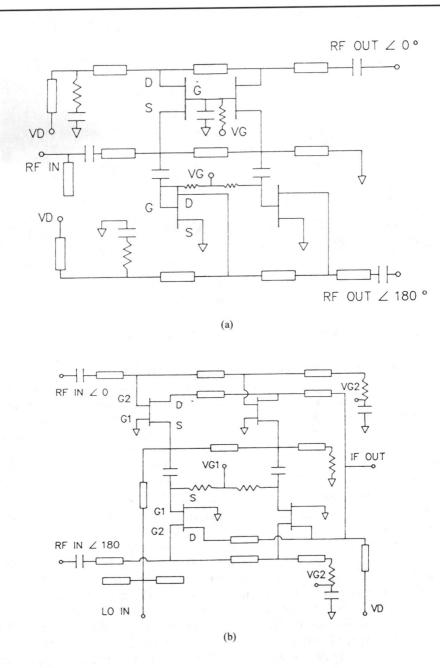

**Fig. 6.54** Dual-gate FET double balanced mixer: (a) transmission line model of distributed active balun; (b) transmission line model of monolithic mixer chip.

by the distributed amplifier-like sections that can be designed to operate over extremely large bandwidths. With the addition of an IF amplifier (active matching) or bandpass matching network, the IF frequency response could be further broadened. Because of the large number of active nonlinear elements, the dynamic range can be made as good or better than the best conventional diode mixers.

The distributed monolithic balun and mixer, shown in Figure 6.55, was designed using 0.5 × 150 $\mu$m dual-gate FETs fabricated on a 0.15 mm thick GaAs substrate. The FETs were modeled as cascade connected single-gate FETs with the linear model elements determined from S-parameter measurements. The nonlinear drain current and transconductance characteristics were obtained from I-V curve data obtained at 1 mHz [12]. The active balun used with the mixer also employed both distributed common source and common gate amplifier sections, in order to obtain a broadband differential phase output with good amplitude balance.

**Fig. 6.55** Monolithic double balanced mixer and active balun ICs.

The mixer-balun combination was evaluated as a conventional double balanced mixer with the LO drive applied to the gate 1 circuit. The RF signal was applied to the active balun, which in turn drives the gate 2 circuit. The dc bias on both gates was adjusted for optimum conversion loss; however, since the mixer performance was very insensitive to bias, the second gate voltage was set to zero and the first gate was biased for a drain current of $I_{dss}/2$. Using these bias conditions, the conversion loss characteristic shown in Figure 6.56 was obtained. The RF-to-IF and LO-to-IF isolations, which demonstrate the excellent balance obtained in the design, are shown in Figure 6.57. Conversion loss performance as a function of LO power and the RF compression characteristics are shown in Figures 6.58 and 6.59. As we can see, the mixer's performance is comparable to hybrid diode designs.

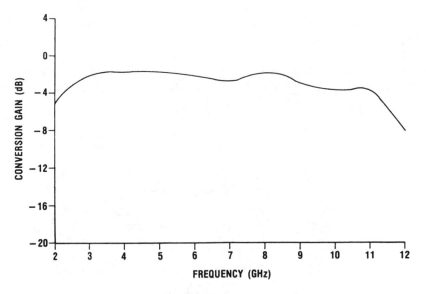

**Fig. 6.56** Monolithic double-balanced FET mixer conversion loss performance as a function of frequency.

This type of structure, with its unique balanced characteristics, can be used as a broadband up-converter as well as a conventional mixer in a variety of receiver applications. In addition, because the mixer is completely balanced, the IF frequency response can overlap the LO or RF responses, which usually can be accomplished only with double-double balanced structures. Fabrication of these balun and mixer structures was accomplished using Texas Instruments baseline ion-implanted GaAs process for 3-inch diameter, 0.15 mm thick semiconductor wafers. The FET channel doping profile was optimized for high transconductance and low noise performance, yielding the best mixer performance. The TiPtAu 0.5 $\mu$m gates were defined with E-beam lithography, the ohmic metal employed was AuGeNiAu, and the MIM capacitors were constructed with a 2000 Å thick layer of $Si_3N_4$. The same baseline low-noise FET process also was used to fabricate the diode structures.

By using this dual-mode characteristic of distributed broadband baluns in

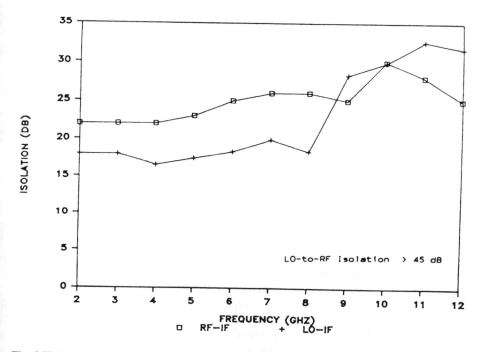

**Fig. 6.57** Isolation performance of double balanced FET mixer.

diode mixer topologies, a very compact monolithic circuit could be designed to operate over a frequency range several octaves wide with performance comparable to conventional passive diode mixers. An alternative all-FET structure, with its unique balanced characteristics, also can be used as a broadband up-converter as well as a conventional down-converting mixer in a variety of receiver applications. Since this mixer is completely balanced, the IF frequency response can overlap the LO or RF responses, which usually can be accomplished only with a double-double balanced structure.

Although the above described mixers are complex, they provide superior performance to other types of balanced structures. In particular, single balanced mixers are easy to design and to fabricate monolithically, but they provide the user with little more than a mixing function. These structures are termination-sensitive, typically narrowband, and offer the user little isolation or spurious-free performance. However, at millimeter-wave frequencies, single balanced mixers usually offer the best noise figure performance, typically the most important system parameter. Several excellent examples of well designed monolithic single balanced mixers [13, 14] are shown in Figure 6.60.

Numerous other types of balanced mixer structures can be fabricated in

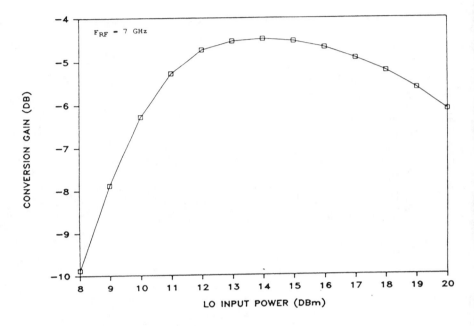

**Fig. 6.58** Conversion loss as a function of LO power.

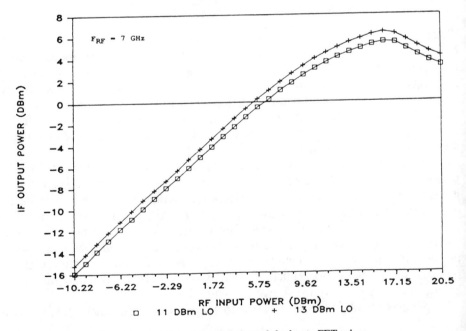

**Fig. 6.59** RF compression characteristics of double balanced dual-gate FET mixer.

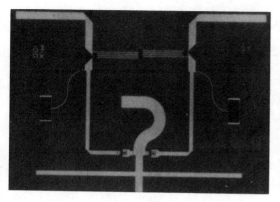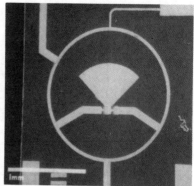

Fig. 6.60 Typical single balanced monolithic mixers.

hybrid form, but few are practical candidates for monolithic implementation due to balun realization problems. As an example, a typical four-diode double balanced mixer, the low frequency equivalent circuit for which is shown in Figure 6.61, requires center-tapped transformer baluns. Such balanced structures are very difficult to build, but can be realized by employing interlaced spiral or rectangular inductors. Transformers of this type can be designed to operate over frequency ranges greater than an octave, but make backside chip processing nearly impossible. The star mixer, another popular form of a double balanced structure (Figure 6.62), is also plagued with similar fabrication problems, although there may be ways to fabricate such a structure if a suspended chip can be utilized.

Completely active double balanced mixers, similar to what has been done with linear ICs in the low frequency world, are now beginning to appear at microwave frequencies. As device gains improve, and as HBTs become available, the frequency operation of such designs will rapidly extend well into the microwave spectrum. A convenient monolithic GaAs FET double balanced structure [15] is shown in Figure 6.63. The circuit (Figure 6.64) consists of two differential pairs with their associated current sources and an external quadrature coupler. The modulation input signal was also supplied from an external source with the proper sine-cosine phase relationship.

## 6.4 VARIABLE ATTENUATOR AND SWITCH DESIGN

### 6.4.1 Introduction

*Electronic warfare* (EW) system requirements include broadband, high dynamic range, and the ability to respond to multiple pulsed and CW signals. Sometimes these requirements are contradictory and result in overly complicated,

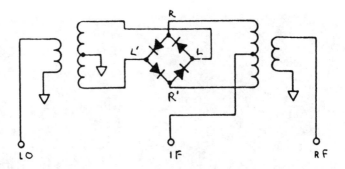

**Fig. 6.61** Circuit diagram of double balanced mixer with transformer hybrids.

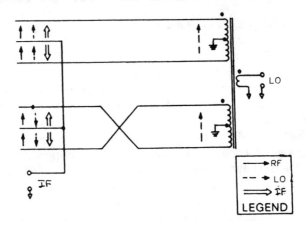

**Fig. 6.62** Circuit diagram of star mixer with multiple secondary transformer hybrids.

expensive designs that will not be bought by the military. Therefore, system architectures with signal control both within and ahead of the IF subsystem are being implemented using the MMIC technology. Two key components for this front-end microwave signal processing are broadband variable attenuators and switches. Of course, similar designs can be used in the IF for additional control at moderate frequencies.

Prior to the development of MMIC, variable attenuators were implemented in MIC technology with either a network of fixed resistors switched in with PIN diodes or using the PIN diode's intrinsic resistance, which varies as a function of the forward bias current. The first method required a large number of components (at least three resistors and two PIN diodes per attenuation level) and did not provide continuous variability; the second method required that the PIN diodes be spaced a quarter-wavelength apart to achieve broadband performance. Because PIN diodes are not readily available in MMIC GaAs processing, whereas FETs

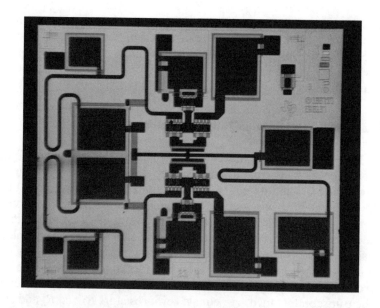

**Fig. 6.63** Monolithic GaAs FET single-sideband modulator.

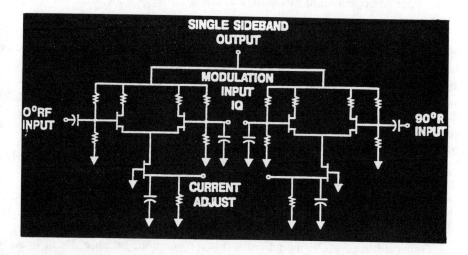

**Fig. 6.64** Modulator circuit configuration.

are, both attenuator and switch designs utilizing these devices are considered. Because the FET can be operated over a wide range of bias conditions and because it is desirable to minimize the number of control points for either the switch or the attenuator, various circuit architectures were evaluated via CAD techniques. This procedure assured the design would come close to meeting the requirements with a minimum number of redesign-fabrication-test cycles. Also due to the small size of its circuits, the FET can be thought as a parallel RC network whose resistance varies with bias; therefore, the designs can be treated in lumped-equivalent terms rather than in a distributed fashion.

### 6.4.2 Nonlinear FET Operation and Selection

The FET is a three terminal device where the gate voltage controls the amount of current flowing from drain to source. Figure 6.65 shows a three-dimensional view of a FET and its equivalent circuit. The elements inside the dashed box are the intrinsic FET parameters. Figure 6.66a shows a typical I-V curve for a FET; and as Figure 6.66b ($V_{ds} = 0$ V region) shows, the I-V curve looks like lines of resistance that are a function of gate bias. The gate voltage varies the drain-to-source resistance, $R_{ds}$. The resistance is changed by depleting the region underneath the gate of free charge carriers thereby increasing the resistance path from drain to source. In this region of operation, several intrinsic parameter values change: the transconductance, $g_m$, equals zero; resistors $R_1$ and $R_2$ and capacitors $C_{gs}$ and $C_{gd}$ are set equal to each other. The other parameters remain constant to a first-order approximation since they are not strongly dependent on the gate voltage.

In the selection of a FET geometry two parameters must be considered: insertion loss (lowest impedance state); and isolation (maximum impedance state). Low insertion loss requires large FET widths to minimize $R_{ds}$; conversely, high isolation necessitates small FET widths to reduce parallel drain-to-source capacitance, $C_{ds}$, which acts as a bypass for the RF signal you are trying to attenuate; or in the case of a switch, the open circuit leg required for good isolation. $C_{ds}$ is a function of the FET's doping density and the physical spacing of the drain and source fingers; it remains almost constant as the gate bias is varied. A determination as to which of the two specifications is important, will play a role in deciding the FET geometry. The design procedure is as follows:

1. Characterize the MESFET; both dc and ac (S-parameters).
2. Use S-parameters to derive capacitance values for the dc model and the remaining elements in the ac model.
3. Generate the SPICE model from this information; because it can linearly scale device parameters, it generates the S-parameter for varying dc bias conditions.
4. Characterize MESFET in both the shunt and series configuration for different bias conditions, as shown in Figure 6.67.

For the design of both the variable attenuator and SPDT switch a 300 $\mu$m FET was chosen.

In most control device functions, the series FET is used in the common gate

# FET PARAMETERS

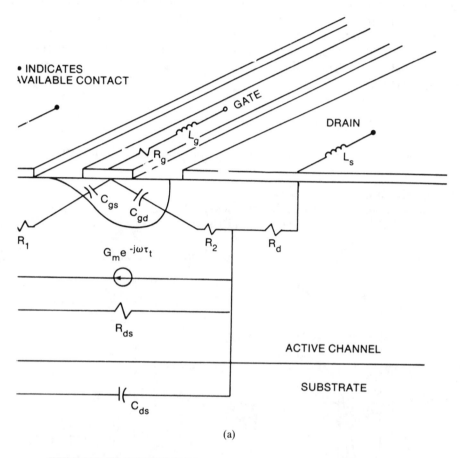

(a)

## FET EQUIVALENT CIRCUIT

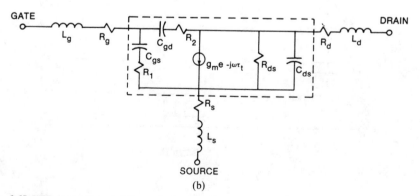

(b)

**Fig. 6.65** FET parameters and its equivalent circuit.

configuration; therefore, care must be taken to ensure that the RF energy you are trying to route from input to output does not leak into the gate port. Although this is not a problem in the isolation mode it will increase your insertion loss in the through mode. The leakage occurs via $C_{gd}$ or $C_{gs}$, depending how you designate the drain and source. This leakage can be remedied by placing a 1–2 kΩ resistor at the gate port [2].

### 6.4.3 Variable Attenuator Design

In the design of the variable attenuator the following electrical specifications were used as our goals.

Bandwidth: 0.5–10 GHz (0.5–18 GHz design goal)
Insertion loss: 1.9 dB design goal at 10 GHz
2.8 dB maximum at 10 GHz

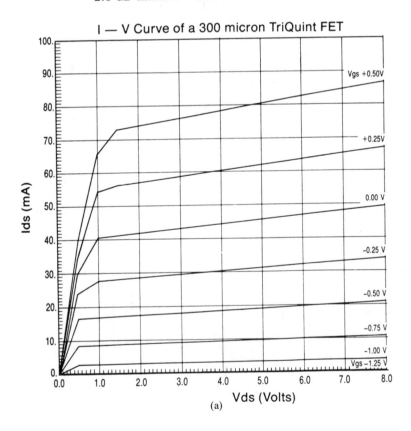

(a)

**Fig. 6.66** I-V curve for a FET (a) typical curve; (b) $V_{ds} = 0$ region.

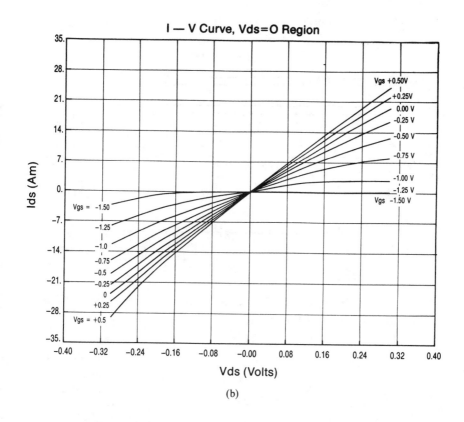

### I — V Curve, Vds=O Region

(b)

Flatness: ± 0.6 dB at 30 dB attenuation
Isolation: 42 dB minimum
Attenuation range: 0–30 dB in 3 dB steps
Attenuation accuracy: 0–9 dB, ± 1 dB
   9–18 dB, ± 2 dB
   18–30 dB, ± 3 dB
OCP: +22 dBm minimum
OIP3: +32 dBm minimum
Return loss: 10 dB minimum at 30 dB attenuation

A circuit must be selected that provides the largest range of attenuation and flatness over frequency while also providing a good source and load match regardless of attenuation value. This match requirement dictates that the attenuator be absorptive. Three such circuits were investigated: bridged-T; symmetrical-T; and symmetrical-Pi. The series FET provides the attenuation and, by the proper selection of the shunt FET's gate voltage ($R_{ds}$), good VSWR can be ensured at any

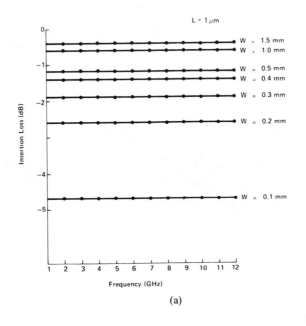

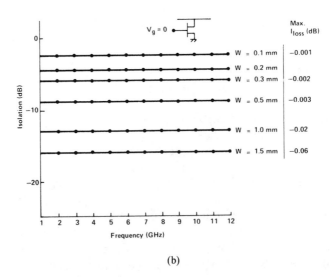

**Fig. 6.67** Insertion loss compared with frequency for an SPST series-FET switch (a). Isolation compared with frequency for (b) SPST parallel-FET and (c) SPST series-FET switches.

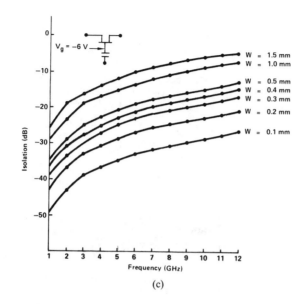

(c)

attenuation level. Figure 6.68 shows the three circuits and the relating equation [3] for resistance that the series and shunt FETs need to achieve. The selection of the proper gate voltage on the shunt FET can be made by using feedback, which will be invariant to process control or different FET geometries [4]. Figure 6.69 shows a reference attenuator cell whose input is terminated in the characteristic impedance of the system (50 Ω); an external operational amplifier, Op Amp, selects the shunt FET gate voltage proper to both the reference and the RF cell shunt FET. For any arbitrary series FET gate voltage, a 50 Ω environment is maintained. This is accomplished by the Op Amp, which in the open loop state will vary its output voltage and therefore the shunt FET's gate voltage by forcing $S_{22}$ of the reference cell to 50 Ω ($Z_0$) to maintain a voltage difference of zero at its input. This technique of feedback can be used for any circuit topology.

The variable attenuator's insertion loss is governed by the minimum $R_{ds}$ that can be achieved for the FET geometry, approximately 10 Ω for the 300 μm FET. The maximum attenuation obtainable is a function of both the maximum $R_{ds}$ and the minimum $C_{ds}$ (95 fF for the 300 μm FET). Figures 6.70–6.72 show the swept frequency response of the three circuits for both return loss and insertion loss as the gate voltage is varied from 0 V to three times pinch-off (−5 V). The symmetrical-T offers the highest attenuation, flattest response, best return loss at maximum attenuation; its electrical performance can be extended to 18.5 GHz with just a 0.8 dB and 2 dB degradation in insertion loss and attenuation, respectively. At these

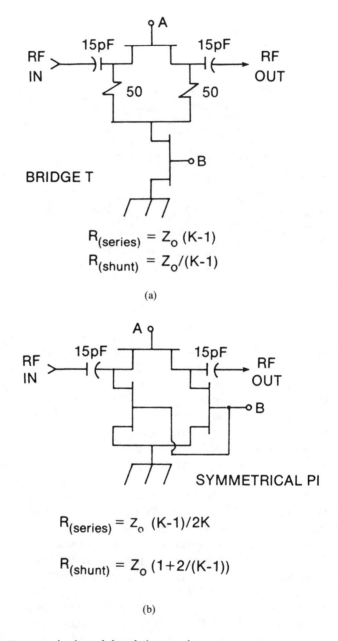

$$R_{(series)} = Z_o (K-1)$$
$$R_{(shunt)} = Z_o/(K-1)$$

(a)

$$R_{(series)} = Z_o (K-1)/2K$$

$$R_{(shunt)} = Z_o (1+2/(K-1))$$

(b)

**Fig. 6.68** Variable attenuator circuits and the relating equation.

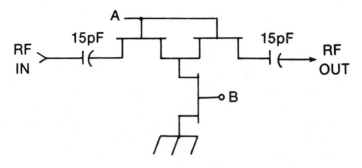

SYMMETRICAL T

$$R_{(series)} = Z_o (1-2/(K+1))$$

$$R_{(shunt)} = 2 Z_o K / (K^2-1)$$

WHERE $K = SQRT [10^{(ATTEN IN DB)/10}]$

$Z_o$ = CHARACTERISTIC IMPEDANCE (50 ohm)

(c)

**Fig. 6.68** cont.

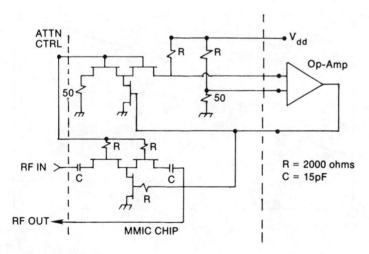

**Fig. 6.69** Reference attenuator cell.

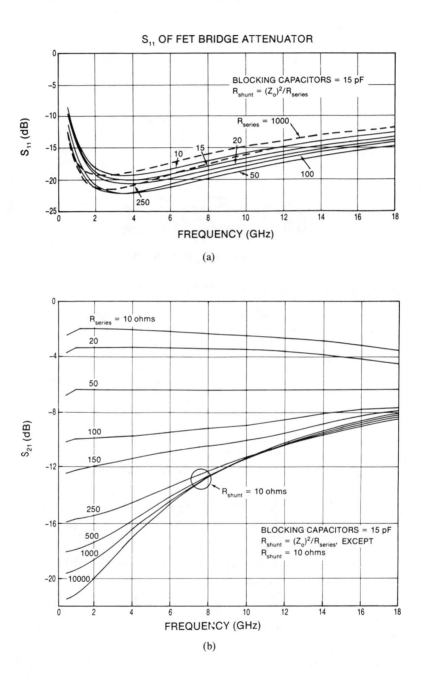

**Fig. 6.70** $S_{11}$ and $S_{21}$ of FET bridge-T attenuator.

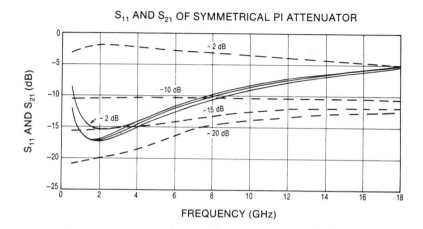

**Fig. 6.71** $S_{11}$ and $S_{21}$ of FET symmetrical-pi attenuator.

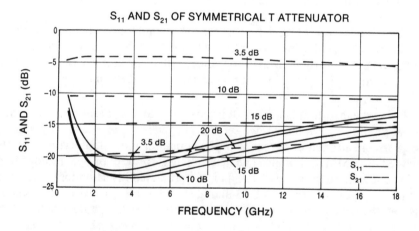

**Fig. 6.72** $S_{11}$ and $S_{21}$ of symmetrical-T attenuator.

extended frequencies its return loss still meets specification. The bridge-T exhibits minimum insertion loss at 10 GHz and a good match at its maximum attenuation. The symmetrical-Pi's return loss degrades with frequency regardless of attenuation level, probably due to $C_{ds}$. Table 6.5 tabulates the computer results of insertion loss, attenuation, and return loss for the three circuits. The symmetrical-T has the highest attenuation because the two series FETs, when turned off, reduce by approximately a half the effective $C_{ds}$; the two series FETs also increase the circuit's insertion loss. Since the symmetrical-pi offers no advantage in electrical performance it was not considered further.

**Table 6.5** Insertion Loss, Attenuation, and Return Loss for Circuits (frequency 10 GHz)

|  | Bridge-T | Symmetrical-T | Symmetrical-pi |
|---|---|---|---|
| Insertion loss maximum | 2.5 dB | 4.2 dB | 3.5 dB |
| Attenuation minimum | 11 dB | 19 dB | 14 dB |
| Return loss | 14 dB | 16 dB | 8 dB |

### 6.4.4 Measurement *versus* Simulation

Table 6.6 shows the design goals for the attenuators and compares the computer simulation results for the circuit with the actual measured performance of the chip in the Tektronix MMIC package. All tests utilized the TriQuint test fixture, which was limited to measurements below 12 GHz due to possible moding problems.

Using the insertion loss plots of Figures 6.70 and 6.71 and comparing them to those of Figures 6.73 and 6.74, we see that, in general, the measured performance of the variable attenuator agrees with the computer predicted performance based upon the GaAs MMIC models. The insertion loss for both bridge-T and symmetrical-T variable attenuators was less than the predicted insertion loss values across the band, as shown in Figures 6.73 and 6.74 and Table 6.6. Also, the attenuation range for both components was somewhat better than the model predicted. Some gain ripple occurs at the high frequency end of the band (above 12 GHz).

### 6.4.5 Switch

The design goals for the SPDT switch are as follows:

Bandwidth: 0.5–10 GHz
Insertion loss: 1.6 dB design goal at 10 GHz
           2.6 dB maximum at 10 GHz
Flatness: ±0.6 dB maximum over full band
Isolation: 45 dB minimum at 10 GHz
Return loss: 10 dB minimum at 10 GHz at the input

**Table 6.6 Variable Attenuator Performance**

| Performance Parameter | MMIC Design Goals | Computer Simulation Results of MMIC Component Performance | | Measured MMIC Component Performance | |
|---|---|---|---|---|---|
| | | Bridge-T | Symmetrical-T | Bridge-T | Symmetrical-T |
| Bandwidth Insertion loss | 0.5–10 GHz 1.9 dB at 10 GHz 2.8 dB max at 10 GHz | dc–18 GHz 3.2 dB at 10 GHz | dc–18 GHz 4.5 dB at 10 GHz | dc–12 GHz 1.2 dB at 1 GHz 2.3 dB at 10 GHz | dc–12 GHz 2.1 dB at 1 GHz 3.2 dB at 10 GHz |
| Flatness | ±0.6 dB max at 10 GHz | ±0.25 dB up to 10 GHz | ±0.25 dB up to 10 GHz | ±0.55 dB up to 10 GHz | ±0.55 dB up to 10 GHz |
| Isolation | 42 dB min across band | Same as attenuation range | Same as attenuation range | Same as attenuation range | Same as attenuation range |
| Attenuation range | 0–30 dB in 3 dB steps | 15.5 dB at 500 MHz 10 dB at 10 GHz step size function of driver | 19 dB at 500 MHz 16.5 dB at 10 GHz step size function of driver | 22 dB at 500 MHz 14 dB at 10 GHz step size function of driver* | 29 dB at 500 MHz 22 dB at 10 GHz step size function of driver* |
| 1 dB output compression point | +22 dBm min | +15 dBm (calculated) | +15 dBm (calculated) | +25.5 dBm | +16.0 dBm |
| Third-order intercept point | +32 dBm min | +25 dBm (calculated) | +25 dBm (calculated) | +35.5 dBm | +26.0 dBm |
| VSWR | 2.0:1 max | 1.32:1 at 10 GHz | 1.29:1 at 10 GHz | 1.29:1 at 10 GHz | 1.29:1 at 10 GHz |

* *Note:* Attenuator testing was done in 3 dB step sizes.

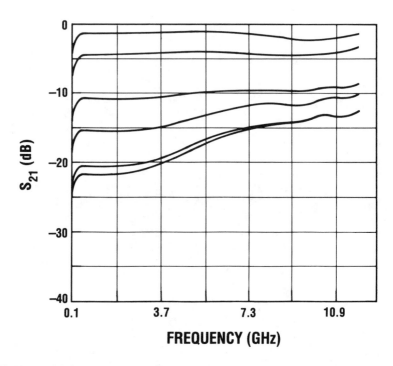

**Fig. 6.73** Measured frequency response of bridge T-attenuator.

The SPDT switch uses two identical "switching legs," consisting of both series and shunt FETs. The shunt FET's $C_{ds}$ is used to form an artificial 50 $\Omega$ transmission line with the addition of series inductors. The operation of the switching leg is to have the series and shunt FETs conversely biased. The series FET improves the isolation at the low frequencies whereas the shunt FETs improve isolation at the high frequencies (by shunting the RF to ground through the low resistance). Further improvements can be made in isolation at the high frequencies by continually adding shunt FETS, but at the expense of insertion loss due to the low $Q$ (resistive) inductors—this is one of the deficiencies of the MMIC technology. One series FET followed by two shunt FETs are used for a switching leg. Figure 6.75 shows an equivalent circuit representation of the switching leg in its two states, and Figure 6.76 is a schematic of the SPDT switch.

The switch's insertion loss is proportional to the $R_{ds}$, but its isolation depends on the series FET's $C_{ds}$ and the shunt FET's minimum $R_{ds}$. Figure 6.77 is a computer simulation of the frequency response for insertion loss, isolation, and return loss of all three ports of both the on and off states of the SPDT switch. The insertion loss and isolation at 10 GHz, respectively, are 3 and 28 dB. An improvement in isolation was achieved by using a second series FET to reduce the effective $C_{ds}$ capacitance at no cost to the input-output match, as illustrated in Figure

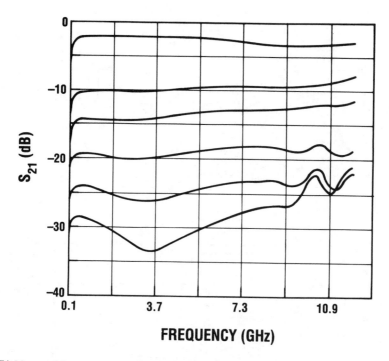

**Fig. 6.74** Measured frequency response of symmetrical T-attenuator.

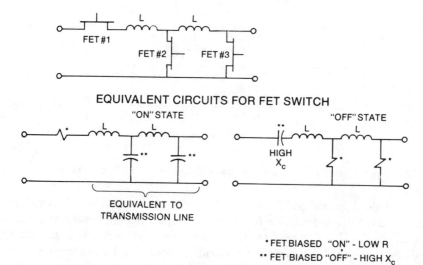

EQUIVALENT CIRCUITS FOR FET SWITCH

"ON" STATE

EQUIVALENT TO
TRANSMISSION LINE

"OFF" STATE

HIGH
$X_c$

\* FET BIASED "ON" - LOW R
\*\* FET BIASED "OFF" - HIGH $X_c$

**Fig. 6.75** Switching leg and its equivalent circuit.

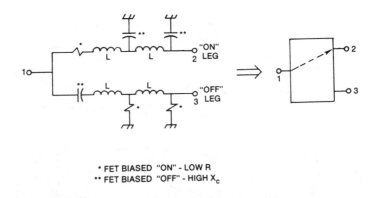

* FET BIASED "ON" - LOW R
** FET BIASED "OFF" - HIGH $X_c$

**Fig. 6.76** Basic SPDT switch.

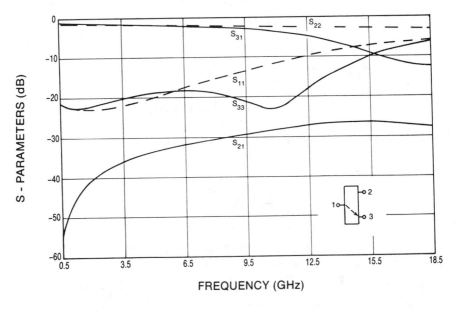

**Fig. 6.77** Computer simulated frequency response of a single series FET SPDT switch.

6.78. The insertion loss and isolation at 10 GHz are 3.6 and 35 dB, respectively; hence, for a 0.6 dB degradation in insertion loss, a 7 dB improvement in isolation was achieved; this is the configuration we chose. Optimization of the inductors revealed that we could extend the frequency of operation by decreasing the series inductor, as illustrated in Figures 6.79 and 6.80 for insertion loss and isolation.

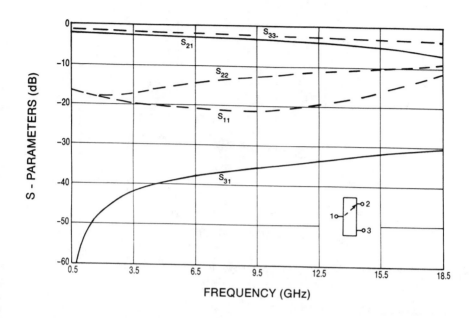

**Fig. 6.78** Computer simulated frequency response of a double series FET SPDT switch.

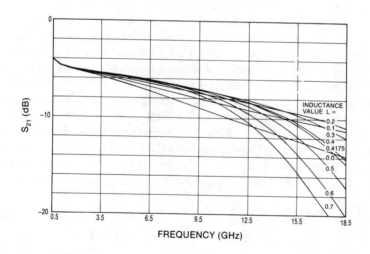

**Fig. 6.79** Insertion loss *versus* series *L* for a FET SPDT switch.

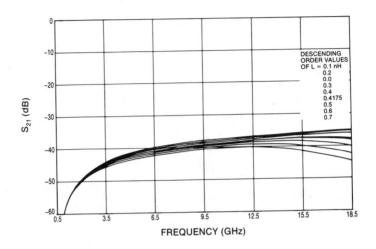

**Fig. 6.80** Isolation *versus* series *L* for a FET SPDT switch.

### 6.4.6 Measurement *versus* Theoretical

Table 6.7 tabulates the design goals, computer simulation results, and measured results. Figure 6.81 is the measured frequency response plot to 12.1 GHz. Again, as in the variable attenuator design, the measured performance of the switch agreed very well with the computer predicted performance. The insertion loss of the switch was actually less than that predicted across the band, as shown in Table 6.7; but the switch isolation was not quite as good as the model predicted at higher frequencies (15 dB measured *versus* 25 dB predicted at 10 GHz). This was due to a parasitic capacitive cross-coupling effect between the series FET pairs at the input of the switch caused by their close proximity on the chip—a layout error. If these input series FETs had been spaced more widely on the chip, the isolation probably would have exceeded the computer prediction as it did for the attenuator design. Resimulation of the switch circuit to include this parasitic capacitance has yielded results similar to those actually measured for the switch.

**Table 6.7 SPDT Switch Performance**

| Performance Parameter | MMIC Design Goals | Computer Simulation Results of MMIC Component Performance | Measured MMIC Component Performance |
|---|---|---|---|
| Bandwidth | dc–10 GHz | dc–18 GHz | dc–18 GHz |
| Insertion Loss | 1.6 dB at 10 GHz<br>2.6 dB max at 10 GHz | 4.5 dB at 10 GHz | dc: 2.0 dB<br>10 GHz: 3.5 dB<br>16 GHz: 2.5 dB |
| Flatness | ±0.6 dB max across band | ±0.5 dB typical over 1 GHz bandwidth | ±0.5 dB up to 10 GHz<br>±1.0 dB 10–18 GHz |
| 1 dB output compression point | +22 dBm min | +15 dBm (calculated) | +23.5 dBm |
| Third-order intercept point | +32 dBm min | +25 dBm (calculated) | +33.5 dBm |
| Isolation | 45 dB min at 10 GHz | 30 dB min at 10 GHz | >35 dB at 1 GHz<br>>16 dB up to 10 GHz<br>>10 dB 10–18 GHz |
| VSWR | Input-Output:<br>2.0 : 1 max at 10 GHz | Input:<br>1.38 : 1 (on) at 10 GHz<br>Output:<br>1.22 : 1 (on) at 10 GHz | Input:<br>1.67 : 1 typical<br>2.3 : 1 worst case<br>Output:<br>1.43 : 1 typical<br>1.9 : 1 worst case |

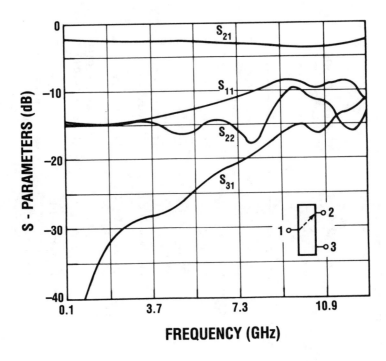

**Fig. 6.81** Measured frequency response of a FET SPDT switch.

# REFERENCES

## Section 6.1

1. W. Titus, Y. Tajima, R. A. Pucel, and A. Morris, "Distributed Monolithic Image Rejection Mixer," *GaAs IC Symp. Tech. Digest*, 1986, pp. 191–194.
2. W. Titus and M. Miller, "2–26 GHz MMIC Frequency Converter," *GaAs IC Symp. Tech. Digest*, 1988, pp. 181–184.

## Section 6.2

1. R. V. Garver, "Broad-Band Diode Phase Shifters," *IEEE Trans. Microwave Theory Tech.*, Vol. MTT-20, May 1972, pp. 314–323.
2. Y. Ayasli, "Microwave Switching with GaAs FETs," *Microwave J.* November 1982, pp. 61–74.
3. G. Matthaei et al., *Microwave Filters, Impedance—Matching Networks, and Coupling Structures*, McGraw-Hill, New York, 1964.
4. R. J. P. Douville and D. S. James, "Experimental Study of Symmetric Microstrip Bends and Their Compensation," *IEEE Trans. Microwave Theory Tech.*, Vol. MTT-26, March 1978, pp. 175–182.

## Section 6.3

1. R. B. Culbertson and A. M. Pavio, "An Analytic Design Approach for 2–18 GHz Planar Mixer Circuits," *IEEE MTT-S Int. Microwave Symp. Dig.*, 1982, pp. 425–427.
2. M. A. Smith, A. M. Pavio, and B. Kim, "A Ka-Band Dual Channel Tracking Receiver Converter," *IEEE MTT-S Int. Microwave Symp. Dig.*, June 1986, pp. 643–644.
3. M. A. Smith, K. J. Anderson, and A. M. Pavio, "Decade-Band Mixer Covers 3.5 to 35 GHz," *Microwave J.*, February 1986, pp. 163–171.
4. S. Egami, "Nonlinear, Linear Analysis and Computer-Aided Design of Resistive Mixers," *IEEE Trans. Microwave Theory Tech.*, March 1973.
5. G. Begemann and A. Hecht, "The Conversion Gain and Stability of MESFET Gate Mixers," *Proc. of the 9th European Microwave Conf.*, 1979.
6. P. Bura and R. Dikshit, "FET Mixers for Communication Satellite Transponders," *IEEE MTT-S Int. Microwave Symp. Dig.*, 1976.
7. R. A. Pucel, D. Masse, and R. Bera, "Performance of GaAs MESFET Mixers at X-Band," *IEEE Trans. Microwave Theory Tech.*, June 1976, pp. 351–360.
8. A. D. Evans, *Designing with Field-Effect Transistors*, McGraw-Hill, New York, 1981.
9. T. S. Howard and A. M. Pavio, "A Distributed Monolithic Dual-Gate FET Mixer," *IEEE Microwave and Millimeterwave Monolithic Circuits Symp. Dig.*, June 1987, pp. 27–30.
10. R. Tsironis and R. Meierer, "Microwave Wide-band Model of Dual-Gate MESFETs," *IEEE Trans. Microwave Theory Tech.*, Vol. MTT-30, 1982, p. 243.
11. R. Tsironis, R. Meierer, and R. Stahlmann, "Dual-Gate MESFET Mixers," *IEEE Trans. Microwave Theory Tech.*, Vol. MTT-32, 1984, p. 248.

12. M. A. Smith, T. S. Howard, K. J. Anderson, and A. M. Pavio, "RF Nonlinear Device Characterisation Yields Improved Modeling Accuracy," *IEEE MTT-S Int. Microwave Symp. Dig.*, 1986, pp. 381–384.

13. L. C. T. Liu, C. S. Liu, J. R. Kessler, and S. K. Wang, "A 30 GHz Monolithic Receiver," *IEEE Microwave and Millimeter-wave Monolithic Circuits Symposium Digest*, June 1986, pp. 41–44.

14. A. Chu, L. Chu, D. Sloat, M. Theobald, J. Teunas, T. Litchfield and W. Moroney, "Low Cost Millimeter Wave Monolithic Receivers," *IEEE Microwave and Millimeter-wave Monolithic Circuits Symposium Digest*, June 1987, pp. 63–67.

## Section 6.4

1. S. S. Bharj and R. Goyal, "MESFET Switch Design," *MSN and CT*, November 1987, pp. 78–85.

2. Y. Tajima et al., "Broadband GaAs FET 2 × 1 Switches," *IEEE GaAs Integrated Circuits Symp. Dig.*, 1984, pp. 81–84.

3. *Reference Data for Radio Engineers*, 5th ed., ITT/Howard W. Sams and Co., 1968, Chapter 10.

4. G. Barta et al., "Surface-Mounted GaAs Active Splitter and Attenuator MMIC's Used in a 1–10 GHz Leveling Loop," *IEEE Trans. Microwave Theory Tech.*, Vol. MTT-34, December 1986.

# Chapter 7
# MMIC-Based Subsystem Case Study: The TVRO Chip

## *A. Podell*

Most MMICs designed today for any system application generally replace existing hybrids to improve the size, weight, and reliability of the overall system. This restricts the MMIC designer to minor repartitioning of the system to insert practical, high yield MMICs and still maintain the overall performance of the system. Hybrids, which have been "tweaked" to get the last bit of performance from the design and technology, may not be good candidates for MMIC insertion. At the present time, few production systems are designed with MMIC capabilities and limitations in mind from the very inception of the system development.

## 7.1 TVRO

An MMIC centered system was designed for the *television receive-only* (TVRO) commercial marketplace. The designers had the complete freedom to partition the system for a custom MMIC, allowing high yield without loss in performance. This program began with setting up specifications acceptable to the TVRO market. Initially, it was planned to put everything on the chip: low-noise preamplifier, RF amplifier, converter, possibly the IF amplifier, and certainly the local oscillator and buffer.

In the first month of the design process, the noise figure requirement dropped by $\frac{1}{2}$ dB, and it was clear that development of an MMIC chip that included the low-noise amplifier could not keep pace with the rate of change in FET technology. The question then became whether the MMIC front-end chip should be preceded by one or two stages of external low noise amplifier in discrete form.

There was no similar question for the local oscillator, because the 1 $\mu$m, low-cost process that was planned would work quite well. Similarly, if incorporated, an IF amplifier would perform adequately with the standard 1 $\mu$m process, and its performance would improve only slightly with an improved process. Also, it was important to be careful that the IF amplifier did not have too much gain with improved process. Figure 7.1 shows a block diagram of the MMIC chip, and Figure 7.2, a top view of the chip in its package.

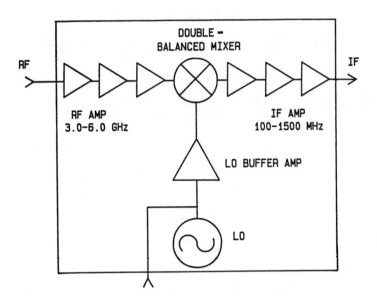

**Fig. 7.1** Block diagram of a MMIC chip.

At the beginning of the project, MIM capacitors were not fully characterized in Tri Quint's analog process, which was considered the most suitable process for circuit fabrication. Therefore, the designer decided to use their digital process and to use diodes for blocking, bypass, and coupling capacitors where necessary. With only a 3 V breakdown, two series diodes would be required for most blocking and bypass applications, as 5 V was the supply voltage. However, in the IF amplifier it would not be necessary to stack diodes if the interstage drop could be controlled to less than 3 V.

Without MIM capacitors, transformer coupling between stages seemed a good choice. However, too many stages of amplification with inductive loads causing out-of-band oscillation was a major concern. For this reason the first stage was RC-coupled to the second stage, which was transformer-coupled to the third. The third stage was transformer-coupled to the diode mixer, which essentially provided a resistive load to the third gain stage. This ensured that the three-stage RF amplifier, with approximately 20 dB of gain, would still be stable when fed from a one- or two-stage low-noise preamplifier.

**Fig. 7.2** MMIC chip in its package (top view).

The FETs were placed inside the transformers to maximize space [1]. The push-pull RF and LO amplifiers minimized the effect of grounding inductance, allowing use of 20 mil thick wafers and inexpensive digital packages. Utilizing unthinned GaAs was advantageous when testing the wafers, handling and die attaching the chips, and in wire bonding. All of the FETs were biased around 25 mA per mm of gate periphery, so that on-chip temperature rise was less than 20°C at the hottest point in this circuit, despite the 20 mil thickness.

The block diagram for the MMIC front end is shown in Figure 7.1. In the layout of the front end, as shown in Figure 7.3, on the left side are (in order from bottom to top) the RF amplifier first stage, second stage, and third stage. As mentioned earlier, the first stage is resistance-capacitance–coupled to the second stage, and the three light colored fingers almost at the bottom of the chip are the input protection diodes to protect against reverse biasing the RF input stage due to static discharge. Because the packaged parts were assembled under uncontrolled conditions, static protection was used at all sensitive inputs on the chip. Because the gate of the RF amplifier stage was biased to approximately −1.0 V it was necessary to use three diodes in series from gate to ground to prevent the negative bias from turning on the protection diodes. The local oscillator static protection diodes are buried underneath the transformers and difficult to point out in a photograph. Such a static protection circuit on this chip has been very useful, in that no identifiable static discharge failure was observed.

The second stage of RF is tuned by a drain-to-drain varactor pair whose bias is provided by the third pad up on the lefthand side. This varactor bias tunes the gain peak center frequency at approximately 1 GHz and helps to compensate for

# 3 - 6 GHz RECEIVER FRONT END

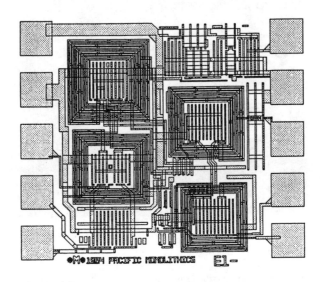

**Fig. 7.3** Layout of MMIC front end.

process-induced variations. The third gain stage is transformer-coupled to a diode mixer, which heavily loads that amplifier, resulting in a very broad bandwidth requiring no varactor tuning. The local oscillator bonding pads are on the lower righthand side of the chip with IF bypass and bias control, the middle pad. The next pad up is ground, and the top pad on the righthand side is the IF output. There are two ground pads on this chip, in the fourth position up from the bottom of the chip on either side.

Comparing single-ended and push-pull amplifier performance including their by-pass capacitor requirements showed that, by far, the smallest chip would result from a push-pull design throughout the LO and RF section and a single-ended IF amplifier. However, the bypass capacitors' area and additional grounding-pad requirements for the single-ended design probably would increase the chip size by a factor of three relative to an optimum push-pull design (Figures 7.3 and 7.4). Even with via hole technology and 4 mil thin GaAs chips, the size of the chip would be 1.5 to 2 times the size of the optimum push-pull design. Moreover, the yield of a thin wafer with via holes would be much lower than that of a 20 mil thick push-pull chip. Last, the processing cost of the much more complex, thinned, and via-holed wafer would add 50–100% to the total cost. It was concluded that the push-pull design would be a factor of three, or even four, times less expensive than a single-ended design under the best conditions.

## 3 - 6 GHz RECEIVER FRONT END

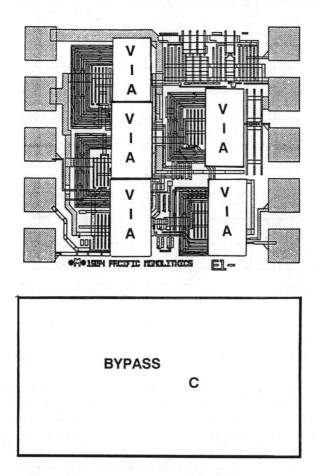

**Fig. 7.4** An optimum push-pull design for the LO and RF section.

Figure 7.5 contrasts the effect of ground lead inductance on the RF amplifier response for push-pull and single-ended designs. We can see that a single-ended RF amplifier would have return gain and potential instability for ground-lead inductances greater than 0.1 nH. Package choices, chip thickness and mounting, and surface mounting procedures all would be limited with a single-ended design.

Clearly, every effort was made to keep the chip size as small as possible, with coupling between the inductors either minimized or utilized. Experimentally, it was found that coupling between adjacent spiral inductors is quite weak, as shown in Figure 7.6. As an experiment, insulated wire was glued to a wooden

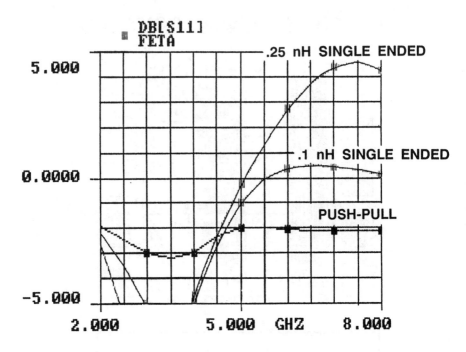

**Fig. 7.5** The effect of ground lead inductance on RF amplifier response.

block and scaled measurements were made at 4 MHz to obtain data for multiturn bifilar spirals scaled to 4 GHz.

The local oscillator was connected directly to the buffer amplifier that drives the diode mixer through a transformer. The simulations indicated that the oscillator could directly drive the mixer, but caution prevailed and a buffer stage was added between the oscillator and the mixer diodes. Computer simulations showed that the oscillator could be bias-tuned from about 4.2 to 5.5 GHz, which indicated that the oscillator would lock up on the external dielectric resonator at 5.15 GHz. Also, the current consumption could be reduced while maintaining an adequate LO drive by reducing the size of the buffer FETs from 800 to 600 $\mu$m gate periphery. Simply setting the gate bias to the appropriate voltage tuned the oscillator to the correct frequency for dielectric stabilization for a wide variety of wafers.

In the RF amplifier, on the other hand, process tolerance for gate-source capacitance variations was accomplished by a low-pass amplifier design for the first stage and a bandpass amplifier for the second stage. Suppose, for example, devices on a wafer had large gate-source capacitance, resulting in reduced gain at 4 GHz. However, in the second stage, which is tuned slightly above the frequency range of interest, additional gate-source capacitance tuned the amplifier to slightly lower frequency, raising the gain in the band of interest from 3.7 to 4.2 GHz.

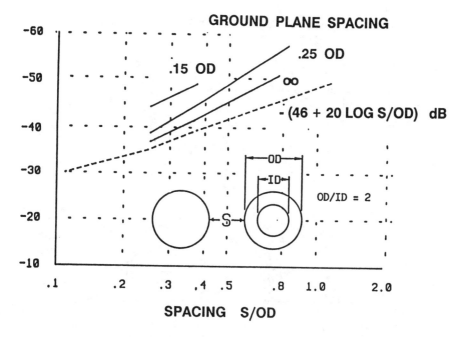

**Fig. 7.6** Coupling between adjacent spiral inductors.

Thus, the reduced gain, for 3.7 to 4.2 GHz, caused by the performance of the first amplifier stage was compensated by an increased gain in the second stage, with the net result that the overall gain of the RF amplifier was essentially independent of the gate-source capacitance of the FETs over a range of ±20%, as shown pictorially in Figure 7.7.

A similar technique, used to optimize the performance of the RF amplifier and the IF amplifier, is shown in Figure 7.8. The RF amplifier was tuned for minimum noise figure over the 3.7 to 4.2 GHz range. Consequently, the gain rolled off 5 dB across this range. Rather than trying to compensate for this effect in the RF amplifier by mistuning, the designer took advantage of the fact that the converter was an inverting down-converter; that is, increasing the input RF and mixing with the local oscillator at 5.15 GHz resulted in a decreasing IF.

A similar problem occurred in the IF amplifier: the amplifier gain rolled off with increasing frequency unless positive feedback or lumped inductors were used to flatten the response. Positive feedback would increase the process sensitivity of the design and passive inductors would be large, and hence expensive to realize on GaAs. It seemed inevitable that the IF amplifier response would roll off by 3 dB or so from 0.9 to 1.5 GHz. Because of the nature of the inverting down-converter, the gain change in the IF amplifier and that in the RF amplifiers over the frequency

## EXAMPLE OF Cgs TOLERANT DESIGN

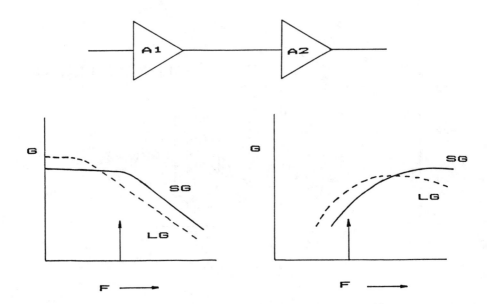

**Fig. 7.7** Gain of the RF amplifier is essentially independent of the gate-source capacitance.

## EXAMPLE OF A SNEAKY DESIGN

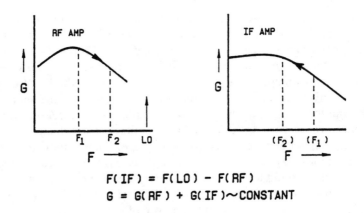

$$F(IF) = F(LO) - F(RF)$$
$$G = G(RF) + G(IF) \sim CONSTANT$$

**Fig. 7.8** RF amplifier tuned for minimum noise figure.

range essentially compensated for one another, and as shown in Figure 7.9, the net frequency response from 3.7 to 4.2 GHz was essentially flat, even though amplifiers with gain slopes across the band were used. This is an excellent example of higher level subsystem integration resulting in relaxed specifications for its component parts.

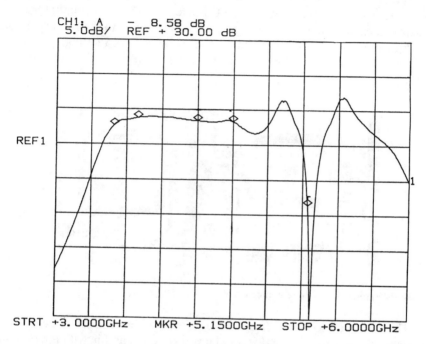

**Fig. 7.9** Compensation of RF and IF amplifiers' change in gain.

As Tri Quint's process matured and circuit design data base was accumulated, wafer yields improved until typical wafers yielded 60% of RF-good parts packaged from a total of 3100 potential devices per wafer (actual yields ranged from 93 to 15%). Only three of more than a hundred wafers processed exhibited poor yield that could not be traced to some process or device characteristic that was far out of the process window. The wafers that measured low RF yield generally were discarded in case something pathologically wrong put them far out of our distribution window.

The most serious problem was the instability of the single-ended IF amplifier, which caused a ripple in the frequency response. This eventually was eliminated by a production change: another power connection was added to the IF amplifier to separate the feedback through the supply lead that was poorly by-passed.

It was necessary to convert the 3.7–4.2 GHz band to 0.95–1.45 GHz IF to mate with existing receiver capabilities. The LO frequency chosen was 5.15 GHz, rather than 2.75 GHz, to minimize the dielectric puck size and allow for frequency inversion, which improved the flatness. The 0.44-in. diameter puck, mounted on the "mother" board with a large (0.180 in.) spacer, stabilized the frequency to better than ±500 kHz over −30 to +60°C range. Symmetrical coupling with 50 Ω microstrip lines on either side of the puck matched well with the push-pull oscillator port of the chip. Figure 7.10 shows the packaged chip sitting on top of the printed circuit board with the puck in its proper location and the top cover removed.

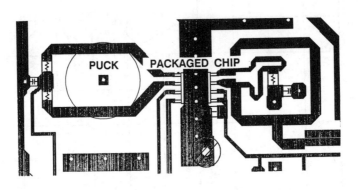

**Fig. 7.10** Packaged chip on top of printed circuit board.

The push-pull local oscillator caused few problems. Results achieved with this push-pull dielectric resonator oscillator were comparable with the best single-ended FET oscillators at this frequency. At the RF input, however, the push-pull design required an external balun for maximum gain and lowest noise figure. This balun was realized on the circuit board, as shown in the right side of Figure 7.10.

The push-pull circuitry offered other advantages, such as easy neutralization of capacitances, convenient coupling to balanced mixers, and broadband positive feedback for the oscillator [1,2]. Other components were required external to the MMIC chip that were impractical to put on GaAs. A 100 pF bypass capacitor, essential for stabilizing the IF amplifier, was located outside of the package, as close as practical to the converter. The lead to this external capacitor caused problems in IF amplifier's stability because the IF gain extended beyond 6 GHz.

There was no indication of instability in the RF amplifier or in the local oscillator, although the bypassing for the power supply was located far from those stages. This was confirmed by evaluating the performance of the individual cells that were constituents of the receiver front end. It was necessary to have these break-out cells available for testing, because it was difficult to have complete confidence that the fully integrated chip would function properly in all respects.

Dual-gate FET mixers described in the literature typically show 3 dB gain and 10–11 dB noise figures with 2–3 V peak-to-peak LO drive voltage [3,4]. The double balanced diode mixer required 2 V peak-to-peak LO drive and had 7 dB conversion loss. The IF amplifier stage following it had 3 dB noise figure, 10 dB gain, and a current drain of only 12 mA. The diode mixer-IF amplifier combination requiring 2 V push-pull LO drive had 3 dB gain and a 10 dB noise figure. A diode mixer followed by an FET IF amplifier was chosen over a dual-gate FET mixer because the combination is more tolerant of LO drive and bias and has essentially the same gain and noise figure. The diode mixer also provided a wideband resistive load which helped to stabilize the high gain RF amplifier. Figure 7.11 shows the input return loss of a transformer-coupled RF amplifier stage with diode (resistive) and dual-gate FET (capacitive) loading. Note the 2 dB return gain of the RF amplifier with FET load. A single-ended IF amplifier was used because the frequencies were lower and the interface to the external circuitry was simpler.

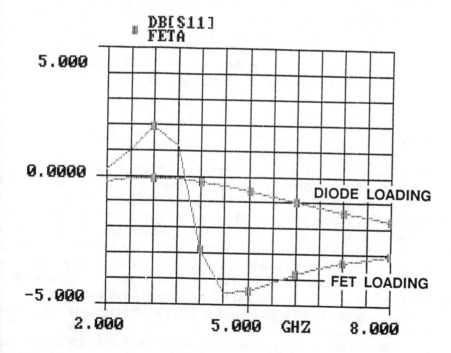

**Fig. 7.11** Input return loss of a transformer-coupled RF amplifier stage with resistive and capacitive loading.

Other required components external to the chip were the bias resistor and the decoupling capacitor at the IF output. By having these components external, the current consumption and the 1 dB compression point of the IF amplifier could be varied in accordance with the specifications. This last stage was a source follower that made it convenient to have an external bias resistor. Again, had the blocking capacitor been on-chip, its capacitance would have had to be quite large (at least 20 pF), which would have been expensive to integrate monolithically, whereas an external 100 pF blocking capacitor cost 3 cents.

Biasing of the RF amplifiers and the IF amplifier initially was handled by an external fixed-bias circuit that was set by selected resistors. Because all of the amplifying devices RC-coupled on the chip used the same ratio of device gate width to load resistance, all of these amplifiers required the same gate bias voltage. This meant that a single bias set point could be used to supply the negative bias for all the amplifier stages on the chip. The oscillator, on the other hand, had a separate bias, which was set for optimum stability and was generally a few tenths of a volt different from the set point of the RF and IF amplifier bias. Later on, as we found that the chips were very consistent and the results were predictable from wafer to wafer, it was possible to use an external feedback circuit that sensed the output voltage from the source follower at the IF port and used this voltage to fix the bias points for the RC-coupled amplifiers on the chip. This eliminated all bias adjustment requirements except for the local oscillator bias. Figure 7.12 shows the final bias circuit.

## 7.2 TESTING

The dc testability was also considered in the design of the bias circuits and the connection of the amplifiers. For example, in order to determine whether there are any broken gates on the chip, it should be possible to pinch off all of the FETs in the circuit. This consideration resulted in some modifications to the RF amplifier because of the importance of dc testability. Cascade Microtech probes do a very adequate job of RF testing. However, the work hours required for RF testing, the cost of the equipment, and the fragility of the probes called for an intensive dc testability program. Throughout the program the desire for dc testability, requiring correlation of the dc test and RF parameters, was always in the forefront, as dc testing with conventional low-frequency needle probes is a well established, inexpensive process.

RF wafer probing at 1% level was used to establish that the wafers were RF "good." Then 100% dc wafer probing identified good chips for packaging. After sawing, approximately 90% of these chips were RF good, results that held when they were tested after packaging. The production test fixtures were manually loaded; the RF, IF, and LO biases manually set; and the 3.7–4.2 GHz gain data

## POTENTIAL BIAS CIRCUITS

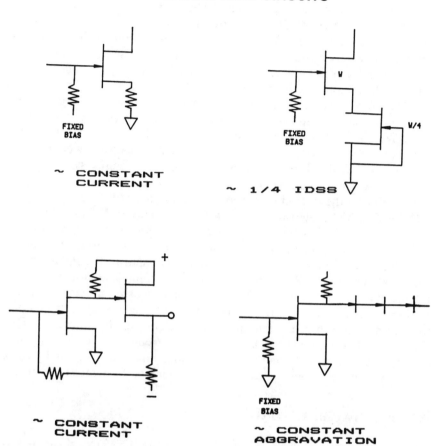

Fig. 7.12 External fixed-bias circuit.

were captured by a personal computer. Over 10,000 packaged parts could be tested per month by one test station. The production test is a fully operational RF test, with fixed tuned RF matching network and 5.15 GHz dielectric stabilization. Figure 7.9 shows the measured gain of the 3–6 GHz MMIC receiver front end with a fixed 5.15 GHz LO.

Small hermetic packages, originally intended for digital applications, were used to package the devices. These have worked satisfactorily up to 9 GHz for our 35 dB gain receiver chips. Measurements on the package alone indicate potential problems above 10 GHz, such as seal ring resonance. Taking advantage of monolithic design, the impedance of the circuitry was scaled on-chip to minimize the effects of the inexpensive packaging, where $L_p = 1.0$ nH, and $C_p = 0.1$ pF.

The Mini Systems package chosen had lead impedances feeding through the package wall higher than 50 $\Omega$. This allowed us to use smaller FETs than those optimum for the traditional 50 $\Omega$ at the RF input without having the package parasitics significantly affect the RF amplifier gain and noise figure. Since external matching was used at this stage, package capacitance directly paralleled the FET capacitance, leading to a higher $Q$, narrower bandwidth, and poorer RF performance. Therefore, this high-impedance package was an ideal choice, particularly when the push-pull RF and LO minimized the effect of poor grounding in the package. We found that ceramic cofired packages with a sufficient number of leads, say eight or ten, had a higher feedthrough capacitance and were much more expensive than the Mini Systems package. It took a few iterations of FET width in the amplifiers to arrive at an optimum matching of the FET size with the package parasitics to make the overall chip easy to use and insert on a circuit board.

The use of distributed transmission line elements rather than lumped elements, in the circuit was never seriously considered for this application. The tradeoff between performance and area utilization appeared to be a strong argument for lumped circuits everywhere but the very input to the RF amplifier. In fact, the matching to the RF amplifier input is accomplished largely outside the monolithic chip, where the losses are very low and the size of matching elements is of no importance. Figure 7.13 (bottom) portrays the tradeoff between performance and cost in lumped *versus* distributed elements. In order to put in an absolute scale, there are many variables, but the trends are very clear: quarter-wavelength lines form extremely expensive matching networks and overcompacted lumped circuits have too low $Q$ and too poor a performance to be practical.

## 7.3 SAMPLE WAFER TESTING

With 3100 chips per wafer, sample RF testing at the 1% level supplies valuable statistical data at low cost. Cascade Microtech probe cards are mounted in a Pacific Western Model SP1B automatic wafer prober. Chips from all areas of the wafer are sampled. Conversion gain is measured with a standard HP 8757A scalar network analyzer, corrected for the effects of the 3 ft connecting cables and 3 dB pads. Although the ceramic thin-film Cascade Microtech probes are extremely fragile, their performance was found very satisfactory. It is important to maintain parallelism between the probe tips and the wafers, so that no particles are between the prober chuck and the wafer under test and the wafer itself must be very flat.

To keep the wafer level test equivalent to actual application, the local oscillator port is coupled to a dielectric resonator in a small cavity. Moreover, the RF and IF ports are connected to circuitry similar to that which would be used in actual application, with several inches of semirigid coaxial cable and the probe

# LUMPED AND DISTRIBUTED CIRCUITS

- FUNDAMENTALLY, LUMPED CIRCUITS ARE WIDER BANDWIDTH.

- A TRANSMISSION LINE IS AN INDUCTOR WITH PARASITIC C,

## BUT

- THE "Q" OF AN INDUCTOR INCREASES WITH ITS VOLUME.

    \* WIDER LINE INCREASES CURRENT CARRYING AREA, BUT NEED LONGER LINE TO MAINTAIN INDUCTANCE VALUE.

    \* PARASITIC C INCREASES ALSO.

    \* AND ITS COST.

    \*\*   MONOLITHIC DESIGN IS A COMPROMISE  \*\*

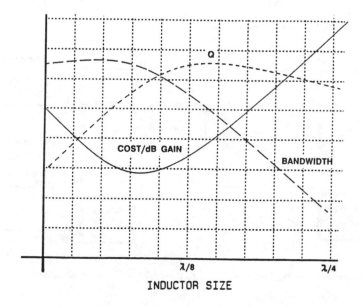

Fig. 7.13 Comparison of performance and cost for lumped and distributed elements.

card traces separating the chip under test from the test circuitry (see Figure 7.14a). Monolithic receivers with gain as high as 45 dB have been tested from 3 to 10 GHz in this RF probe test set. The test data is in close agreement with the packaged IC measurements. The gain is recorded with a high-speed plotter; current and bias voltages are added manually. RF wafer probing at 1% level is an invaluable aid in identifying the occasional wafer that passes dc yet fails RF tests.

An external local oscillator was used to test both probe and receiver repeatability, and a second mixer added, so that amplitude and phase data could be measured with an HP 8510 system (see Figure 7.14b). With chips selected from all areas of the wafer, over 90% of the receivers fell within a 1 dB gain and 3° phase window from 3.5 to 6 GHz (see Figure 7.15).

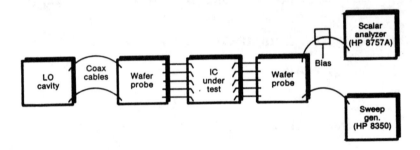

**Fig. 7.14a** Chip under test and the test circuitry.

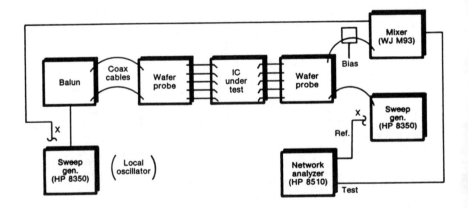

**Fig. 7.14b** Measurement of amplitude and phase to test probe and receiver repeatability.

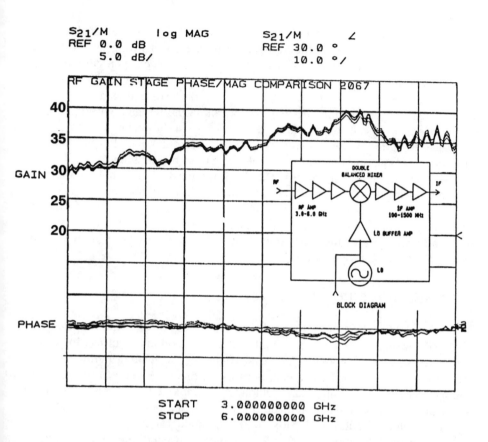

**Fig. 7.15** Receiver chip gain and phase results.

## 7.4 dc WAFER PROBING

RF good wafers are 100% dc probed, and the bad devices inked. The wafers are mounted on an expandable plastic backing then sawn. Subsequently, the unmarked (good) devices are die attached and wire bonded in their flatpack packages. Establishing a dc test program with a high correlation to RF performance required close attention to device and circuit modeling. Equally important and often overlooked is the need to design for dc testability. "Built-in" tests were designed to strongly indicate the gain of the RF and IF amplifiers. Moreover, simple pinch-off and *I versus V* tests for the entire chip indicated broken gates, leaky devices, and weak short circuits. RF, LO, and mixer stages are transformer-coupled, so a short circuit in the transformer appears as a strong short circuit from $V_{dd}$ to ground. On the other hand, shorted inductor turns or open-coupling capacitors are virtually impossible to detect in dc tests. Fortunately, such problems are unlikely and occur globally if at all, so that test patterns can establish their presence. The dc testing is not sufficient to determine the RF gain within 5 dB or the performance of the local oscillator. However, correlation of wafer dc good to packaged RF good is now running about 90%. For much simpler circuits such as one- or two-stage amplifiers, the correlation could be very high.

For some high-gain designs, the dc probe cards could cause the circuit to oscillate during testing. Ferrite beads reduce this problem, but frequently very large on-chip bypass capacitors are required to stabilize a 30 dB gain amplifier. These large capacitors raise the cost of manufacturing the chip. Low-cost, higher-frequency response probe card equivalent in parasitics to 20 or 30 mil long wire bonds would be highly desirable.

## 7.5 DESIGNING FOR TESTABILITY

As a rule, the more pads the circuit has, the more information that can be obtained through dc testing. Tying all the circuit grounds together, for example, eliminates some stage-by-stage dc tests. High-gain, wideband, multistage amplifiers present a probe card challenge, because of their tendency to oscillate on the common power or ground lead inductances. Extensive use of push-pull circuitry in the present case study minimized these problems, both in testing and packaging. The value of additional pads solely intended for more comprehensive dc testing can be determined by simple economics: wasted chip area *versus* fewer assembly rejects. Even self-testing circuitry will become practical as the complexity of monolithic subassemblies increases to the system level.

## 7.6 PACKAGED UNIT TESTING

Physical realization of a test fixture that would provide for accurate data acquisition at high production rates proved challenging. One approach to fixture design addressed the issues of electrical performance, mechanical integrity, and production efficiency. The resulting fixture is shown in Figure 7.16. To achieve full functional testing of the packaged MMIC required a hybrid softboard support circuit with *dielectric resonating oscillator* (DRO) tuning capability. In addition, the printed circuit board had to be able to withstand tens of thousands of test cycles without reflecting mechanical wear in RF measurement inaccuracies. The circuit board material selected was 0.31 in. thick polyimide glass (Howe Industries, HI-3003). A metalization scheme of 1 oz. copper plated up with 50 $\mu$in. nickel and 30 $\mu$in. gold was chosen for its ruggedness and conductivity. The wear resistance of these circuits has been outstanding: none has needed replacement or replating although more than 25,000 units have been product tested. We should also note that the large trace geometry needed on the polyimide glass make designing support circuitry at these frequencies very comfortable.

Repeatable device location on the circuit was achieved by building a pocket on the surface of the board (no cutouts in the PCB were made to maintain its structural integrity under cyclic loading). The three-sided pocket is formed of a metal backstop flanked by nylon line siding. The surface-mounted LNB device slides directly into the pocket and against the backstop with its "gull-winged" leads fitting tightly over the nylon line for alignment. In this way, very quick tweezer-aided insertion and extraction of the device is possible with highly repeatable location (see Figure 7.16).

A clamp machined of low dielectric constant *polyphenylene oxide* (PPO) applied contact pressure at the tips of the device leads for reliable operation of each device under test. The package itself also was pressed down to ensure a reliable RF ground. The clamp is actuated by a toggle clamp and ball bearing slide mechanism having positional repeatability of better than 1 mil. Clamping force is fully adjustable with a single screw and can be locked readily in when calibrated. The body of the fixture is a simple construction of two machined aluminum pieces. The assembly is mounted on a plastic base and fully connected for ease of set up and mobility.

Some automation is added to the final test, both in timing and data acquisition. Under operator control, the test frequency range is 3.7–5.2 GHz for the initial dc bias adjustment, and 3.7–4.2 GHz for final adjustment. The transfer gain is measured in 100 MHz increments and stored in an IBM PC. Automatic computerized bias adjustment was attempted but not incorporated into this set up. Due to the consistency of performance over a wafer, this was unnecessary and human testers often detected irregular chip performance that would not have been detected by our program. Simple Pascal programs that run on inexpensive PCs appear quite capable of performing tests at very reasonable cost. On the other hand, it was with great relief that we abandoned the automatic computerized bias adjustment.

The style of fixturing designed for the LNB production line provides the basis for developing an entire family of fixtures for future subsystems in small packages. Slight modifications to the support circuitry, the pocket, and the PPO clamp along with the adjustability already achieved allow ready conversion of existing fixture parts into functioning testers for new products.

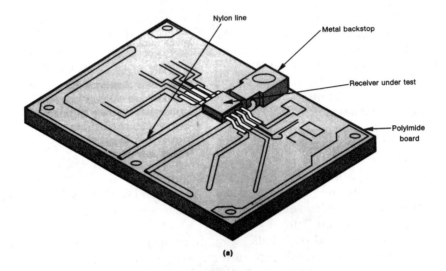

(a)

(b)

Fig. 7.16 Test fixture design.

## 7.7 CONCLUSION

Testing thousands of monolithic receiver front ends per week was not as formidable as expected. Designing the chip for testability then distributing the test burden between RF and dc level wafer and package testing provided a cost-effective high-volume MMIC test procedure. A simple, manually loaded test fixture with PC based automatic data acquisition has been used successfully to test 7000 parts per week.

The TVRO application required high-gain and low-noise figure performance over the RF input band of 3.7–4.2 GHz. However, dynamic range and output power capability were not particularly important and power consumption of approximately 0.5 W was acceptable. By aiming for modest performance, process tolerance, and dense high integration, the designers managed to produce low-cost monolithic receiver parts. Without friends, dedicated engineers, and cooperative customers, this enormous task could not have been completed in time to be nullified by signal scrambling in late 1985.

## REFERENCES

1. S. A. Jamison et al., "Inductively Coupled Push-Pull Amplifiers for Low Cost Monolithic Microwave ICs," *GaAs IC Symp. Tech. Dig.*, November 1982.
2. J. Culp et al., "A 1.0 × 1.6 mm 4 GHz GaAs Receiver," *IEEE Int. Solid-State Circuits Conf. Dig.*, February 1983.
3. K. Honjo, Y. Hosono, and T. Sugiura, "X-Band Low-Noise GaAs Monolithic Frequency Converter," *GaAs IC Symp. Tech. Dig.*, October 1984.
4. C. Kermarrec, P. Harrop, C. Tsironis, and J. Faguet, "Monolithic Circuits for 12 GHz Direct Broadcasting Reception," *IEEE Microwave and Millimeter-Wave Monolithic Circuits Symp. Dig.*, 1982.

# Chapter 8
# Design Automation Tools for MMIC Design

## R. Goyal

## 8.1 INTRODUCTION

To a microwave system designer, MMICs offer the same benefits over the *monolithic integrated circuits* (MICs) or hybrids that silicon ICs have given designers of lower-frequency digital systems: small size, low weight, high reliability, lower power consumption, improved electrical performance, and optimum design. In addition, large numbers of circuits can be produced within a given specification window, by batch processing several wafers and thousands of circuits on each wafer, which reduces the manufacturing cost. However, these benefits will not be fully realized until all aspects of the technology, including design techniques, have reached a suitable maturity.

Design automation tools play an important role in reducing the design cycle time [1,2] from concept to final product introduction of an integrated circuit. The roots of the advanced CAD systems currently used for IC design can be traced back to the 1960s, when CAD placement and routing tools were developed for printed circuit board layouts. Today, a variety of tools exist to aid the low-frequency silicon design engineers. These tools help them create successful IC designs in a minimal amount of time. However, in the microwave IC design area, the development and use of monolithic microwave and millimeter wave ICs has moved so fast in the recent years that design automation tool development has not kept pace with it. In fact, there is a need for a conceptual overhaul of design automation philosophy and system to adapt it to MMIC development.

Today, because of the increased complexity of circuits, shorter product lives, and longer design cycles, computer-based design automation tools have become an absolute necessity for silicon-based VLSI circuit designs. The advent of low-cost workstations has put the equivalent of a 32 bit mainframe computer within reach of almost every IC design engineer. This surge in hardware price-performance has greatly expanded the potential for *electronic design automation* (EDA). As GaAs and high-speed silicon processes are becoming available at more economical prices and requirements for more complex, high-speed, reliable, small-size microwave circuits is increasing, there is an increased demand for MMIC design automation tools that can provide shorter, accurate, more efficient design cycles. This need is making EDA a critical part of the microwave circuit design development, from concept to prototyping and manufacturing. However, the effectiveness and efficiency of the EDA tools depends heavily not only on the quality of the tools but also on their completeness and integration through all stages of product development. It is important to view the IC product development cycle as an unbroken chain of different but highly interdependent tasks. This requires a set of software tools for microwave IC design that are tightly integrated through a common interface and data base, so that information can flow smoothly and quickly throughout the various stages of the design cycle.

Historically, even the first generation of design automation systems had a significant impact on the productivity of those who design low-speed, conventional digital ICs. In this design environment, we find heterogeneous platforms, a mix of unrelated design tools, and many different unrelated design methodologies—which mirrors the status of MMIC design tools today. A variety of MMIC design software tools are available from various vendors, which have their own strengths and weaknesses and are quite unrelated from user interface point of view.

In today's realm of LSI and VLSI silicon digital IC design, an MMIC would appear to present a fairly simple design effort. Superficially, today's MMICs appear to be relatively simple circuits, where the component count in any typical MMIC circuit rarely exceeds a few hundred. However, in reality, the design task even at this level of integration is far more involved and complex than it appears to be [3]. The main reasons for this are as follows:

- The analog nature of MMICs requires the involvement of both active and passive components.
- The frequency of operation of the circuits is in the range of 0.1–40 GHz, which results in prominent parasitic effects of the device layout and the interconnects.
- Component proximity and the interconnect coupling effects are prominent, too.
- Material and process variations in both active and passive devices are prominent as GaAs material and process technology are less mature than the silicon technology.
- Circuit analysis and the measurement techniques are complex at microwave frequencies.

- MMICs are designed to get the maximum benefit from the technology, which requires accurate modeling and better optimization techniques, along with the integrated design tools environment.

## 8.1.1 Design Perspective: MMICs *versus* Hybrid

At first glance, the only difference between hybrid and MMIC design appears to be the difference in the substrate material; for example, from alumina or soft substrate to GaAs. However, some significant differences in the design philosophy demand consideration. The most practical consideration is the cost of design iterations. Hybrid designs can easily be iterated on-board many times until they achieve the desired performance specifications, which is prohibitively expensive and time consuming with MMIC designs. As a result, MMIC designs must meet performance specifications within a minimum number of iterations. With MMIC technology maturing rapidly and becoming available through various foundry services, there has been a surge in designing of MMICs. However, most of these MMIC designs, layouts, and topologies are based on traditional hybrid design techniques. These MMIC designs are copies or extensions of hybrid-based concepts, which has resulted in a large number of research papers describing laboratory MMIC performance achievements. Unfortunately, very few of these have escaped the labs and reached the marketplace. The main reason for such a small success rate for MMICs is the lack of tuning capability that hybrid circuit designers have grown so accustomed to. Further, the lack of suitable MMIC design tools lowers the certainty of success in the design cycle. At the present time, the initial cost of developing an MMIC is considerably higher than that of an equivalent hybrid circuit. To reduce the cost per chip produced will require reducing the developmental cost of MMICs. This is particularly important for microwave ICs, whose major application at this time is in a wide variety of custom products manufactured in small quantities.

In hybrid design, the emphasis is placed on converting the paper design to the board level, ready for testing as early as possible. This is because it is always possible to fix the performance by tuning the component values while making measurements. Quite contrary to this, in MMIC design, the most detailed paper design possible is necessary to ensure the success of the fabricated circuit. This includes detailed modeling of passive and active components, layout parasitics, and process tolerances. Hence, there is a basic difference in the design philosophy between MICs and MMICs. MMIC design places tremendous demands on the design automation tools' accuracy, availability, and integrated working environment. Designing a reproducible process-tolerant MMIC is significantly more difficult than designing a prototype MMIC.

### 8.1.2 Digital-Analog *versus* MMIC Design

Design automation tools have become an essential part of the design cycle in the area of low-frequency analog and digital circuit designs. The question arises, why can these design automation tools not be used directly or adapted to microwave circuit designs, since the design cycle for both types of circuits seems to be the same—superficially at least? In spite of their similarities, designing a manufacturable circuit at microwave frequencies, which looks simple from the point of view of component count, is a much bigger challenge than it appears to be. This is because high frequency design techniques are quite different from conventional low-frequency analog-digital design techniques. Not only are active and lumped passive devices involved in microwave circuit design but distributed elements as well. Moreover, at microwave frequencies, layout parasitics are critical to the extent of limiting the circuit performance if not treated carefully. An advanced digital IC design, for example, may incorporate tens of thousands of gates or flip-flops that must be laid out and connected. A bigger challenge then for a CAD environment in digital design is to help create an incredibly complex road map of circuit traces and ensure its integrity as well as the integrity of the data base. On the other hand, an advanced microwave circuit may contain only 100 components, but each must be characterized and modeled thoroughly and with detailed accuracy. Also, manufacturability is a bigger issue in GaAs; much more so than in conventional CMOS silicon because of material properties and variation in electrical properties. Typically, a digital design station simulates just one instrument (a logic analyzer) or, in an analog low-frequency circuit design, function generators, oscilloscope, frequency sweeper, *et cetera*. However, in microwave design we deal with voltage and current along with incident and reflected powers and hence need to simulate a network analyzer, power meter, spectrum analyzer, and power compression meter along with such conventional instruments as oscilloscope, function generator, and frequency sweeper. Moreover, because of the different types of microwave circuits, we need three different types of circuit simulation programs: linear, nonlinear, and mixed mode.

The iterative nature of the design procedure for microwave integrated circuits, as described later, makes it essential to have a design environment with all the design tools available on the same platform; that is, an integrated workstation and all the design tools working under the same data base, to make all the design information available to every design tool whenever necessary.

### 8.1.3 Typical MMIC Design Cycle

As more engineers become involved in MMIC design, with real system applications in mind, they will expect to design circuits that work the first time. In fact, most of the tools used today in designing an MMIC were originally used for hybrid circuit design or for silicon IC design. These tools have been quite inadequate to address the needs of MMIC design.

A typical procedure followed by an MMIC designer is shown in Figure 8.1. It depicts the interrelations of desired specifications, circuit topology, circuit simulation, sensitivity analysis, layout, and other considerations [2]. After defining the specifications for a monolithic circuit, we choose a preliminary topology for the circuit. The topology should be considered in terms of its suitability to monolithic implementation in view of the process availability. Unless a cell library of different circuits available is at the disposal of the circuit designer, the choice of topology depends mainly on the designer's past experience. Predictions of performance, optimization, sensitivity analysis, and yield are an important portion of the MMIC design cycle. The chosen topology, performance window, and the element values are iterated using available microwave circuit simulation and optimization tools, until the designer reaches an acceptable yield. After the design optimization, we must physically lay out the generation of mask tooling for the MMIC fabrication. The layout is checked for correctness relative to the defined design and electrical rules and the connectivity is checked for conformance with the process rules and the schematic of the circuit. Next, parasitic effects, induced by the layout, are extracted and the proximity and coupling effects are all taken into account and put back in the circuit description file for circuit analysis. The circuit file representing the layout-dependent components is now simulated for electrical performance and compared with the target specifications. If the results are satisfactory, the design is sent to manufacturing; otherwise, the whole exercise is repeated, which may include considering an entirely different topology of the circuit. Most often the adjustment of circuit components values is sufficient to achieve the desired performance. The iterative nature of the design flow described here is a considerable task for designers. Software and computer hardware available today do not easily couple the functions of simulation, optimization, yield analysis, layout, back-end design verification, and parasitic extraction. This makes the task of an MMIC designer difficult and tedious, resulting in many expensive iterations of fabrication before a design produces an MMIC worthy of manufacturing.

## 8.1.4 Design Approaches

In the broad sense, circuit design approaches [4,5] for MMICs can be classified into custom design, standard cell-semicustom design, and gate array. Due to the analog nature of MMICs and the need for high-level performance, gate array does not seem to be an appropriate approach in the near future for designing MMICs. The gate array design approach is more applicable to digital circuit

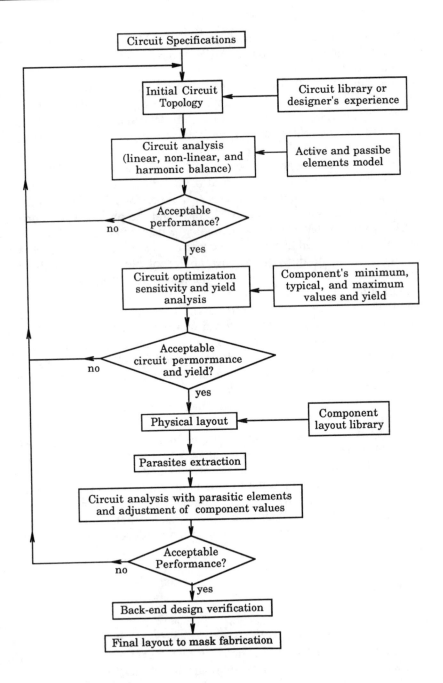

**Fig. 8.1** Typical MMIC design flow chart.

design. Hence, it is appropriate to elaborate on the other two approaches.

Currently, the full custom approach is the method most commonly used in designing MMICs. Of course, custom design provides the optimum design, circuit yield, and the performance for an application; however, this approach is expensive and the design cycle is longer. In other words, the best circuit performance is at the expense of increased developmental cost. Lack of user friendly, efficient design tools make this approach even more expensive and time consuming, particularly for MMIC designs.

An integrated software package for custom MMIC design, when run on contemporary workstation hardware, provides an MMIC designer the capability to create from concept to final mask generation on a single station; it puts all the design software at the finger tips of the MMIC designer. This improves the efficiency and the accuracy of the design and thus cuts the total design time to half the time it takes to design a typical MMIC today. At the same time, design can be more accurate and have higher probability of working right the first time. Presently, an MMIC designer spends more time looking for appropriate design tools and learning different computer systems than concentrating on designing circuits.

As systems engineers start using MMICs and look toward implementing multiple functions on the same chip, the standard cell approach to MMIC design seems to be the most attractive. Much of the cost and time associated with designing MMICs can be reduced by establishing a library of predesigned and characterized standard cells that the system designer can use to build complex subsystem designs. The microwave designer maintains a cell library of the most commonly used building blocks, such as amplifiers, mixers, switches, and several microwave passive components. These cells are well designed, simulated, and characterized. According to our experience 80 percent of the system engineers' needs could be met by an appropriate cell library, leaving the rest to custom design. Most of the remaining 20% typically would be the front ends of the systems or the driver output stages, which push the performance limits of the MMIC technology. A list of typical standard cells includes

Generic gain blocks
Special purpose gain blocks
Mixers
Switches
Phase shifters
Oscillators
Attenuators
Isolators
Baluns

However, the standard cell approach to MMIC design is quite different from an implementation point of view in what is available in the digital circuit design area.

This is because of the following:

Analog nature of MMICs.

Intricate issues of impedances and so forth that must be addressed while designing at microwave frequencies.

Different frequency bands of operation of the circuits.

Application specific performance requirements.

Some factors will require standardization during the development of a standard cell library. These include power supply (as well as bus and ground line distribution); impedance levels at the input and the output of the cell; compatible cell dimensions; input and output pads (location of supply pads, dc and RF pads); single- versus double-ended configurations; integration of the cells. Different types of circuits, such as low noise, power, linear, and nonlinear, require quite different circuit analysis tools. Moreover, it will be necessary to develop software tools to address the use of a standard cell library design approach for a subsystem-level MMIC design. The sets of standard cells designed, fabricated, and characterized will not be sufficient and will not make MMIC design efficient unless specific application tools are developed to address the simulation of standard cells, interface effects when two or more standard cells are connected, and the layout issues.

## 8.2 DESIGN AUTOMATION TOOLS

Traditionally speaking, computer integrated design manufacturing includes the following specific steps: design, engineering, testing, manufacturing, and assembly. Different components of this system are presented in Figure 8.2. The whole conceptual system can be divided into two main components: the traditionally termed CAD, which includes computer-aided engineering and testing; and computer-aided manufacturing, which includes computer-assisted process control, robotics, manufacturing resource planning, and other wafer-fabrication issues. The tight coupling of these disciplines is essential in achieving the producible, affordable, and reliable MMICs. It is quite common to address these issues in a modern automated or semiautomated silicon design and wafer fabrication facility. In fact, one reason for the high rate of success of silicon products is attributed to taking these issues seriously.

Even though GaAs technology is becoming popular, its production requirements are not as high as those for digital integrated circuits in silicon. At the present time it is not unreasonable to say that the design tools for GaAs and high-speed silicon microwave ICs are at the level of sophistication silicon digital IC design tools were in during the mid-1970s. Computer-aided manufacturing and testing issues have been discussed in the earlier chapters; it is now appropriate to define the integrated design work environment for IC design and its main advantages.

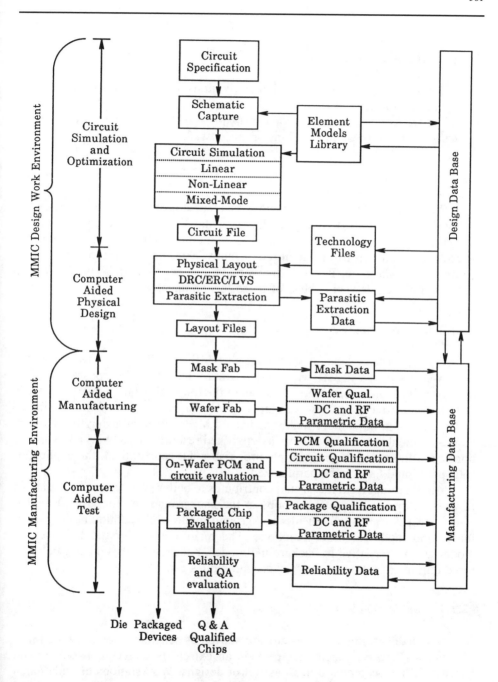

**Fig. 8.2** Components of a computer integrated manufacturing system.

### 8.2.1 Workstation Environment

Designers have many and varied motivations for using a workstation based (CAE) design automation system. The CAE-based concept, accepted by a majority of design engineers, has become a major thrust within the semiconductor industry. Some of these motivations are discussed.

#### 8.2.1.1 Availability

Workstations have become so universal that almost all software vendors feel at home with them, particularly in silicon digital VLSI design area; as a result, software is much more readily available for workstations than for mainframes. As a result, workstations can be acquired instantly instead of the arduous process of porting to a mainframe. Because the workstations are so ubiquitous, application software for them is years ahead of those for the mainframe.

#### 8.2.1.2 Response Time

On the surface it might seem that a 25 MIP mainframe gives much better response time than a 10 MIP workstation. But for all practical purposes this is not true, because many more users try to share the total MIPs on the mainframes. In fact, users are lucky if they can get one MIP to themselves. With distributed MIPs, users can have more highly dedicated MIPs either by using the workstations as standalone machines or as a networked resource. If networked, workstations can run single jobs on multiple processors. Of course, then, the application software would have to be parallelized.

Another important factor that contributes to better performance is the ability of workstations to run different jobs in the background and foreground modes. Users can safely run jobs in the background mode while continuing to use the foreground mode for other functions. The mainframes offer batch processing modes as an alternative to background modes, but these modes cannot be monitored during execution as the background modes can.

#### 8.2.1.3 Graphic Capability

Universally, almost all designs are started as schematics or translated into schematics. Because schematics provide an extremely convenient tool for the designer, it has become a critical aspect of designs. Workstations provide much better graphics than mainframes in terms of the software, response time, and resolution. In the mainframe environment, the terminals are located remote from the processors, hence the drawback in the graphics capability. Response time becomes a function of the speed of the connection between the processor and the remote terminal.

Currently, microwave design practice makes use almost exclusively of text entry for creating the netlist of the circuit description file. This circuit text file, created by a text editor, is read by the circuit analysis software. The text file approach to creating circuit description files has many severe limitations over graphical schematic entry, such as keeping track of several versions of the circuit, making modifications in the circuit topology by applying sections of one circuit onto the other, and interface between designers when several engineers are working on the sections of circuit design.

### 8.2.1.4 Top to Bottom Simultaneous Access

Throughout the design cycle, the designer needs simultaneous access to all the design automation tools that might be needed. This means that, at the highest level of abstraction, the designer might need access to a low-level simulation tool. This is very easy in a workstation environment. Data base and design data management is much easier in the workstations because of the superior file management techniques possible in this environment. Design data can be easily shared among individuals and groups, which leads to enhanced productivity.

### 8.2.1.5 Mouse and Menu Driven Software

An important aspect of the workstation is the ability to drive the software via a mouse. At the front end of the design cycle, where the process is highly interactive, a mouse enhances user productivity. It is very difficult to conceive of mainframes providing mouses for remote terminals. The mouse has revolutionized design entry and manipulation. By placing the mouse at the proper place and clicking, the user can call up anything in the application software. It speeds up entering each command manually on a command line.

Almost all design automation software for workstations is more or less menu driven. This means that all software operations can be controlled and operated from within a menu. This, coupled with the mouse, has become a very powerful tool for designers.

### 8.2.1.6 Cost and Local Control

Because mainframes are very large and targeted toward many users, they naturally are very expensive. In almost all cases, designers have to fight for mainframe resources with other branches of the organization, including accounting and payroll. The total cost of the mainframes thus can be huge in cost per designer. Workstations provide a much cheaper option for the designer. Workstations also provide localized control of software and hardware and remove the possibility of operators' killing jobs in the middle through overload or ignorance. System crashes can be fixed much faster, and system utilization can be optimized.

### 8.2.1.7 Networking

Modern workstation-based design automation systems provide an excellent environment for networking across software and hardware platforms. The ability to access files and data across different machine boundaries is difficult on a mainframe but can be achieved very easily on a workstation. Networking has given rise to the concept of diskless nodes, whereby the workstations themselves have no disk storage but are connected to a common storage area, to which the user has access. Diskless node configuration is transparent to the user and much more efficient than local disk storage.

Networking has given users computing power akin to that of mainframes in certain cases. It also brings up interesting concepts of parallel computing, and many vendors are working on that concept. The concept of networking has really made the whole world very small. Users do not have to worry about the location of actual computers or workstations and can transparently access any or all of available computer resources.

### 8.2.1.8 Custom Chips

With more and more workstation vendors opting for custom chips, software and hardware are being optimized for higher performance. *Reduced instruction set computer* (RISC) technology is giving desk top computers more and more MIPs while optimizing the performance of the design automation tools. Many workstations are based on general purpose microprocessors, but as demands for higher performance workstations increase, more and more vendors are going to custom microprocessors, tailored to their individual environments. Today, 10 MIP workstations are available, and who knows how many workstations will be available at the turn of the century.

### 8.2.1.9 Operating System, Connectivity and Windows

An important issue in integrating the CAE environment is the portability of software from one machine to another across a range of hardware platforms. The most popular commercial workstation hardware today is described in Table 8.1. Various workstation standards have existed for many years, but their acceptance has been hampered for two reasons. It takes time for workable standards to evolve and become accepted, and standard implementations are usually slower than nonstandard approaches. High performance is often the foremost criterion for choosing workstation hardware. Typically, to get the highest level of performance we have to pass layers of software standards and interact directly with the hardware. At long last, workable standards have evolved, and powerful, affordable workstations are available that use those standards. In today's workstations the de facto operating system standard is Unix Berkeley 4.X merging with AT&T's System V, resulting in a hybrid Unix standard. Unix provides a very flexible operating system for design engineers. Most of the workstations and CAE vendors are Unix based and differ significantly with operating systems on the mainframes, where file management is a real problem. The operating system is one area where CAE-CAD vendors have been able to standardize with Unix. The current generation of powerful engineering workstations all offer Unix, and more vendors are adopting Unix.

*De facto* standards are evolving in the networking of workstations, too. For example, TCP-IP protocol with Ethernet is now accepted by most workstation vendors. This connects various workstations into a distributed computing environment. Sun's *network file system* (NFS) also is supported by many workstation vendors for convenient access to files distributed in a computing environment made up of workstations from different vendors.

The ability to simultaneously use multiple windows is available only at a workstation. Using multiple windows, the designer might view graphics in one window and modify the netlist (text) on the next window. The windows are updated when the user wishes. The use of multiple windows has proven a significant motivator for designers to use workstations. Window managers have been available on many workstations for some time now. These windows allow simultaneous access to multiple applications from several overlapping text and graphics windows. We can manipulate these windows in various ways: moving, sizing, overlapping, hiding, and iconizing. Window managers are powerful tools but until recently suffered from some serious problems. For one thing, there have been too many different window systems. A user had many different commands to remember. Also, the windows were very dependent on the machine applications. Learning one machine's windowing system was not adequate, as other machines could have entirely different windowing commands. Also it was difficult to effectively access a network from within the windowing environment.

In an ideal situation, all workstations, hence, all applications, will have a common window manager. Also, if the window manager provides effective access to the network, we could run the application software on any machine on the

**Table 8.1** Most Popular Commercial Workstation Hardware

| Company | Product | CPU | Clock Frequency (MHz) | Operating System | Memory (min/max) (bytes) |
|---|---|---|---|---|---|
| Apollo Computer | Domain Series 3000 | MC68020 | 12 | Unix System V and Berkeley 4.2 and Aegis | 2M/8M |
| | Domain Series 4000 | MC68020 | 25 | Unix System V and Berkeley 4.2 and Aegis | 4M/32M |
| | DN5XX Turbo Workstations | MC68020 | 20 | Unix System V and Berkeley 4.2 and Aegis | 8M/32M |
| Apple Computer | Macintosh II | MC68020 | 15.7 | Macintosh OS A/UX (Unix System V and Berkeley 4.2) | 1M/8M |
| Digital Equipment Corp. | Vaxstation 2000 | Microvax II | 40 | Ultrix (Unix) or VMS | 4M/6M |
| | Vaxstation II | Microvax II | 40 | Ultrix (Unix) or VMS | 2M/16M |
| | Vaxstation II/GPX | Microvax II | 40 | Ultrix (Unix) or VMS | 3M/16M |
| | Vaxstation 3200 | CMOS CPU (VAX based) | 22 | Ultrix (Unix) or VMS | 8M/16M |
| | Vaxstation 3500 | CMOS CPU (VAX based) | 22 | Ultrix (Unix) or VMS | 16M/32M |
| Hewlett-Packard | HP 9000 Model 318M | MC68020 | 16.67 | HP-UX | 4M/4M |
| | HP 9000 Model 330 | MC68020 | 16.67 | HP-UX | 4M/8M |
| | HP 9000 Model 350 | MC68020 | 25 | HP-UX | 4M/48M |
| | HP 9000 Model 825SRX | HP Precision architecture | 12.5 | HP-UX | 8M/48M |

| | Graphics | | | Storage (bytes) | | | |
|---|---|---|---|---|---|---|---|
| Monitor | Resolution (pixels) | Color/ Palette | Floating-point Processor | Floppy Disk | Hard Disk | Tape | Local-Area Network |
| 15-in. monochrome | 1024 × 800 | | MC68881 | 5¼ in. 1.2M | 72M-348M | 60M | Apollo token ring or Ethernet |
| 19-in. monochrome | 1280 × 1024 | | | | | | |
| 15- or 19-in. color | 1024 × 800 | 16/4096 or 256/16.7M | | | | | |
| 19-in. monochrome | 1280 × 1024 | | MC68881 | 5¼ in. 1.2M | 155M-348M | 60M | Apollo token ring or Ethernet |
| 15- or 19-in. color | 1024 × 800 | 256/16.7M | | | | | |
| 19-in. color | 1280 × 1024 | 256/16.7M or 1.3M/16.7M | MC68881 optional FPX | | 155M-696M | 60M | Apollo token ring or Ethernet |
| 12-in.monochrome | 640 × 480 | | MC68881 | 3½ in. 800k | 20M-80M | | Appletalk, Ethernet |
| 13-in. color | 640 × 480 | | | | | | |
| 15- or 19-in. gray scale | 1024 × 864 | | 78132 (proprietary) | 5¼ in. 1.2M | 44M-159M | 95M | Ethernet |
| 15- or 19-in. mono-chrome | 1024 × 864 | | | | | | |
| 15- or 19-in. color | 1024 × 864 | 16/16.7M | | | | | |
| 19-in. monochrome | 1024 × 864 | | 78132 (proprietary) | Dual 5¼ in. 400k | 71M-159M | 95M | Ethernet |
| 19-in. gray scale | 1024 × 864 | | 78132 (proprietary) | Dual 5¼ in. 400k | 71M-159M | 95M | Ethernet |
| 19-in. color | 1024 × 864 | 16 or 256/16.7M | | | | | |
| 19-in. gray scale | 1024 × 864 | | CMOS floating-point unit | | 71M-318M | 95M | Ethernet |
| 19-in. color | 1024 × 864 | 256/16.7M | | | | | |
| 19-in. gray scale | 1024 × 864 | | CMOS floating-point unit | | 280M-560M | 296M | Ethernet |
| 19-in. color | 1024 × 864 | 256/16.7M | | | | | |
| 17-in. monochrome | 1028 × 768 | | MC68881 | 3½ in. 270k-630k 5¼ in. 270k | 101M-571M | 15M-140M | Ethernet |
| 19-in. monochrome | 1280 × 1024 | | MC68881 optional floating-point accelerator | 3½ in. 270k-630k 5¼ in. 270k | 101M-571M | 15M-140M | Ethernet |
| 16- or 19-in. color | 1280 × 1024 | 256/16.7M | | | | | |
| 19-in. monochrome | 1280 × 1024 | | MC68881 optional floating-point accelerator | 3½ in. 270k-630k 5¼ in. 270k | 10M-571M | 15M-140M | Ethernet |
| 16- or 19-in. color | 1280 × 1024 | 256/16.7M | | | | | |
| 19-in. color | 1280 × 1024 | 16/16.7M | HP Proprietary | | 132M-571M | 15M-140M | Ethernet |

**Table 8.1** cont.

| Company | Product | CPU | Clock Frequency (MHz) | Operating System | Memory (min max) (bytes) |
|---|---|---|---|---|---|
| Silicon Graphics | Iris 4D/60 Turbo | RISC $\mu$P (MIPS COMPUTERS) | 12.5 | Unix System V and Berkeley 4.3 | 4M/12M |
| | Iris 3130 | MC68020 | 16 | Unix System V and Berkeley 4.3 | 8M/16M |
| Sun Microsystems | Sun 3/50 | MC68020 | 15 | Sun OS (Unix System V and Berkeley 4.2) | 4M/4M |
| | Sun 3/60 | MC68020 | 20 | Sun OS | 4M/24M |
| | Sun 3/160 | MC68020 | 16.67 | Sun OS | 4M/16M |
| | Sun 3/260 | MC68020 | 25 | Sun OS | 8M/32M |
| | Sun 4/260 | MB86900 (RISC) | 16.67 | Sun OS | 8M/128M |
| Sony Microsystems | News 711 | Sony | 16.67 | 4.2 BSD/C, Bourne | 4M |
| | News 841 | Sony | 16.67 | 4.2 BSD/C, Bourne | 8M/16M |
| MIPs | M/120 RISC | R3000 | 16.7 | SV, 4.3 BSD | 8M/48M |
| | M/1000 RISC | R2000 | 15 | SV, 4.3 BSD | 16M/80M |

| | Graphics | | | Storage (bytes) | | | |
| Monitor | Resolution (pixels) | Color/ Palette | Floating-point Processor | Floppy Disk | Hard Disk | Tape | Local-Area Network |
|---|---|---|---|---|---|---|---|
| 19-in. color | 1280 × 1024 | 4096/16.7M or 16.7M/16.7M | Proprietary | | 170M-340M | 60M | Ethernet |
| 19-in. color | 1024 × 768 | 4096/16.7M or 16.7M/16.7M | Weitek 1064 | | 170M | 60M | Ethernet |
| 19-in. monochrome | 1152 × 900 | | | | 71M-141M | 60M | Ethernet, GM-MAP |
| 19-in. gray scale | 1152 × 900 | | MC68881 | | 71M-282M | 60M | Ethernet, GM-MAP |
| 19-in. high-resolution monochrome | 1600 × 1280 | | | | | | |
| 19-in. monochrome | | | | | | | |
| 16- or 19-in. color | 1152 × 900 | | | | | | |
| | 1152 × 900 | 256/16.7M | | | | | |
| 19-in. monochrome | 1152 × 900 | | MC68881 | | 141M-1.2G | 60M | Ethernet, GM-MAP |
| 19-in. gray scale | 1152 × 900 | | | | | | |
| 19-in. color | 1152 × 900 | 256/16.7M | | | | | |
| 19-in. monochrome | 1152 × 900 | | MC68881 | | 280M-2.3G | 60M | Ethernet |
| 19-in. high-resolution monochrome | 1600 × 1280 | | | | | | |
| 19-in. gray scale | 1152 × 900 | | | | | | |
| 19-in. color | 1152 × 900 | 256/16.7M | | | | | |
| 19-in. monochrome | 1152 × 900 | | Weitek 1164/1165 | | 280M-2.3G | 60M | Ethernet |
| 19-in. high-resolution monochrome | 1600 × 1280 | | | | | | |
| 19-in. gray scale | 1152 × 900 | | | | | | |
| 19-in. color | 1152 × 900 | 256/16.7M | | | | | |
| | | | | — | — | — | Ethernet |
| | | | | | 286M-1.4G | | Ethernet |
| | | | | — | — | — | Ethernet, TCP/IP, DECnet |
| | | | | — | — | — | Ethernet, TCP/IP, DECnet |

network using the same commands. This would be possible only after setting some standards for window manager software. X windows, developed at MIT, provides high-performance, high-level, device-independent graphics. Almost all of today's workstation manufacturers have accepted this X windowing as a standard. Software development tool kits are developed around this windowing system, greatly simplifying sophisticated user interfaces. With X windows, programs can run on different workstations or even on mini or mainframe computers, and the results can be displayed on any workstation in the network. X window also offers a device-independent environment. Without such a standard, it will be necessary to alter the software for each new type of display. Under X window architecture, a server does all the display manipulations. The server receives the requests from all the application programs using the display and executes them. Thus, entire networks of the workstations with different display hardware can utilize the same executable application code. As new displays are added, the only requirement is a new server that understands X windows system.

Today's bottlenecks in silicon digital VLSI design is the coordination of tools and data for all phases of the IC design process, which has resulted in the concept of workstation. Main features of an integrated workstation are as follows:

1. The whole system must have a common user interface. All the application software packages (e.g., schematic capture, circuit simulation, physical layout, back-end verification, parameter extraction, and testing software) run from a common user interface. This should support a consistent methodology for invoking and manipulating all design tools and design data. The interface must be heavily menu driven, visual, and intuitive to the microwave circuit designer; that is, very user friendly. This allows the designer to use any other piece of application software, after learning to use one package user interface. Most important, this enables the designer to work as an IC design engineer without the need for computer programming as well.

2. There must be a common, unified data base for all the design tools. This is a concept for the new generation of design tools developers. Most of the traditional workstations for digital silicon IC design do not use this concept. Also, they provide the interface between different components of the software through file translation, which is inefficient, slow, and inflexible. To build a more interactive and efficient design environment requires a unified data base approach that supports direct procedural access for any tool and provides a closely coupled design environment. Data exchange among all components of the design software should be improved by reducing the number of traditional interfaces and data base translators.

3. Data base and the design environment must be opened, by providing documented procedural access to the data and services. This allows the user to easily integrate other software packages, in-house developed or from other vendors, within one design environment.

The bright picture painted for workstations is not always so bright. Mainframes will remain the workhorses of the computing industry. Workstations are good engineering and development tools, but they take a long time to run big jobs. If mainframes are available during off peak hours, then they should be used for bigger jobs. The workstation industry is relatively young and still expanding. This has led to unstable software in many cases. It takes a long time of continuously debugging and enhancing to make a particular piece of software completely stable.

Networking seems to be a very good thing to use but it, too, has many problems. In order to have a totally satisfactory handshake between two machines flawless software and hardware is required. All data must be accessible to all users over a network. There are many problems with graphic data over a network and individual solutions have to be developed for particular situations.

On the whole, the use of workstations has proven very effective and productive in the design process. This does not mean that mainframes have no use for designers and other users. They will still remain the workhorses, providing vital run time and archiving resources to users. On the other hand, workstations are not very useful in a standalone mode, where the user is faced with, among other things, memory and disk storage problems. Unless networked to other workstations and to file and computer services, their efficiency drops drastically.

The simplistic view provided for the design tools hardware is not at all simple and straightforward. A host of other issues really make the life of design engineers very complex. Issues such as difference in language among various simulators, translation between schematic and netlists, turnaround time for simulation, lack of correlation between the simulation results of various levels of simulation, and nonstandard hardware description language (described later in Section 8.2.7) between vendors and customers makes the job of design entry, verification, and final fabrication that much more difficult.

## 8.2.2 Schematic Capture

Text entry of the circuit using text editor to create a circuit file has been a tradition in all microwave circuit designs, whether for hybrids or MMICs. Such a method for creating a circuit file for circuit simulation is suitable for hybrid circuit design, where the designs are tweaked more on the actual circuit board than during the design cycle. However, designing an MMIC is more of an iterative procedure, as described in Figure 8.1. The designers have to iterate the circuit topology several times to include the parasitics and so forth that make text entry very inefficient and prone to errors.

Schematic entry, or the graphical representation of the circuit schematic, is almost essential and widely accepted in all VLSI designs. In such a method, a symbol library is created of the basic components, such as MESFET, diode, transmission lines, inductors, capacitors, distributed elements, and junctions. Such a component library is used to create the circuit schematic with proper connectivity. It is quite convenient to delete, or introduce, any components once an initial schematic is developed. The designer need not remember, or manipulate, the node numbers as required in netlist file generation using text editor. Moreover, when the whole subsystem is subdivided into several MMIC chips, it is convenient, using schematic editor, to interface all chips and perform circuit simulation on the composite subsystem.

The schematic symbol library will be quite different from that for traditional VLSI circuits because of the use of distributed elements in MMIC designs. Open and shorted stubs, distributed transmission lines, *et cetera* will need special attention in schematic editor program for MMICs.

Schematic capture programs for MMIC designs are now in their infancy. Most of the commercial programs are the natural extensions of existing schematic programs for silicon VLSI design. Also, these packages are offered by third party vendors, and there is no tight coupling between the circuit simulation, schematic editor, and the component cell library, which limits their usefulness.

The definition of electrical parameters of the MMIC components is quite different from the traditional silicon VLSI components, where an active FET device is the main component and the required parameters generally are only the gate lengths and gate widths. In other passive elements, such as resistors, inductors, and capacitors, their lumped electrical value sufficiently represents the device at low frequencies in silicon VLSI design. At microwave frequencies, physical parameters such as substrate thickness, physical length and width of the interconnect lines, physical separation between the coupled lines, need to be specified. Hence, the specific electrical properties of the components used in MMIC design are quite different from those used in silicon VLSI design, which would require special attention. Also, in MMIC designs, as described in Section 8.2.3, a variety of circuit simulation programs are used, which require different information about the circuit components. To have a general interface between the schematic capture and all these circuit simulation programs, would require further special attention.

The main features of a typical schematic capture program for MMIC design are the following:

Symbol library for active, passive, and distributed elements of MMIC design.
Proper definition of electrical parameters for the components.
Capability to create user-defined symbols and components, and a method to define electrical properties for these black-box components.
Hierarchical design capabilities.
Cell library for commercial and foundry components.

Interface netlist extraction for commonly used circuit simulators.
Wires to connect components.
Text for better user interface with the schematic diagram.
Capability of moving, copying, deleting, rotating, mirroring, *et cetera*.
Zoom-in and zoom-out.
Interface with commonly used plotters and printers.
Interface with commonly used circuit simulation programs.

Almost all workstation-based VLSI design tools have the capability of schematic capture, although at this time they are not fully equipped to address MMIC design needs. The most popular VLSI design workstation vendors are Mentor, Daisy, Cadence, Caeco, Silvar-Lisco, and Valid Logic, who offer schematic capture programs on a range of host computers, from single user workstations to mainframe computers. Also several PC-based standalone schematic capture programs are commercially available, though with limited capabilities and closed architecture. This makes them even less sufficient for MMIC design, because they are written primarily to address digital circuit designs. Besides, the public domain VLSI design software OC Tools from the University of California-Berkeley has schematic capture capabilities. The MMIC design workstation from Hewlett-Packard also offers schematic capture capabilities, which are also linked to their circuit simulation programs.

## 8.2.3 Circuit Simulation

Circuit simulation programs have almost completely replaced the traditional breadboarding or fabrication and testing of integrated circuits as a means of design acceptability. This is particularly true in silicon digital VLSI circuits. As the simulation software addresses the needs of MMIC designs, the trend is similar to that in silicon VLSI design. In fact, a breadboard approach may give results that have little or no resemblance to integrated circuit performance because of very different parasitics in both cases, especially at high frequencies in GHz range. Also, fabrication and testing of an MMIC to verify a design is quite expensive and time consuming. Moreover, it might not be easily possible to probe all the internal nodes in an MMIC to debug the design. On the other hand, depending on the accuracy of the device models available in the circuit simulators, these models can provide accurate predictions of the circuit performance and information impossible to obtain from actual measurements.

Simulation is an extremely wide area to cover. Simulation provides the ability to verify designs at all levels of abstraction. Different simulators can be used for different levels of design and for different types of circuits. It is difficult to find any one simulator that is sufficient for all levels of abstraction and all types of circuits. At the very highest level, the designer will be dealing with the architecture of the system itself. At this level, the designer deals only with blocks with the elements inside the blocks not defined yet. The software packages for this level are system performance analysis tools. In microwave system design environment, such tools either do not exist or are in their infancy, and their capabilities are severely limited.

After defining the architecture, the designer needs to define what will be inside the blocks. The components inside the blocks can be defined by their behavior or by their functions. By defining the behavior, the functionality of the components is put in abeyance. Behavioral level simulators to define the system and verify the design are quite common in silicon digital VLSI systems. However, such models do not exist in microwave systems.

Finally, the designer needs to define the individual components and elements within the block, the most detailed level of circuit simulation. At this level, designs are supposed to be verified for accuracy and functionality. Most of the vendor-supplied simulators address this level most vigorously. The circuits of individual components and elements are more detailed, and hence, circuit analysis needs more time to run. This is the lowest level at which the designer works, the level of individual active and passive components along with distributed transmission lines.

In microwave IC design, even for comparatively small circuits, the speed of simulation becomes a critical issue; hence, the performance of the workstations is important to consider. Circuit simulation software for chip-level IC design is the most developed software commercially available in the MMIC design cycle. However, no single software is capable of addressing all types of circuit designs: low-noise, small-signal, and high-power amplifiers, mixers, switches, phase shifters. Three different types of circuit analysis tools commonly are used: linear-frequency domain, nonlinear-time domain, and mixed mode-harmonic balance simulators. Linear-frequency domain simulators are used specifically for analyzing small-signal linear circuits, such as small-signal and low-noise amplifiers. Time domain nonlinear simulators are used for designing switching and other large-signal circuits, such as limiting amplifiers, power amplifiers, and phase shifters. Harmonic balance simulators are extremely useful for designing circuits with nonlinear devices operating in a large-signal mode combined with distributed microwave components, such as transmission lines and microwave junctions (e.g., mixers, distributed amplifiers). Essentially, as the number of nonlinear devices operating in a large-signal mode increases in a circuit, harmonic balance techniques of circuit analysis have little or no advantage over nonlinear-time domain techniques, provided there are no distributed elements in the circuit.

### 8.2.4 Linear-Frequency Domain Simulation

Linear or frequency domain circuit analysis, synthesis, and optimization techniques are the most developed and described [1–3] concept in microwave circuit simulation. In this text, it is appropriate to avoid the details of the mathematical concepts on which frequency domain simulators operate. Numerous linear microwave simulators are developed in-house or available commercially to the microwave designers.

The most popular and widely used linear circuit simulators are Super Compact℗ from Compact Software, and Touchstone℗ from EESof. Both of these programs are available on several computer hardware platforms, ranging from personal computers to main frames. Also, the MMIC design workstation from Hewlett-Packard has its own microwave circuit simulation package. All microwave circuit analysis programs have very similar features [4]:

- *Circuit elements:* Both passive and active circuit elements are defined as building blocks in the circuit netlist file and are cascaded either on a ladder network or in nodal form. Nodal analysis provides the most flexibility in interconnections, particularly for feedback networks and during analysis procedures. Passive elements can be defined as lumped or distributed elements. S-parameter and optimum noise figure data can be entered into the data file in separate blocks.

- *Optimization:* Several optimization methods are available in any software package. The least-path optimization is often used in addition to direct search and conjugate-gradient algorithms. Optimization requires entering an error function to the program to which it attempts to converge. Usually, the total error function, which is a sum of squares of individual error functions, is weighed. Each of the optimized parameters (e.g., VSWR, gain, noise figure) can be weighed as a function of frequency. The circuit elements to be optimized are defined as varying between the upper and lower limits defined in the input circuit description file.

- *Sensitivity analysis:* Monte Carlo analysis is commonly used for sensitivity analysis of the circuit performance as a function of variations in individual circuit components. Particular component values are altered in random fashion within their specified limits providing valuable information about circuit yield within specific performance. In Monte Carlo analysis, the circuit element values are varied independently, without any correlation between the different circuit elements. In practice, several circuit element values are dependent on each other; for example, if greater than average, the MESFET channel doping increases the $g_m$ as well as $C_{gs}$ values and decreases the sheet resistivity of monolithic resistors, as the same implant is commonly used for MESFET channel and N-monolithic resistors. Hence, the Monte Carlo analysis results provide the conservative view of the yield.

To estimate more realistic yield, it is important to develop tools that take into account the correlation between components. In real applications, Monte Carlo analysis indicates only the qualitative estimates of the circuit yield and the general trends, such as whether the circuit yield is 10 or 80%. A yield of 10% does not necessarily indicate the true estimated yield to be 10% but definitely points out that the circuit topology needs serious considerations before committing to fabrication. Similarly, an 80% yield does not mean such high yields but indicates that the circuit is quite process tolerant.

Bias-dependent models for nonlinear devices such as FET and diode are becoming a common part of these circuit simulators. However, we must note that by themselves these programs are not capable of evaluating the dc operating point of the circuit, as they have only frequency domain information. The user needs some other nonlinear circuit analysis program to determine the dc bias point of the circuit. Once the bias point is determined, such models can identify the small-signal equivalent circuit element values of the nonlinear device at the specified bias point. It is definitely a user friendly feature; however, it does not eliminate the need for a separate program to determine the bias or the dc operating point of the active devices in the circuit. A comparison of the features of two commercial linear programs is provided in Table 8.2. One of the severest limitations of both programs is the validity of the models for distributed elements. These models have questionable validity above 12 GHz and are most likely unusable at millimeter-wave frequencies. Another limitation is the closed nature of these programs. Although the capability of implementing user-defined models is an added benefit, this feature is very restricted and requires the circuit designer to also be a computer programmer and circuit simulation techniques expert.

### 8.2.5 Nonlinear-Time Domain Simulator

Time domain circuit analysis has been an essential part of the digital and low-frequency analog design. Simulation tools based on nonlinear analysis have allowed designers a detailed and in-depth knowledge of the circuit before building prototypes [5,6]. Lack of such circuit analysis software would have made it difficult for silicon VLSI designs to advance so far in such a short period of time. Methods of simulating the nonlinear behavior of microwave circuits is becoming more important and has received substantial attention in a MMIC designer community. Nonlinear quantities of particular interest include gain compression, power saturation, intermodulation distortion, and harmonic content in switching circuits, mixers, amplifiers, and other large signal analog circuits. Time domain

**Table 8.2** Comparison of Two Linear Programs

| Feature | Super Compact | Touchstone |
|---|---|---|
| Maximum no. of ports | 4 | 4 (9 for subnetworks) |
| Maximum no. of nodes | 50 per block | — |
| Analysis method | Ladder, nodal | Nodal |
| Hierarchical circuit definition | No | No |
| User-defined models | Yes | Yes |
| Optimization | Gradient (Fletcher-Powell-Davidson) random | Gradient (first order), Quasi Newton (Fletcher-Powell-Davidson) random |
| Error function | Least squares | Least squares, mini-max, least $p$th worst case |
| Yield analysis | Monte Carlo | Monte Carlo |
| Noise analysis | Nodal | Two-port–Nodal |
| Tune mode optimization | Yes | Yes |
| User interface | Command driven | Menu driven |
| Graphics | Yes | Yes, advanced |
| Network analyzer interface | Yes | Yes |
| Synthesis capability | Yes | Yes |

analysis is commonly used in silicon digital and high precision analog circuit design, and SPICE is the program most widely used for time domain analysis. A main reason for the popularity of SPICE is that it works for most circuit analysis cases, and the University of California, Berkeley, charges only for the cost of magnetic media and shipping. Further, by providing access to the source code of the program, the designer can make changes, such as implementation of various device models, although such changes in the program are hampered by the absence of any support.

Because the time domain circuit analysis techniques are comparatively new to microwave designers, compared with the linear circuit analysis software, it is appropriate to describe such techniques in detail, particularly in view of SPICE [7–9] program. Time domain techniques result in the most complete analysis of the circuit, because they provide the transient as well as the steady-state responses. The analysis consists of establishing the system of equations that represent each component in the circuit and solving them in time domain. The equations are obtained by applying the *Kirchhoff current law* (KCL) involving branch currents and the *Kirchhoff voltage law* (KVL) involving branch and node voltages. A nonlinear circuit with elements such as MESFET and diode results in a set of nonlinear integro-differential equations. These equations are solved in time domain by applying various numerical techniques, such as equation formulation, linear equation solution, nonlinear equation solution, and integration. In frequency domain linear circuit analysis, techniques such as partial differential equations are represented in frequency domain. Such a method converts nonlinear partial differential equations into algebraic equations, which makes the solution procedure relatively easy.

A "generic" SPICE program contains built-in models for the following linear and nonlinear circuit elements:

Resistor
Capacitor
Inductor
Coupled inductor
Ideal transmission line
Independent voltage and current sources
Voltage controlled voltage and current sources
Current controlled voltage and current sources
Diode
BJT (bipolar junction transistor)
JFET (junction field effect transistor)
MOSFET (metal-oxide-semiconductor field effect transistor)
MESFET (metal-semiconductor field effect transistor)

Diode and MESFET models are described in detail in Chapter 4. The mathematical representation of other components available in SPICE is shown in Table 8.3 [10].

Convergence is a common problem in using SPICE: during the dc operating point solution or the transient solution, the numerical techniques fail to find a unique solution or converge to an incorrect solution for the differential equations representing the circuit. Several options provided in SPICE allow the user to choose one numerical method over another and some other user-controlled parameters can improve convergence in a particular situation. Also, an internal algorithm of source stepping can be used to find the correct solution. In this method, the independent sources are turned-off as an initial condition then increased by a certain amount to obtain convergence. At each increment, the starting solution is that obtained at the previous step. The method is repeated until all sources are completely turned on. Moreover, if the circuit designer has an initial idea of the voltage of some critically unstable nodes, convergence can be obtained easily by setting nodes to the specific values. A good initial idea of some node

**Table 8.3** Mathematical Representation of Various Components Available in a SPICE Program

| Two terminal elements: | Symbol | Linear branch relations for DC and Transient Analysis | | Relations for AC analysis | |
|---|---|---|---|---|---|
| | | Voltage controlled | Current controlled | Voltage controlled | Current controlled |
| **Resistors:** | | | | | |
| Linear | | $i = V/R$ | $V = RI$ | $i = v/R$ | $v = Ri$ |
| Non-Linear | | $I = I(V)$ | $V = V(I)$ | $I = i(v)$ | $v = v(i)$ |
| **Capacitors:** | | | | | |
| Linear | | $q = Cv; i = dq/dt \Rightarrow i = C\, dv/dt$ | | $i = j\omega Cv$ | |
| Non-Linear | | $q = q(v); i = dq/dt$ $\Rightarrow i = (dq/dv)(dv/dt) = C(v)dv/dt$ | | $i = j\omega C(v)v$ | |
| **Inductors:** | | | | | |
| Linear | | $\phi = Li; v = d\phi/dt \Rightarrow v = L(di/dt)$ | | $v = j\omega L\, i$ | |
| Non-Linear | | $\phi = \phi(i); v = d\phi/dt$ $\Rightarrow v = (d\phi/di)(di/dt) = L(i)di/dt$ | | $v = j\omega L(i)\, i$ | |

**Table 8.3** cont.

| Two terminal elements: | Symbol | Linear branch relations for DC and Transient Analysis | | Relations for AC analysis | |
|---|---|---|---|---|---|
| | | Voltage controlled | Current controlled | Voltage controlled | Current controlled |
| Independant Sources<br><br>Voltage Source | $+ \bullet \downarrow i$<br>$v \ \pm$<br>$- \bullet$ | $v(t) = Es(t)$ | | $v = Ve^{j\phi}$ | |
| Current Source | $+ \bullet \downarrow i$<br>$v \ \bigcirc$<br>$- \bullet$ | $i(t) = Is(t)$ | | $i = Ie^{j\phi}$ | |
| Transformer | $I_1 \ \ I_2$<br>$V_1 \ \ V_2$ | $M = k\sqrt{L_1 L_2}$<br><br>$V_1 = L_1\dfrac{dI_1}{dt} + M\dfrac{dI_2}{dt}$<br><br>$V_2 = M\dfrac{dI_1}{dt} + L_2\dfrac{dI_2}{dt}$ | | $M = k\sqrt{L_1 L_2}$<br>$V_1 = (j\omega L_1)i_1 + (j\omega M)i_2$<br>$V_2 = (j\omega M)i_1 + (j\omega L_2)i_2$ | |
| Transmission Line | $i_1(t) \ V_1^+(t) \ \ V_2^+(t) \ i_2(t)$<br>$V_1(t) \ V_1^-(t) \ V_2^-(t) \ V_2(t)$ | $V_1(t) = V_1^+(t) + V_1^-(t)$<br>$V_2(t) = V_2^+(t) + V_2^-(t)$<br>$i_1(t) = \dfrac{1}{Z_0}\left[V_2^+(t) - V_2^-(t)\right]$<br>$i_2(t) = \dfrac{1}{Z_0}\left[V_2^+(t) - V_2^-(t)\right]$<br>$T = \dfrac{1}{c}$<br>where, $V_1^+(t) = V_2^-(t+T)$<br>$V_2^+(t) = V_1^-(t+T)$<br>l = length of the line<br>c = speed of propogation<br>T = time delay on the line | | | |

Table 8.3 cont.

| Two port elements: Controlled sources | Symbol | Linear branch relations for DC and Transient analysis | | Relations for AC analysis | |
|---|---|---|---|---|---|
| | | Linear | Non-linear | Linear | Non-linear |
| VCVS (Voltage Controlled Voltage Source) | | $v_k = E_k v_C$ | $v_k = v_k(v_C)$ | $v_k = E_k e^{-j\omega\tau} v_C$ | $v_k = e^{-j\omega\tau} v_k(v_C)$ |
| VCCS (Voltage Controlled Voltage Source) | | $i_k = G_k v_C$ | $i_k = i_k(v_C)$ | $i_k = G_k e^{-j\omega\tau} v_C$ | $i = e^{-j\omega\tau} i_k(v_C)$ |
| CCVS (Current Controlled Voltage Source) | | $v_k = H_k i_C$ | $v_k = v_k(i_C)$ | $v_k = H_k e^{-j\omega\tau} i_C$ | $v_k = e^{-j\omega\tau} v_k(i_C)$ |
| CCCS (Current Controlled Current Source) | | $i_k = F_k i_C$ | $i_k = i_k(i_C)$ | $i_k = E_k e^{-j\omega\tau} i_C$ | $i_k = e^{-j\omega\tau} i_k(i_C)$ |

voltages helps obtain the correct final solution and hence avoid convergence problem. Common reasons for nonconvergence are ill-defined nonlinear device model equations or inappropriate model parameters for these devices [11,12].

Another problem faced in using time domain analysis tools is the long simulation time. Most applications in MMIC design use steady-state simulations, which requires all transients in the circuit to decay before the steady state is reached. This can result in prohibitively long simulation runs, particularly in circuits where picosecond circuit response must be observed in the presence of much larger (e.g., millisecond) time constants, often the case in very high $Q$ circuits. However, this limitation is imposed by the speed of the computer and not necessarily the simulation software itself.

SPICE or other time domain simulators lack distributed element models valid at microwave frequencies because the structure of these programs and the numerical techniques used prohibit the implementation of models of such components as lossy transmission line, open and shorted stubs, and different bends and junctions. The partial solution to such a problem is to use lumped-element models for the distributed elements. However, such models are not sufficient and accurate at higher frequencies and certainly are quite inaccurate at millimeter frequencies.

SPICE2, a generic version of the SPICE program consists of seven major overlays, shown in Figure 8.3. The root segment, SPICE2, is the main control portion of the program. The root calls in the various overlays in the sequence required for the specific simulation. The main control loop of the program is shown in the flow chart in Figure 8.4. The main program begins by initializing some program constants and reading the job title card. If an end-of-file is encountered on the input file, the program terminates. Otherwise, the READIN overlay is called to read the remainder of the input file. The READIN program stops rending after a .END card or an input end-of-file is encountered. After complete circuit description file is read, it is checked for errors and consistency and internal data is set up to define the circuit nodal and element information. Element values are evaluated if temperature analysis is requested and certain analyses are proposed depending on the type requested in the input file. Finally, the circuit input and analysis output are printed out in a file.

The READIN overlay consists of several subroutines. This overlay reads the input file, checks each line of input for syntax, and constructs the data-structure for the circuit. The ERRCHK overlay checks further for inconsistency in the input circuit description file, such as incomplete defined elements and model parameters of the incompletely defined elements. Model parameters of the device models are checked and printed in tabular form. Then, ERRCHK constructs an ordered list of the user node numbers, the hierarchy of the circuit is completely flattened, and extra node numbers are created internally to correspond to the hierarchical circuit sections. It also creates a node table and checks that every

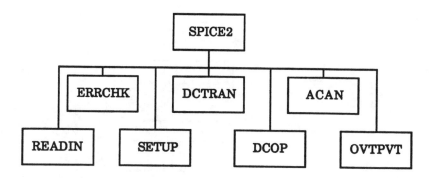

**Fig. 8.3** Block diagram of SPICE2.

node in the circuit has at least two elements connected to it and a dc path to ground. In microwave circuits such as filters, this might not be true if a node is connected to ground through a capacitor. To satisfy this requirement for SPICE, it is necessary to introduce a high value (100 KΩ) resistance in parallel with such a capacitor. The independent sources and coupled inductors in the circuit also are processed at this time.

In the SETUP overlay, the admittance matrix is set up and matrix coefficients that are nonzero are identified. Finally, the executable machine code is generated that solves the circuit equations for dc and transient analysis.

DCTRAN is the largest and most complicated overlay in SPICE2 program. This overlay performs the dc operating point analysis, the transient initial-condition analysis, the dc transfer-curve analysis, and the transient analysis. The DCOP overlay prints out the nonlinear device operating points and the linearized device model parameters. The dc transfer-function analysis and dc, small-signal sensitivity analysis are also printed.

The ACAN overlay determines the small-signal frequency-domain analysis. A complex system of linear equations is constructed for each frequency point and then solved. The ACAN overlay also includes noise analysis and distortion analysis.

The OVTPVT overlay interpolates the output variables to match the specified printing increment and generates the tabular listings specified by the user. It generates the line-printer plots of the simulation. Finally, the Fourier analysis for transient analysis is preformed in this overlay.

A detailed description of these overlays with their subroutines and the flow of data through different subroutines is shown in Figure 8.5. The simulation of a circuit usually requires a combination of three basic analyses: dc analysis, time-domain transient analysis, and small-signal ac analysis. SPICE contains all three analyses, as well as several subanalyses. The types of analysis available in SPICE

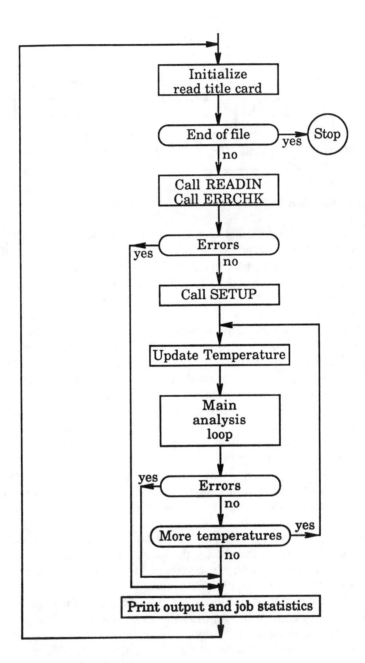

**Fig. 8.4** Main control flowchart of SPICE.

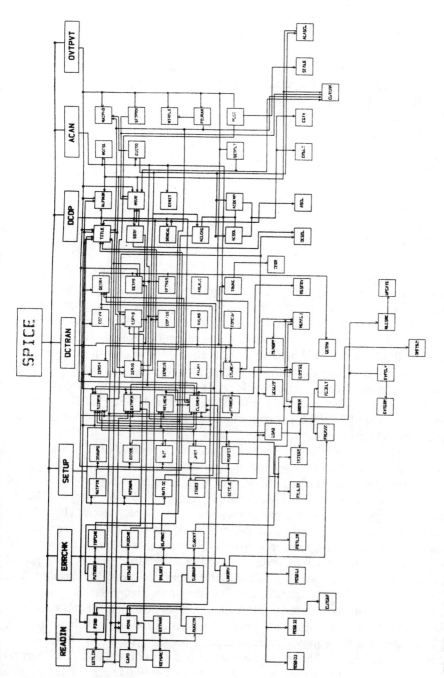

**Fig. 8.5** Detailed description of SPICE overlays with their subroutines and data flow.

are as follows (see Table 8.3b):

    *dc Analysis:*
    dc operating point
    Linearized device model parameterization
    Small-signal transfer function
    Small-signal sensitivity
    dc transfer curve

    *Transient Analysis:*
    Time-domain response
    Fourier analysis

    *ac Analysis:*
    Small-signal frequency-domain response
    Noise analysis
    Distortion analysis

A dc analysis determines the quiescent operating point of the circuit. All energy-storage elements in the circuit are ignored in a dc analysis by treating capacitors as open circuits and inductors as short circuits. At the conclusion of the operating point analysis, the program prints the circuit node voltages, independent voltage source currents, and the total quiescent power dissipation of the circuit. The flow chart for calculating dc operating point is shown in Figure 8.6.

For small-signal operation, the perturbational response of the circuit can be determined by modeling the nonlinear elements in the circuit by equivalent linearized models, such as diode, BJT, JFET, MOSFET, and MESFET, available in most versions of SPICE. The parameters of the linearized models depend upon the quiescent operating point.

If an output variable and an input source are specified, SPICE also determines the dc, small-signal transfer function value; that is, the ratio of the output variable to the input source. In addition, the program determines the small-signal input resistance and the small-signal output resistance of the circuit. The dc sensitivities of specified output variables with respect to every circuit element can also be evaluated. The dc transfer characteristics of the circuit are determined by evaluating the dc operating point repetitively for successive values of a designated input source. Such dc transfer curves are used to determine the noise margins of digital circuits as well as the large-signal characteristics of analog circuits. The flow chart for evaluating dc transfer curve is shown in Figure 8.7.

Transient analysis determines the time-domain response of the circuit to specified time-domain inputs. The initial timepoint, arbitrarily defined as time zero, is determined by a previous dc operating point solution. The independent sources in the circuit may have constant or time-dependent values. The interval between starting and final times specified by the user is divided into internal time

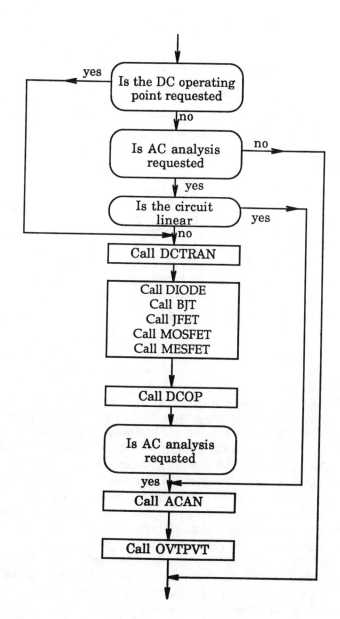

**Fig. 8.6** dc operating point and ac analysis flowchart.

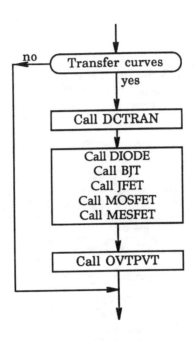

**Fig. 8.7** dc transfer-curve flowchart.

steps. The spacing between successive timepoints is controlled by the program to ensure an accurate solution, and the program determines the circuit solution at each successive timepoint starting from time zero. Voltage or current output variables are stored at each timepoint and can be listed in tabular form or plotted at the conclusion of the analysis. The output routines contain a Fourier analysis subprogram that determines the first nine Fourier coefficients of a specified output. This capability is useful for estimating the Fourier harmonic distortion components of a near-sinusoidal waveform. However, the inherent inaccuracies of fitting a Fourier series to a time-domain waveform may limit the usefulness of this subanalysis to relatively large values of harmonic distortion. The flowchart for transient analysis is shown in Figure 8.8.

The small-signal, frequency-domain analysis portion of SPICE is useful for designing of analog circuits that operate in small-signal modes. The perturbational response of the circuit is obtained using linearized models for the nonlinear elements. The parameters for these linearized models are determined by the dc operating point analysis. The small-signal linear equivalent circuit is analyzed in the frequency domain with the phasor method. All circuit voltages and currents including the independent sources are complex variables, usually expressed in terms of magnitude and phase. An ac analysis is performed at successive values of

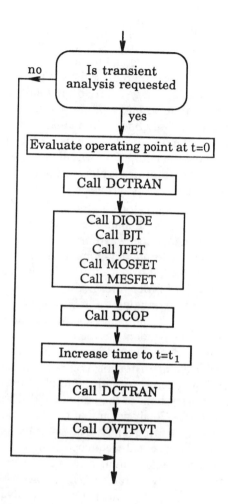

**Fig. 8.8** Transient analysis flowchart.

frequency specified by the user to determine the frequency response of circuit transfer functions. The specified voltage and current outputs are stored at each frequency point. At the conclusion of the analysis, these outputs can be listed or plotted as magnitude, magnitude in dB, phase, real part, or imaginary part.

The ac analysis portion of SPICE contains the added capability of evaluating the noise characteristics of the circuit. Detailed noise models for MMIC components are described in Chapter 4. Each resistor in a circuit generates thermal noise, and each semiconductor device generates both shot noise and flicker noise in addition to the thermal noise generated by the device's ohmic resistances. This noise generation is modeled by the equivalent, independent sources associated with each noise-generating element. However, the accuracy of the noise analysis depends on the accuracy of noise models for the active devices.

Because the equivalent noise sources in the circuit are not statistically correlated, the contribution of each noise source to the total output noise at a specified output is computed separately. The total output noise is the rms sum of the individual noise contributions. Only one additional ac solution at each frequency point is required to evaluate the rms sum of the noise of the circuit. After the noise analysis is complete, both the total output noise and the equivalent input noise can be presented either in tabular form or as plots. In addition, the contribution of each noise source in the circuit can be printed at selected frequency points.

For relatively large levels of distortion, the method of fitting a Fourier series of a periodic waveform to the time-domain response yields a good approximation of the distortion components of the waveform. Small levels of distortion are determined more accurately by frequency-domain distortion analysis. The distortion analysis can evaluate second-order and third-order harmonic distortion as well as second-order and third-order intermodulation distortion. The distortion analysis is performed by approximating the model equations for each nonlinear element by a Volterra series. Each nonlinear element is modeled by the linearized equivalent element and an independent source that represents the distortion generated by that particular element. Distortion analysis, therefore, is analogous to noise analysis. For a particular distortion component, the small-signal circuit contains several distortion generators, and the total distortion component is the vector sum, instead of the rms sum, of the individual distortion contributions. At the conclusion of the distortion analysis, any of the distortion components can be listed or plotted as a function of frequency. The contributions of every distortion source also can be printed at selected frequency points.

Recently, several commercial nonlinear circuit analysis programs have appeared specifically for microwave circuit designs. Most of these programs are variations on the generic SPICE program. The main disadvantage of the commercial versions of the SPICE program, besides their high price tags, is the lack of control on the device models available in the program. The user is generally dependent on the models offered by the vendor, such as for MESFET and diodes. At the same time, using the generic SPICE program and modifying it to implement customer models needs specialized skills, such as understanding of SPICE program flowchart (SPICE2G.6 has approximately 10,000 lines of Fortran code), its internal workings, and programming language. Different procedures have been developed in Appendix A8 to provide capabilities to SPICE users working in a microwave circuit analysis environment. These include getting S-parameters and developing device models without making any changes in the SPICE program itself.

### 8.2.6 Mixed-Mode–Harmonic-Balance Simulator

*Harmonic-balance* (HB) simulators are a particularly important set of mixed-mode simulators, whereby the nonlinear devices are described in the time domain, whereas the linear part of the network is represented by frequency-domain equations at a finite number of discrete frequencies. The basic assumption of HB analysis is that the only interesting aspects of circuit behavior are related to its steady state of operation, so that attention can be restricted *a priori* to the steady-state regime. The immediate consequences of this viewpoint are that only constant (dc) and time-periodic excitations are allowed in harmonic-balance simulation and no transient information, such as turn-on or turn-off, can be obtained from this analysis. Thus, the HB approach implies a consistent loss of understanding of the circuit physical operation with respect to an ideal time-domain simulator, mainly for two reasons: the transient behavior itself is of interest for some applications, such as the determination of the settling time of a VCO; and it usually can be assumed that during a transient all the natural frequencies of a circuit are excited, so that, in general, any unstable ones would show up; thus instabilities can be detected by time-domain analysis. In exchange for these limitations, harmonic-balance simulators offer a few significant advantages with respect to practical time-domain programs, the most important of which are numerical efficiency and physical accuracy in the modeling of high-frequency microwave circuits.

The time-domain simulation of a nonlinear circuit typically requires the circuit to be analyzed at a sequence of time instants starting from some suitable initial situation, usually the result of a dc analysis, until the transient dies out and the required steady state is reached. This procedure may be very lengthy, because the number of RF cycles required for the transient to die out may be large. A particularly long transient may occur in some circumstances, as when the circuit contains time constants that are large with respect to the RF period (large capacitors) or in the presence of high-$Q$ components such as dielectric resonators. Also, in many cases of practical interest (e.g., multitone excitation with incommensurate source frequencies), the steady-state waveforms may not be periodic, so that it may be difficult even to establish whether a steady state has been achieved. Thus, if the user's interest is focused on the steady state, which is most often the case in microwave applications, the time-domain approach may not be cost effective. In such cases a dramatic increase of computational efficiency can be obtained by resorting to analysis strategies that directly address the steady state, taking for granted that a sufficient amount of time has elapsed since circuit turn-on for all kinds of transient signals to die out.

A major aspect of the nonlinear simulation problem is represented by the evaluation of the circuit response to a *multitone* excitation, consisting of a composite input signal arising from the superposition of several sinusoids of arbitrary frequencies, amplitudes, and phases. It is important to note that the terms *excitation* and *input* should be intended here in the broadest possible sense, to include any signal that might be injected into the circuit.

From this standpoint, many situations of practical interest may be viewed as cases of multitone excitation, be they purposely created by the user of the circuit to perform specific electrical functions, or generated by some unwanted circuit behavior. Typical examples of the former are modulation and frequency conversion, of the latter, the generation of spurious signals in oscillators.

In all such cases, the exciting signals intermodulate in the device nonlinearities, so that all possible intermodulation (IM) products of their frequencies are generated. Any time-dependent electrical quantity such as a voltage or a current may then be represented by means of a generalized Fourier series of the form

$$x(t) = \text{Re}\left[\sum_{\mathbf{k}} X_{\mathbf{k}} \exp\left(j \sum_{i=1}^{N} k_i \omega_i t\right)\right] \tag{8.1}$$

where $\omega_i$ is the angular frequency of the $i$th exciting tone, and $\mathbf{k}$ is a vector of harmonic numbers $k_i$. The summation in (8.1) is extended to all possible vectors $\mathbf{k}$. A function of the form (8.1) is usually referred to as an *almost periodic function* [13] and is *quasiperiodic* [13] when the infinite summation is truncated to a finite one for numerical reasons. A function is termed as *strictly periodic* when it is represented by a conventional Fourier series; that is, when $N = 1$ in Equation (8.1). Note that (8.1), in fact, is strictly periodic when all of the exciting tones have commensurate frequencies.

Although it is taken for granted that (8.1) must be truncated in some way for practical purposes, several different truncation criteria can be devised and—what is more important—the effects of truncation cannot be established except on a case-by-case, empirical basis. Nevertheless, a most common choice is to restrict the summation in (8.1) to those IM products satisfying the inequality

$$0 \le \sum_{i=1}^{N} |k_i| \le M \tag{8.2}$$

The quantity $\Sigma |k_i|$ is usually referred to as the *order of the corresponding IM product*. Thus, (8.2) is equivalent to considering all IM products up to a given maximum order M. Note, however, that (8.2) does not always represent the best choice from a physical viewpoint: for instance, for mixer analysis it is usually convenient to select a spectrum consisting of groups of lines symmetrically placed around the local-oscillator harmonics.

Until 1983, virtually the only available frequency-domain technique for analyzing a nonlinear circuit operating in a quasiperiodic electrical regime was represented by the Volterra series [14,15]. Since then, with the advent of general-purpose harmonic-balance simulators, much work has been done to solve the multitone analysis problem. From the HB viewpoint, the difficulty essentially consists in finding the spectral components of the nonlinear subnetwork response to an excitation of the form (8.1). As a matter of fact, once such an algorithm has been established, the remaining part of the analysis would not differ substantially from the conventional strictly periodic case.

The CPU time required by a harmonic-balance analysis making use of the Newton iteration to solve the nonlinear system is usually dominated by the time required to compute the Jacobian matrix, especially when the partial derivatives are evaluated numerically. In turn, a parameter playing a major influence on this computational cost is represented by the number of sampling points used to carry out time-to-frequency-domain conversions via the *discrete Fourier transform* (DFT).

Let us consider a signal represented by a suitably truncated series of the general form (8.1). The spectrum of this signal consists of a dc component plus a finite number of discrete spectral lines, referred to as *harmonic components,* whose total number is denoted by $N_H$. Since the dc component is identified by a real number and each harmonic by a complex number, the total number of degrees of freedom of the signal is given by

$$N_D = 2N_H + 1 \tag{8.3}$$

Now, the amount of information contained in the set of time-domain samples must be large enough to completely identify the signal; thus, the total number of samples of the time-domain waveforms to be computed for a correct application of the DFT must satisfy the inequality

$$N_P > 2N_H \tag{8.4}$$

A sampling mechanism making use of the minimum number of points defined by (8.3) is referred to as a *nonredundant sampling;* otherwise is called *redundant sampling*. The numerical efficiency of any HB analysis method depends essentially on the degree of redundancy of the sampling mechanism adopted and the kind of algorithm used to perform the DFT.

### 8.2.6.1 Reduction to a Strictly Periodic Regime

The most obvious way of dealing with a multitone excitation is to try to reduce the associated quasi-periodic electrical regime to a strictly periodic one and then to apply the conventional HB technique. This can be done in a conceptually straightforward way when the exciting tones have commensurate frequencies; that is:

$$\omega_i = n_i \omega_0 \qquad (8.5)$$
$$i = 1, 2, \ldots, N$$

where the $n_i$'s are integers and $\omega_0$ is the largest common submultiple of the exciting frequencies. In such cases, all frequencies appearing in (8.1) are integer multiples of a same fundamental $\omega_0$, so that the electrical system, indeed, is strictly periodic. Note that, because of numerical truncation, (8.5) is always satisfied in a digital computer, so that the numerical representation of a quasi-periodic regime, in fact, is always a strictly periodic one. Thus, in principle, a trivial solution to the multitone problem is always available.

In practice, the applicability of this simple approach depends on the actual values of the exciting frequencies $\omega_i$. To understand this point, let us denote by $n_M$ the largest of the integers $n_i$ appearing in (8.5), and let us consider a quasi-periodic regime described by (8.1) with the truncation criterion (8.2). Then the highest harmonic of $\omega_0$ appearing in (8.1) has the order $Mn_M$, which may be much larger than $N_H$ if $\omega_0$ is small with respect to at least one of the $\omega_i$'s. By Shannon's theorem, a strictly periodic function must be sampled at least at the Nyquist rate, which means that the minimum number of sampling points is $2Mn_M + 1$. As a consequence, the sampling mechanism may turn out to be extremely redundant, depending on the frequency values, and thus the computation may be extremely inefficient. This is because the spectrum typically consists of clusters of lines separated by comparatively broad gaps, where no significant harmonics exist, but is treated as if it were dense for sampling purposes.

For example, let us consider a two-tone intermodulation problem with $M = 4$, $\omega_1/2\pi = 5$ GHz, $\omega_2/2\pi = 5.01$ GHz. In this case the fundamental is 10 MHz and the highest harmonic is 20.04 GHz, so that the minimum number of sampling points is 4010, which usually is unreasonable from the viewpoints of both numerical efficiency and memory. As a matter of fact, it has been shown that working with a number of sampling points larger than 1000 is substantially impractical, even when using large mainframes such as vector supercomputers [16]. This example demonstrates quantitatively the practical limitations of such a method.

Despite its limitations, this simple method just described has been used successfully to solve a number of typical two-tone excitation problems, such as mixer analysis with relatively large IF values [17–19]. It is worth mentioning that this approach was used to implement the first direct numerical optimization of a FET mixer [18].

The reduction of a quasiperiodic regime to a strictly periodic one can be accomplished in a much more sophisticated and efficient way by resorting to the mapping technique [20]. In this method, a map is established between the original spectrum and an auxiliary one (the so-called describing spectrum) in such a way that the describing spectrum can be efficiently treated by the DFT and the Fourier coefficients of the nonlinear subnetwork response remain the same.

To understand how this can be done, consider a two-tone excitation problem defined by the truncation criterion (8.2). In the mathematical representation (8.1) of a generic waveform, replace the original (angular) frequencies $\omega_1$, $\omega_2$ by a corresponding set of describing frequencies $\omega_1'$, $\omega_2'$ defined as follows:

$$\omega_1' = M\omega_0'$$
$$\omega_2' = (M + 1)\,\omega_0' \tag{8.6}$$

where $\omega_0'$ represents the fundamental frequency of the describing spectrum. The actual correspondence established by (8.6) between the two frequency spectra is best illustrated by the example reported in Table 8.4, where the two sets of frequencies are listed for the $M = 3$ case.

**Table 8.4** Actual Correspondence between Two Frequency Spectra

| Original Frequency | Describing Frequency |
|---|---|
| $\omega_2 - \omega_1$ | $\omega_0'$ |
| $2\omega_1 - \omega_2$ | $2\omega_0'$ |
| $\omega_1$ | $3\omega_0'$ |
| $\omega_2$ | $4\omega_0'$ |
| $2\omega_2 - \omega_1$ | $5\omega_0'$ |
| $2\omega_1$ | $6\omega_0'$ |
| $\omega_1 + \omega_2$ | $7\omega_0'$ |
| $2\omega_2$ | $8\omega_0'$ |
| $3\omega_1$ | $9\omega_0'$ |
| $2\omega_1 + \omega_2$ | $10\omega_0'$ |
| $2\omega_2 + \omega_1$ | $11\omega_0'$ |
| $3\omega_2$ | $12\omega_0'$ |

In general, the spectral lines of the quasi-periodic signal are mapped by (8.6) onto the first $M(M + 1)$ harmonics of the fundamental $\omega_0'$ of the describing spectrum, no matter what are the actual values of $\omega_1$, $\omega_2$. Thus the transformed spectrum is always dense, with no gaps embedded between consecutive lines, and can be treated most efficiently by the DFT. Because the minimum required number of sampling points is $2M(M + 1) + 1$, sampling is nonredundant and the maximum possible numerical efficiency is achieved for the given number of spectral components.

The invariance of the Fourier coefficients under the mapping (8.6) is shown in [20] for the case of a memoryless (resistive) nonlinearity of the form $f[x(t)]$. The basic idea is that, if the given nonlinearity generates only a finite number of nonnegligible IM products (such as those listed in Table 8.4 for $M = 3$) under the applied two-tone excitation, it is equivalent to a polynomial nonlinearity of order $M$ (*generating function*). The invariance then follows from the one-to-one correspondence between the spectral lines produced by the generating function under the same excitation in the original and in the transformed frequency domains.

When the nonlinearity is not purely resistive, some care must be taken in the application of this technique, since the frequency-dependent operations, such as differentiation with respect to time, must be performed in the frequency domain using the actual frequencies [20]. For example, let us consider a very simple reactive nonlinearity of the form

$$i(t) = \frac{d}{dt} q[x(t)] \tag{8.7}$$

where $q(x)$ is memoryless. In two-tone excitation conditions the harmonics $Q_k$ of $q[x(t)]$ can be computed by the mapping technique; however, the harmonics of $i(t)$ must be computed in the original frequency domain as

$$l_k = j(k_1\omega_1 + k_2\omega_2)Q_k \tag{8.8}$$

In practice, this means that each nonlinear term in the device model must be decomposed into its frequency-dependent and frequency-independent parts, so that the frequency-dependent steps of the model description can be performed in the frequency domain at the actual frequencies. Although conceptually this need not imply any restriction of the nonlinear device modeling capability, it does, nevertheless, result in considerably greater difficulty in handling nonlinear models. This may be particularly annoying from the viewpoint of general-purpose HB simulators allowing the introduction of user-defined device models and probably is why the mapping technique, although very appealing in principle, has not found widespread practical application until recently. It is worth noting that in the case of more than two exciting tones, a simple mapping law similar to (8.6) is not readily available: for a given order $M$, the mapping has to be found by solving numerically a constrained diophantine problem [20].

An important advantage of the approaches described in this subsection is that they can make use of very efficient numerical methods such as the *fast Fourier transform* (FFT) or the Hartley transform to compute the DFT.

### 8.2.6.2 Matrix Methods

The spectral components of the nonlinear subnetwork response to a quasi-periodic excitation of the form (8.1) can be computed directly by relating the harmonics to the time-domain samples through a set of linear equations. Equation (8.1) can be rewritten as

$$X(t) = X_0 + \mathrm{Re}\left[\sum_{n=1}^{N_H} X_n \exp(j\Omega_n t)\right] = Y_0 + \sum_{n=1}^{N_H} (Y_{2n-1} \cos\Omega_n t + Y_{2n} \sin\Omega_n t) \quad (8.9)$$

where $\Omega_n$ represents the frequency of any one of the IM products, arbitrarily numbered from 1 to $N_H$, and thus is a quantity of the form $\sum k_i \omega_i$. By comparing the two expressions on the righthand side of (8.9), we get

$$\begin{aligned} Y_0 &= X_0 \\ Y_{2n-1} &= \mathrm{Re}(X_n) \\ Y_{2n} &= \mathrm{Im}(X_n) \\ n &= 1,2, \ldots , N_H \end{aligned} \quad (8.10)$$

Let us define an arbitrary set of sampling instants $t_r$ $(r = 1,2, \ldots , N_P)$, and compute both sides of (8.9) at each $t_r$. Then introduce the following vectors:

$$\begin{aligned} \mathbf{Y} &= [Y_0 \ldots , Y_{2n-1} \, Y_{2n} \ldots]^T \\ \mathbf{S} &= [x(t_1) \ldots x(t_r) \ldots]^T \end{aligned} \quad (8.11)$$

and a matrix $\mathbf{F}$ whose generic ($r$th) row is given by

$$\mathbf{f}_n = [1 \cos\Omega_1 t_r \, \sin\Omega_1 t_r \ldots \cos\Omega_n t_r \, \sin\Omega_n t_r \ldots] \quad (8.12)$$

The relationship between harmonic components and time-domain samples may then be expressed by the linear system:

$$\mathbf{F}\,\mathbf{Y} = \mathbf{S} \quad (8.13)$$

The matrix $\mathbf{F}$ has dimensions $N_P x N_D$, so that in principle we have only to choose $N_P = N_D$ and to solve the system (8.13) by conventional methods in order to solve the problem. In practice, some care must be taken in selecting the sampling points, to ensure that (8.13) be well conditioned from the numerical viewpoint.

As a matter of fact, numerical ill-conditioning occurs for the most obvious choice (i.e., uniform sampling $(t_r = r \, \Delta t)$. As shown by Kundert et al. [21], in this case, some of the rows of $\mathbf{F}$ usually are almost linearly dependent, so that the matrix is nearly singular. Then the method is not directly applicable because the accuracy becomes poor. Nevertheless, uniform sampling can be used [22], provided that (8.13) is solved by least-squares, rather than in a conventional sense. To do so, a redundant sampling is used with a total number of points 1.5 to 3 times larger than the theoretical minimum given by (8.3), so that $\mathbf{F}$ becomes a tall rectangular matrix. Both sides of (8.13) are then premultiplied by $\mathbf{F}^T$ to obtain

$$\mathbf{Y} = (\mathbf{F}^T \, \mathbf{F})^{-1} \, \mathbf{F}^T \, \mathbf{S} \tag{8.14}$$

It is shown [22] that the matrix $\mathbf{F}^T \, \mathbf{F}$ is singular if and only if one of the following conditions is satisfied:

$$|\Omega_i \pm \Omega_k|\Delta t = 2n\pi \qquad \begin{matrix} i,k = 1,2, \ldots N \\ n = 0,1,2, \ldots \end{matrix} \tag{8.15}$$

Thus, in practice, it is sufficient to choose the sampling interval far enough from all of the values defined by (8.15) to ensure a good numerical behavior of the algorithm.

An important particular case of (8.15) is encountered when two of the IM products—say, the $i$th and the $k$th—are very close to one another. Then, since the lefthand side of (8.15) should not be small in order to avoid ill-conditioning, $\Delta t$ must satisfy the constraint

$$\Delta t \gg \frac{1}{|\Omega_i - \Omega_k|} \tag{8.16}$$

which means that the sampling interval is usually large.

An obvious drawback of this procedure is the need for redundant sampling, which results in a considerable loss of numerical efficiency. This need is generated by the use of a set of uniformly spaced sampling instants, a condition that must be eliminated if a nonredundant sampling is desired. This requires an alternate method for choosing the sampling instants in order to ensure a good numerical behavior of the algorithm. A possible solution is given by the "near-orthogonal selection algorithm'' described by Kundert et al. [21]. The basic idea is to select an optimum set of sampling instants within a larger initial set in such a way that the rows of the matrix **F** are orthogonal to each other as much as possible. This results in the best possible numerical conditioning of the problem for the given initial set of sampling instants. The criterion for choosing the latter is heuristic and established on the basis of experience. The method suggested in [21] is to pick up, by a conventional random generator, twice as many points as required (i.e., $2N_D$) within the time interval $[0, 3T_{max}]$, where $T_{max}$ is the period associated with the smallest of the exciting frequencies.

For every sampling instant $t_r$ found in this way, a corresponding row of the system matrix can be computed by (8.12). A nearly orthogonal set of rows (and thus a well-conditioned matrix) is then obtained in the following way. One of the rows, say, $\mathbf{f}_1$, is arbitrarily selected as the first row of the final matrix. The component of $\mathbf{f}_r$ orthogonal to the direction of $\mathbf{f}_1$ is then computed as

$$\mathbf{f}'_r = \mathbf{f}_r - \frac{\mathbf{f}_1^T \mathbf{f}_r}{\mathbf{f}_1^T \mathbf{f}_1} \mathbf{f}_1 \qquad r = 2,3,2N_D \tag{8.17}$$

Note that the row vector (8.12) always has the same Euclidean norm $(N_H + 1)^{1/2}$. Thus the value of $r$ yielding the maximum norm of $\mathbf{f}'_r$ identifies the row vector $\mathbf{f}_r$ that is most nearly orthogonal to $\mathbf{f}_1$. This is chosen as the second row of the final matrix and used in place of $\mathbf{f}_1$ to perform the second step of the algorithm. The same procedure is repeated until a set of $N_D$ nearly orthogonal rows has been selected.

It is noteworthy that the point selection procedure as well as the inversion of the system matrix **F** have to be carried out only once prior to starting the Newton iteration. After that, the extraction of the harmonic components from the time-domain samples requires only the multiplication of a square matrix of size $N_D$ by a vector of the same size. Thus, the overhead required by the initialization, although relatively large [21], is not practically significant with respect to the overall analysis cost.

Nevertheless, it is quite clear that the speedup provided by the FFT or by similar techniques is not available in this case. Thus, the matrix method even in its nonredundant version is considerably less efficient than other (equally nonredundant) FFT-based algorithms, such as the mapping technique described in the previous subsection. As an example, let us consider a two-tone excitation problem defined by the truncation criterion (8.2) with $M = 7$. In this case, $N_D = 113$ and the matrix method takes about 20 ms to carry out a transform (after initialization) on a VAX 8800. On the other hand, the mapping technique can perform the same transform by an FFT of size 128, requiring about 2.5 ms on the same machine.

### 8.2.6.3 The Multiple Fourier Transform Method

The analysis methods described earlier make use of a one-dimensional array of sampling instants to describe the time-domain response of the nonlinear sub-network. Thus, they are referred to as *one-dimensional methods*. An effective alternative to the one-dimensional philosophy is to make use of a multidimensional set of time-domain samples followed by a multiple Fourier transformation [23]. To illustrate this concept, let us consider a typical nonlinearity of the parallel conductance-capacitance type, whose response $i(t)$ to a time-domain excitation $v(t)$ is defined by

$$i(t) = G[v(t)] + C[v(t)] \frac{dv}{dt} \tag{8.18}$$

$G$ and $C$ are assumed to be nonlinear memoryless functions of the arguments indicated in brackets. The extension to more general types of nonlinearities is immediate [23].

Let $v(t)$ be a multitone excitation of the form (8.1). When a suitable truncation criterion is adopted, $v(t)$ is completely identified by the finite set of its harmonics. Let

$$v(t) = \mathrm{Re}\left[ \sum_{\mathbf{k}} V_{\mathbf{k}} \exp\!\left(j \sum_{i=1}^{N} k_i \omega_i t\right) \right] \tag{8.19}$$

then, for the time derivative, the expression

$$\frac{dv}{dt} = \mathrm{Re}\left[ \sum_{\mathbf{k}} \left(j \sum_{i=1}^{N} k_i \omega_i\right) V_{\mathbf{k}} \exp\!\left(j \sum_{i=1}^{N} k_i \omega_i t\right) \right] \tag{8.20}$$

It is quite clear that (8.19) and (8.20) are structurally very similar and differ only in the actual values of the complex coefficients.

The key step toward the application of the multiple-transform approach is to consider the quantities

$$z_i = \omega_i t \tag{8.21}$$

appearing in (8.19) and (8.20) as *independent* variables. By means of (8.21), $v(t)$ is changed into a function of several variables $z_i$ and is $2\pi$-periodic with respect to each one. Assuming that all of the variables except one, say, $z_i$, are assigned constant values, $v(t)$ is reduced to a $2\pi$-periodic function of one variable, which can be sampled at the usual Nyquist rate, with uniformly spaced sampling points defined by

$$z_i = (r_i - 1) \frac{2\pi}{N_{Pi}} \qquad i = 1,2, \ldots N_{Pi} \tag{8.22}$$

If the highest-order harmonic number of $\omega_i$ appearing in (8.19) is denoted by $K_i$, then the number of sampling points in the $i$th dimension must satisfy the inequality (similar to (8.4))

$$N_{Pi} > 2K_i \tag{8.23}$$

For $i = 1, 2, \ldots, N$, Equation (8.22) defines an $N$-dimensional grid of sampling points in the $z$-space.

To compute the harmonics of (8.18) the following procedure is now adopted. Assume that the harmonics of $v(t)$ have been assigned in some way; say, by the generic iteration of a nonlinear solving algorithm. The harmonics of the time derivative are readily obtained by inspection of (8.20). Both $v(t)$ and its time derivative may now be sampled by computing through (8.21) the values they take at all points of the $N$-dimensional $z$-space defined by (8.22) for $i = 1,2, \ldots, N$. The sampled values of the linear subnetwork response are then directly computed by (8.18).

Let us now consider the one-dimensional string of samples of $i(t)$ corresponding to a set of predetermined values of $z_2, z_3, \ldots, z_N$, and to all values of $z_1$ defined by (8.22). These are the time-domain samples of a $2\pi$-periodic function of $z_1$ and can be treated in the usual way. The application of the FFT to these samples yields the harmonics of such function. The same procedure is repeated for all possible combinations of $z_2, z_3, \ldots, z_N$. The resulting $N$-dimensional matrix of harmonics is then treated in exactly the same way as the original matrix of time-domain samples, but with $z_2$ now playing the role of $z_1$. The same procedure is repeated cyclically with respect to all the independent variables $z_i$ until $i(t)$ has been Fourier-transformed with respect to each one. At this stage, the required harmonics are available.

The preceding discussion is simply a description of the conceptually easiest implementation of the *multidimensional Fourier transform* (MFT). It shows that an MFT can always be reduced to a sequence of one-dimensional transforms. Of course, this is usually not the most efficient solution from the numerical viewpoint. Specialized MFT algorithms are available in most mathematical libraries, which can reduce the CPU time required to carry out the transform even by one order of magnitude or more in the case of high-dimensional problems.

As discussed earlier, the numerical efficiency of a harmonic-balance method is heavily dependent on the required number of sampling points. In the case of a multitone excitation defined by the truncation criterion (8.2), we have $K_i = M$ for all values of $i$, so that the minimum required number of samples is

$$N_P = (2M + 1)^N \tag{8.24}$$

This implies that the MFT method generally makes use of redundant sampling with respect to the criterion (8.2). As an example, for a two-tone excitation the total number of samples used by a nonredundant mapping technique is

$$N_P = 2M(M + 1) + 1 = 2M^2 + 2M + 1 \tag{8.25}$$

and thus is about one-half of the number given by (8.24) for $N = 2$.

On the other hand, the MFT method has a number of distinctive advantages over other approaches described. If library routines for performing the MFT are available, this method is easy to implement and does not require any major modification to the architecture of an existing harmonic-balance simulator. Also, the subroutines for nonlinear device description remain strictly the same, which is very desirable for user-oriented simulators allowing user-defined device models [24]. Unlike the mapping technique, the MFT can easily be extended from the commonplace two-dimensional excitation to cover cases of more than two intermodulating tones. For instance, the only full numerical analysis of two-tone IM distortion in microwave mixers reported so far [25] makes use of a three-dimensional MFT.

For these reasons, the MFT method is believed to offer good trade-offs between ease of implementation and numerical efficiency and is preferred for general-purpose harmonic-balance simulators [24].

### 8.2.6.4 Harmonic Balance Solution Approach

In the general case of multitone excitation, the starting point for an HB analysis is the assumption that all electrical quantities related to circuit operation such as voltages, currents, and charges, may be represented by means of generalized Fourier expansions. This may be considered the quantitative interpretation of the qualitative concept of a circuit operating in steady-state conditions. Out of the whole set of previously mentioned electrical quantities, a suitable number are chosen as the state variables to be used in the mathematical description of the circuit behavior. Quite obviously the number of state variables depends on the strategy adopted for solving the circuit equations. In the following, we shall be concerned mainly with the so-called piecewise harmonic-balance technique (PHB), whereby the circuit is subdivided into linear and nonlinear subnetworks [25], both of which are multiports and share a common number of ports $n_D$. The nonlinear are referred to as the *device ports*. When using the PHB method, the number of state variables usually is equal to the number of device ports. A generic state variable is denoted by $x(t)$.

For numerical purposes we have to deal with systems having a finite number of degrees of freedom. This can be obtained by truncating the generalized Fourier expansions for the state variables according to some suitable criterion. After truncating, the circuit steady state is completely identified in the state space by a state vector $\mathbf{X}$, whose elements are the complex amplitudes of all spectral (sinusoidal) components of the state variables. Analyzing a nonlinear circuit by the HB technique means finding the state vector, once the linear subnetwork topology, the nonlinear device models, and the exciting sources have been specified. A peculiar advantage of this approach is that the problem can be reduced to the solution of a nonlinear system of algebraic equations for which fast numerical techniques are available.

The typical flow diagram of an HB analysis is shown in Figure 8.9. As for any mixed-mode simulation approach, the flow is bifurcated into a time-domain and a frequency-domain branch. The exchange of information between the two branches is made possible by the Fourier transform. In a standard analysis the frequency-domain branch is passed through only once, in the sense that the linear subnetwork is analyzed once at all frequencies of interest (those appearing in the generalized Fourier expansions of the state variables), and the results are stored for subsequent use. The analysis consists in computing one of the circuit matrices of the linear multiport, such as the admittance matrix, so that the linear subnetwork equations can be formulated as

$$\mathbf{I}_k = \mathbf{Y}(\omega_k)\mathbf{V}_k + \mathbf{J}_k \tag{8.26}$$

Where, $\omega_k$ represents the angular frequency of a generic spectral component and $\mathbf{I}_k$, $\mathbf{V}_k$ are vectors of current and voltage phasors at the linear subnetwork ports at frequency $\omega_k$. The admittance matrix is denoted by $\mathbf{Y}(\omega)$ and $\mathbf{J}_k$ represents a generic forcing term.

In the time-domain branch, the state-variable waveforms as well as their derivatives of any required order are computed for any given state vector $\mathbf{X}$. The nonlinear subnetwork usually consists of a set of nonlinear devices that are described by time-domain equations yielding the voltages and currents at the device ports as functions of the state variables and of their time derivatives. As an example, if we assume that the voltages at the device ports are chosen as the state variables ($x(t) = v(t)$), a typical formulation could be of the form:

$$\mathbf{I}(t) = \mathbf{G}[v(t)] + \mathbf{C}[v(t)] \cdot dv/dt \tag{8.27}$$

where $\mathbf{G}$ is a vector-valued nonlinear function, and $\mathbf{C}$ is a square matrix, all dimensions being equal to $n_D$. Although (8.27) is most common, more general

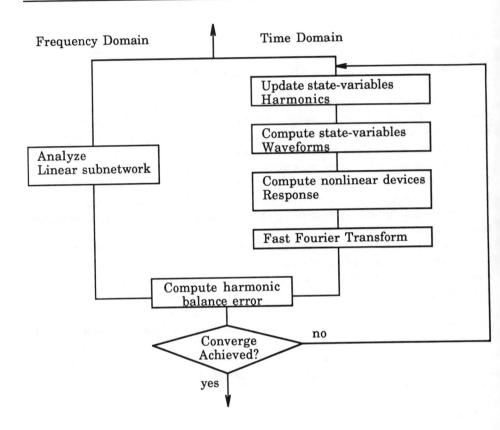

Fig. 8.9 Flowchart of harmonic-balance analysis.

formulations of the device equations may be necessary, or simply useful, for complicated device models [26].

The state-variable waveforms can be used through (8.27) or similar equations to compute the current and, if required, the voltage waveforms at the device ports. A Fourier transformation then yields a set of current and voltage phasors $\mathbf{I}_{Nk}$, $\mathbf{V}_k$, where the subscript $N$, for "nonlinear," is used to recall that the $I_{Nk}$ are generated through the nonlinear subnetwork equations. If the linear subnetwork equations are available in the form (8.26), the voltage phasors can be used to obtain a further evaluation of the current phasors, which in general will be different from the previous one. Such differences represent the *harmonic-balance errors*. For the description to be physically consistent, the harmonic-balance errors must be zero, which leads to the following formulation of the solution:

$$\mathbf{E}_k(\mathbf{X}) = \mathbf{I}_{Nk}(\mathbf{X}) - \mathbf{Y}(\omega_k)\mathbf{X}_k - \mathbf{J}_k = 0 \tag{8.28}$$

for all **k,** or more synthetically:

$$E(\mathbf{X}) = 0 \tag{8.29}$$

where **E** is the set of all HB errors.

According to the flowchart of Figure 8.9, the system (8.29) usually is solved by means of some iterative procedure. The Newton iteration is a common choice because of its numerical efficiency; however, the Newton algorithm may fail to converge when the circuit is strongly nonlinear and in general when the starting point is too far from the desired solution. For this reason, some HB simulators [24] make use of a more robust iteration scheme, such as a variable metric algorithm, that may be used to reach a point in the state space from which the Newton iteration may be safely started. In any case, an important aspect of the HB technique is that all the state-variable harmonics are determined simultaneously, with no need for a preliminary dc analysis.

### 8.2.6.5 Comparison with Time Domain

Let us denote by $N_H$ the number of spectral components that must be controlled in the analysis. When using the PHB technique, the total number of degrees of freedom of a steady state (i.e., the number of scalar quantities that must be known to completely determine it) is expressed by

$$N = n_D(2N_H + 1) \tag{8.30}$$

When using a time-domain simulator, the *minimum* number of time steps to be taken within a period of the steady state is $2N_H + 1$, still assuming that the number of harmonics to be kept under control is $N_H$. However, for purely numerical reasons, to control the propagation of numerical error from one time step to another, the actual number of time steps per period is usually much larger. Furthermore, the PHB technique makes use of a number of state variables equal to the number of device ports. In time-domain analysis, the state variables are the voltages across all capacitors, plus the currents through all inductors, plus the voltages at the transmission line ports [27], and thus is much larger, possibly by one order of magnitude or more, depending on circuit topology. Finally, in time-domain analysis it may be necessary to extend the integration over several RF cycles to achieve steady state, whereas only one cycle need be considered in HB analysis. The conclusion is that a time-domain analysis usually is extremely redundant with respect to an HB one, because it requires a much larger amount of information to achieve the same results. Although this argument is only qualitative, it gives a clear feeling of the superior efficiency that HB simulation can provide, within the previously mentioned limitations.

Another important advantage of HB simulation is related to the fact that it allows the linear subnetwork to be analyzed in the frequency domain. The linear subnetwork usually is treated as generated by the interconnection of a number of building blocks or circuit components. In a typical time-domain simulator, the only available circuit components are simple, idealized lumped elements, such as resistors, inductors, capacitors, and possibly elementary transmission-line models [27,28]. Anything more complex must be simulated by an equivalent circuit consisting of interconnected ideal components. This represents a serious limitation especially for MMIC analysis, it rules out all modern field-theoretical approaches to the simulation of MMIC components, because these methods invariably work in the frequency domain [29]. Also, the increased number of reactive components and, thus, of state variables may result in a very consistent loss of numerical efficiency. Commercially available mixed-mode simulators Microwave Harmonica™ [30] from Compact Software Libra™ [31] from EEsof Inc., and the mixed-mode simulator on an MMIC workstation offered by Hewlett-Packard have many distributed elements models, including microstrip discontinuities, radial stubs, lossy and dispersive coupled-microstrip sections, and so on, many of which do not admit a reliable equivalent-circuit representation. Furthermore, a harmonic-balance simulator will accept "black box" components defined by means of measured scattering (or other) parameters, whereas a time-domain simulator will not. Also, a public domain mixed-mode microwave simulator is available almost free from the University of California at Berkeley.

One final point is worth mentioning. The most modern trend in MMIC analysis is to overcome the circuit component concept and treat a chip as a whole; that is, as a unique electrical system to be analyzed globally by electromagnetic techniques [32]. In this case, the admittance matrix of the multiport linear subnetwork is directly obtained as the result of a mathematical procedure simultaneously involving the entire circuit pattern. The circuit description is layout- rather than component-oriented, and all possible types of couplings among different circuit parts are automatically taken into account, including proximity, surface-wave, and radiation effects. Although at present these techniques are not yet sufficiently developed to make them generally available, in the future they are likely to provide the ultimate accuracy in MMIC simulation. When it comes to nonlinear circuit analysis, these new techniques are fully compatible with the piecewise harmonic-balance concept but not with the time-domain approach.

### 8.2.7 Physical Layout

Physical design or chip layout involves converting the schematic design of a MMIC into a multilayer physical layout consistent with the process dependent design rules. During the last several years, the physical layout process has evolved from an entirely manual one of cutting Rubylith sheets to represent each mask layer to a completely automatic procedure using the compilers [33–37], particularly for silicon VLSI digital designs. However, on-line interactive IC layout remains one of the most widely used approaches to designing full custom ICs, and full custom still is the method of MMIC layout most commonly used. Four different approaches are used in silicon VLSI layout design [38]:

Geometric layout;
Symbolic layout;
Cell-based layout;
Procedural layout.

Geometric layout rules refer to the procedure in which the design creates the exact shapes of the circuit components on an IC mask. This system is usually a "polygon pusher" drawing system, and the polygons have no electrical significance compared to the circuit schematic. Ensuring the validity of such a layout requires back-end design verification as DRC-ERC-LVS. A symbolic layout is based on the "stick layout" concept developed for digital IC design [39]. Details of the layout are hidden to simplify the complex layout design. However, ultimately, symbolic layout techniques produce complete geometric layout details. Cell-based layout is a hierarchical system where the basic cells are layed out using one of these methods and stored in a cell library. These cells do not necessarily have all layers of layout visible to make the chip layout easier. Most often the externals or the visible shape of the cell is a single polygon representing the input, output, power, and ground pad locations. These cells are then used to complete the detailed layout by connecting them according to the circuit schematic. Procedural layout is generally used for structured designs, mostly digital circuit designs such as PLAs, PLDs, and ROMs. Such a method is not applicable to MMIC physical designs.

A hierarchical physical layout system suitable for MMIC designs will have the following main features:

- Ability to draw rectangular, 45°, 90°, and all angles as well as arcs and circles.
- User-defined colors and fill patterns for all layers.
- Boolean editing operations on layers such as logical AND, OR, NOT, XOR, and resizing to achieve increased editing flexibility and data entry for complex geometries.
- Conventional component and polygon moving, copying, deleting, mirroring and rotating capabilities.
- Window manipulations such as zoom-in and zoom-out.
- Edit-in-place to improve productivity in hierarchical design.
- Undo capability.

Layout systems are commercially available from Applicon, Valid Logic,

Silver-Lisco, Mentor, Daisy and Cadence. These software packages run on a variety of computer hardware from single-user workstations to mainframe computers. Several PC-based layout systems are available, but with limited capabilities. Also, several public domain physical layout systems are available, such as KICS, MAGIC, and OC Tools from the University of California-Berkeley. KICS is an older program and has limited capabilities; however, MAGIC has well-developed layout software to satisfy the needs of MMIC design. Different layout systems have their own internal data bases to represent various shapes; however, the industry standard data formats for layout are GDS, Applicon, and CIF (Caltech Interface Format). Almost every commercial IC layout system has interface to one or more of these standard data formats.

### 8.2.8 Back-End Design Verification

The *design verification software* (DVS) system has its origin in the development and evolution of integrated circuit technology in the 1970s. As the circuit size grew to the point where it was not reliable to check the layout manually, an imperative need arose for some form of computer-aided IC layout [40,41]. As the layout data already was handled by computers, the task of verification became to examine this mass of data for mistakes and unusual feature constructions, which might make the chip fail after fabrication.

Design verification software should be independent of technology; otherwise, it is extremely difficult to maintain the software, because the underlying fabrication technology changes so quickly. Moreover, there are many different technologies, such as NMOS, CMOS, bipolar, and GaAs, and each one would require a specialized design verification software package. An efficient approach is to keep verification rules in a technology file that uses a language easily understood by users. The user will have to define these technology files to completely describe a particular process in a way that is understood by, and compatible with, the generic verification software.

Layout design verification consists of three functions: design-rule, functional, and performance verifications. Design rules are a set of geometric constraints enforced on the physical layout [42]. The software package that checks the design for the critical violations of geometrical design rules imposed by the fabrication technology is called the *design-rules checker* (DRC). It detects situations where the layout geometries fail to conform to feature tolerance specifications, such as minimum width of conductor lines, spacing between them, or whether contact openings are enclosed by conductor lines with sufficient margins. These rules guard against catastrophic failures due to the limitations inherent in the manufacturing processes, such as alignment tolerances and photolithography, etching, and plating.

If design rules are violated, the chip yield will decrease, adversely affecting the chip production costs. For example, a metal line that is too narrow will have a greater possibility of opening up during the manufacturing process. Or the chip may be affected by electromigration, which causes reliability problems. Design rule verification checks these geometric design rules.

Although DRC provides basic and critical functions for checking the layout data, designers are still left with the daunting task of checking whether the chip represents the same circuit as the original design, usually given in the form of schematic diagram. In the case of manually generated layouts or full-custom designs, which is the method of designing MMICs most commonly used today, it becomes very difficult to manually verify a design with more than a few hundred devices. To relieve designers of this tedious and time-consuming task, *electrical rules checkers* (ERCs) and network consistency checkers (NCCs) or layout versus schematics (LVSs) are developed.

Comparing the physical layout with the logical schematic design involves two steps. First, the logical network needs to be extracted from the layout, and then it is compared with the original network. The first part of network extraction is the function of the ERC, and while accomplishing this, it can also perform some important checks that are not covered by the DRC. These checks might be called *electrical rules checks,* in the sense that they are not merely geometrical rules but deal with electrical properties, such as devices and nets. For example, any transistor should have the proper set of terminals: emitter, base, and collector for bipolar transistors and gate, source, and drain for MESFETs. The other obvious electrical rule error is power-to-ground shorts. Violations of these rules are reported by the ERC before the NCC starts its comparison. Once the layout is found to be free of electrical rule violations, it is compared with the schematic. This is the final check step to conclude that there are no functional discrepancies between the logical and physical design.

DRC, ERC, and LVS have become indispensable tools in the design cycle. By the late 1970s, most IC manufacturers and design houses were using these tools in one way or another. The importance and popularity of the layout verification products lie in the high cost of fabrication process. Running a design through the fabrication line is so costly in time and money that it is imperative to catch all the possible defects before a design is committed for mask fabrication or for manufacturing.

Another significant area in back-end verification is called *electrical parameters checker* (EPC) or the *performance verification software*. This is the logical extension of the ERC and LVS products. Once ERC verifies that all device and net constructions are error-free and LVS makes sure that the layout network is a faithful representation of the original logical design, we can be sure that the fabricated chip will meet the functional specifications. The next major concern is whether the chip will meet the performance specifications, such as appropriate frequency bandwidth, gain, flatness, and noise figure, which depends on the quantitative values of the components either layed out intentionally or inserted into the circuit as parasitics (e.g., interconnects).

As the layout data is the closest we can get to the physical realization of the chip in fabrication, it may also be the most accessible source of data for electrical parameter calculation. EPC, or the performance verification software, extracts parameters such as device dimension, parasitic capacitance, and resistance. And by extension, some packages offer the capability of comparing these extracted values against user-specified tolerances. EPC is a late-comer to the layout verification area and probably the most advanced product. It has not been accepted as readily as other products but is gaining increased recognition. With its close interface to simulation packages, it may become as useful as the other products. For MMIC design, such software is in its infancy and cannot address the high-frequency coupling effects, distributed elements effect, and the discontinuities generated in the layout.

A more accepted parameter extraction software in MMIC design is based on the solution of electromagnetic field equations. However, such software is very slow in execution because of the complexity of the mathematical problem for irregular shapes of layout interconnections. Moreover, the interface of such a parasitic extraction software with the layout software and the schematic and simulation software will be almost essential for any efficient design environment. Such an interface is almost nonexistent at this time.

In an MMIC parameter extraction, it is not sufficient to know the parasitic resistance and capacitance in interconnect lines. More important is the evaluation of microwave coupling between closely layed out interconnect lines and coupling between active and passive components. The most commonly used commercial packages for back-end design verification are Dracula™ [43] from Cadence, and Design Verification Software (DVS) from Silvar-Lisco. Also, there are some back-end design verification programs in the public domain software OCTools and MAGIC [44], which were described earlier.

DVS is a back-end process in the chip design, which means that by the time the design projects get to DVS, the design is already complete and the remaining task is to make sure there are no errors. Because of this back-end nature, the DVS system cannot take advantage of any type of interaction with the design process itself. The DVS software should be able to process almost all kinds of technologies and designs techniques without imposing any extra burden on the designers. This leads to great sacrifice in the efficiency and performance of the product. For example, if the design is orthogonal (right-angle geometries only), the processing algorithm can be tremendously simplified. If the input cell hierarchy is constructed like the standard cell designs, a great amount of CPU and disk resources can be saved by employing the hierarchical method in the processing. These extra constraints on the design methodology often are not acceptable to circuit designers. The DVS vendor, thus, is left with the most general cases to deal with, which makes the design of the product complicated and less efficient than it otherwise could be. Proprietary DVS systems developed by individual semiconductor manufacturers have less of a problem of this kind. This might be the most compelling reason why many companies have developed their own DVS tools [40].

### 8.2.8.1 Verification Techniques

IC physical designs consist of graphical polygons that are placed at different levels or layers depending on a particular process. These polygons must be presented in a coordinate form to be understood by the DVS software. The representation of these polygons and the types of functions defined to operate on them distinguish edge-based from area-based verification systems, the two commonly used techniques in design verification software. Area-based verification systems represent the areas defined by the polygons, either by the polygons themselves or by a subset of the edges of the polygons. These areas represent the minimum geometric unit on which verification functions can operate. The only functions that can be defined are those that operate on the areas to produce other areas. Examples of such functions include Boolean functions as ANDing, ORing, ANDNOT, etc., merge, oversizing, and undersizing operations.

Edge-based verification systems decompose each polygon of a layer into its constituent edges. Each edge is represented by the coordinates of the end points and the direction of the polygon interior. The generating layer is also retained for each edge. The ability to define and access the layer of each edge is of key importance for an edge-based verification system.

Since the edge is the fundamental unit on which functions can operate in an edge-based system, each edge must be present in the data base. The edge is represented by the coordinates of the end points and the direction of the polygon interior. Other information includes the generating layer, the node identification number, and the cell number (for hierarchical analysis). A polygon area can be represented as a series of enclosing edges.

Area-based systems employ a variety of techniques to represent polygons. One method is to save the sequence of polygon vertices [45]. Another method defines polygon areas using only the nonhorizontal edges. Still other techniques include the use of trapezoids to represent areas. In all area-based systems, individual edges are not important by themselves but only as parts of the entire polygon. It might be possible to define an access method that could reconstruct and return a certain amount of edge-related information from the area data base. However, unless the data base contains information about each edge, it is impossible to obtain all the necessary information about each individual edge.

### 8.2.8.2 Flat, Hierarchical, and Incremental Verifications

In the flat approach, all the hierarchy is exploded onto one flat level and all geometric figures are treated equivalently. This equivalence makes the programming task a little simpler than it is with other approaches. When using the hierarchical approach, the regularity of the design is taken into account. Cells that are repeated many times will be checked only once, because they have the same intracell geometric relations. The cell-composite interactions will be checked only once if the cell has the same environment. In order for the program to remember the checked interactions and skip the check if it is encountered a second time, more bookkeeping is included in the checking processes.

With the cell-incremental approach, the program examines the previous verification history. Previously verified subcells are not verified again if they have not been revised. The data base should contain information about the modification date and verification date of cells (or obtain information from the file system). In the hierarchical design methodology, this cell-incremental capability is quite desirable. Because designers usually have to wait for the results before pursuing further, a fast turnaround will increase the productivity significantly.

Whereas the cell-incremental approach skips reverifying unchanged subcells, the figure-incremental approach skips reverifying unchanged figures. This capability requires very close coupling between the design data base and the checkers so that the checkers know which figures were revised after the previous verification. The figure-incremental approach is useful when only a small portion of a flat design is changed. If a fair amount of changes are made, we might as well do a complete check.

The figure-incremental approach facilitates interactive checking in which checking is invoked every time the designer enters a figure. Interactive checking may be cost-effective in the personal workstation environment, because the job can be performed in the background and, presumably, the graphic editing will require only a small fraction of the computing power of the workstation. However, in order to be efficient, the interactive checker can only check local design rules.

Designers may not like the interactive feature because sometimes they want to cut corners to expedite the design process. For example, designers may want to create an illegal figure, knowing that it will be remodeled later on. The interactive DRC option is often turned off, even though it is available.

The main features of a typical design verification software package are as follows:

- *Graphics processing:*
  Graphics data format interface conversion of layout data into information understandable to DVS software (e.g., for edge-based systems, the polygon information is converted into edge information).

Hierarchy expansion, if needed.
Conversion into main data base format.
Removing redundant lines or cleaning.
Logical area merging.
Sizing.
Polygon association, such as touching or not.

- *DRC:*

  Basic conventional checks—width, spacing, enclosure, overlap, area.
  More advanced features—checks based on conditions, operations on edge
  data, directional checks, and checks on results of other checks.

- *ERC:*

  Extraction of net and device information from physical layout—heavy use
  of logical merges to create component layers, polygon association for
  connectivity processing, and definition of device flag and terminals and
  terminal count.
  Errors in construction.
  Device related errors—more or fewer than proper number of terminals,
  unusual connection, terminal shorts.
  Node related errors—current path to $V_{DD}$, $V_{SS}$, isolated node check,
  nodes not connecting two or more devices.
  Logic gate check—group of devices forming logic gate but output is not
  used.

- *LVS:*

  Net list format—interface to other formats, clarity and flexibility, hierarchy capability.
  Handling special situations—parallel devices, multifinger MESFETs, re-
  versibility check.
  Error representation—listing and graphics, option to limit number of errors, flexibility in defining starting points.
  Back annotation—identification annotated back in the design.

- *Parameter Extraction:*

  Device dimension—MESFET width and length, resistor width and length,
  intensional capacitor dimension, bipolar devices (user defined dimension
  calculation).
  Parasitic capacitance—conductor line parasitics, vertical overlaps, planar
  line-to-line.
  Resistance—derived from the dimension for explicit definitions only.
  Distributed RC—pi structure approximation, user-defined models.
  Other derived parameters—load-to-driver ratio.

- *Limitations of Current DVS System:*

  General—user interface not very advanced, many are written in old-style

Fortran code, hierarchy handling is not satisfactory.

For MMIC designs—most limitations are in ERC and parameter extraction programs; device recognition for inductors, transmission lines, open and shorted stubs, *et cetera,* currently not possible; in parasitic extraction, much more sophisticated formula-driven calculation is necessary.

### 8.2.8.3 Future Trends

Although verification software has improved in recent years, further improvements are desired for greater efficiency in speed and memory usage. In this respect, hierarchical verification is definitely the approach to select. It is faster by one order of magnitude than flat verification, and it also reduces disk storage by one order of magnitude. However, actual performance varies according to the extent of hierarchy and regularity in a circuit.

By taking the design hierarchy into account, the layout-*versus*-logic interconnection verification is more effective in locating discrepancies. It also can handle much larger circuits, because the design is analyzed level-by-level in the hierarchy instead of on one totally expanded flat level. Parameter extraction is a software package that merits the greatest attention at this time for MMIC designs. Much progress has been made in the area of artificial intelligence. Knowledge-based expert systems can be used to do in-depth analysis of rule violations.

### 8.2.9 Design Languages

As the software for MMIC designs proliferates to more and more workstation vendors, there will be a need to define standards for design data bases [45a]. This will be necessary for two main reasons: first, to communicate the design data base between various design workstations, and second, to communicate the design data at different steps of design from schematic capture to circuit simulation, then to physical layout and back-end design verification. This is particularly important when the designers are working with various MMIC foundries and an MMIC based subsystem is designed in cooperation with several companies using different design environments. Such standards for design data bases will provide a user friendly, efficient interface between design workstations from different vendors as well as between different components of the design software [46]. As these standards are accepted more in the CAD-CAE industry, a workstation vendor will not be able to lock a customer to one system, as the data relevant to the customer can be easily ported to any other system. Also, the ASIC vendor will be able to supply design libraries of their cells to a variety of CAE packages without an extra effort to port the cell libraries to various workstations individually.

Two representations for such standards commonly used in silicon digital IC design are the *electronic design interface format* [47] (EDIF) and *VHSIC hardware description language* (VHDL) [48,49]. Evolution of EDIF began in and has been adopted by many commercial workstation vendors for silicon digital IC design since then. EDIF has been adopted as an EIA standard. EDIF standard is a design data exchange format rather than a tool to be used directly for circuit design. Such a standard serves as a universal format to exchange data among designers and foundries. EDIF is a public domain data format that can transfer information about libraries, schematic capture, behavior, functional and logic structures, circuits, geometric layout, and test definitions for generation and simulation. EDIF can handle information on library and cell organization, cell interfaces, cell details, and processing technologies as well.

The cell interface includes logic symbols, ports, parameters, boundaries, port-to-port timing, feedthroughs, and functional test patterns. The cell-detail description includes netlists, schematic diagrams, geometric layouts, gate arrays, logic models, symbolic layouts, and simulation parameters. The processing-technology description includes layer definitions, device-size scaling information, and simulation values.

The main features of EDIF can be described as follows:

- EDIF has capability to define simulation-model transfer.
- EDIF currently supports structural and low-level functional description of the circuits.
- This is a nonprocedural language; that is, the order in which the statements appear is not intended to suggest the order in which events occur in the model. This approach makes the description of parallelism in the model very natural.
- The structure is separated from the behavior in defining the circuit: the structure information of the circuit resides in a netlist, the behavior information resides in the logic model view.
- Communication among different entities are handled via the netlist view of EDIF.
- EDIF provides for easy and general expression of timing behavior.

Major vendors of workstations who have indicated their committment to EDIF are Daisy, Hewlett-Packard, Mentor, Valid Logic, and Cadence. VHDL evolved as a part of the *very high speed integrated circuits* (VHSIC) program by the US Department of Defense and is being adopted as a standard by IEEE. Unlike EDIF, VHDL is a common design language that is execute oriented. It defines a digital design at a very high level, such as *register transfer level* (RTL), in its behavioral, architectural, or structural terms. Finally, VHDL generates the response of the circuit to given inputs. VHDL provides circuit descriptions that are oriented toward the use of a time-domain, event-driven simulator. VHDL is based upon the concepts of the ADA language and itself is a higher-order language such as Fortran, Pascal, or C. Just as these programming languages are user readable for describing and communicating the flow of a particular mathematical algorithm, VHDL is a user-readable language that describes the flow of logic operations in a circuit. It provides an integrating framework for design tools with:

- An unambiguous schematic representation of hardware components.
- A structure language for early detection of many errors of description.
- An extensive system to which additional tools can be attached in a straight-forward manner.

At this time, based on the developmental status of both languages, one language does not significantly outperform the other. On the one hand, VHDL is very strong in representing the hierarchical design methodology, making it easier for the system level designers. However, at this time VHDL does not have the capability to allow the easy flow of information at the chip-level designs. EDIF is inherently better at enabling data flow at the chip designer level and the flow of data between CAD and CAM. However, EDIF cannot handle data flow at the system level using hierarchical design methodology.

The major drawback of using EDIF in describing MMICs is its inability to express electrical models for analog or microwave circuits. In this situation, the alternative is to represent these circuits in Y-, Z-, or S-parameters in EDIF. However, it is difficult to represent nonlinear microwave circuits. A major effort will be required to improve the capabilities of EDIF to represent such circuits.

## 8.3 MASK FABRICATION

Since the late 1970s, electron beam (E-beam) exposure systems have played an increasing role in the fabrication of masks for the semiconductor industry. E-beam systems offered some significant advantages over the optical pattern generators and steppers that were the mask-making standards in the 1960s and 1970s [1]. E-beam data flow starting from layout data base format to converting into E-beam format is shown in Figure 8.10. A comparison of optical *versus* E-beam mask fabrication is shown in Table 8.5. In GaAs MMIC fabrication, where the gate lengths are always pushed to the limits of wafer fabrication capabilities, there are demanding requirements on minimum line widths and layer overlay tolerances. Particularly for one or two critical masks, the E-beam mask fabrication technique is almost universally used. Moreover, because of the problems in overlay tolerances of mixing E-beam and optically generated mask sets, E-beam fabricated mask sets are most commonly used in GaAs MMIC manufacturing.

The specifications limit of MEBES III, one of the most commonly used pieces of E-beam mask fabrication equipment, are shown in Table 8.6 [2,3]. Besides the mask-making equipment, the other critical aspect in mask fabrication is

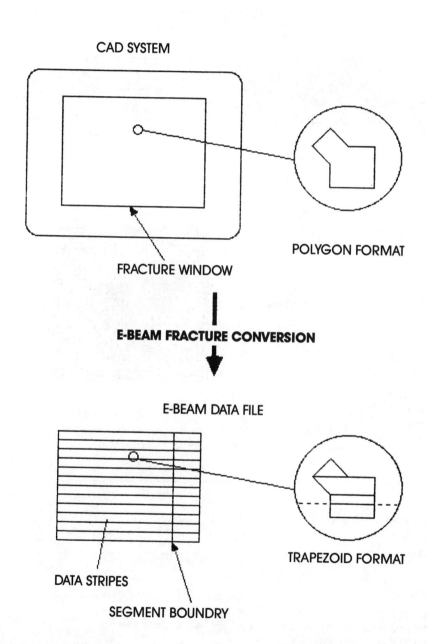

CAD SYSTEM

POLYGON FORMAT

FRACTURE WINDOW

**E-BEAM FRACTURE CONVERSION**

E-BEAM DATA FILE

TRAPEZOID FORMAT

DATA STRIPES

SEGMENT BOUNDRY

**Fig. 8.10** E-Beam data flow.

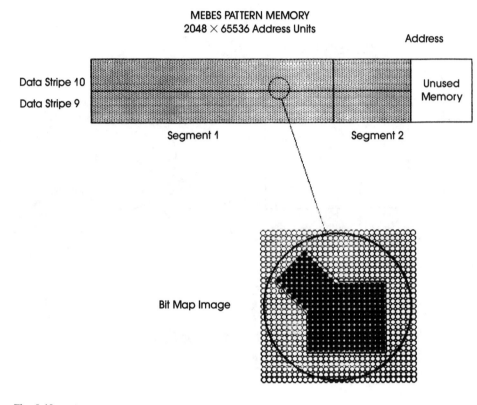

MEBES PATTERN MEMORY
2048 × 65536 Address Units

**Fig. 8.10** cont.

**Table 8.5** Optical *versus* E-Beam for 1X Projection Masks

*Optically Generated Masks*

Advantages:

Lower cost
Nonscalloped edge quality on angles, circles, and both X- and Y-parameters
Design grid of 0.1 $\mu$m resolvable

**Table 8.5** cont.

*Optically Generated Masks*

Disadvantages:

Die size and density (flash count) are limited. Die size limit is imposed by the quality and range
of view of the optical lense used in the optical system
Interdie critical dimension uniformity is a function of light uniformity
Array complexity is limited by reticle cost and manufacturability
Reticle-to-reticle critical dimension and registration variation
Larger number of processing steps limit theoretical throughput

*E-Beam Generated Masks*

Advantages:

Improved registration capability
Array size and density flexibility
Improved throughput due to fewer process steps
Die size is limited only by 6-in. stage travel

Disadvantages:

Minimum design grid 0.1 $\mu$m, which is the limit of the address size or the spot size
Cost, especially when utilizing design grid < 0.25 $\mu$m is higher, as it considerably increases the
write time
Scalloping on angled, circular, and horizontal edges

the choice of mask plate blanks and the thin film coating the blanks. Four principal
materials are used for mask plates: green and white soda lime, low expansion
borosilicate, and synthetic quartz. The main considerations in the choice of a
material are the thermal expansion coefficient, refractive index, transmittance

**Table 8.6** MEBES III Typical Mask Specifications (3 $\sigma$ or 99.7%)

| | |
|---|---:|
| Line edge roughness | $\pm 0.05 \ \mu$m |
| Pattern butting accuracy | $\pm 0.08 \ \mu$m |
| Level-to-level overlay accuracy | $\pm 0.12 \ \mu$m |
| Line width control | $\pm 0.10 \ \mu$m |
| Line width accuracy | $\pm 0.10 \ \mu$m |
| Machine-to-machine overlay accuracy | $\pm 0.15 \ \mu$m |

coefficient, and material durability. A comparison of material characteristics relevant to the mask fabrication is shown in Table 8.7. Transmission coefficient of light at different wavelengths from the UV to the deep-UV spectrum is shown in Figure 8.11 [4] for different materials. Even though quartz is the most expensive material mentioned, it is the one most widely used for mask fabrication today, especially when larger plate sizes and extremely small line widths (submicron range) are required. The single most important reason for this choice is the extremely low coefficient of thermal expansion for quartz. Thermal coefficient of expansion of quartz is plotted as a function of temperature in Figure 8.12 [4]. In photolithography, a low thermal coefficient of expansion is directly related to the registration; the smallest linewidth realistically possible in wafer processing depends not only on resolution but also on the accuracy with which successive masks can be aligned. The runout between different mask layers, from errors in the successive mask or reticle levels, can be caused either by thermal expansion effects or by insufficient flatness in the quartz plate.

Plate flatness is controlled in the manufacturing process, which should be able to produce 2–5 $\mu$m flat plates, and in the use of properly thick plates (e.g., 90 mils for 4 in., 120 mils for 5 in., and 150 mils for 6 in.) to reduce sag.

In exposure systems a mask temperature rises more than the temperature of the wafer. Even if both mask and wafer could be maintained at the same rise in temperature, runout still would occur because mask materials are not available that match the thermal expansion coefficient of GaAs (6.86 ppm/°C). The control of runout was a leading reason for increased use of LE borosilicate glass blanks in the past. With the use of larger wafers and finer lines, quartz is now often the material of choice. The use of quartz reduces the thermally produced runout to about 0.1 $\mu$m. Runout as a function of array size on the mask plate for different materials is shown in Figure 8.13 [4].

Besides material of the plates, the other critical choice is the material coating the plates. Commonly used materials are low-reflectance or medium-reflectance

**Table 8.7 Plate Blanks Material Composition**

| | White Soda Lime | Green Soda Lime | LE Borosilicate | Synthetic Quartz |
|---|---|---|---|---|
| *Composition:* | | | | |
| $SiO_2$ | 73% | 70% | 60% | 100% |
| BO | | | 5% | |
| AlO | 1% | | 15% | |
| NaO | 15% | 8% | 1% | |
| KO | 1% | 9% | 1% | |
| RO | 10% | 13% | 18% | |
| *Material characteristics:* | An economical substrate which is widely used for producing working (copy) masks. | | Possesses excellent thermal properties very similar to the silicon wafer. | A superior glass substrate for producing photomasks with very fine patterns. Exhibits extremely low expansion and excellent chemical properties. |
| *Material Properties:* | | | | |
| Thermal expansion coefficient 25 ~ 100° ppm/°C | 9.4 | | 3.7 | 0.52 |
| Refractive index | 1.52 | | 1.53 | 1.47 |
| Transmittance 200 nm | 0 | | 0 | 90% |
| 436 nm | 90% | | 91% | 92% |
| Specific gravity | 2.5 | | 2.6 | 2.2 |
| Softening point (°C) | 740 | | 900 | 1,870 |
| Young's modulus (kg/mm²) | 7,300 | | 8,450 | 7,413 |
| Vickers microhardness (kg/mm²) | 420 | | 640 | 870 |
| Chemical durabilities (Å/min.) NaOH 10% 80°C | 50 | | 120 | 40 |

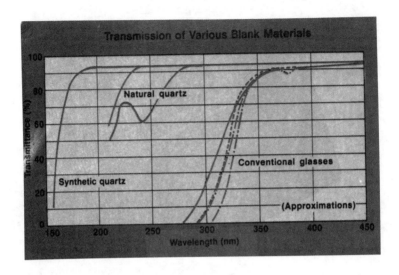

**Fig. 8.11** Transmission of various blank materials (from [4]).

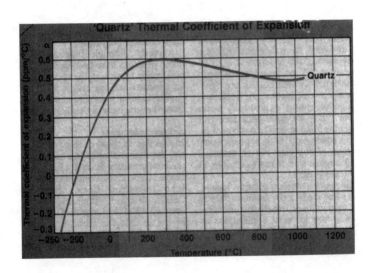

**Fig. 8.12** Quartz thermal coefficient of expansion (from [4]).

chrome and see-through chrome and iron oxide. See-through chrome, an alternative to conventional blanks coated with iron oxide, is particularly valuable where manual alignment to existing circuitry is still practiced. See-through chrome combines the transparency of iron oxide for alignment with the edge quality of

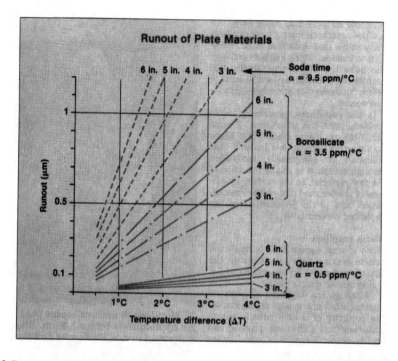

**Fig. 8.13** Runout as a function of wafer size, mask material, and temperature difference between mask and wafer (from [4]).

chrome. Relevent characteristics of the coating materials are shown in Table 8.8. LR, MR, or HR coatings of chrome are chosen to optimize the photolithographic process in terms of wavelength of light, exposure time, and the type of photoresist used. See-through chrome or see-through iron oxide coated plates are generally used for backside processing alignment in GaAs MMIC fabrication, such as for substrate via holes.

Choice of spot size during E-beam mask fabrication is an important consideration. During the E-beam writing process, the whole reticle is divided into smaller strips of total address size of 2048 × 65536, where the absolute length and width of the strip depends on the spot size or the exposure size, as shown in Table 8.9. By reducing the spot size, we can achieve better overall line definition and smaller line widths, but at a cost in the total number of strips into which the E-beam layout reticle is to be divided. This increases the number of buttings (for example, in a gate finger), which increases the butting problems if extreme care is not taken during exposures and stage travel of E-beam system. The roughness of acute angle lines (other than 45°) also depends on the spot size as shown in Figure 8.14. This fabrication limitation of the E-beam system requires all gate fingers or

**Table 8.8 Comparison of Coating Materials**

| Name | Application | Optical Density (visual) | Reflectance (436 nm) | Film Thickness | Transmittance |
|---|---|---|---|---|---|
| Standard: | | | | | |
| Low-reflective (LR) chrome (double layer) | | 2.6 | 11% | 1050Å | |
| Medium-reflective (MR) chrome | Reticle and mask (optical and EB, especially EB) | 2.9 | 31% | 900Å | |
| High-reflective (HR) chrome | | 2.9 | 51% | 800Å | |
| See through: | | | | | |
| See-through chrome | Mask for high precision mask aligners | 450 nm : 2.1 Visual : 0.6 | 22% | 1800Å | 550 nm : 15% |
| See-through iron oxide (FeO) | | 450 nm : 2.3 Visual : 0.5 | 22% | 3100Å | 550 nm : 30% |
| Double-sided, low-reflectance: | | | | | |
| Low-reflective chrome (triple layer) | Kasper aligner | 2.5 | 11% | 1200Å | |

**Table 8.9** Strip Length and Width *versus* Spot Size (address size 2048 × 65536)

| Spot Size (μm) | Strip Size (μm) |
|---|---|
| 0.125 | 256 × 8192 |
| 0.25 | 512 × 16384 |
| 0.375 | 768 × 24576 |
| 0.50 | 1024 × 32768 |
| 0.75 | 1536 × 49152 |
| 1.0 | 2048 × 65536 |

small line widths to be laid out in the same orientation. If 0.5 μm gate fingers are layed out horizontally as well as vertically, the horizontal lines will be quite rough and thus not useful for wafer fabrication or photolithography printing on the wafer. Typical E-beam 1X mask specifications [5] for an MMIC process using contact printing are shown in Table 8.10. Defect density of 1 defect per square inch of no more than 1.0 μm size and pinhole size of no more than 1.0 μm can be achieved during E-beam mask fabrication.

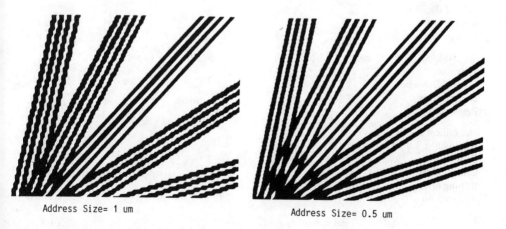

Address Size= 1 um

Address Size= 0.5 um

**Fig. 8.14** Effect of spot size on roughness of acute angle lines.

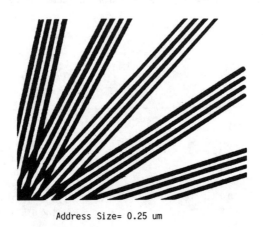

Address Size= 0.25 um

**Fig. 8.14** cont.

**Table 8.10** Typical E-Beam 1X Mask Plate Specifications

Critical layers
(line widths 0.5 $\mu$m or less):

| | |
|---|---|
| Spot size | 0.125 $\mu$m |
| Critical dimension | X $\pm$ 0.1 $\mu$m |
| Butting error | < 0.1 $\mu$m |
| Registration | $\pm$ 0.15 $\mu$m |

(line widths > 0.5 $\mu$m and < 1.0 $\mu$m):

| | |
|---|---|
| Spot size | 0.25 $\mu$m |
| Critical dimension | X $\pm$ 0.15 $\mu$m |
| Butting error | $\pm$ 0.1 $\mu$m |
| Registration | $\pm$ 0.15 $\mu$m |

Noncritical layers
(line widths 3.0 $\mu$m or more):

| | |
|---|---|
| Spot size | 0.5 $\mu$m |
| Critical dimension | X $\pm$ 0.2 $\mu$m |
| Butting error | $\pm$ 0.1 $\mu$m |
| Registration | $\pm$ 0.25 $\mu$m |

# APPENDIX 8A[1]

## 8A.1 Network Parameters Output

Traditionally, circuit simulation programs such as SPICE have the capability of analyzing circuits in time domain as well as frequency domain. However, in frequency-domain analysis, the output from these generic programs has been in the amplitude and phase of voltage and currents format, whereas microwave circuit designers are more accustomed to analyzing the output in terms of S-parameters. To implement the format to obtain the S-parameters out of SPICE is an involved process and requires a knowledge of programming language, the data structure of SPICE, and algorithms to convert voltages and currents into S-parameters for a generalized circuit.

We present a procedure to evaluate the S-parameters of a circuit by using any generic or modified version of SPICE without making any internal modifications in SPICE. This method is qualified for a two-port network, but the concept can be extended to an $n$-port network. The method is generalized for analyzing any circuit, active or passive, with input, output, and ground terminals. The circuit to be analyzed is considered to be a black box.

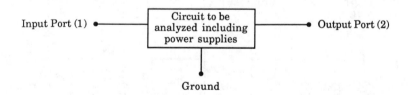

Input Port (1) — Circuit to be analyzed including power supplies — Output Port (2)

Ground

**Fig. 8A.A**

---

[1] This appendix was written by R. Goyal and W. Thomann.

To obtain the S-parameters for the circuit, extra input and output circuitry needs to be attached. Moreover, either $S_{22}$ and $S_{12}$ or $S_{11}$ and $S_{21}$ can be estimated in once SPICE analysis. Hence, to get all four S-parameters of the circuit, SPICE would need to be run twice. The circuit to be analyzed is defined in the SPICE input file as a subcircuit.

**Table 8A.A**

| RS | 1 | 0 | 50.0 ohm | | |
|----|-----|-----|----------|---|---|
| I1 | 1 | 0 | AC -20M | | |
| E11 | 10 | 0 | 1 | 0 | 2 |
| V11 | 10 | 11 | AC 1 | | |
| R11 | 11 | 0 | 1.0 ohm | | |
| Xckt | 1 | 2 | ckt | | |
| RL | 2 | 0 | 50.0 ohm | | |
| R21 | 21 | 0 | 1.0 ohm | | |
| E21 | 21 | 0 | 2 | 0 | 2 |
| .AC | LIN | 20 | 1 GHz | 20 GHz | |

```
.PRINT   AC   VM(21) VDB(21) VP(21) VM(11) VDB(11) VP(11)
.PLOT    AC   VDB(21) VDB(11)
.END
```

**Table 8A.B**

| RS | 1 | 0 | 50.0 ohm | | |
|----|-----|-----|----------|---|---|
| E12 | 12 | 0 | 1 | 0 | 2 |
| R12 | 12 | 0 | 1.0 ohm | | |
| Xckt | 1 | 2 | ckt | | |
| RL | 2 | 0 | 50.0 ohm | | |
| I2 | 2 | 0 | AC -20M | | |
| V22 | 20 | 22 | AC 1 | | |
| E22 | 20 | 0 | 2 | 0 | 2 |
| R22 | 22 | 0 | 1 | | |
| .AC | LIN | 20 | 1 GHz | 20 GHz | |

```
.PRINT   AC   VM(12) VDB(12) VP(12) VM(22) VDB(22) VP(22)
.PLOT    AC   VDB(12) VDB(22)
.END
```

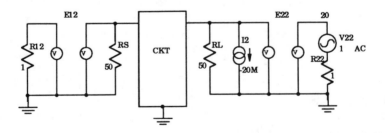

**Fig. 8A.B**

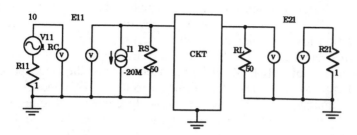

**Fig. 8A.C**

As mentioned earlier, before performing ac small-signal analysis, the dc operating point is evaluated, for which all nodes in the circuit are to have a dc path to ground. In some matching RLC circuits, a large resistor (on the order of 100 Ω) might be required across capacitors. Also, after the dc operating point is evaluated, all voltage power supplies are short-circuited, and the current sources are open-circuited before performing ac analysis. In such a situation, a large inductance (on the order of 100–200 nH) must be inserted between the voltage supply and circuit. This will ensure that the input is not shorted directly to ground during ac analysis. A large value of inductance would not significantly affect the performance of the circuit, even at lower frequencies. This large inductance serves only to make SPICE algorithms work properly and does not reflect a need in real circuit.

The main advantage of using this technique to estimate S-parameters with SPICE, as opposed to linear–frequency-domain circuit analysis programs, is that the circuit can be optimized for active device size and the bias condition without evaluating the device's equivalent circuit element values at every bias point (as would be necessary if frequency domain programs were used). A phase shifter circuit is analyzed using AISPICE™,[2] which is essentially based on SPICE 2G.6 (developed by the University of California-Berkeley) with implementation of an accurate GaAs MESFET model.

The lumped element phase shifter essentially consists of a low-pass filter structure comprising inductors L1 and L2 and drain-to-source capacitance of M3. The high-pass filter comprises inductor L3 and capacitor C2. To achieve the low-pass filter structure, the MESFETs M2 and M3 are switched off and MESFETs M4, M5 are switched on. To achieve a high-pass filter structure, the bias is reversed.

---

[2] AISPICE is the Anadigics, Inc. proprietary version of SPICE.

The phase shifter is designed to produce a phase shift of 45° between the two bias states. The SPICE file is detailed in Table 8A.1(a,b) and the computed phase shift and insertion loss are shown in Table 8A.2(a,b). The circuit element described by Z in Table 8A.1(a,b) is the reserved character to identify the MESFET model in this version of SPICE. Z can be replaced by J and the model parameters in the .Model card can be adjusted accordingly to use a JFET model in generic SPICE. This technique has also been used to analyze filters and to design MESFET switches.[3] The accuracy of the active circuit analysis is, of course, limited by the accuracy of the active device models implemented in SPICE.

**Table 8A.1(a)**

*PHASE SHIFTER*                                    *TEMPERATURE = 27.000 DEG C*

```
.WIDTH OUT=80
.OPTIONS LIMPTS=3000

.SUBCKT PS1 30 37 45

Z2 30 34 35 BX .5
RB2 34 39 2000
LP1 30 38 0.25NH
LP2 38 35 0.25NH
Z3 35 36 37 BX 1.0
RB3 36 33 2000
CLP2 35 37 .5P
Z4 38 40 41 BX 0.1
RB 39 40 2000
Z5 41 42 45 BX .5
RB5 42 33 2000
LHP 41 45 1.0NH
LB1 39 46 .1NH
LB2 33 47 .1NH
*VLP 47 45 DC 0
*VHP 46 45 DC -5
VLP 47 45 DC -5
VHP 46 45 DC 0
.ENDS PS1
```

[3] "MESFET Switch Design," S. S. Bharj and Ravender Goyal, *MSN & CT*, November 1987.

**Table 8A.1(a)**

```
XPS1 1 2 0 PS1
RS 1 0 50
RL 2 0 50
I1 1 0 AC −20M
E11 10 0 1 0 2
V11 10 11 AC 1
R11 11 0 1
E21 21 0 2 0 2
R21 21 0 1
.AC LIN 20 8GHZ 12GHZ
.PRINT AC VDB(21) VP(21) VDB(11)
.MODEL BX GASFET(VTO=−1.4 VBI=.8 RG=0 ALPHA=1.5 BETA=.09
+LAMBDA=.06 CGS0=.67PF CGD0=9.5E−14 CDS=.18PF IS=1E−15)
.PLOT AC VDB(21) VP(21)
.END
```

**Table 8A.1(b)**

```
.WIDTH OUT=80
.OPTIONS LIMPTS=3000

.SUBCKT PS1 30 37 45

Z2 30 34 35 BX .5
RB2 34 39 2000
LP1 30 38 0.25NH
LP2 38 35 0.25NH
```

**Table 8A.1(b)**

```
Z3 35 36 37 BX 1.0
RB3 36 33 2000
CLP2 35 37 .5P
Z4 38 40 41 BX 0.1
RB 39 40 2000
Z5 41 42 45 BX .5
RB5 42 33 2000
LHP 41 45 1.0NH
LB1 39 46 .1NH
LB2 33 47 .1NH
VLP 47 45 DC 0
VHP 46 45 DC −5
*VLP 47 45 DC −5
*VHP 46 45 DC 0
.ENDS PS1
XPS1 1 2 0 PS1
RS 1 0 50
RL 2 0 50
I1 1 0 AC −20M
E11 10 0 1 0 2
V11 10 11 AC 1
R11 11 0 1
E21 21 0 2 0 2
R21 21 0 1
.AC LIN 20 8GHZ 12GHZ
.PRINT AC VDB(21) VP(21) VDB(11)
.MODEL BX GASFET(VTO=−1.4 VBI=.8 RG=0 ALPHA=1.5 BETA=.09
+LAMBDA=.06 CGS0=.67PF CGD0=9.5E−14 CDS=.18PF IS=1E−15)
.PLOT AC VDB(21) VP(21)
.END
```

**Table 8A.2(a)**

*AC ANALYSIS*                                    *TEMPERATURE = 27.000 DEG C*

LEGEND:

 *: VDB(21)
+: VP(21)

```
     FREQ        VDB(21)

*) ---------- −2.500D+00    −2.000D+00    −1.500D+00    −1.000D+00    −5.000D−01
                             - - - - - - - - - - - - - - - - - - - - - - - - -
+) ---------    1.500D+01     2.000D+01     2.500D+01     3.000D+01     3.500D+01
                             - - - - - - - - - - - - - - - - - - - - - - - - -
8.000D+09 −2.262D+00 .       *         .           .           .        +      .
8.211D+09 −2.170D+00 .     *           .           .           .  +
8.421D+09 −2.082D+00 .       *         .           .         +  .
8.632D+09 −1.999D+00 .        *        .           .       +  .
8.842D+09 −1.920D+00 .          *      .           .     +    .
9.053D+09 −1.846D+00 .           *     .           .   +      .
9.263D+09 −1.776D+00 .            *    .          +           .
9.474D+09 −1.710D+00 .             *   .        +            .
9.684D+09 −1.648D+00 .              *  .      +              .
9.895D+09 −1.589D+00 .               *+ .                     .
1.011D+10 −1.535D+00 .              + *.                      .
1.032D+10 −1.483D+00 .           +    *.                      .
1.053D+10 −1.435D+00 .         +       .*                     .
1.074D+10 −1.391D+00 .       +         .  *                   .
1.095D+10 −1.349D+00 .     +           .   *                  .
1.116D+10 −1.311D+00 .    +            .     *                .
1.137D+10 −1.276D+00 .  +  .           .       *              .
1.158D+10 −1.243D+00 . +   .           .        *             .
1.179D+10 −1.213D+00 .+    .           .          *           .
1.200D+10 −1.186D+00 .  +  .           .           *          .
                     - - - - - - - - - - - - - - - - - - - - - - - - - - - - -
```

**Table 8A.2(b)**

---

*AC ANALYSIS*                                       *TEMPERATURE = 27.000 DEG C*

---

LEGEND:

 *: VDB(21)
 +: VP(21)

```
    FREQ       VDB(21)

*) ---------  -1.600D+00    -1.400D+00    -1.200D+00    -1.000D+00    -8.000D-01
              - - - - - - - - - - - - - - - - - - - - - - - - - - - - - - - - -
+) ---------  -3.500D+01    -3.000D+01    -2.500D+01    -2.000D+01    -1.500D+01
              - - - - - - - - - - - - - - - - - - - - - - - - - - - - - - - - -
8.000D+09 -8.039D-01 .            .             .             .          +       *
8.211D+09 -8.222D-01 .            .             .             .            +    *  .
8.421D+09 -8.415D-01 .            .             .             .           +    *   .
8.632D+09 -8.619D-01 .            .             .             .  +        *        .
8.842D+09 -8.835D-01 .            .             .            .+        *           .
9.053D+09 -9.062D-01 .            .             .          +.        *             .
9.263D+09 -9.303D-01 .            .             .         +  .     *               .
9.474D+09 -9.557D-01 .            .             .        +    .  *                 .
9.684D+09 -9.826D-01 .            .             .     +      .*                    .
9.895D+09 -1.011D+00 .            .             .   +      *.                      .
1.011D+10 -1.041D+00 .            .             . +      *                         .
1.032D+10 -1.073D+00 .            .           .+      *                            .
1.053D+10 -1.106D+00 .            .           .+    *                              .
1.074D+10 -1.142D+00 .            .         +  . *                                 .
1.095D+10 -1.180D+00 .            .        +   . *                                 .
1.116D+10 -1.220D+00 .            .      +    *.                                   .
1.137D+10 -1.262D+00 .            .    +    *                                      .
1.158D+10 -1.307D+00 .            .  +    *    .                                   .
1.179D+10 -1.354D+00 .            . +*          .                                  .
1.200D+10 -1.404D+00 .            X             .             .                    .
              - - - - - - - - - - - - - - - - - - - - - - - - - - - - - - - - -
```

Similar techniques are used for analyzing the circuit to obtain two-port Z, Y, G, and H parameters. Again, considering the circuit to be a black box and defined as a subcircuit in SPICE, Table 8A.3 (parameter files 1–12) can be used to obtain these parameters, which also include the alternative methods for S-parameters.

## 8A.2  Device Model Development

Device models can also be developed by using similar methods without making any changes in the source code of the program. Nonlinear voltage-current-controlled voltage-current or current-voltage sources, such as available in SPICE, are the most useful components in developing the device models. These dependent sources are defined as

$i = f(v)$ voltage-controlled current source (VCCS)
$v = f(v)$ voltage-controlled voltage source (VCVS)
$i = f(i)$ current-controlled current source (CCCS)
$v = f(i)$ current-controlled voltage source (CCVS)

where the functions are polynomials and the arguments may be multidimensional.

Suppose that the dependent function is one-dimensional (i.e., it depends on only one independent variable), the functional value $f$ is defined as

$$f = p_0 + (p_1 f_a) + (p_2 f_a^2) + (p_3 f_a^3) + (p_4 f_a^4) + (p_5 f_a^5) + \cdots \tag{8A.1}$$

When the dependent function is two-dimensional, depending on two independent variables $f_a$ and $f_b$, $f$ is defined as

$$\begin{aligned}
f = {} & p_0 + p_1 f_a + p_2 f_b + p_3 f_a^2 + p_4 f_a f_b + p_5 f_b^2 + p_6 f_a^3 \\
& + p_7 f_a^2 f_b + p_8 f_a f_b^2 + p_9 f_b^3 + p_{10} f_a^4 + p_{11} f_a^3 f_b + p_{12} f_a^2 f_b^2 \\
& + p_{13} f_a f_b^3 + p_{14} f_b^4 + \cdots
\end{aligned} \tag{8A.2}$$

Similarly, a three-dimensional functional value $f$ is defined as

$$\begin{aligned}
f = {} & p_0 + p_1 f_a + p_2 f_b + p_3 f_c + p_4 f_a^2 + p_5 f_a f_b + p_6 f_a f_c + p_7 f_b^2 \\
& + p_8 f_b f_c + p_9 f_c^2 + p_{10} f_a^3 + p_{11} f_a^2 f_b + p_{12} f_a^2 f_c + p_{13} f_a f_b^2 + p_{14} f_a f_b f_c \\
& + p_{15} f_a f_c^2 + p_{16} f_b^3 + p_{17} f_b^2 f_c + p_{18} f_b f_c^2 + p_{19} f_c^3 + p_{20} f_a^4 + \cdots
\end{aligned} \tag{8A.3}$$

**Table 8A.3** Circuit Parameter Files 1–12

FILE 1

This file gives the two-port open circuit impedance parameters $Z_{11}$, $Z_{21}$.

```
I1 0 1 AC 1
I2 0 2 AC 0
X  1 2 CKT
.AC LIN 1 1GHz 1GHz
.PRINT AC VM(1) VP(1) VM(2) VP(2)
.END
```

FILE 2

This file gives the two-port open circuit impedance parameters $Z_{12}$, $Z_{22}$.

```
I1 0 1 AC 0
I2 0 2 AC 1
X  1 2 CKT
.AC LIN 1 1GHz 1GHz
.PRINT AC VM(1) VP(1) VM(2) VP(2)
.END
```

FILE 3

This file gives the two-port short circuit admittance parameters $Y_{11}$, $Y_{21}$.

```
V1 1 0 AC 1
V2 0 2 AC 0
F1 3 0 V11
R1 3 0 1
```

FILE 6

This file gives the two-port hybrid parameters $H_{12}$, $H_{22}$.

```
I1 0 1 AC 0
V2 2 0 AC 1
F1 3 0 V21
R  3 0 1
X  1 2 CKT
.AC LIN 1 1GHz 1GHz
.PRINT AC VM(1) VP(1) VM(3) VP(3)
.END
```

FILE 7

This file gives the two-port inverse hybrid parameters $G_{11}$, $G_{21}$.

```
V1 1 0 AC 1
I2 0 2 AC 0
R  3 0 1
F1 3 0 V11
X  1 2 CKT
.AC LIN 1 1GHz 1GHz
.PRINT AC VM(3) VP(3) VM(2) VP(2)
.END
```

FILE 8

This file gives the two-port inverse hybrid parameters $G_{12}$, $G_{22}$.

FILE 10

This file gives the two-port S-parameters $S_{12}$, $S_{22}$.

```
X  1 2 CKT
R1 10 2 50
R2 40 0 1
R3 50 0 1
R4 1 0 50
G1 0 40 2 0 2
G2 0 50 1 0 2
VIN 10 0 AC 1
.AC LIN 1 1GHz 1GHz
.PRINT AC VM(50) VP(50) VM(40,10)
                  VP(40,10)
.END
```

FILE 11

This file gives the two-port S-parameters $S_{11}$, $S_{21}$.

```
X  1 2 CKT
R1 1 10 50
R2 0 40 1
R3 0 50 1
R4 2 0  50
G1 0 40 1 0 2
G2 0 50 2 0 2
IIN 40 0 AC 1
VIN 10 0 AC 1
```

X 1 2 CKT
.AC LIN 1 1GHz 1GHz
.PRINT AC VM(3) VP(3) IM(V2) IP(V2)
.END

FILE 4

This file gives the two-port short circuit admittance parameters $Y_{12}$, $Y_{22}$.

V1 0 1 AC 0
V2 2 0 AC 1
F1 3 0 V2 1
R1 3 0 1
X 1 2 CKT
.AC LIN 1 1GHz 1GHz
.PRINT AC IM(V1) IP(V1) VM(3) VP(3)
.END

FILE 5

This file gives the two-port hybrid parameters $H_{11}$, $H_{21}$.

I1 0 1 AC 1
V2 0 2 AC 0
X 1 2 CKT
.AC LIN 1 1GHz 1GHz
.PRINT AC VM(1) VP(1) IM(V2) IP(V2)
.END

V1 0 1 AC 0
I2 0 2 AC 1
X 1 2 CKT
.AC LIN 1 1GHz 1GHz
.PRINT AC IM(V1) IP(V1) VM(2) VP(2)
.END

In addition to previously described input files, the following files (File 9–12) can also be used to get the small-signal S-parameters.

FILE 9

This file gives the two-port S-parameters $S_{11}$, $S_{21}$.

X 1 2 CKT
R1 1 10 50
R2 0 40 1
R3 0 50 1
R4 2 0 50
G1 0 40 1 0 2
G2 0 50 2 0 2
VIN 10 0 AC 1
.AC LIN 1 1GHz 1GHz
.PRINT AC VM(40,10) VP(40,10) VM(50)
VP(50)
.END

.AC LIN 1 1GHz 1GHz
.PRINT AC VM(40) VP(40) VM(50) VP(50)
.END

FILE 12

This file gives the two-port S-parameters $S_{12}$, $S_{22}$.

X 1 2 CKT
R1 10 2 50
R2 40 0 1
R3 50 0 1
R4 1 0 50
G1 0 40 2 0 2
G2 0 50 1 0 2
IIN 40 0 AC 1
VIN 10 0 AC 1
.AC LIN 1 1GHz 1GHz
.PRINT AC VM(50) VP(50) VM(40) VP(40)
.END

These nonlinear sources are defined in SPICE as follows:

VCCS:

GXXXXXX   N+ N− ⟨POLY(ND)⟩ NC1+ NC1− . . . $p_0$ ⟨$p_1, p_2, p_3, . . .$⟩
⟨IC = . . .⟩

where GXXXXXX represents the VCCS element and N+ and N− are the positive and negative nodes of the controlled source. Current flow is from the positive node through the source to the negative node. POLY(ND) is specified only if the source is multidimensional (i.e., the controlled source depends on more than one controlling voltage source). ND is the dimension of the polynomial (the default value is 1), which must be a positive number (i.e., if no (POLY ⟨ND⟩) defined, one-dimensional polynomial is assumed). NC1+ and NC1− are the positive and negative controlling nodes. One pair of nodes must be specified for each dimension. The polynomial coefficients are $p_0, p_1, p_2, . . . , p_n$. The order of specifying the polynomial coefficients must be strictly followed as shown in equations (8A.1) to (8A.3). The optional initial condition IC is the initial guess of the source at the value of the controlling voltage.

Similarly, the other nonlinear sources are defined as follows:

VCVS:

EXXXXXX   N+ N− POLY(ND) NC1+ NC1− . . . $p_0, p_1, p_2, . . .$ IC = . . .

CCCS:

FXXXXXX   N+ N− POLY(ND) NC1+ NC1− . . . $p_0, p_1, p_2, . . .$ IC = . . .

CCVS:

HXXXXXX   N+ N− POLY(ND) NC1+ NC1− . . . $p_0, p_1, p_2, . . .$ IC = . . .

The following examples will help to develop a model for a three-terminal MESFET device. The objective is to obtain the drain current as a function of drain-to-source ($V_{ds}$) and gate-to-source ($V_{gs}$) voltage (i.e., the function $f(v)$ is $f(V_{gs}, V_{ds})$ in this case, and the dependent variable is the drain current).

*Case I:* Voltage source dependent on the addition or subtraction of two voltages $V_1$ and $V_2$ as shown in Figure 8A.1. These functions are defined as follows:

E1 3 0 POLY(2) $\underbrace{1\ 0}_{V_1}\ \underbrace{2\ 0}_{V_2}\ \underbrace{0\ 1}_{\substack{\text{polynomial}\\\text{coefficients}}} \pm 1$

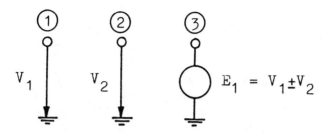

**Fig. 8A.1** Voltage source for Case I.

*Case II:* Voltage source as a multiplication of two voltages $V_1$ and $V_2$ as shown in Figure 8A.2. We define these functions as follows:

E1 3 0 POLY(2) 1 0 2 0 0 0 0 0 1

                       $V_1$  $V_2$  polynomial
                                coefficients

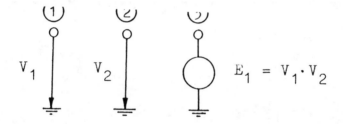

**Fig. 8A.2** Voltage source for Case II as the multiplication of two voltages.

Similarly, the voltage source can be defined as a square of one voltage $V_1$ as shown in Figure 8A.3. Thus,

E1 2 0 POLY(2) 1 0 1 0 0 0 0 0 1

                       $V_1$  $V_2$  polynomial
                                coefficients

This can also be defined as

E1 2 0 POLY(1) 1 0 0 0 1

                       $V_1$ polynomial
                           coefficients

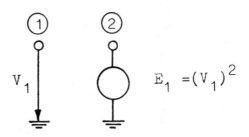

**Fig. 8A.3** Voltage source for Case II as the square of one voltage.

*Case III:* Nonlinear current sources can be defined as the division of two voltages $V_1$ and $V_2$ as shown in Figure 8A.4. We define

$$I_1 = g_1 \cdot V_1 = 1\left(\frac{A}{V}\right) \cdot V_1$$

However, as $I_1 = I_2 = g_2 \cdot V_2 \cdot V_3$ with $g_2 = 1(A/V)$, if $R$ is very large,

$$g_1 \cdot V_1 = g_2 \cdot V_2 \cdot V_3$$

and, hence,

$$V_3 = \frac{g_1}{g_2} \cdot \frac{V_1}{V_2} = \frac{V_1}{V_2} (V)$$

This is defined in SPICE as

G1 0 3 1 0 1

G2 3 0 POLY(2) 2 0 3 0 <u>0 0 0 0 1</u>

                                    polynomial
coefficients

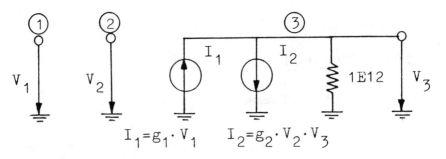

**Fig. 8A.4** Voltage source for Case III.

*Case IV:* Case III can be extended to achieve a nonlinear current source as a function of some power of one or two voltages, as shown in Figure 8A.5. For example, let us define

$$I_1 = g_1(V_1)^{11}$$

$$I_2 = g_2(V_2)^{10}$$

if $R \to \infty$, $I_1 = I_2$

and, hence,

$$g_2(V_2)^{10} = g_1(V_1)^{11}$$

i.e.,

$$V_2 = \left(\frac{g_1}{g_2}\right)^{1/10} \cdot (V_1)^{1.1}$$

The value of $R$ is between 100 M$\Omega$ and $10^{12}$ $\Omega$. This circuit in SPICE is defined as

$$\text{G1 0 2 1 0 } \underbrace{\text{0 0 0 0 0 0 0 0 0 0 0 1}}_{(V_1)^{11}}$$

$$\text{G2 3 0 3 0 } \underbrace{\text{0 0 0 0 0 0 0 0 0 0 1}}_{(V_1)^{10}}$$

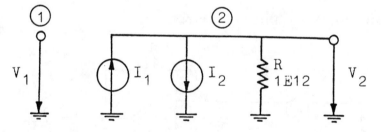

**Fig. 8A.5** Nonlinear current source for Case IV.

*Case V:* The hyperbolic tangent (tanh) function of $V_{ds}$ is commonly used to represent the drain current of GaAs MESFETs. The diode current equation available in SPICE can be used to generate this function as shown in Figure 8A.6(a–d). First, we generate the required exponential function, as shown in Figure 8A.6(a). The diode current is given by the equation:

$$I_D = I_S(e^{V_D/(N \cdot V_T)} - 1) = (e^{V_D} - 1) \text{ (amperes)}$$

assuming $I_S = 1$ A and $N \cdot V_T = 1$, i.e., $N = 38.6698$; hence,

$$I_D + 1 \text{ A} = e^{V_D} \text{ (amperes)}$$

In the above circuit, $V_2$ is represented by

$$V_2 = -(I_D + 1 \text{ A}) \times 1 \,\Omega = -e^{V_1} \text{ (volts)}$$

with $V_1 = V_D$ due to the ideal operational amplifier (OPAMP) in the circuit. The tanh function can be expressed with the exponential function as

$$\tanh(x) = \frac{e^x - e^{-x}}{e^x + e^{-x}} = \frac{1 - e^{-2x}}{1 + e^{-2x}}$$

Thus, we must generate the function $e^{-2x}$ with the ideal OPAMP modeled as a voltage-controlled voltage source and with a gain of two for the voltage source E1. In SPICE, the circuit and notation is as shown in Figure 8A.6(b):

E1 2 0 1 0 0 −2

D1 2 3 DMOD

IB 0 3 1AMP

R1 3 4 10HM

E2 4 0 3 0 1E7

.MODEL DMOD D(IS=1 N=38.6698)

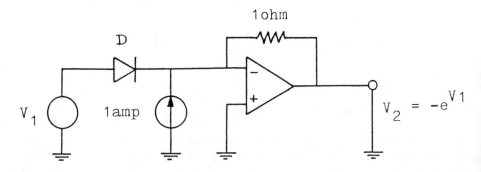

**Fig. 8A.6(a) Exponential function for Case V.**

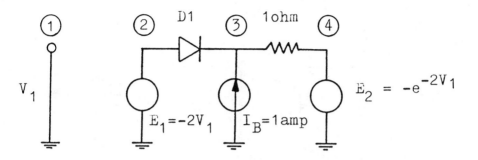

**Fig. 8A.6(b)** Circuit for Case V.

*Case VI:* For the complete tanh function, we perform another addition and subtraction, and a division as described in Case III. Thus, the tanh function is modeled with the circuits shown in Figure 8A.6(c) and (d).

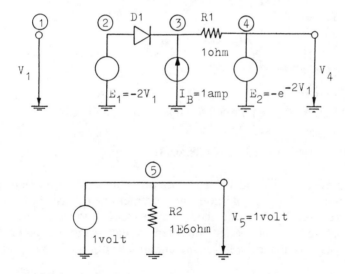

**Fig. 8A.6(c)** Model circuit for Case VI.

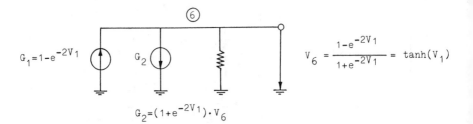

**Fig. 8A.6(d)** Model circuit for Case VI with tanh function.

The notation in SPICE is

$$
\left.\begin{array}{l}
\text{E1 2 0 1 0 } -2 \\
\text{D1 2 3 DMOD} \\
\text{I1 0 3 1} \\
\text{R1 3 4 1} \\
\text{E2 4 0 3 0 1E7}
\end{array}\right\} \text{E2} = -e^{-2V_1} \text{ as in Case V}
$$

$$
\left.\begin{array}{l}
\text{V2 5 0 DC 1} \\
\text{R2 5 0 1E6}
\end{array}\right\} V_2 = 1 \text{ volt}
$$

$$
\left.\begin{array}{l}
\text{G1 0 6 POLY(2) 5 0 4 0 1 1} \\
\text{G2 6 0 POLY(3) 5 0 4 0 6 0 0 0 0 0 0 0 1 0 } -1 \\
\text{R6 6 0 1E12}
\end{array}\right\} \text{division}
$$

.MODEL DMOD D(IS=1 N=38.6698)

The terms $1 - e^{-2V_1}$ and $1 + e^{-2V_1}$ are not generated by separate controlled voltage sources as for the addition and subtraction in Case I. Instead, these sources are realized by the controlled sources of the divider circuit to minimize the number of nodes. This is accomplished by increasing the dimension of the current sources. Compare the following equations with the currents in Case III:

$$I_1 = 1 - e^{-2V_1} = V_2 + V_4 \qquad \text{(two-dimensions, voltages)}$$

$$I_2 = (1 + e^{-2V_1}) \cdot V_6 = V_5 \cdot V_6 - V_4 \cdot V_6 \text{ (three-dimensions, voltages)}$$

with $V_2 = 1$ volt, hence,

$$\text{G2 6 0 POLY(3) } \underbrace{5\ 0}\ \underbrace{4\ 0}\ \underbrace{6\ 0}\ 0\ 0\ 0\ 0\ 0\ 0\ 0\ \underset{\uparrow}{1}\ 0\ \underset{\uparrow}{-1}$$

$$\qquad\qquad\qquad\quad V_5\ \ V_4\ \ V_6 \qquad\quad (V_5V_6)\ (-V_4V_6)$$

*Case VII:* The MESFET model development also requires a $C_{gs}$ capacitor dependent on the drain-to-source voltage $V_{ds}$. This voltage, however, is not across the capacitor:

$$C_{gs} = C'_{gs}V_{ds}$$

This equation can be implemented by using the Miller effect, as in the circuit of Figure 8A.7. The capacitor in the feedback of the OPAMP appears at the input node, and is given by

$$C_{gs} = C'_{gs}(A_v + 1)$$

where $A_v$ is the voltage gain. Controlling the gain $(A_v + 1)$ by the voltage $V_{ds}$ results in the desired function and the circuit of Figure 8A.8 with the OPAMP modeled as an ideal voltage-controlled voltage source.

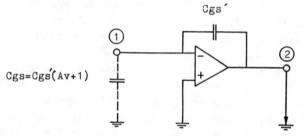

**Fig. 8A.7** Voltage-dependent capacitor for Case VII.

Calculating the input impedance proves the equivalence of the two circuits in Figures 8A.7 and 8A.8:

$$V_{gs} = V'_{Cgs} - V_{gs}(V_{ds} - 1) = V'_{Cgs} + V_{gs} - V_{gs} \cdot V_{ds}$$

Thus,

$$V'_{gs} = V_{gs} \cdot V_{ds}$$

and with

$$V'_{gs} = \frac{1}{j\omega C'_{gs}} I$$

$$\frac{I}{j\omega C'_{gs}} = V_{gs}V_{ds}$$

$$Z_{IN} = \frac{V_{gs}}{I} = \frac{1}{j\omega(C'_{gs}V_{ds})}$$

and it follows that

$$C_{gs} = C'_{gs} V_{ds}$$

**Fig. 8A.8** Model circuit for Case VII with OPAMP as a voltage-controlled voltage source.

With the previously described circuits, we can now implement a complete nonlinear MESFET model. The equations for $I_{ds}$ and the capacitors are to be described in equations (8A.4)–(8A.6). Functions $\tanh(V_{ds})$ and $(V_{gs} - V_{po})^N$ have already been discussed. $V_{po}$ represents the pinch-off voltage and $\lambda$ models the output resistance of the MESFET. The gate-to-source capacitance ($C_{gs}$) and the gate-to-drain capacitance ($C_{gd}$) are divided into two parts for more accurate results. A complete model to implement these equations (8A.4)–(8A.6) is shown in Figure 8A.9. The first part is modeled as the capacitance of the gate-to-source diode ($D_{gs}$) and the gate-to-drain diode ($D_{gd}$). The second part of $C_{gs}$ is the capacitance dependent on the drain-to-source voltage as described in Case VII, and the second part of the gate-to-drain capacitance is the fixed capacitor $C_{gd1}$. In addition, the gate, drain, and source parasitic resistances are taken into account as $R_g$, $R_d$, and $R_s$, respectively.

$$I_{ds} = \beta \cdot W \cdot [(V_{gs} - V_{po})^N + \lambda(V_{gs} - V_{po})^M V_{ds}] \tanh(\alpha V_{ds}) \tag{8A.4}$$

$$C_{gs} = W \left[ \frac{C_{gs0}}{\left(1 - \dfrac{V_{gs}}{V_{bi}}\right)^{M_{gs}}} + C'_{gs1} V_{ds} \right] \tag{8A.5}$$

$$C_{gd} = W \left[ \frac{C_{gd0}}{\left(1 - \dfrac{V_{gd}}{V_{bi}}\right)^{M_{gd}}} + C_{gd1} \right] \tag{8A.6}$$

where

$I_{ds}$ = drain-to-source current;
$\beta$ = transconductance parameter;
$W$ = width of the device.

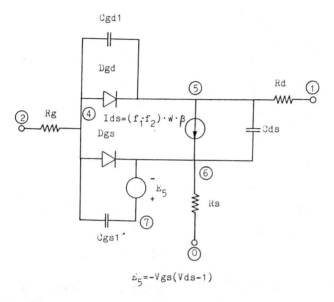

Fig. 8A.9 Model circuit to implement equations (8A.4) to (8A.6).

A set of parameters of a typical 1 mm width GaAs MESFET with 0.5 $\mu$m gate length is shown in Table 8A.4.

**Table 8A.4**

| | |
|---|---|
| $W = 1.0$ | $C_{gs0} = 6.7\text{E}-13$ |
| $\beta = 0.09$ | $V_{bi} = 0.8$ |
| $V_{po} = -1.4$ | $M_{gs} = 0.6$ |
| $N = 1.1$ | $C'_{gs1} = 4.5\text{E}-14$ |
| $\lambda = 0.06$ | $C_{gd0} = 9.5\text{E}-14$ |
| $M = 1.5$ | $M_{gd} = 0.7$ |
| $\alpha = 1.5$ | $C_{gd1} = 5.0\text{E}-14$ |
| $R_s = 0.1\ \Omega$ | $C_{ds} = 1.8\text{E}-13$ |
| $R_g = 0.1\ \Omega$ | |
| $R_d = 0.1\ \Omega$ | |

The realization of the MESFET model shown in Figure 8A.9, described by (8A.4)–(8A.6), with model parameters detailed by Table 8A.4, is shown in Figure 8A.10. The SPICE "netlist" to describe such a model is given in Table 8A.5. The

668

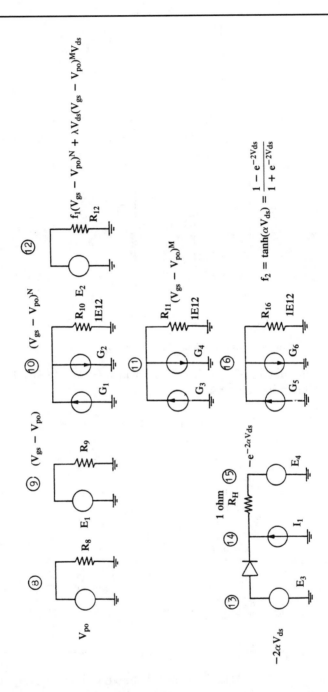

**Fig. 8A.10** Realization of MESFET circuit model.

netlist is compatible with PSPICE™ (MicroSim Corporation). A similar netlist file can also be compatible with a generic or any other version of SPICE. The only difference may be the method of varying $V_{gs}$ and $V_{ds}$ to simulate the $I_{ds}$-$V_{ds}$ characteristics for varying $V_{gs}$. Thus, VGS, VDS, .DC, and .PLOT lines in Table 8A.5 may be different for other versions of SPICE. A plot of $I$-$V$ characteristics for different $V_{gs}$ obtained from the netlist in Table 8A.5 is shown in Figure 8A.11. The large value resistors R8, R9, R10, R11, R12, and R16 may also need to be adjusted to achieve convergence in SPICE. However, these values must be kept as large as possible.

**Table 8A.5**

```
*GaAs FET MODEL 1*
VGS 2 0 DC −0.5
VDS 1 0 DC 3
RG 2 4 0.01
RD 1 5 0.01
RS 6 0 0.01
DGS 4 6 DMOD1 OFF
CGS1 4 7 4.5E−14
DGD 4 5 DMOD2 OFF
CGD1 4 5 5.0E−14
GDS 6 5 POLY(2) 12 0 16 0 0 0 0 0 9E−2 ⟶ Ids = β · W · (f1)f2
CDS 5 6 1.8E−13
VDUM 8 0 DC 1.0
R8 8 0 1E6
E1 9 0 POLY(2) 4 6 8 0 0 1 1.4
R9 9 0 1E6
G1 0 10 POLY(1) 9 0 0 0 0 0 0 0 0 0 0 0 1
G2 10 0 POLY(1) 10 0 0 0 0 0 0 0 0 0 0 0 1
R10 10 0 1E12
G3 0 11 POLY(1) 9 0 0 0 0 1
G4 11 0 POLY(1) 11 0 0 0 1
R11 11 0 1E12
E2 12 0 POLY(3) 10 0 11 0 5 6 0 1 0 0 0 0 0 0 0.06
R12 12 0 1E6
E3 13 0 5 6 −3
D1 13 14 DMOD3
I1 0 14 1
RH 14 15 1
E4 15 0 14 0 1E12
G5 0 16 POLY(2) 8 0 15 0 0 1 1
G6 16 0 POLY(3) 8 0 15 0 16 0 0 0 0 0 0 0 10 −1
R16 16 0 1E12
```

pinch-off voltage
$(V_{gs} - V_{po})$

$(V_{gs} - V_{po})^N$

$(V_{gs} - V_{po})^M$

$f_1 = (V_{gs} - V_{po})^N + \lambda(V_{gs} - V_{po})^M \cdot V_{ds}$

$f_2 = \tanh(\alpha V_{ds})$

**Table 8A.5**

```
E5 7 6 POLY(2) 4 6 5 6 0 1 0 0 −1
.MODEL DMOD1 D(IS=1E−15 CJO=6.7E−13 VJ=0.8 M=0.5)
.MODEL DMOD2 D(IS=1E−15 CJO=9.5E−14 VJ=0.8 M=0.7)
.MODEL DMOD3 D(IS=1 N=38.6698)
.NODESET V(10)=1 V(11)=1 V(16)=1
.DC VDS 0.0 8.0 0.25 VGS 0.4 −1.4 0.2
.PLOT DC I(VDS)
.OPTIONS LIMPTS=390
.END
```

In some cases, it might be necessary to use an even more accurate MESFET model. The previously described model does not correctly represent the output resistance. The model shows a decreasing output resistance by increased drain-to-source current, similar to the output-characteristic of a bipolar transistor, whereas an actual MESFET shows the opposite behavior. The modified equation (8A.5a) accurately models this phenomenon, but it is more complicated and requires more lines of SPICE code. The associated parameters are described in Table 8A.6. The often important gate-to-source resistance $R_{IN}$ is also included in the modified model:

$$
I_{ds} = W\beta \left[ \underbrace{\frac{(V_{gs} - V_{po})^N}{C}}_{f_1} \underbrace{\tanh(\alpha V_{ds})}_{f_2} \right.
$$

$$
\left. + \overbrace{\frac{\lambda \cdot B \cdot [\tanh(A \cdot V_{gs}) + 1]}{\rho(-V_{po})^M}}^{f_3} \underbrace{[V_{ds} - \tanh(V_{ds})]^\rho}_{f_4} \right] \tag{8A.5a}
$$

An additional effect is the drain-lag effect, which is modeled as a parallel current source $G_{HB}$ in addition to the $I_{ds}$ current source. A voltage-controlled current source controlled by its own terminal (the bulk terminal) acts as a resistor. However, this resistor is only in parallel at higher frequencies due to the high-pass filter for the controlling voltage ($C_{BD}$, $R_{BS}$). Such a backgate model is described in detail in chapter 4. This detailed model is shown in Figure 8A.12. Such a model can be described in SPICE using techniques described for Cases I–VII and the

671

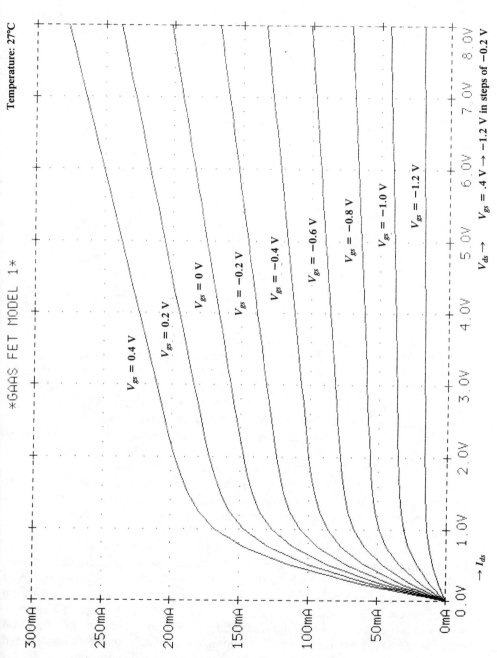

**Fig. 8A.11** GaAs FET model 1.

previous MESFET model. The details of describing such a model are presented in Figure 8A.13. The netlist of the SPICE program to describe this model is provided in Table 8A.7.

**Table 8A.6**

| | | |
|---|---|---|
| $\beta = 0.092$ | $C_{gs0} = 6.7\text{E}{-}13$ | $R_{\text{IN}} = 0.7$ |
| $V_{po} = -1.4$ | $C'_{gs1} = 4.5\text{E}{-}14$ | $C_{BD} = 0.03$ pF |
| $N = 1.5$ | $V_{bi} = 0.8$ | $R_{BS} = 1\text{E}9$ |
| $M = 1.5$ | $M_{gs} = 0.6$ | $G_{HB} = 4\text{E}{-}3$ |
| $\alpha = 1.5$ | | |
| $\lambda = 0.06$ | $C_{gd0} = 5.0\text{E}{-}13$ | $R_d = .01\ \Omega$ |
| $\rho = 0.6$ | $C_{gd1} = 7.0\text{E}{-}14$ | $R_g = .01\ \Omega$ |
| $A = -2.2$ | $M_{gd} = 1.8$ | $R_s = .01\ \Omega$ |
| $B = 1.2$ | | |
| $C = 1.0$ | $C_{ds} = 1.9\text{E}{-}13$ | |

Figure 8A.14 shows the simulation of $I$-$V$ characteristics using (8A.5a). Also plotted in this figure are the measured values of the corresponding current at a particular $V_{ds}$ and $V_{gs}$. As discussed in the previous simplified model, the resistors R9, R10, R11, E2, R15, E4, R19, E6, R23, R24, R25, R26, and R27 in Table 8A.7 may need to be adjusted if the same SPICE file is run on a version of SPICE other than PSPICE. However, their values must be kept as large as possible.

The main disadvantage of using the above approaches to model the MES-FETs is that we would need to define a similar subcircuit for each size of device, which could lead to a long SPICE netlist. However, for microwave circuits, where the number of different sized devices used in a circuit is small, such a method is very attractive as it does not require modification of the SPICE program. Such a model has been successfully used for accurately simulating a circuit with more than 10–15 active devices.

We should stress that those models are for dc and transient analysis only, because the dc operating voltages $V_{gs}$ and $V_{ds}$ used in the equation for the drain-to-source current are set to zero by SPICE for the ac analysis. An ac analysis can be performed with the SPICE FET-MODEL by specifying the parameters with the help of the dc analysis and data sheets. We suggest designing the circuit with an ac analysis first, because this requires less computation time. Then, make a final run with the MESFET-MODEL, which is a complete nonlinear model, to confirm the results and to refine the design, especially for large-signal conditions.

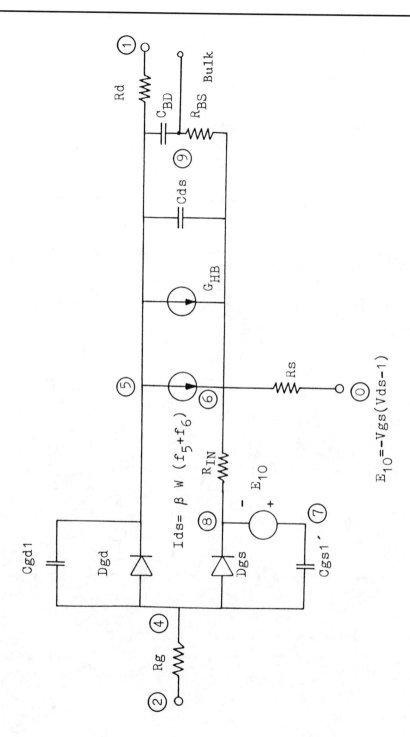

**Fig. 8A.12** Backgate model.

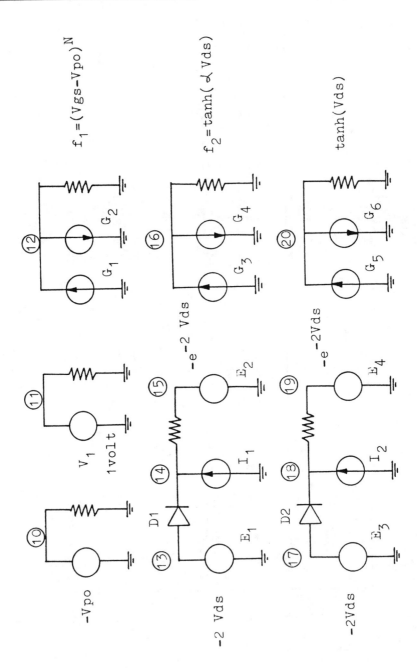

**Fig. 8A.13** Description of backgate model.

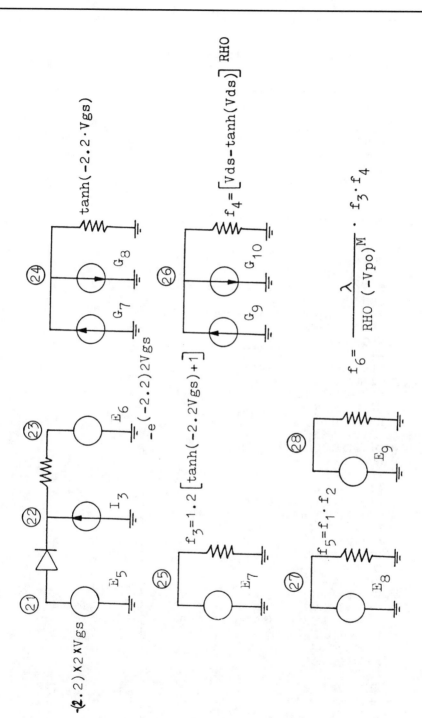

**Fig. 8A.13** (cont'd).

**Table 8A.7**

```
*GaAs FET MODEL 2 ·
VGS 2 0 DC −0.5
VDS 1 0 DC 3
RG 2 4 0.01
RD 1 5 0.01
RS 6 0 0.01
DGS 4 8 DMOD1 OFF
DGD 4 5 DMOD2 OFF
RIN 8 6 0.7
CGD1 4 5 7.0E−14
CDS 5 6 1.9E−13
CGS1 4 7 4.5E−14
GDS 6 5 POLY(2) 26 0 27 0 0 9.2E−2 9.2E−2
CBD 5 29 0.03PF
RBS 29 6 1E9
GMB 5 6 6 29 4E−3
VPO 9 0 1.4
R9 9 0 1E6
G1 0 10 POLY(2) 4 8 9 0 0 0 0 0 0 0 1 3 3 1
G2 10 0 POLY(1) 10 0 0 0 1
R10 10 0 1E12
V1 11 0 DC 1.0
R11 11 0 1E6
E1 12 0 5 6 −3
D1 12 13 DMOD3
I1 0 13 1
RH1 13 14 1
E2 14 0 13 0 1E12
G3 0 15 POLY(2) 11 0 14 0 0 1 1
G4 15 0 POLY(3) 11 0 14 0 15 0 0 0 0 0 0 0 1 0 −1
R15 15 0 1E9
E3 16 0 5 6 −2
D2 16 17 DMOD3
I2 0 17 1
RH2 17 18 1
E4 18 0 17 0 1E12
G5 0 19 POLY(2) 11 0 18 0 0 1 1
G6 19 0 POLY(3) 11 0 18 0 19 0 0 0 0 0 0 0 1 0 −1
R19 19 0 1E12
E5 20 0 4 8 4.4
D3 20 21 DMOD3
I3 0 21 1
RH3 21 22 1
E6 22 0 21 0 1E12
G7 0 23 POLY(2) 11 0 22 0 0 1 1
G8 23 0 POLY(3) 11 0 22 0 23 0 0 0 0 0 0 0 1 0 −1
R23 23 0 1E9
```

Annotations:

$\longrightarrow I_{ds} = \beta \cdot W(f_5 + f_6)$

$\}$ backgate model

$\}$ pinch-off voltage

$\}\, f_1 = (V_{gs} - V_{po})^{1.5}$

$\}$ auxiliary voltage = 1 volt

$\}\, \tanh(1.5 \times V_{ds}) = f_2$

$\}\, \tanh(V_{ds})$

$\}\, \tanh(-2.2 \times V_{gs})$

**Table 8A.7** cont.

E7 24 0 POLY(2) 23 0 11 0 0 1.2 1.2
R24 24 0 1E6 $\Bigg\}$ $f_3 = 1.2[\tanh(-2.2 \times V_{gs}) + 1]$

G9 0 25 POLY(2) 5 6 19 0 0 0 0 0 0 0 1 $-3$ 3 $-1$
G10 25 0 POLY(1) 25 0 0 0 0 0 0 1 $\Bigg\}$ $f_4 = [V_{ds} - \tanh(V_{ds})]^\rho$
R25 25 0 1E9

E8 26 0 POLY(2) 10 0 15 0 0 0 0 0 1
R26 26 0 1E6 $\Bigg\}$ $f_5 = f_1 \times f_2$

E9 27 0 POLY(2) 24 0 25 0 0 0 0 0 60.368E$-$3
R27 27 0 1E6 $\Bigg\}$ $f_6 = \dfrac{\lambda}{\rho(-V_{po})^M} \times f_3 \times f_4$

E10 7 8 POLY(2) 4 8 5 6 0 1 0 0 $-1 \longrightarrow -V_{gs}(V_{ds} - 1)$
.MODEL DMOD1 D(IS=1E$-$15 CJO=6.7E$-$13 VJ=.08 M=0.6)
.MODEL DMOD2 D(IS=1E$-$15 CJO=5.0E$-$13 VJ=0.8 M=1.8)
.MODEL DMOD3 D(IS=1 N=38.6698)
.NODESET V(10)=1 V(15)=1 V(19)=1 V(23)=1 V(25)=1
*END  MODEL 2 *
.DC VDS 0.0  8.00  0.2 VGS 0.4 $-1.4$ 0.2
.PLOT DC I(VDS)
.OPTIONS LIMPTS=500
.PROBE
.END

## ACKNOWLEDGEMENTS

The editor wishes to thank the following for their contributions to this chapter: Mr. M. Mannan of National Semiconductor, Santa Clara, California for contributing Sections 8.2.1 and 8.2.2; Prof. Vittorio Rizzoli of the Department of Electronics and Information Systems, University of Bologna, Italy for contributing Section 8.2.6; Mr. Tony Sison of Silvar Lisco, Menlo Park, California for his help toward the completion of Section 8.2.8.

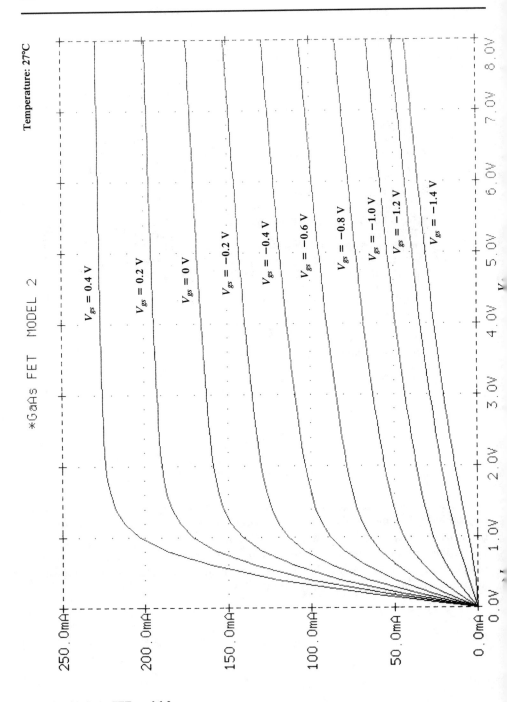

**Fig. 8A.14** GaAs FET model 2.

# REFERENCES

## *Section 8.1*

1.  J. Tenedorio, "Techniques Further the Design and Realization of MMICs," *MSN & CT*, March 1986, p. 66–75.
2.  C. W. Suckling, "Procedures Examined for Successful GaAs MMIC Design," *MSN & CT*, March 1986, pp. 79–87.
3.  R. Goyal, "What We Need for MMIC Design Automation," *Microwave S & RF*, June 1987, p. 77.
4.  A. R. Newton et al., "CAD Tools for ASIC Design," *Proce. IEEE*, Vol. 75, June 1987, pp. 765–776.
5.  D. M. Caughey et al., "Rapid Implementation of a Custom IC Design System Using Acquired CAD,"; Custom Integrated Circuit Conf., May 1982.

## *Section 8.2*

1.  K. C. Gupta, Ramesh Garg, and Rakesh Chadha, *Computer-Aided Design of Microwave Circuits*, Artech House, Dedham, MA, 1981.
2.  R. H. Jansen, "Computer-Aided Design of Hybrid and Monolithic Microwave Integrated Circuits—State of the Art, Problems and Trends," *Proc. 13th European Microwave Conf.*, September 1983, pp. 67–78.
3.  G. R. Hoffman, "Introduction to the Computer Aided Design of Microwave Circuits," *Proc. 14th European Microwave Conf.*, September 1984, pp. 731–737.
4.  D. Trubitt et al., "Desktop CAD Offers Alternatives to the Microwave Design Engineer," *Microwave Systems Design Handbook*, 1984, pp. 72–91.
5.  P. Antognetti et al., *Computer Design Aides for VLSI Circuits*, NATO ASI Series, Martinus Nijhoff Publishers, Hingham, MA, 1986.
6.  L. O. Chua et al., *Computer Aided Analysis of Electronic Circuits: Algorithms and Computational Techniques*, Prentice-Hall, Englewood Cliffs, NJ, 1975.
7.  L. W. Nagel, *SPICE2: A Computer Program to Simulate Semiconductor Circuits*, Electronics Research Laboratory Rep. No. ERL-M520, University of California, Berkeley, 1975.
8.  E. Cohen, *Program Reference For SPICE2*, Electronics Research Laboratory Rep. No. ERL-M592 University of California, Berkeley, 1976.
9.  *SPICE Version 2G User's Guide*, University of California, Berkeley.
10. W. J. McCalla, *Fundamentals of Computer-Aided Circuit Simulation*, Kluwer Academic Publishers, Hingham, MA, 1988.
11. D. Divekar, *FET Modeling and Circuit Simulation*, Kluwer Academic Publishers, Hingham, MA, 1988.
12. P. Antognetti et al., *Semiconductor Device Modeling with SPICE*, McGraw-Hill, New York, 1988.
13. G. Ioos and D. D. Joseph, *Elementary Stability and Bifurcation Theory*, Springer-Verlag, New York, 1980.
14. D. D. Weiner and J. R. Spina, *Sinusoidal Analysis and Modeling of Weakly Nonlinear Circuits*. Van Nostrand-Rheinhold, New York, 1980.
15. I. W. Sandberg, *IEEE Trans. CAS-30*, February 1983, pp. 61–77.

16. V. Rizzoli et al., *Dig. IEE Colloquium CAD Microwave Circuits,* London, November 1985, pp. 13/1–13/5.

17. M. A. Smith et al., *IEEE MTT-S Int. Microwave Symp. Dig.,* June 1986, pp. 381–384.

18. V. Rizzoli, C. Cecchetti, and A. Neri, *Proc. 16th European Microwave Conf.,* September 1986, Dublin, pp. 692–697.

19. W. R. Curtice, *IEEE Trans. Microwave Theory Tech.,* Vol. MTT-35, April 1987, pp. 441–447.

20. D. Hente and R. H. Jansen, *Proc. Institute Electrical Engineers,* Part H, October 1986, pp. 351–362.

21. K. S. Kundert et al., *IEEE Trans. Microwave Theory Tech.,* Vol. MTT-36, February 1988, pp. 366–378.

22. A. Ushida and L. O. Chua, *IEEE Trans. Circuits and Systems,* Vol. CAS-31, September 1984, pp. 766–779.

23. V. Rizzoli, C. Cecchetti, and A. Lipparini, *Proc. 17th European Microwave Conf.,* September 1987, Rome, pp. 635–640.

24. U. L. Rohde, *Microwave J.,* October 1987, pp. 203–210.

25. M. S. Nakhla and J. Vlach, *IEEE Trans. Circuits and Systems,* Vol. CAS-23, February 1976, pp. 85–91.

26. V. Rizzoli et al., *Proc. Institute Electrical Engineers,* Pt. H, Vol. 133, October 1986, pp. 385–391.

27. M. I. Sobhy and A. K. Jastrzobski., *Proc. 15th European Microwave Conf.,* September 1985, Paris, pp. 1110–1118.

28. L. W. Nogel, "Memo ERL-M520," Electronic Research Laboratory, University of California-Berkeley, 1975.

29. R. H. Jansen, R. G. Arnold, and I. G. Eddison, *IEEE Trans. Microwave Theory Tech.,* Vol. MTT-36, February 1988, pp. 208–219.

30. *Microwave Harmonica User Manual,* Compact Software Inc., 1988.

31. *Libra User Manual,* EEsof Inc., 1988.

32. Y. L. Chow et al., *Proc. 16th European Microwave Conf.,* September 1986, Dublin, pp. 625–630.

33. M. D. Warlick et al., "Compiled GaAs ASICs Reduce Development Time," *Computer Design,* 1987, pp. 81–86.

34. G. Janae et al., "A Knowledge Based GaAs Design System," *VLSI System Design,* April 1987, pp. 68–75.

35. A. C. Parker et al., "Automating the VLSI Design Process Using Expert Systems and Silicon Compilation," *Proc. IEEE,* Vol. 75, June 1987, pp. 777–785.

36. W. Birmingham et al., "Knowledge-Based Expert Systems and their Application"; *Proc. 23rd Design Automation Conf.,* June 1986, pp. 531–539.

37. "Bristle Blocks: A Silicon Compiler," *Proc. 16th Design Automation Conf.,* June 1979, pp. 310–313.

38. A. R. Newton, "A Survey of Computer Aids for VLSI Layout," *Dig. Tech. Papers, Symp. on VLSI Technology,* September 1982, pp. 72–74.

39. C. Mead and L. Conway, *Introduction to VLSI Systems,* Addison-Wesley, 1980.

40. S. Chao et al., "A Hierarchical Approach for Layout versus Circuit Consistency Check"; *17th Design Automation Conf.,* 1980, Minneapolis.

41. M. T. Yin, "Layout Verification of VLSI Designs," *VLSI Design,* July 1985, pp. 30–35.

42. "Speeding IC Design Verification," *Test and Measurement World,* September 1988.

43. S. Perry et al., "Edge-Based Layout Verification," *VLSI Systems Design,* September 1985, pp. 106–114.

44. G. S. Taylor et al., "MAGIC's Incremental Design-Rule Checker," *21st Design Automation Conf.,* 1984, Albuquerque.

45.  T. G. Szymanski et al., "GOALIE: A Space-Efficient System for VLSI Artwork Analysis," *Int. Conf. Computer Aided Design*, 1980, Santa Clara, CA.

45a. B. R. Epstein et al., "Circuit Descriptive Languages for Microwave CAD," *Microwave J.*, May 1988, pp. 357–363.

46.  R. Goering, "Implementations May Blossom for Design Interface Standards," *Computer Design*, February 1987.

47.  "EDIF: Electronic Design Interchange Format," Version 200, May 1987, EIA Interim Standard No. 44, Washington, D.C. For further information concerning microwave applications contact Dan Nash, Raytheon Co., Bedford, MA.

48.  J. Hines, "Where VHDL Fits in the CAD Environment," *24th ACM/IEEE Design Automation Conf.*, June 1987.

49.  "VHDL Language Reference Manual," Draft Standard 1076/B, CAD Language Systems, Rockville, MD, May 1987.

## Section 8.3

1.  G. Skinner, "Progress Toward a Computerized Mask Shop," *Solid State Technology*, May 1988, pp. 131–134.

2.  K.M. Wishnuff, "Realistic Mask Specifications," *Solid State Technology*, May 1988, pp. 123–125.

3.  A.J. Serafino, *et al.,* "Technique for Repair of Mask Defects," *Semiconductor International*, June 1988, pp. 132–135.

4.  —"The Limits of Maskmaking," *Solid State Technology*, May 1988, pp. 141–153.

5.  P. Burggraff, "Photomask and Reticle Blanks," *Semiconductor International,* December 1987, pp. 38–42.

# Chapter 9
# On Wafer Testing of MMICs

*I. G. Eddison and R. Goyal*

## 9.1 INTRODUCTION

During the early development of the silicon integrated circuit industry, the twin disciplines of harsh competition and large volume production led to many painful lessons in such areas as process control, production engineering, testing technology, and packaging techniques. From these experiences grew an impressive array of high technology equipment designed to provide the integrated circuit manufacturer with consistent and accurate handling of semiconductor wafers and chips. Today highly automated, computer controlled wafer chip handling stations at every manufacturing stage are common in the silicon industry. This predominance of automation is particularly true in the circuit test area, where the need to reduce costs to a minimum has given rise to comprehensive on-wafer IC test facilities. Much of the silicon IC technology, at least in principle, can be transferred to GaAs IC production, but difficulties arise in some areas because of the higher frequencies (generally above 1 GHz) at which GaAs circuits operate. There are particular difficulties, however, in the case of on-wafer testing.

Until recently, the size and parasitic inductances associated with conventional on-wafer probers prevented component testing at frequencies above 1 GHz. During the early development of GaAs FET devices and ICs, before dicing chips, the only means of checking for device functionality was by using a dc/1 MHz autoprober test. Although this technique screens out gross failures and provides valuable statistical data for discrete FET devices, it has serious shortcomings in testing ICs. In particular, the more complex circuitry and wide variety of circuit elements inherent to an IC cannot always be evaluated adequately at dc/1 MHz. For example, conventional 1 MHz measurements cannot resolve low-value capacitances and many capacitor coupled circuits cannot be verified functionally at such low frequencies. These disadvantages were originally overcome by using a mixture of dc probe screening and detailed microwave testing on a small sample of packaged chips.

Traditional microwave characterization of GaAs MMIC involves dicing the fully processed wafers, bonding and wiring individual chips onto subcarriers, mounting these into test jigs, and measuring the microwave performance of the resultant assembly using conventional test facilities. However, the process is slow, labor intensive, wasteful of resources, and involves complicated computer processing to extract the true IC performance (an operation known as *deembedding*). More important, as GaAs IC technology matures, questions of circuit complexity, cost, and throughput become crucial. Because the cost of microwave packaging and testing is a significant fraction of an IC development and production costs, faulty chips have to be identified as early as possible in the production cycle in order to minimize the final cost of the MMIC. Further, even modest production runs of several thousand GaAs IC units per annum cannot be handled with the traditional characterization techniques. Also, the advent of GaAs foundry processes now calls for extensive microwave characterization so that the foundry can provide statistically meaningful IC design data.

These considerations were important factors in the drive to develop techniques for on-wafer characterization of microwave components. Such needs are now being satisfied by the emergence of *automated test systems* (ATE) based on probes capable of operating at microwave frequencies. Thus, the RF functional operation of an IC can be verified prior to dicing from the wafer, providing a rapid feedback of circuit functional information to process and design engineers without the added uncertainties of dicing, die bonding, wiring, and packaging effects.

## 9.2 MICROWAVE PROBES

For many silicon ICs with operating frequencies below 30 MHz, individual circuits can be accurately tested on wafer, using a traditional probe card technique. Here, the interface between conventional electronic test equipment and the small IC contact pads is achieved using tungsten needles mounted on a printed circuit board (the probe card). A wide variety of circuits can be tested quickly by producing a probe card for each new IC design. In this way, after testing one type of IC, a completely different IC wafer design can be evaluated simply by replacing one probe card with another and calling up the required ATE test software routine. An example of a typical Si IC wafer under test using a conventional probe is shown in Figure 9.1.

This test methodology is used by many semiconductor manufacturers, benefitting production costs, timescales, and data monitoring. However, the conventional probe card has severe limitations when ICs with higher operating

**Fig. 9.1** A typical probe card used in silicon IC production testing (courtesy of Plessey plc.).

frequencies have to be tested, due to the physical size of the probe needles, which are approximately 18 mm long. As the frequency is raised, the probe needles behave increasingly as antennae and the coupling between them masks the input to and output from the IC's contact pads. If frequencies above 30 MHz are involved, the simplest solution is to significantly reduce the physical length of the probe needles. Such a solution has been reported [1] with a claimed highest frequency capability of approximately 4–5 GHz.

This particular probe (Figure 9.2) consists of a 0.25 mm (10 mil) sapphire substrate on which 50 Ω impedance microstrip lines are fabricated. These microstrip lines radiate from a 2 mm (80 mil) diameter hole in the center of the substrate. Metal probe needles of beryllium copper are connected to the microstrip lines at the hole through the substrate center. These metal fingers contact the probe pads of the IC under test, while the microstrip lines provide a well-controlled impedance interface between the miniature probe needles and conventional microwave coaxial connectors. Although the probe needles give a resilient, low-resistance contact to the wafer, they are difficult to assemble and align with the contact pads of the device under test. In addition, the probes are electrically long at microwave frequencies and their resultant 0.7 nH inductance can be a serious limitation. To illustrate the difficulties Figure 9.3 shows the measured input reflection coefficient ($S_{11}$) of the probes connected to a 50Ω through line. Large, rapid variations in return loss with frequency are obvious, and these problems are caused by the self-resonant, radiating behavior of the probe fingers. As a result this probe card is unlikely to be useful above 4 GHz. However, at lower frequencies the probe system is a low-cost solution to production testing that offers good correlation

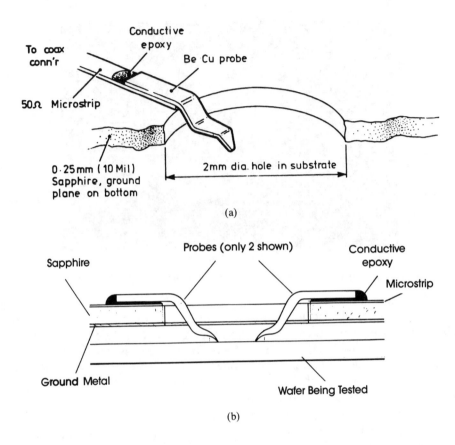

Fig. 9.2 Microwave probe card (a) detail and (b) side view (courtesy of Hewlett-Packard).

between probed and packaged results, as shown in Figure 9.4. The opportunities for further miniaturization of these probes for higher frequency performance are limited, and an alternative probe technique must be found for microwave frequencies.

Before examining alternative probe techniques, it is essential to consider the particular needs of testing MMICs at microwave frequencies. There are two main generic classes of on-wafer probe testing: in-circuit testing for diagnostic purposes, and terminal testing for either new component characterization or production tests against a product specification. The in-circuit testing application is rarely used at the moment, and probes for this purpose are discussed later in the chapter. In the majority of cases the designers and users of the MMICs want to know the terminal characteristics of their ICs or test devices. In this case, if meaningful on-wafer microwave measurements are to be made, a probe system

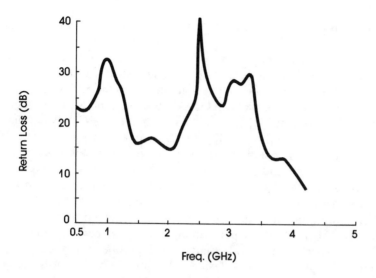

**Fig. 9.3** Measured $S_{11}$ of probes connected by a 50 $\Omega$ through line (courtesy of Hewlett-Packard).

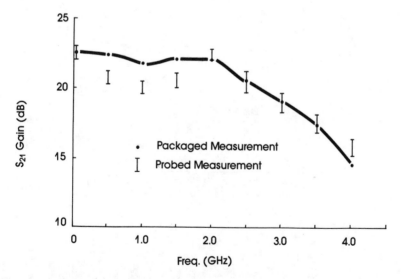

**Fig. 9.4** Correlation between on-wafer probed and packaged amplifier performance (courtesy of Hewlett-Packard).

must be devised with a path that combines low-insertion loss, low-interchannel crosstalk, and good matching between the measuring equipment and the high-frequency input and output pads. These requirements are dictated by the need to minimize uncertainties in the measured signals caused by losses, reflections, and crosstalk.

To meet the probe requirements, a transmission medium is needed that is readily interfaceable with a standard coaxial to microstrip transition (e.g., SMA, K-connector or 2.4 mm connector) and yet can transform the 50 $\Omega$ line dimensions down to the 100–150 $\mu$m probe pads on GaAs MESFETs and MMICs. A convenient solution to the problem is the *coplanar waveguide* (CPW) system first proposed by Gleason et al. [2], shown in Figure 9.5. The basic properties of CPW [3] allow the fabrication of constant impedance tapered lines by keeping a fixed conductor line width to ground space ratio (A/B). Indeed, CPW is the basis of the microwave probes produced commercially by both Cascade Microtech and Design Technique International. A schematic representation of the simplest of these probes is shown in Figure 9.6, and a photograph of a typical probe head is shown in Figure 9.7. The high-frequency performance of coplanar waveguide probes is exceptional when compared to the alternative needle probe systems. Indeed, with special care in the coplanar probe's design, fabrication, and use, it is possible to achieve acceptable performance up to 50 GHz.

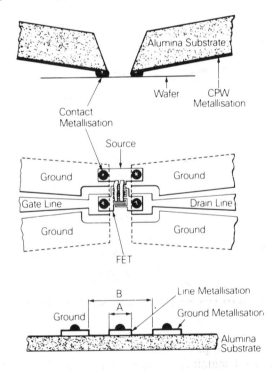

**Fig. 9.5** Coplanar waveguide probe system.

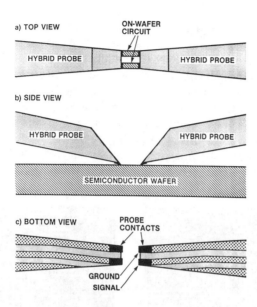

a) TOP VIEW

ON-WAFER CIRCUIT

HYBRID PROBE

HYBRID PROBE

b) SIDE VIEW

HYBRID PROBE

HYBRID PROBE

SEMICONDUCTOR WAFER

c) BOTTOM VIEW

PROBE CONTACTS

GROUND
SIGNAL

**Fig. 9.6** Schematic representation of simplified coplanar waveguide probe (ground signal).

**Fig. 9.7** Photograph of typical coplanar waveguide probe head (courtesy of Cascade Microtech).

A convenient means of examining the probe performance is through the use of *time domain reflectometry* (TDR). In this technique, a fast rise time pulse is injected into the unknown device and the reflected signals are displayed as a function of time. The time delay associated with each reflection can be calibrated against distance, and the imperfections and their physical locations can be identified easily. Figure 9.8 shows the results from a 40 ps rise time TDR measurement on two CPW probes contacting a short length of through line. The discontinuity at A is typical of standard SMA connectors and the region A-B is the length of semirigid coaxial cable connected to the first probe head. The region B-C is the first probe, with point C being the discontinuity at the probe tips; region C-D is the second probe, with point D being another SMA connector and E being an SMA matched load. Throughout this system the maximum reflection coefficient is 20 m$\rho$ and the major discontinuities are at the connector interfaces and the probe tips. Also, the CPW probe transmission lines introduce little or no reflection.

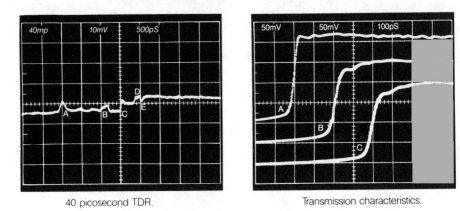

40 picosecond TDR.                                    Transmission characteristics.

**Fig. 9.8** A 4CPS TDR response of two probes connected by a through line (courtesy of Cascade Microtech).

Figure 9.9 shows the measured transmission and reflection loss characteristics of 50 GHz single CPW probe from 45 MHz to 50 GHz. This probe is of a ground-signal-ground tip configuration for optimum probe performance. The results correspond to a single probe insertion loss of better than 3.0 dB up to 50 GHz. Return loss is better than 20 dB to 35 GHz and better than 12 dB to 50 GHz. The undoubted benefits of the new generation probes over the conventional needle probes can be easily seen by comparing the return loss performances shown in Figures 9.3 and 9.9, respectively.

Interprobe isolation is another primary parameter of high-frequency probes. High isolation between input and output probes is essential, because any crosstalk between the probes cannot be accounted for in any network analyzer 12-term error model. Figure 9.10 shows the crosstalk between two probes over 0.4 GHz to 40 GHz with both probes open circuited. In this open circuit configuration, both probes are in the air approximately 2.5 mm (100 mils) above the work stage and separated horizontally by 5 mm (200 mils). Over this band, the worst case isolation is 50 dB; that is, the output probe detects any signal on the input probe at a

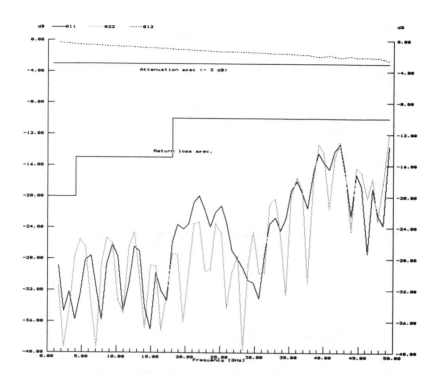

**Fig. 9.9a** 45 MHz to 26.5 GHz.

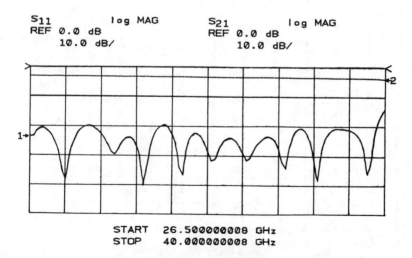

START 26.500000008 GHz
STOP 40.000000008 GHz

**Fig. 9.9b** 26.5 to 40 GHz.

**Fig. 9.9** Measured $S_{11}$ and $S_{21}$ characteristics of a 50 GHz coplanar waveguide probe (courtesy of Cascade Microtech).

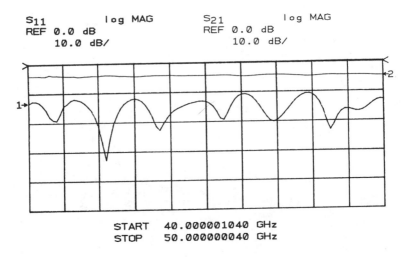

**Fig. 9.9c** 40 to 50 GHz.

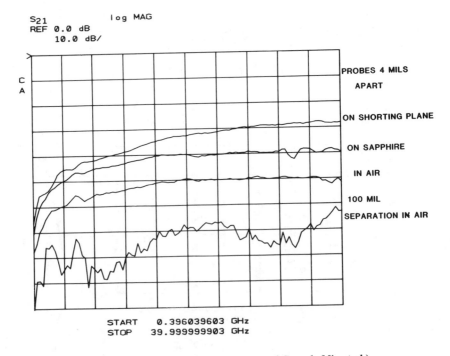

**Fig. 9.10** Probe crosstalk measurements (isolation) (courtesy of Cascade Microtech).

level of −50 dB. This low crosstalk, in effect, is the network analyzer system noise floor.

Such excellent microwave performances can be achieved only through careful and precise design and control of CPW transmission line manufacturing. A possible probe design problem is that of radiation from the probe substrate. Most of the electromagnetic field in coplanar waveguide is confined between the signal and ground electrodes and within the substrate dielectric. However, some electromagnetic fields can stray outside the immediate metal conductor areas, particularly in the area of the backface of the dielectric substrate. Consequently, radiation of energy from the back of the CPW can result in unwanted coupling between probes and the devices under test. Clearly, this unwanted radiation has to be minimized. A simple and effective solution is to apply lossy material along the back surface of the probe to absorb any stray radiation. Such a technique is used on most commercially available probes [4].

Another serious probe problem arises from ground-lead inductance. Basically, the probe system must provide a low-inductance RF ground return path. Figure 9.11 illustrates the difficulty when probing a MMIC amplifier chip. The requirement is for an accurate measurement referenced to the input and output bond pads. At first sight, the probe satisfies this need but the ground return path reveals a likely deviation from the ideal.

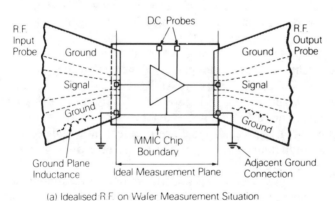

(a) Idealised R.F. on Wafer Measurement Situation
(Adjacent R.F. Ground)

**Fig. 9.11 Possible measurement error caused by common lead inductance. Idealized RF on wafer measurement situation (adjacent RF ground).**

Because the best quality RF ground potential is from the coaxial to CPW transition, the ground path from the probe tips to the transition can be considered inductive in nature, see Figure 9.12. The greatest contribution to this inductance is close to the probe tips, where the coplanar ground strips are physically narrow. A

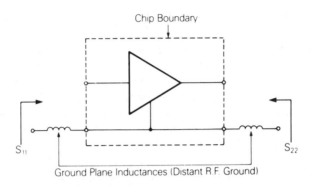

**Fig. 9.12** Measurement uncertainty due to ground plane inductances (distant RF ground).

partial solution may be achieved by welding a metal channel over the front face of the CPW to electrically connect the ground electrodes. Even in this case, the electrically long distance from the probe tips to the coaxial-CPW transition presents a ground-lead inductance that could cause measurement error in some test situations [5].

This ground-lead inductance also is an important consideration in the type of probe contact geometry chosen for microwave testing. Figure 9.6 shows the simplest possible probe implementation, comprising of a signal and one ground electrode in the region of the probe tip; that is, a ground-signal contact configuration. This single ground electrode has an associated ground-lead inductance that, depending on its value, degrades the quality of the microwave measurements. The degradation, for example, in probe crosstalk as a function of ground-lead inductance is graphically illustrated in Figure 9.13. Therefore, for higher microwave frequencies, this ground inductance must be minimized. A more satisfactory probe configuration uses ground-signal-ground electrodes and contact pads. Here, the ground-lead inductances at the probe tips are effectively in parallel, and this cuts the inductance in half.

Both the interprobe coupling and the common lead-inductance problems also can be overcome by the alternative probe design shown in Figure 9.14. This Plessey probe uses a *grounded coplanar waveguide* (GCPW) as the signal transmission medium, because the rear ground metallization inherent to GCPW has the advantage of providing a large, good quality RF ground potential. More important, by incorporating wraparound grounds and via metallizations through the substrate, the front-face ground electrodes are automatically provided with low common inductance paths. In this way a low-impedance ground return can be realized within 250 $\mu$m of the probe tips, and thus, all the ground electrodes can be electrically joined without the complications of metal channel connectors. Further, the large-area ground metallizations intrinsic to GCPW provide excellent RF

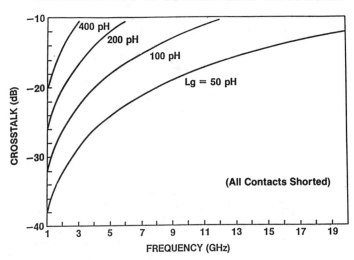

**CROSSTALK DUE TO COMMON LEAD INDUCTANCE**

**Fig. 9.13** Crosstalk due to common lead inductance (courtesy of Hewlett-Packard and Cascade Micro-tech).

shielding properties, which give the probes low interchannel crosstalk and high interprobe isolation.

Using a sensitive 28 ps rise time time domain reflectometry system, GCPW probes have been tested to reveal an intrinsic microwave performance very similar to the CPW probe behavior shown in Figure 9.8. Once again, the coaxial to microstrip-GCPW probe launcher is the limiting factor in achieving satisfactory VSWRs. Introduction of the K-connector, with its direct launch to 0.25 mm alumina gives much improved performances. Using K-connector launchers on tip-to-tip probe pairs, gives the microwave S-parameter performance shown in Figure 9.15. These results show that low connector to probe tip insertion loss (2 dB) and high return loss (18 dB) can be achieved up to at least 20 GHz. The connector probe interface quality is a prime factor in the achievement of microwave probe operation to frequencies as high as 50 GHz [6]. Although special care in fabrication tolerances is necessary to achieve probe performance to 50 GHz, major improvements are achieved through the use of the new 2.4 mm coaxial connector [7].

All of these probes are designed for microwave on-wafer testing purely as a means of accurate terminal characterization. However, there is also a need for probes to perform in-circuit diagnostic testing. The aim here is to provide probes with access to nodes within the circuit to sample the circuit's internal waveforms. Consequently, as many different impedance levels may be experienced, the main

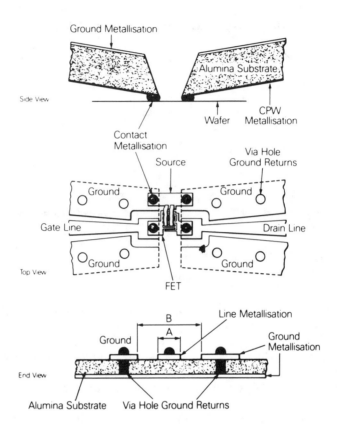

**Fig. 9.14** Grounded coplanar waveguide probe system.

requirement of the probe is that it does not load the circuit and hence distort the circuit's operation. An attractive means of carrying out in-circuit testing is through the use of electro-optic coupling, a technique already being used on MMICs [8]. These electro-optic probes use a very short pulse laser to sample the GaAs circuit node under test. The inherently noninvasive nature of this probe technique does not cause any loading of the GaAs IC, and consequently the circuit operation is not affected by the probes themselves. Some impressive results are reported from electro-optic probes with circuits operating at frequencies up to 100 GHz.

A particularly good example of the uses of in-circuit testing is the examination of signals at several points within an IC. Take for instance a GaAs distributed amplifier MMIC where the circuit relies for its operation on the correct phasing of the traveling wave signals on the FET's gate and drain transmission lines. The circuit designer attempts to achieve optimum operation by selecting the correct

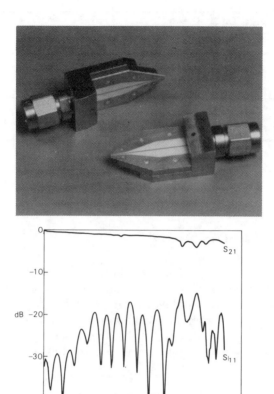

**Fig. 9.15** Pair of probes and their measured performance in back-to-back configuration (courtesy of Plessey plc.).

electrical lengths of the various transmission lines connecting the FET gates (the gate line) and the FET drains (the drain line). It is not always possible to achieve the correct line lengths, and if the completed IC does not perform as expected, then it is very difficult to determine the faulty circuit element. However, an electro-optic prober can be used to sample the waveforms at various locations within the IC. A comparison of these sampled waveforms can readily highlight incorrect signal phases or amplitudes within the circuit, hence, easing the diagnosis of circuit design faults. Figure 9.16 shows the results of probing a traveling wave amplifier, and the interrelation of the gate voltages within the IC can be easily seen [8].

Whereas the attractions of electro-optic probing are self-evident, the lack of commercially available systems is a major drawback. However, lower-cost in-

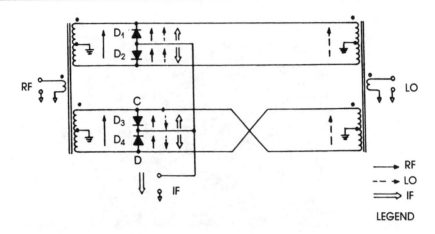

**Fig. 9.16** Schematic of active probe system.

circuit probers are available that use a high-input impedance amplifier within a conventional needle probe (Figures 9.16 and 9.17). The amplifier, which typically has an input impedance of 1 MΩ with a capacitance of below 100 fF, then drives a 50 Ω line feeding the appropriate oscilloscope. These active probes are low cost and readily available (e.g., Picoprobe); but their operating speeds are below a few GHz, and so they are not suitable for many of today's broadband high-frequency MMICs.

**PICOPROBE® MODEL 28** mounted on R & K MODEL 329 positioner.
(Can be mounted on any micropositioner for use with any probing system)

**Fig. 9.17** Commercially available probe head (courtesy of GGB Industries).

## 9.3 ON-WAFER CALIBRATION

The usefulness of any microwave probing technique depends on accurate measurements, which in turn depends on accurate calibration. For scalar measurements it is possible to characterize the probes and feeds from the test equipment for insertion loss and use this information to adjust the results. For sensitive vector S-parameter measurements, however, a more precise model of the imperfections between the test equipment and the device must be determined.

For conventional network analysis in a coaxial or waveguide medium, a wide range of calibration components and techniques is available. By measuring a variety of such components—for, e.g., matched load, short circuit, open circuit— it is possible to construct error models to the measurement ports and thus remove the error terms from subsequent measurements, using an 8- to 12-term error correction [9]. In the case of on-wafer vector measurements, the provision of accurate calibration components and error correction techniques is far more difficult. At the present time, there is much debate as to the optimum form of calibration components, the most accurate calibration philosophy, and the true location of the measurement reference planes.

The most common calibration technique used for vector network analysis measurements is the conventional *short-open-load-through* (SOLT) technique. The use of this technique with microwave probers requires the realization of miniature calibration components that can be accessed at the probe tips. The minimum calibration component set, shown in Figure 9.18, is a short circuit, an open circuit (usually the probes lifted into the air), an accurate 50Ω load resistor, and a short coplanar through transmission line. In addition, it is useful to have some "known" mismatch impedances that can be used after calibration to verify the system accuracy.

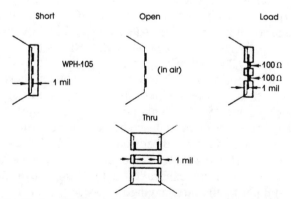

**Fig. 9.18** Basic SOLT reflection calibration of ground-signal-ground prober (courtesy of Hewlett-Packard and Cascade Microtech).

Figure 9.19 shows a set of planar, SOLT calibration and verification components for use with a GCPW probe set. The thin film components are formed on a 25 mm (1 in.) square alumina substrate with a thin resistive layer (NiCr, for example) and plated gold conductors. The design, comprises the following calibration components:

Distributed matched loads incorporating pseudo-T attenuators.
Short circuits.
Through lines.
50 $\Omega$ terminations for isolation measurement and alignment.
Mismatch termination.
Offset open and short circuits.
Large test cell to determine sheet resistivity.

The components are designed to be the same size as the IC under test to remove the need for adjusting the probes between calibration and measurement. Calibration using these components allows the use of an error correction under computer control, resulting in S-parameter measurements with reference planes at the probe tips; that is, the IC RF contact pads. This removes the need for sophisticated but error-prone de-embedding techniques and is particularly valuable in individual IC component element characterization. The design and verification of these "calibration standards," of course, are far from straightforward. Attention has to be taken of the likely effects of the standard's physical dimensions, their behavior with frequency, and the effects of the probes' behavior during the calibration.

The short circuit element must be large enough to ensure good quality contact to the probe yet be small enough to avoid excess capacitances and minimize unwanted coupling to the probes. Here, the aim is to minimize any errors in the measurement reference plane caused by the short circuit's electrical length. In the case of the short on the calibration set of Figure 9.19, the shorting bar is only 100 $\mu$m wide, enough to allow the 50 $\mu$m square probe tips to contact the bar. The shorting bar itself is connected to a good quality RF ground by the via holes at either end of the short. The calibration standard also contains GCPW offset short circuits of three different lengths. Measuring these offset shorts and comparing the $S_{11}$ values with those of the original short allows, by extrapolation, the accurate determination of the electrical length of the original short. It is easily shown that the error in the reference plane caused by the small shorting bar is of the order of 50 $\mu$m or less. Given that the bond pads on a typical MMIC are 100 $\mu$m square it is obvious that the short calibration component is of adequate quality.

At present the open circuit calibration is done simply by lifting the probes off the calibration standard substrate leaving the probe tips in free air. Once again, this technique can be shown to be adequate but there is some debate over the accuracy of the open circuit. In the standards of Figure 9.19, a number of offset GCPW open circuits are included to verify the open calibration.

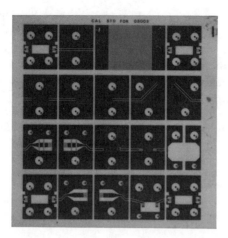

**Fig. 9.19** Integrated calibration components on alumina (courtesy of Plessey plc.).

The 50 Ω matched load is a vital part of the calibration, and it is essential that the resistance value is accurately maintained. The physical size of the load is also significant, especially if broad bandwidths and high frequency performance are required. At higher frequencies the finite physical dimensions of a badly designed load resistor can lead to inductive behavior and hence less than ideal load operation. A possible solution to this problem is to design the load as a broadband, high-performance pseudo-T attenuator [10], which can be optimized to exhibit near ideal performance over multioctave bandwidths. This is the approach adopted in the standards of Figure 9.19. An alternative solution is to produce very small 100 Ω resistors to connect in parallel between the signal and symmetrical ground contacts of the probe. This is the approach adopted in the CPW probe standards of Figure 9.18.

The final calibration component is the through line, which ideally must have zero length. This is approximated by the input and output probes connected through a minimum length CPW or GCPW transmission line. Alternatively, a known electrical length through line can be used, and the known reference plane offset then can be accounted for within the error correction process.

Clearly, at this early development stage, the accuracy and traceability of on-wafer microwave probe standards must be treated with caution. To gauge the quality of these "calibration standards," the individual calibration components can be subjected to conventional jig measurement followed by de-embedding correction routines. The subsequent results backed up by prober calibration runs show that, although the components are inferior to their coaxial counterparts, they form the basis of meaningful and self-consistent microwave network analysis measurements.

Figure 9.20 shows the probed return losses of the mismatch terminations, 17 Ω and 150 Ω nominal 6 dB return loss verification components from the integrated calibration substrate, and 56 Ω erroneous load resistor from another substrate. The last demonstrates the ability to resolve return loss as high as 25 dB with a precision of magnitude of 0.01 at up to 10 GHz and 0.02 at up to 18 GHz. The nominal 6 dB return loss mismatches have been well characterized by the wafer prober in both magnitude and phase. From these, current estimates of measurement uncertainties are ±0.25 dB, ±2.5° to 8 GHz, degrading to ±0.75 dB, ±7.5° at 18 GHz.

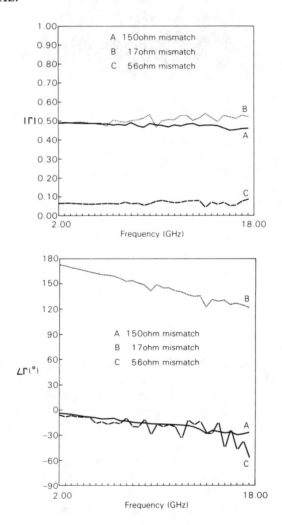

**Fig. 9.20** Reflection coefficient of mismatch loads (courtesy of Plessey plc.).

An alternative calibration approach involves the fabrication of on-wafer calibration components as dropins on the GaAs IC wafer. Figure 9.21 shows a calibration chip containing matched loads, open circuits, short circuits, through lines, test FETs, offset short circuits and alignment loads. (Alignment loads are simply 50 Ω resistors between ground and signal ports to allow rapidly setting up the probes on the contact pads.) The potential advantages of this approach are that each wafer run automatically produces new standards fabricated to semiconductor processing accuracies (1–2 $\mu$ms), and the calibration components inherently have the same shunt capacitances to ground as the components to be measured on wafer. As well as being very accurate, these on-wafer calibration standards allow automatic calibration during wafer testing. This ability to check probe calibration during wafer testing ensures that wafers cannot be diagnosed incorrectly as rejects through a probe failure.

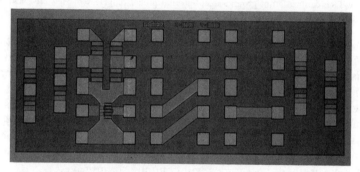

**Fig. 9.21** Integrated calibration of components on GaAs (courtesy of Plessey plc.).

An alternative calibration technique to the SOLT approach is the *through-reflection-line* (TRL) calibration routine. The ''standards'' used in this case are a through connection, a known reflection coefficient, and a transmission line of known length and impedance. For microwave probers, the TRL calibration set can be realized as a short length of coplanar transmission line (T), the probes in free air (R), and a comparatively long length of coplanar transmission line (L), see Figure 9.22. The TRL calibration technique places the reference plane at the center of the short line (T), and then the known delay of the through line is used to move the reference plane back to the probe tips. The TRL approach also relies on the characteristic impedance of the long transmission line to set the system reference impedance level. Therefore, the line used must be realized very accurately with close control of the transmission line's conductor electrodes' dimensions. A potential disadvantage of the SOLT approach is that the probe's separation has to be varied in order to measure the L standard.

Comparisons of results from wafer probers calibrated using both SOLT and TRL techniques reveal very close agreement between the two calibration approaches. The SOLT calibration has the advantage that the system impedance is

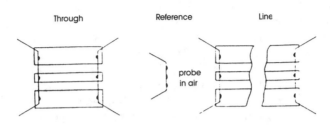

**Fig. 9.22** Basic TRL calibration of ground-signal-ground prober (courtesy of Hewlett-Packard and Cascade Microtech).

set by an accurately trimable resistor. Conversely, although the TRL calibration and transmission S-parameters self-consistent. In both cases, the reference planes are established within a few tens of microns of the probe tips, a distance that is insignificant compared to the normal 100 $\mu$m MMIC or FET bond pad sizes. The choice of calibration approach, therefore, is dominated largely by available calibration standards and the personal preference of the test engineer, since both approaches give very similar, consistent microwave probed results. However, as we gain more experience of microwave probers and their calibration, there is little doubt that measurement accuracies will continue to improve.

## 9.4 ATE SYSTEM ARCHITECTURE

Given the basis of a viable wafer probe technology, serious consideration can be given to the ATE needs of an on-wafer measurement equipment. The choice of ATE architecture is dictated by the tests to be performed. Perhaps the simplest application of the microwave prober is a S-parameter test set for MMIC development and characterization. Here, the simple ATE system of Figure 9.23, comprising probe station, automatic network analyzer, and computer controller, is all that is required. The controller is used to drive the probe station's stage movements, to put the analyzer through the appropriate automatic calibration and measurement routines, to organize data collection and storage, and if appropriate to perform data manipulation, such as equivalent circuit modeling. For development and characterization applications, accuracy often is a prime requirement rather than testing throughput or flexibility. Here, the simple ATE has the major advantage that its direct interface between the probes and the network analyzer demands accurate control of transmission line impedance, it does use fewer calibration components and inherently keeps the reference planes of the reflection

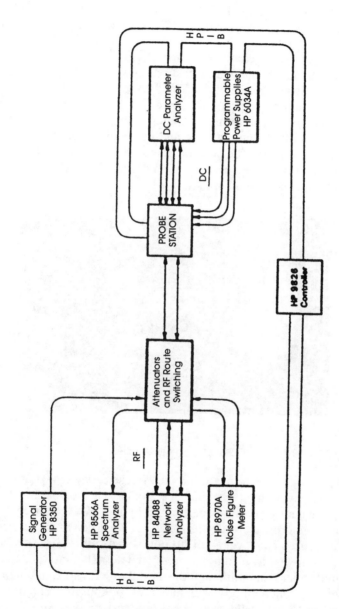

**Fig. 9.23** Simple ATE system.

(i.e., no routing switches) minimizes uncertainties in the measurements caused by interconnect VSWRs and loss. However, if a more production-oriented application is envisaged, then a more complex ATE system with greater flexibility and data throughput is needed.

Truly versatile wafer characterization equipment implies the capability to perform all conceivable functionality tests on any possible MMIC. Figure 9.24 shows the test hardware in operation. The ATE comprises a rack of RF and dc test equipment, an automatic probe station, and a desktop computer. Selection of the RF test equipment is performed by microwave switches within the rack to generate two signal lines, one outgoing and one return, between the test equipment and the probe station. The desktop computer drives all the test equipment, sets the RF signal routing, controls the wafer probe step movements, and manages the large quantity of measurement data.

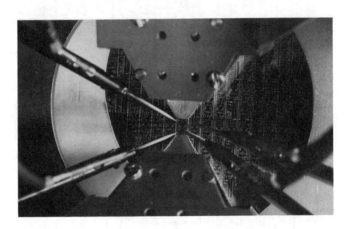

**Fig. 9.24** Microwave probes in operation (courtesy of Plessey plc.).

In an ATE system, care has to be taken in selecting test routines to avoid running into excessively long test times per wafer. The comparatively slow and complex measurement procedures performed by microwave test equipment can quickly dominate microwave wafer prober test times. For example, carrying out S-parameter measurements over a full 801 point bandwidth under several different bias conditions followed by a gain compression determination leads to a test time of several minutes per chip. Although this is still orders of magnitude faster than the conventional test jig methods, with hundreds to thousands of chips per wafer, the total wafer test time could easily be a full working day. In some cases this approach is acceptable but in final production screening, for example, cost and throughput considerations demand much faster wafer test times. Therefore, a speedier test routine is needed, which examines only the most critical MMIC performance parameters. Experience has already shown that MMICs passing a critical test, such as a narrow band $S_{21}$ measurement, normally also pass the more comprehensive test routines. In selecting the most efficient test schedule, the most important parameter measurements and their elapsed times have to be calculated. These measurement times are then added to the wafer probe time overheads of wafer stage movement, connect and disconnect, chip bias, and chip thermal-electrical settling.

## 9.5 ON-WAFER MICROWAVE TESTING

### 9.5.1 MMIC Layout Considerations

In laying out components for the RFOW test it is necessary to imitate, as closely as possible, the electrical environment of the intended application. For example, in the absence of substrate via holes, continuous ground rails as shown in Figure 9.25 are needed to connect the ground contacts of input and output probes and to provide a minimum inductance ground reference to elements within the chip that finally would be grounded by local bond wires. Obviously, there is a frequency-dependent maximum length beyond which this interconnect no longer fulfils its purpose. In particular, because the interconnect effectively forms a transmission line with the superior ground plane of the back face of the wafer, an effective open circuit referenced to the probe ground will exist halfway along a $\lambda/2$ interconnect.

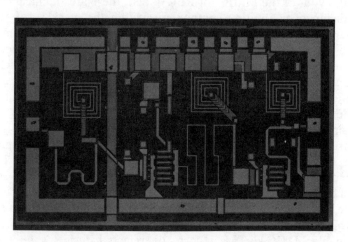

**Fig. 9.25** Photograph of S-band LNA chip (courtesy of Plessey plc.).

The use of through substrate vias overcomes the problems of grounding. Figure 9.26 shows an S/C band LNA that uses via holes for reduced source inductance on 600 $\mu$m FETs, and the only concessions necessary for RFOW testing are the additional probe ground pads with via holes for direct connection to the back of the wafer. Another approach, where possible and appropriate, is the use of entirely balanced circuit configurations [11]. These push-pull circuits have a virtual ground with no net RF current flowing to ground. Here, the provision of baluns between unbalanced test equipment lines and balanced transmission lines on RF probes allows on-wafer testing with little compromise.

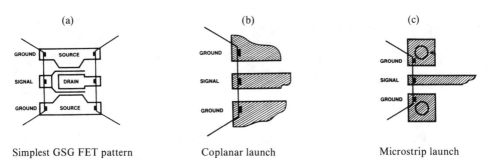

| (a) | (b) | (c) |
|---|---|---|
| Simplest GSG FET pattern | Coplanar launch | Microstrip launch |

**Fig. 9.26** Recommended DUT connection pad layout configurations.

In the layout of the MMIC for RFOW test, there are significant advantages in the inclusion of calibration components within the mask set if S-parameter measurement is intended. Not only does this facilitate the measurements by removing the need to exchange calibration pieces and the wafer under test—and probably the adjustment of probe height due to different substrate thickness—but also in the accuracy of the error correction by guaranteeing identical substrates for calibration and measurement. This is particularly important when testing high-speed Si devices on N$^+$Si substrates, which give rise to large bond pad capacitances. Using on-wafer calibration components, these capacitances are included in the measurements performed during calibration and effectively removed along with other nonidealities of the measurement systems. Otherwise, it would be necessary to estimate their value and perform postmeasurement de-embedding.

Another advantage of including calibration components on the wafer is that they can be used for periodic checking of the probe calibration during wafer test. By doing this, any degradation in probe behavior or drift in calibration is identified before valuable MMIC die are incorrectly rejected during wafer probing. For example, if one probe tip fails without detection during the wafer test, the ATE system will start incorrectly inking chips as faulty. A periodic probe check avoids this costly mistake.

### 9.5.2 Probe-to-IC Interface Issues

For the best measurement accuracy, the device under test should have connection pads laid out to achieve a high-quality, low-mismatch interface to the probe's coplanar transmission lines. Ideally, the connection pads should be in a colinear ground-signal-ground format to ensure a good launch into either microstrip or coplanar transmission lines on the MMIC. Figure 9.27 shows the recommended configurations for a test FET (the minimal MMIC), a coplanar line based MMIC, and a microstrip line based MMIC. The latter configuration is that found on most MMICs with GaAs via holes positioned within the ground contacts to provide a low-inductance path to ground from the probe to the MMIC wafer's backside metalization.

The provision of low-inductance paths to ground is vital in ensuring good quality measurements. At the lower microwave frequencies the via hole is not strictly required for successful on-wafer probing. However, from X-band upward for single-ended circuits, the provision of low-inductance ground interconnects is vital for accurate microwave probe results. Although the ground-signal-ground probe configuration is preferred for high-quality, high-frequency measurements, other options are available to the MMIC designer. As well as having shielded, unshielded, balanced, differential, and single-ended options for RF interconnects, probes also are available for MMIC power supply decoupling. Here, ICs can be powered from very low impedances rather than from the highly inductive impedances inherent to needle probes.

Although the choice of IC contact pad configuration depends very much on the particular MMIC design, the positioning of the contact pads on the IC has to be designed with some care. First, consideration has to be given to the issue of interprobe coupling. Here, the input and output microwave probes should be placed as far apart as possible, because interprobe isolation is a function of probe separation. Hence, it is normal for the IC's microwave input and output pads to be placed at opposite edges of the chip (i.e., the west and east edges of the chip). Consequently, the remaining two chip edges (north and south) are normally used for dc power supply and control line connection points. Second, the coupling of energy from the probes to sensitive circuit elements on the MMIC has to be considered. Care should be taken to ensure that no circuit element crucial to IC performance is placed adjacent to the probe connection points. For example, a critical feedback network inductor should not be placed next to the RF probe connection pads, as the presence of the probe during on-wafer testing could significantly affect chip performance.

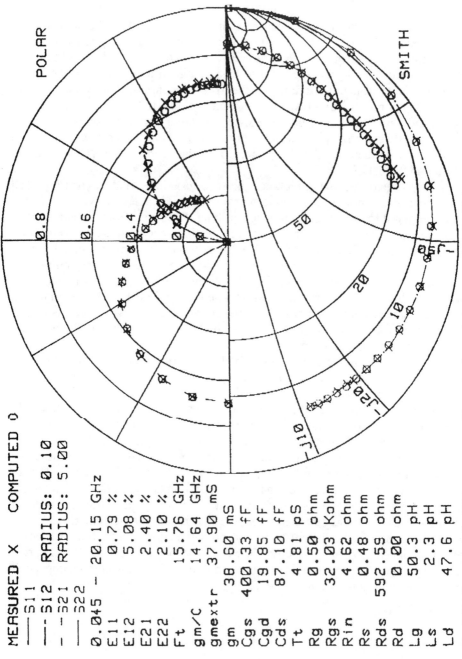

**Fig. 9.27** Microwave prober FET measurements (X) compared to equivalent computed results (O) (courtesy of Cascade Microtech).

### 9.5.3 Examples of MMIC Wafer Testing

The photograph in Figure 9.28 is of a broadband amplifier produced using a 0.5 $\mu$m gatelength process. The chip layout has a conventional distributed amplifier topology with the ground-signal-ground RF on wafer connection pads at the circuit's input and output ports. To ensure low inductance grounds, and hence guarantee accurate measurements over the required wide bandwidths, there is a GaAs via hole under each ground connecting pad. Each wafer run has 74 available die sites for this particular circuit design, as three other circuit designs are included on the multiproject mask set. Figure 9.29 presents the results from a probe measurement performed on a typical wafer of this circuit. It is apparent that 54 of the 73 circuits tested had nearly identical gain performances over the 1–18 GHz bandwidth, demonstrating a yield at the wafer stage of approximately 74%. With each die site carrying its own unique X-Y coordinate code, the identification of operational chips after wafer dicing is straightforward.

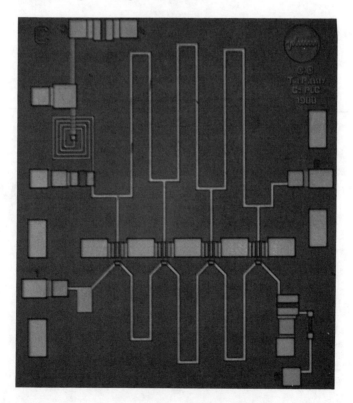

**Fig. 9.28** Photograph of low-noise amplifier MMIC (courtesy of Plessey plc.).

| Number of Die Sites on Wafer | 77 |
| Number of Chips Operational | 54 |
| Yield | 70% |

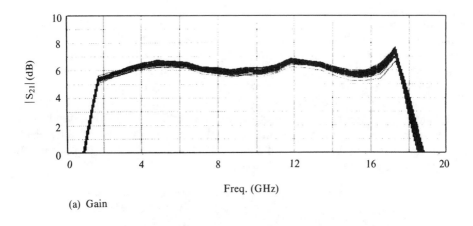

(a) Gain

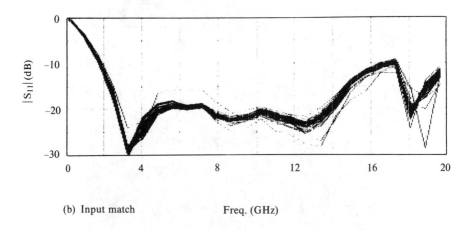

(b) Input match          Freq. (GHz)

**Fig. 9.29** RF on-wafer measurements of a 2 to 18 GHz amplifier MMIC (courtesy of Plessey plc.).

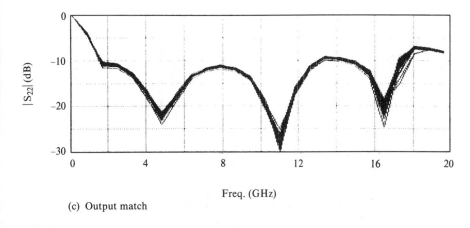

(c) Output match

**Fig. 9.29** cont.

Figure 9.30 compares the measured results from a typical chip and the designed response of the MMIC amplifier circuit. The success of the circuit design is obvious, but the advantage of the microwave prober as a development tool is also worthy of note. Here, within a working day of completing the wafer processing, the microwave prober can test all available chips, and off-line data processing can verify the design accuracy. This close interaction between computer-aided test and design is an intriguing concept, since it opens up many exciting possibilities for MMIC design. For example, if during an early chip development phase the measured response deviates from the desired design response, then the circuit analysis programs could be used to reoptimize the circuit file to fit the measured response. With care, this optimization could identify the design faults, enabling a fast redesign.

The example of MMIC testing is a relatively straightforward measurement to show the basic possibilities for on-wafer testing at microwave frequencies. Of course, the microwave prober combined with a sophisticated ATE system can be a powerful MMIC development or production tool. One example of a more complex application of the prober system is the design and development of a 2–6 GHz feedback amplifier MMIC, in which the prober plays a major role in speeding up the acquisition of IC performance data.

A photograph of the three-stage LNA chip was shown in Figure 9.25. This amplifier occupied 27% of the GaAs wafer area giving 110 chip sites on a single 2 in diameter wafer. A microwave wafer prober was used to evaluate the LNA chips

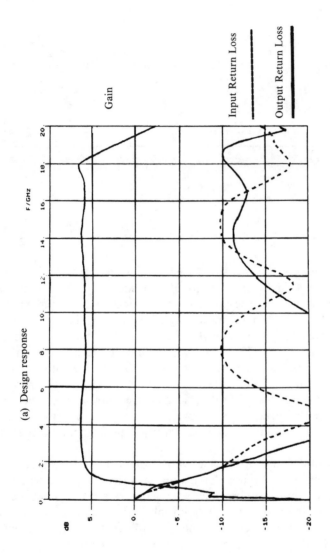

Gain

Input Return Loss

Output Return Loss

(a) Design response

**Fig. 9.30** Comparison of simulated (a) and measured (b) results on a 2 to 18 GHz amplifier MMIC (courtesy of Plessey plc.).

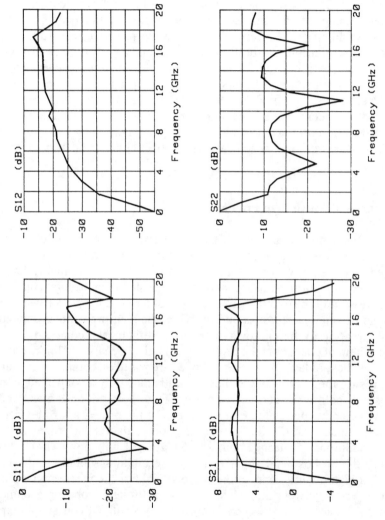

Average S-parameters of 54 out of 77 devices

(b) Average of probe measurements

**Fig. 9.30 cont.**

quickly and cheaply. The large quantity of data that the prober can produce is amply demonstrated by the plots of Figure 9.31. These computer generated plots of gain, noise figure, and input match represent only 50% of the circuits probed on one 2 in wafer. This wealth of data can be difficult to interpret, and therefore off-line data analysis is used to compare the measured chip performances against the demanded chip specifications. In this way, clearer presentations can be derived, such as those of Figure 9.32. Here, the microwave prober gain results from a complete wafer are presented as both a wafer map and a performance histogram. Similar results can be produced for all the measured microwave parameters in turn. By way of a summary, a composite wafer map and histogram of all the microwave results is shown in Figure 9.33. The results are analyzed against the complete product specifications, revealing an overall chip yield of better than 50%. This production-oriented use of microwave probers allows the testing of MMIC wafers against product specifications to select chips and determine manu-facturing yield.

### 9.5.4 Limitations of Microwave Probing

All microwave probes, in either CPW or GCPW format, are required to act as high-quality microwave connectors and go through several hundred connect and disconnect cycles per wafer tested. This is an extreme requirement, demand-ing both the precision normally associated with high-quality metrology connectors and a robustness and repeatability not expected of any ordinary microwave con-nector. Not surprisingly, success in microwave probe measurements is dependent on care and attention to the microwave probes and their vital contact tips.

The first item of concern is the probe's mechanical behavior. Unlike conven-tional probe needles, today's microwave probes are based on ceramic material not metal. As a result, the microwave probes can withstand only a controlled amount of bending stress before the ceramic fractures. This shortcoming is worsened by the normally recommended deliberate overtravel of the probes after first contact with the device under test (DUT). To ensure good contact most probes are sub-jected to an additional 50 $\mu$m (2 mil) to 250 $\mu$m (10 mil) downward motion. Consequently the probes bend slightly and the resultant scrubbing at the probe tips achieves a high-quality connection to the DUT. For the inexperienced user, the margin of error between good contact and a broken probe or a damaged airbridge on the adjacent die site seems depressingly small. Of course, no experi-enced user of probes admits to probe breakage, but all users have broken probes at some time or other. This particular problem is analogous to the overtightened connector syndrome commonly encountered on network analyzers. The solution is analogous: use the minimum overtravel on the microwave prober (e.g., 50 $\mu$m); that is, avoid overtightening the connector.

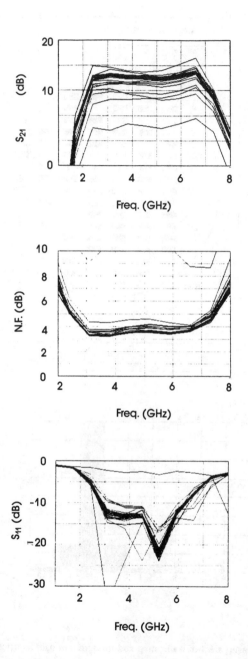

**Fig. 9.31** On-wafer RF measurements of gain, noise figure, and input match (courtesy of Plessey plc.).

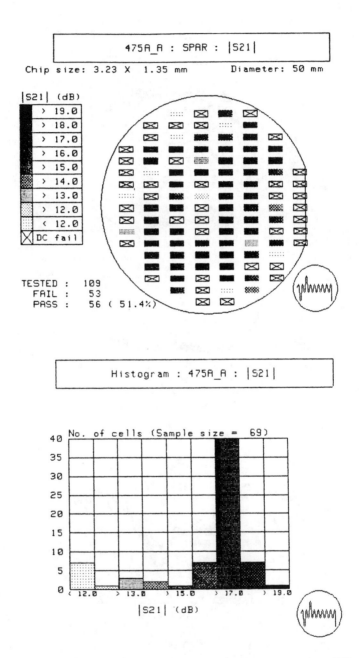

**Fig. 9.32** On-wafer RF measurement of wafer map and histogram of gain (courtesy of Plessey plc.).

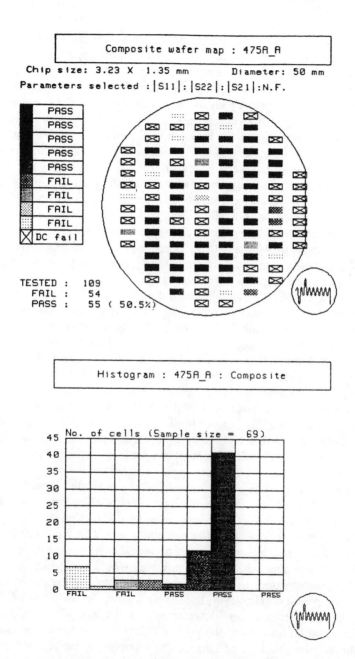

**Fig. 9.33** On-wafer RF measurement of composite wafer map and histogram (courtesy of Plessey plc.).

The other major problem concerns the probe tips themselves. First, since the tips are on the underside of the coplanar structure, they are not always easy to see, and therefore alignment to the contacts of the wafer under test is difficult. The only solution here is care and experience. Second, the probe tips tend to pick up gold shavings from the contact pads on the wafers under test. As a worst case, this gold can become lodged between ground and signal lines, thus shorting out the probe. Alternatively, this gold can build up preferentially on one of the probe tips, degrading contact to the other tips and consequently also degrading measurement quality and repeatability. Checking dc continuity during on-wafer testing can detect the short circuit. The more insidious contact degradation through gold build up can be monitored only by including verification standards on the wafer PCM cells, which are probed periodically during wafer test. Of course, the solution to these probe drawbacks is the regular cleansing of the probe tip areas.

Recently, these problems with CPW probes have been recognized, and some manufacturers such as Design Techniques have developed a new range of CPW probes that have needle point pin contacts.

These new probes are available with the same standard contact pitches (100 $\mu$m to 250 $\mu$m) and mounting configurations as the CPW probes. Hence, the retrofitting of the new, more robust probe heads to existing wafer stations is straightforward. The new contact pins are claimed to be compliant enough to allow 25 $\mu$m of flexure. The contact pins protects both the DUT and the CPW ceramic probe medium, while providing a positive, low-loss connection between probe and DUT. In addition, since the pins protrude slightly from the probe tip, the ATE operator has a clearer view of the DUT contact pads and the probe tips. Even with these more robust, easier to align probes, some breakages may occur. However, these new probes have field replaceable tips. These claims are impressive, and this latest probe development may prove to have cured many of the handling problems some probe users have experienced.

Any microwave probe's operational lifetime is related directly to the over-travel mentioned earlier and the electrical power carried by the probes during test. Instantaneous overload effects can be caused by connecting or disconnecting the probes while power is applied to the die under test. Clearly, very large current spikes could occur in some conditions; and in inductive power supply situations, disconnecting the probes while a die is powered would lead to arcing at the probe tips. Installing an appropriate current limit-sense circuit on the die power supplies can reduce the problem. However, the simplest safeguard is to write into the ATE software a die powering up and down routine during the probe connect time.

Besides the limitations of the probes themselves, a few ATE measurement configurations must be treated cautiously; namely the power and noise figure measurements. In the measurement of MMIC power output, the principal problem is the dissipation of dc power and the consequent thermal effects on device performance. For example, for power amplifiers with power added efficiencies of about 20%, for every 1 W of RF output power the MMIC chip must dissipate as heat 5 W of dc energy. Therefore, the provision of good thermal heatsinking for the wafer under test is a primary concern. Unfortunately, a back-metalized wafer resting on a wafer stage can never have the thermal performance of the finished solder-bonded MMIC. As a result, it is always going to be difficult to accurately test multiwatt output power MMICs on-wafer, since the device's active channel temperatures will be higher than in the final packaged form. However, for MMICs below the 1 W level, sensible 1 dB power compression measurements can be achieved on-wafer.

Another equipment limitation for power measurements is the provision of high-RF drive powers at the input probe tips and the accurate determination of the power level at the probe tips. This measurement equipment problem could demand the use of TWT amplifiers to provide the RF input drive and, possibly, bandpass filters to ensure single-tone operation. Calibration of the system requires the straightforward determination of the S-parameters of the adapters, cables, bias tees, and probes. Similarly, accurate output power measurements require an accurate determination of the load impedance presented to the DUT by the output probe, adaptors, bias tee, attenuator pad, and power sensor. Figure 9.34 shows a typical power measurement set up to illustrate the calibration difficulties. As we can see, the DUT is presented with a load impedance that is a combination of the probe-adaptor and power sensor parameters. The power delivered by the DUT is attenuated by the "gain" ($G$) within the probe-adaptor network before detection at the power sensor. Therefore, we need to know the appropriate network parameters to solve equation 9.1 and derive the likely uncertainties in the DUT output power:

$$\Gamma_L = \frac{S_{11} + S_{12}S_{21}\Gamma_{ps}}{(1 + S_{22}\Gamma_{ps})}$$

$$P_{ps} = P_{dut} + G(\Gamma_{ps})$$

(9.1)

The final difficulty commonly encountered during power MMIC measurements is one of decoupling the dc power lines. The large gatewidth, high-$g_m$ MESFETs required on power ICs need careful bias decoupling to avoid oscillation. Here probes containing high-value bypass capacitors with low-inductance dc lines are needed. Fortunately, such probes can be procured from the microwave prober manufacturers.

Perhaps the most difficult RF measurement that can be made on wafer is the noise figure of low-noise amplifier MMICs. Figure 9.35 shows a typical probe set up for the conventional automatic noise figure measurement. To make an accurate noise measurement, the system must be calibrated so that the losses and VSWRs of the switches, adaptors, and probes can be determined. This can be done readily using the network analyzer and the appropriate known calibration standards.

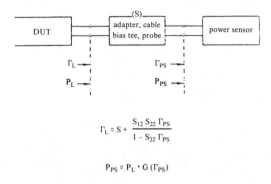

**Fig. 9.34** Block diagram illustrating power measurement corrections for adaptor losses and reflections.

Thus, the losses between the noise source and the probe tips as well as those between the probe tip and the noise meter input are known. These figures are entered in the noise figure meter so that the *excess noise ratio* (ENR) at the probe tips is known, as is the DUT noise output at the probe tips. The corrections involved in the on-wafer noise figure measurement are analogous to those already described for on-wafer power measurements. Having calibrated the system and entered the necessary correction factors, a check calibration is performed with the probes connecting the minimum length through line on the probes' calibration standard substrate. A good calibration provides the confidence of a zero dB noise figure. Finally, the wafer prober is applied to the MMIC wafers for measurement of circuit noise figure.

Measurement uncertainties are caused by the mismatch reflections at the probes, adaptors, switches, and DUT in such a system. The likely measurement errors in the noise figure determination can be calculated [12] to give a lower bound for accurate measurements. Given that the microwave probes exhibit degrading insertion losses and reflection losses with frequency, it is possible to show that the minimum measurable NF rises with frequency. For example, at S-band frequencies 2 dB noise figures are the lowest that can be realistically measured on a wafer prober, whereas at 20 GHz, this figure degrades to 4 dB. However, for today's 0.5 $\mu$m gate MESFET based MMICs these figures are practical, and on-wafer noise measurements can be made as part of the MMIC product testing and selection procedure.

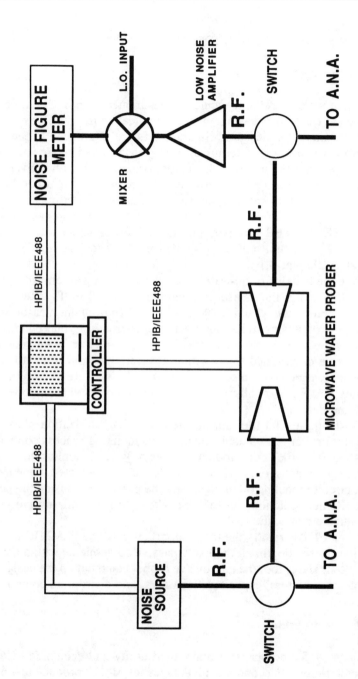

**Fig. 9.35** Schematic of an on-wafer noise figure measurement system.

## 9.6 ON-WAFER PRODUCTION TESTING

### 9.6.1 Foundry Process Design Support

The ability to make rapid and accurate measurements on individual circuit components such as transistors has found major application in MMICs as a design data base tool. Because the measured S-parameters are referenced to the probe tips and therefore to the device terminals, there is no need to de-embed the measured data to extract the inherent device performance, as with conventional testing. Consequently the probe measured data tends to be very smooth in its behavior with frequency thus making equivalent circuit model fitting straightforward and accurate. The use of a microwave prober and postmeasurement analysis software such as FETFITTER ™ provides both much faster device measurements and faster modeling than the traditional jig measurement technique. This is explained in detail in Chapter 4.1.

To illustrate the benefits of on-wafer microwave probing in terms of volume of data, Figure 9.36 shows the measured S-parameters of 40 FETs taken from a complete 3 in. diameter wafer of $0.5 \times 200 \ \mu$m FETs. From similar measurements made over a number of wafer batches and a range of transistor $I_{ds}$ bias conditions a wealth of foundry design data can be accrued. For example, accurate spreads in S-parameters can be determined, both within a wafer and from wafer to wafer. Device bias dependent behavior also can be readily derived. Repetition of these measurements on test FETs of different gate widths can be used to provide accurate device scaling data, too.

As well as being a vital tool during the early characterization phase of a process, the wafer prober is a valuable means of maintaining process control. By continually testing PCM FETs on production wafers, we can monitor the process quality. Any change in measured FET S-parameters can be detected immediately, and equivalent circuit modeling often highlights the parameter causing the performance change. Remedial action can then be taken, with the microwave prober serving as a valuable diagnostic tool.

Although the FET is rightly the most important part of any MMIC process, the accurate characterization of all the likely passive components within a circuit cannot be forgotten. Here, too, the microwave prober can build up the design data base for a range of inductors, capacitors, resistors, and distributed components.

### 9.6.2 MMIC Product Testing

Although the wafer prober is of undoubted utility and economic value for process characterization, its economic advantages for MMIC produce testing are not quite so obvious. Situations can be found where the wafer prober would be

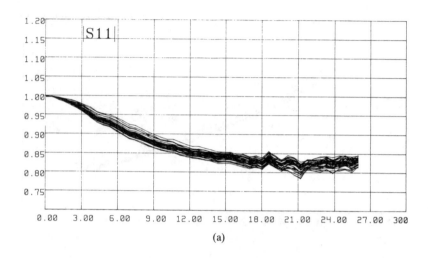

(a)

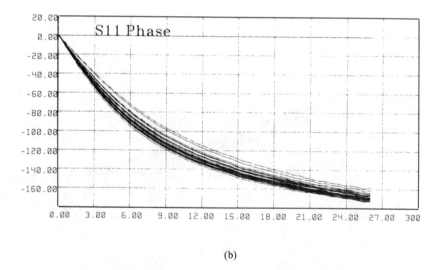

(b)

**Fig. 9.36** Typical probed S-parameter measurements on foundry test FET wafer (courtesy of Anadigics Inc.).

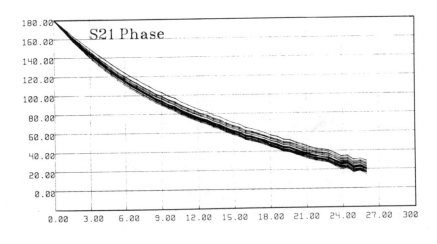

(c)

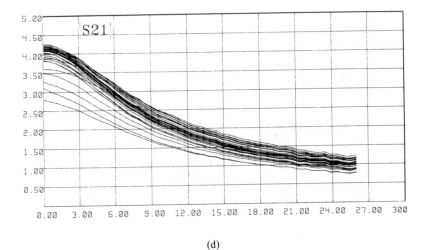

(d)

**Fig. 9.36** cont.

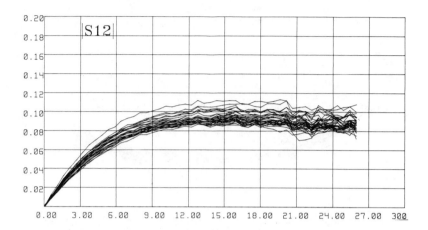

(e)

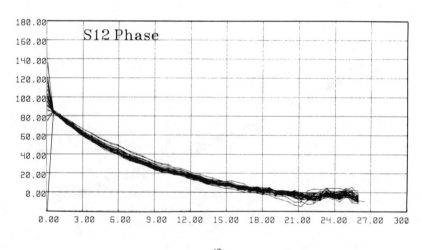

(f)

**Fig. 9.36** cont.

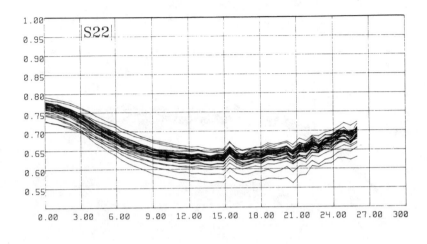

(g)

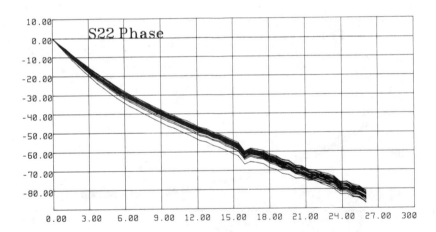

(h)

**Fig. 9.36** cont.

vital to cost-effective MMIC production. However, in some circumstances, the capital and running costs of a microwave prober would dominate the savings in production costs made possible by the prober. Therefore, before looking at the economic aspects let us consider the technical advantages offered by a wafer prober in MMIC system production.

Consider the miniature phase array module shown in Figure 9.37. This module contains three MMIC switches, three MMIC amplifiers, and a phase shifter MMIC, all mounted on a single MIC ceramic tile. Each of these modules can be assembled in one operation using good MMICs, preselected for their on-wafer microwave probe performance. At the assembly stage all the chips are known to be good, and the production is performed with a very high probability that the completed unit will perform to specification. Without the confidence given by the wafer prober, this aggressive production strategy would carry the high risk of producing modules containing faulty chips. The reworking of faulty modules, of course, is expensive and carries the risk of damaging the module during rework. An alternative production strategy is to assemble each MMIC on its own pretestable tile before integrating all the pretested MIC units into the module. This approach does not need a wafer prober, but it is more expensive in assembly effort and tends to produce a relatively large final module.

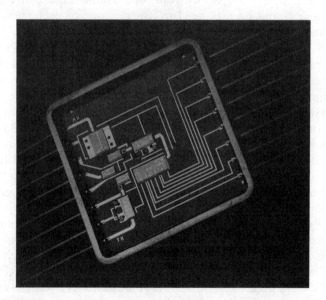

**Fig. 9.37** Packaged miniature phased array radar module (courtesy of Plessey plc.).

These considerations suggest that a microwave wafer prober is a necessity for efficient production of complex MMIC based systems. The module described was produced using the wafer prober as a key part of the manufacturing cycle. As a consequence, good yields of modules were achieved; a typical module response is shown in Figure 9.38. The reasons for the importance of the prober in this assembly can be readily calculated by considering the known yields of the individual chips. The amplifier and phase shifter MMICs have chip yields of approximately 60%, whereas the SPDT switches have yields approaching 90%. Now, assuming an assembly from randomly selected MMICs, the probability, $P$, of assembling three fully functional switches and four fully functional die of the other MMICs is simply given by

$$P = (0.6)^4 \times (0.9)^3 = 0.094 \qquad (9.2)$$

Clearly, with such a low probability (9.4%) of success, preselection of the working die is desirable. However, at some level of chip yield, chip cost, wafer probing cost, and packaging cost, it may be cheaper to package randomly selected chips and simply throw away the nonworking modules. For example, if the system could be built in a cheap package, with only one multifunction chip that had a 90% yield, then there would be less need to use microwave probing as part of the production process (i.e., $P = 0.9$).

Figure 9.39 is the flow diagram for the production of GaAs Tx/Rx modules needed in a phased array radar. This production model starts with the GaAs MMIC processing and flows through dc probe and visual inspection, sawing of wafers for chip separation, and MIC unit assembly to final module test. For the purposes of examining the impact of RF on wafer probing the model can be run for production strategies with or without the RFOW prober. From the flow diagram, a cost model is readily constructed in which all the appropriate input costs, overheads, workhours, wafer diameter, chip size, manufacturing, and testing yields are taken into account. The assumed production rate is 100,000 modules per annum with each module containing three MMIC chips in separate ceramic packages.

Figure 9.40 is a graph showing the results of the cost model as a function of chip yield. The curve represents the results for only one set of input parameters, but the results are typical of many parameter sets, and general conclusions can be drawn from the curve. First, it is quite clear that a microwave (RFOW) prober facility is not always necessary, because in some cases its overhead and operating costs outweigh its benefits. However, for chip yields below 60–70%, the microwave prober offers considerable cost savings. Indeed, at 10–20% yields, the prober facility would reduce costs by millions of dollars a year at the production rates envisaged for array radars.

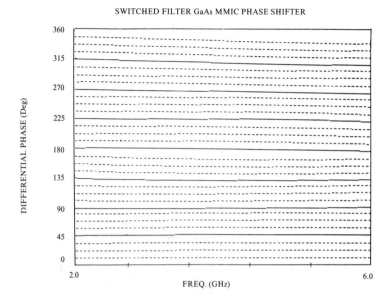

**Fig. 9.38** Measured performance of a typical phase shifter module (courtesy of Plessey plc.).

## 9.7 THE FUTURE STATUS OF MICROWAVE PROBING

During the early 1980s several companies realized that if high-speed GaAs ICs were to be manufactured, then a means had to be found to evaluate the ICs high-speed functionality on-wafer. The earlier work started at Tektronix [2], which later grew to become the commercial prober supplier Cascade Microtech, has had the greatest effect on industry. Indeed, Cascade Microtech probers are now used by nearly all the GaAs IC companies for both analogue and digital testing. Also, the undoubted benefits of microwave probers are now being realized by the high-speed silicon industry. Microwave probes are available with more than six signal lines at 100 $\mu$m pitch, and it is thus possible to measure the functionality of high-speed digital ICs on-wafer. Therefore, it is possible to measure gate delays directly, instead of by the dubious technique of monitoring ring oscillator performance at the low frequencies of conventional probers.

Considering the likely high-frequency limits of microwave probers, we have to recognize that coplanar waveguide, which is the basic transmission line medium for the probes, is inherently broadband and well behaved to frequencies well in excess of 50 GHz. However, the present limitations in high-precision connectors, the difficulties with waveguide to coplanar transitions, the existence of suitable S-parameter measurement equipment, and the need to reduce common lead

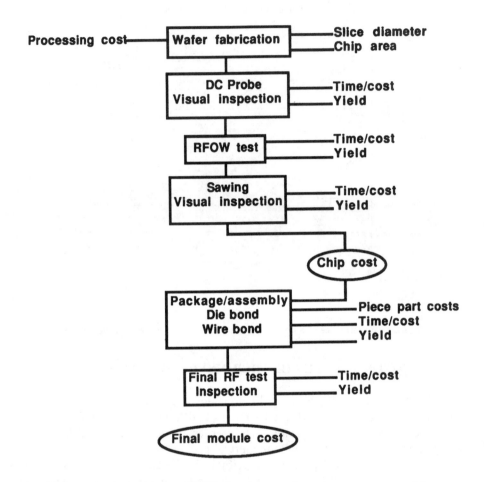

**Fig. 9.39** Production flow model for a MMIC phased array radar system.

inductances all are factors that could slow the development of faster probers. Despite these difficulties, the drive toward mm-wave ICs no doubt will see microwave probes emerge with higher operating frequencies than the present 50 GHz.

Whereas it is relatively easy to predict with confidence the future of microwave probers in terms of frequency and pin count, the question of test philosophy is more contentious. No doubt the use of microwave on-wafer testing for developmental and production purposes is increasing worldwide. What is not so clear is whether the prober will become a standard production test technique or remain as a vital development aid. With the expected reduction in chip processing and packaging costs, a key need to satisfy the market for MMICs, the capital and running costs of operating sophisticated microwave on-wafer ATE systems for 100% chip testing could well become the dominating chip production cost. Fur-

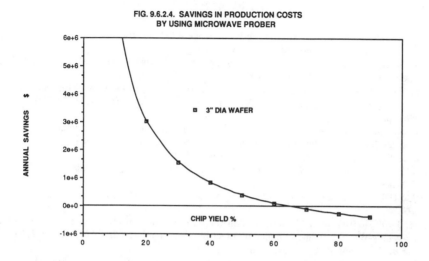

**Fig. 9.40** Estimated production cost of a phased array module as a function of chip yield.

ther, with MMIC yields now improving, chips could be incorporated into subassemblies with a high degree of confidence. Equally, for single-function MMICs reaching production, strong correlations are witnessed between simple dc functionality and RF performance. Hence, chip selection could be performed on the basis of faster, cheaper dc on-wafer testing. Conversely, for multifunction MMICs (e.g., single chip receiver) the yields and dc to RF correlation may be low enough to justify 100% microwave on-wafer production testing.

## ACKNOWLEDGMENTS

The author acknowledges with gratitude the many workers in this field who have freely provided their results for this chapter. In particular, Jeff Buck of Plessey Research has provided a large amount of up-to-the-minute test data, reviewed the manuscript, and offered much valuable advice; Eric Strid of Cascade Microtech has also provided relevant results and figures from his company's probe product lines. Heather Barbour's patient and meticulous typing and Keith Jenkins' artwork preparation are gratefully acknowledged as is the help of all Plessey's MMIC staff. Thanks are also given to Sheo Khetan of Anadigics Inc.

Part of this work has been supported by the procurement executive, Ministry of Defence (Directorate of Components, Valves and Devices) sponsored from the Royal Signals and Research Establishment, Malvern. Some of the MMIC chips described were produced on programs supported by the US Department of Defense.

# REFERENCES

1. D. Hornbuckle et al., "A Microwave Probe System," *Proc. Hewlett-Packard RF and Microwave Measurement Symp.*, Munich, 1983.
2. K. R. Gleason et al., "Precise MMIC Parameters Yielded by 18 GHz Wafer Probe," *Microwave System News*, 1983, pp. 55–65.
3. C. Wen, "Coplanar Waveguide: A Surface Strip Transmission Line suitable for Non-reciprocal Gyromagnetic Device Applications," *IEEE Trans. Microwave Theory Tech.*, Vol. MTT-17, 1969, pp. 1087–1101.
4. L. R. Lockwood et al., "Wafer Probe," US Patent No. 4,697,143, September 1987.
5. Cascade Microtech, "On Wafer Measurements Using the HP8510 Network Analyser and Cascade Microtech Wafer Prober," Product Note 8510-6, 1987.
6. K. E. Jones et al., "MM-Wave Wafer Probes Span 0 to 50 GHz," *Microwave J.*, April 1987, pp. 177–183.
7. "50 GHz Coaxial Connector Developed in Team Effort," *Microwave Systems News*, April 1986.
8. M. J. W. Rodwell et al., "Internal Microwave Propagation and Distortion Characteristics of Travelling-Wave Amplifiers Studied by Electro-Optic Sampling," *IEEE MTT-S Int. Microwave Symp. Dig.*, June 1986, pp. 333–336.
9. J. Fitzpatrick, "Error Models for Systems Measurements," *Microwave J.*, May 1978.
10. H. J. Finlay et al., "Design and Application of Precision Microstrip Multi-Octave Attenuators and Loads," *Proc. 6th European Microwave Conf.*, Rome, 1976.
11. D. Ferguson et al., "Transformer Coupled High Density Circuit Technique for MMIC," *IEEE Microwave and mm-Wave Monolithic Circuits Symp. Dig.*, May 1984, pp. 34–36.
12. "Noise Figure Measurements, Principles and Applications," Hewlett-Packard Application Note 5952-8266.

# Chapter 10
# MMIC Packaging

*B. Berson, F. Rosenbaum and R. A. Sparks*

## 10.1 INTRODUCTION

As the price goals of volume MIMIC production are achieved, it becomes increasingly apparent that packaging concepts borrowed unmodified from the hybrid microwave industry are inadequate long-term solutions to the MIMIC packaging problem. Because of the high cost of the microwave elements of hybrid microwave systems, packaging costs were not driven down to the level that will be required for volume MIMIC production. These packages have not been designed for cost and high volume manufacturability.

As we go higher in frequency, the chip cost increases, as smaller gate length and tighter tolerances take their toll on yield, but packaging costs rise even more steeply. Simple ceramic packages, useful to about 10 GHz, can cost from $7–9 in high volume. More sophisticated ceramic packages, useful to 20 GHz, can cost from $35 to $75 and up. Custom ceramic high-frequency packages invariably will involve nonrecurring engineering and initial tooling charges that may range from $20,000–$40,000.

MMIC packages differ from standard electronic ones in a variety of ways:

- MMIC packages are a part of the circuit not just a housing.
- Electromagnetic waves must be transmitted in and out in controlled ways.
- Package parasitics become more critical as frequency increases.
- Knowledge of mechanical, electrical, and thermal disciplines is required for a successful design. A team effort also is required, and this must overlap "school-taught" disciplines.
- Standards and a degree of standardization will be required for the industry to progress.

Recognition of these differences is critical to creating optimum packages. Packages fulfill multiple functions:

- Physical protection from physical damage, static discharge, mechanical forces, and chemical contamination.
- Thermal environment including heatsinking and heat transfer.
- Electrical environment including controlled impedance levels with minimum insertion loss, high-density connections, and multilayer, multimedia transmission.
- Mechanical environment including storage, manufacturing, and testing.

Only the simultaneous meeting (or optimization) of all of these will result in the desired performance.

Clearly, a major driving force in MMIC packaging is the application being considered. Although a number of commercial insertion opportunities for MMICs are being explored at the present time, the greatest technology thrust currently is in military systems. The *microwave-millimeter wave monolithic integrated circuits* (MIMIC) program [1], initiated by the U.S. Department of Defense in 1986, has as its objective the development of affordable and reliable MMICs for insertion in military electronic systems.

One of the main objectives of this program is to generate an approach to packaging that performs the mechanical function with acceptable electrical performance and reliability, at a cost that is reasonable in the context of extremely inexpensive GaAs IC chips. Highly integrated GaAs MIMIC subsystems are a new technology, and few manufacturers have experience estimating costs in the truly high-volume environment that will become a dominant aspect of MIMIC application. The integration level and the packaging approach both will contribute to determining system costs for MIMICs.

MMIC packaging occurs at a number of levels, as illustrated in Figure 10.1. At level 1, the MMIC chip that may include several active devices (FETs, diodes, *et cetera*), passive circuits, and bias lines is mounted into a container that, for simple illustration, is shown as a ceramic package. This package can be mounted on a substrate that also may include other active devices, passive circuits, *et cetera*, which is then put in a housing with coaxial cable or other connectors. This is component-level packaging, level 2. Finally, that housing can be mounted on a board, which is put into another housing. This is module or system-level packaging, level 3. The degree of complexity at each level is dependent upon the complexity of the initial chip and its intended application. As the technology advances, the complexity at each level tends to advance.

Package design can be approached in several ways. Commonly, the package designer begins by specifying the dissipated power that must be removed from the MMIC die together with the worst-case ambient temperature during operation. The overriding concern here is that the maximum junction temperature in the

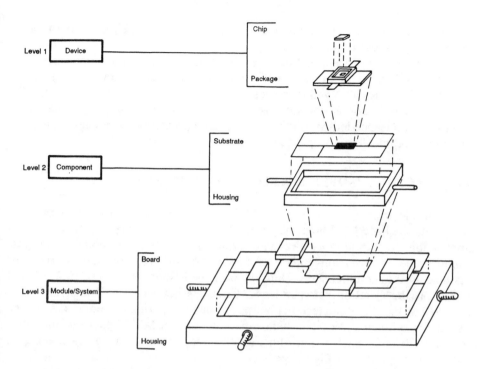

**Fig. 10.1** Multilevel MMIC packaging.

MMIC should not exceed 150°C or the rated temperature for continuous operation. The die attach process, the package base, and the scheme for mounting the package in the customer's system have to be engineered to meet this maximum temperature specification.

The second important parameter in a list of package design priorities is the range of microwave frequencies over which the device must operate. At microwave frequencies, the package causes signal reflections, signal loss, and unwanted coupling between circuit components. The electrical feedthrough and package dimensions must be carefully designed to minimize these effects.

The third broad area of package design includes mechanical and environmental considerations. The package must protect the MIMIC and its support circuitry from moisture, corrosive materials, and particulates that might degrade microwave performance or reduce the useful product life. It also must protect against mechanical forces (shock or vibrations) that would damage the circuitry or the package itself.

Finally, the package must be easy to install into a customer's system. It is also sometimes desirable to have the package easily removable for maintenance.

A system as complex as an active array antenna certainly should have thought given to this aspect of package design.

Despite the high cost of packaging, the amount of research being carried out in chip manufacture far exceeds that of package manufacture. In this chapter, we delineate the issues involved in the design and fabrication of packages for MMICs; we consider the various types of packages on the market today; and finally, we discuss what is needed in standards and standardization to ensure that cost-effective, technically appropriate packages are available.

## 10.2 ELECTRICAL DESIGN AND MEASUREMENT

The electrical performance of a MMIC chip is influenced by the characteristics of its package. Although performance specifications depend on the chip's application, we can identify a range of typical parameters for ceramic packages. For example, the insertion- and return-loss per transition through a feedthrough or via hole used to conduct signals into the package is shown in Figure 10.2. Also of interest is the isolation between leads. Crosstalk between signal lines and from signal-to-dc bias lines can lead to poor performance, spurious responses, and even oscillation problems in certain high-gain situations. A MMIC package may have from 2 to 20 leads, each a signal line, control line, or a bias line. Typical isolation data also are plotted in Figure 10.2.

Note that measured performance, such as that for isolation, will be strongly affected by the impedance terminations used in the measurement, as well as the mounting configuration or test fixture used to characterize the package. The excitation of external RF currents due to improper grounding must be avoided by using proper techniques. The actual results obtained with an active chip mounted in the package may differ considerably from those obtained with a straight through microstrip line or internal resistive loads. Thus, the package specifications must be written to describe the performance under actual conditions. For this reason, it is desirable to model the package to predict its performance under best- and worst-case conditions.

In a MMIC device or component, microwave and control signals must be transmitted to and from the chip, and dc power must be brought to it. The electrical design problem is to control the impedance levels of the transmission lines used for these purposes to obtain the desired electrical performance from the packaged chip. In principle, this is not a difficult task. In practice, however, some complicating factors make the problem more interesting. Because the transmission lines must be brought through walls or substrata of the package, physical discontinuities at these junctions scatter the signal energy.

The discontinuities resulting from geometrical variations in different parts of the package give rise to local changes in the electromagnetic modes. Other discontinuities are presented by changes in line widths, substrate heights, and bonding

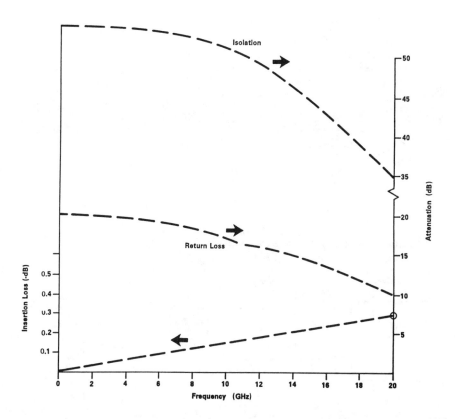

**Fig. 10.2** Nominal specifications for a ceramic package (insertion loss and return loss are per transition).

pads. Furthermore, signal energy is not confined to the transmission lines but may be coupled from one line to others, including the dc power lines. Coupling generally occurs in an uncontrolled manner. The energy also can be coupled to propagating modes in the complicated waveguide structure provided by the package itself. Self resonances of the structure must be located above the useful bandwidth of the package or eliminated entirely. Resonance-free performance beyond 30 GHz has been demonstrated recently [2] in a ceramic-frame package by metalizing all sides of the individual layers as shown in Figure 10.3.

The primary transmission lines at the MMIC designers disposal are microstrip and coplanar waveguides. Other transmission media encountered in packages are stripline regions, found where the planar transmission lines penetrate the dielectric sidewalls of metal-lidded packages, as shown in Figure 10.4, and balanced microstrip, in ceramic-lidded packages. A typical MMIC ceramic package is shown schematically in Figure 10.5.

MMICs lend themselves readily to use with coaxial feedthroughs, too, so

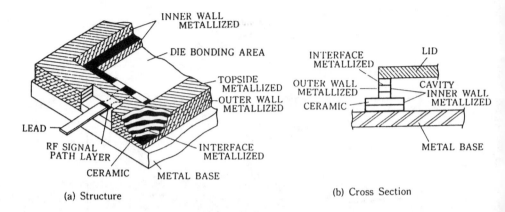

(a) Structure            (b) Cross Section

**Fig. 10.3** Metalized multilayer ceramic package (from [2]).

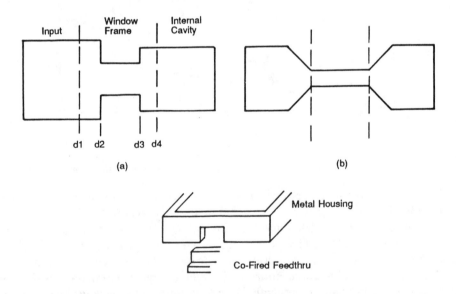

**Fig. 10.4** Feedthrough (courtesy Motorola Government Electronics Group).

that the wide body of knowledge and experience from hybrid microwave integrated circuit technology can be employed. Very careful assembly steps have been developed for soldering or brazing the coaxial glass beads that establish hermeticity. Figure 10.6 shows a typical glass bead installation cross section and sparkplug-launcher connector provided by commercial suppliers. Wire bonds are used to make the transition from traces on the substrate to the MMIC chip.

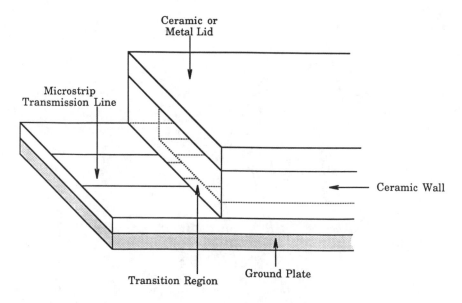

**Fig. 10.5** Typical MMIC ceramic package.

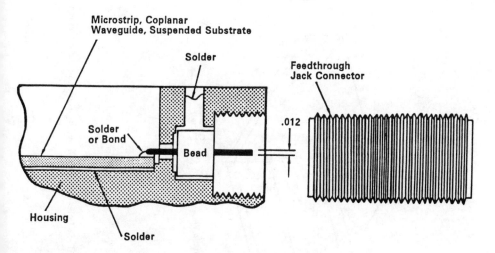

**Fig. 10.6** Glass bead installation and sparkplug-launcher coaxial connector.

Wirelike via holes often are used to carry signals through substrates or to bring traces to ground, as depicted in Figure 10.7. To deal with this complex electromagnetic environment, computer-aided design tools are extremely helpful. The

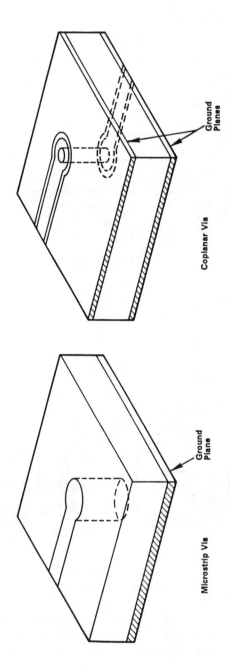

**Ground Planes**

**Coplanar Via**

**Ground Plane**

**Microstrip Via**

**Fig. 10.7** Signal via holes for grounding, transmission, or transition from microstrip to coplanar wave-guide.

modeling needs for MMIC packaging includes the following:

- Electrical properties, including impedances, coupling, modes, multilayer, multimedia, multiconductors, and three-dimensional analysis.
- Package-chip interaction (again, accuracy to Figure 10.1), including port-to port analysis, bandwidth isolation, and compatibility with circuit models.
- Thermal analysis, including temperature profiles, heat transfer, performance degradation, and thermal expansion stress.
- Mechanical design, including strength, stress, forces, deformations, and microcracks.

Thermal and mechanical issues will be discussed later.

The designer's job is to model the various elements of the package to obtain an overall representation of the package and so predict its electrical performance. Several distinct steps are involved in the process. First, the package needs to be separated into regions that can be modeled appropriately. These may include transmission line segments; walls and via holes, and bonding pads. Suitable models for each region must be developed and verified experimentally. The results of the modeling must be expressed in terms useful to the designer. For example, a microstrip transition at a sidewall, like that shown in Figure 10.4, can be represented as two 50 $\Omega$ transmission lines, each of a dimension appropriate to the open and dielectrically loaded regions, with the sidewall discontinuities modeled as shunt capacitances. Now, a frequency-dependent equivalent circuit can be constructed and used in a linear circuit analysis program, such as Touchstone or SuperCompact, to allow an overall simulation of the chip and package performance.

An interesting example [3] of ceramic package feedthrough modeling and subsequent measurement is provided in Figures 10.8 and 10.9, showing the electrical equivalent circuit and predicted performance. Measurements were taken of a network analyzer and the results de-embedded to the physical beginning and end of the feedthrough, marked A and B. The model component values were adjusted using Touchstone to fit the measured data.

Figure 10.10 illustrates in a conceptual way the multitransmission line nature of the interconnection problem. Signal energy can be coupled between lines. The amount of coupling depends on the geometry of the lines, their lengths, and their load terminations. The coupled lines support different modes of electromagnetic propagation and signals in different modes may have different propagation velocities. For TEM-like modes, two coupled lines can be analyzed in terms of two orthogonal sets, called the even and odd modes. An example of the differing propagation velocities for these modes is shown in Figure 10.11: $v = c/\lambda$, where $c$ is the speed of light. There are no general computer design aids to predict coupling and propagation effects for general problems, such as that shown in Figure 10.10.

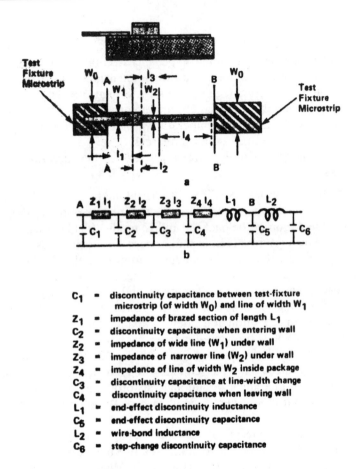

$C_1$ = discontinuity capacitance between test-fixture microstrip (of width $W_0$) and line of width $W_1$
$Z_1$ = impedance of brazed section of length $L_1$
$C_2$ = discontinuity capacitance when entering wall
$Z_2$ = impedance of wide line ($W_1$) under wall
$Z_3$ = impedance of narrower line ($W_2$) under wall
$Z_4$ = impedance of line of width $W_2$ inside package
$C_3$ = discontinuity capacitance at line-width change
$C_4$ = discontinuity capacitance when leaving wall
$L_1$ = end-effect discontinuity inductance
$C_5$ = end-effect discontinuity capacitance
$L_2$ = wire-bond inductance
$C_6$ = step-change discontinuity capacitance

**Fig. 10.8** Ceramic package feedthrough and electrical equivalent circuit (from [3]).

However, some use can be made of the coupled line elements in Touchstone and Super Compact to simulate certain geometries.

Electromagnetic simulation of package performance is in its infancy. Three-dimensional modeling is needed, because some geometric discontinuities require higher-order modes to match boundary conditions. Some commercial computer design aids exist including Linmic (spectral domain) and Ansoft (finite element method) but more comprehensive design aids (thermal, mechanical, electrical) are needed. Figure 10.12 illustrates what such aids might include.

Considerable effort has been expended on measurement techniques for on-chip evaluation of MMICs, and workable methods for it have been developed. Thus, MMIC manufacturers can track their results prior to dicing and packaging the product. However, final testing of the packaged device is necessary to determine if all the specifications have been met. Thus chip, package, labor, and test

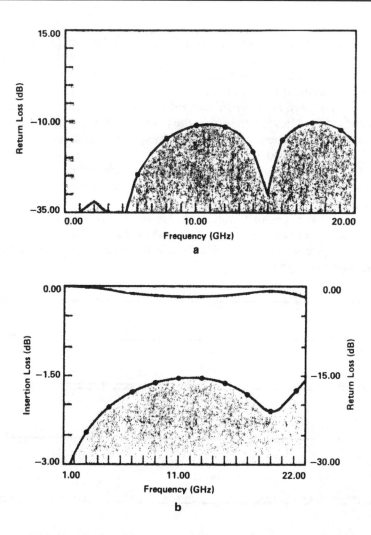

**Fig. 10.9** (a) predicted performance of a compensated ceramic package feedthrough, and (b) the calculated return loss of two feedthroughs (from [3]).

costs must be absorbed to reach the final result. The weak link in this chain is package performance. At this stage of MMIC development, it is not clear what methods are available to pretest packages before chip insertion. If all chips are to be tested, should all packages be tested as well? Four major issues present themselves.

1. *Lids:* What is the effect of the lid on level-1 packages? Can measurements be done systematically with and without lids and a method developed to correct the unlidded results?

2. *Terminations:* What termination conditions should be applied at chip bonding pads in order to accurately characterize the package in the given application? Are 50 Ω loads the appropriate choice or will a one-port measurement with open-circuited pads suffice?

3. *One- or multiport measurements:* How much interaction between signal lines and power lines is obtained in the package? This is important to reduce oscillation and frequency response problems. The coupling and mismatch environment of a bonded chip in a lidded package will be quite different from that of an empty package. Are multiport measurements necessary? It will be necessary to develop measurement techniques to assess the significance of this situation.

4. *Fixtures:* What types of fixtures are needed to give correct measurement results? Fixtures will be needed to address the questions just mentioned. And the answers to those questions will have an impact on the design of production package test fixtures.

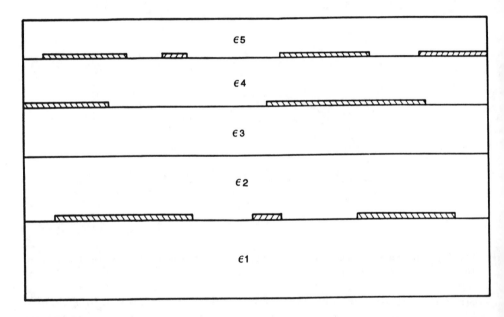

**Fig. 10.10** Multiconductor package.

## 10.3 THERMAL DESIGN

Consideration of the temperature range of operation of a MMIC package is fundamental to preventing device failures due to heat that is not properly dissipated. GaAs chips can be divided into two general classes: small signal chips in

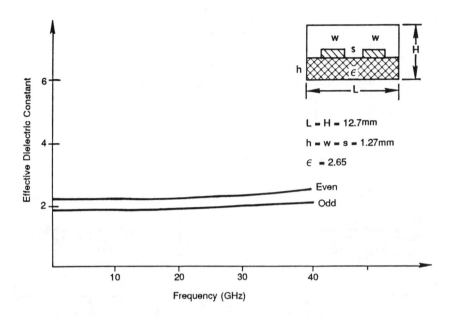

**Fig. 10.11** Effective dielectric constant for even and odd modes of two coupled microstrips.

which the dissipated device power does not contribute significantly to the overall channel temperature, and large signal chips in which the dissipated power in the chip is the major contributor to channel temperature. The thermal design of packages for the first class generally is not critical; therefore, the latter class will be addressed in this section.

Simple modeling is a good start for understanding the thermal environment of MMIC packages. Figure 10.13 shows a cross section of a level-1 device that illustrates the media boundaries between the source of heat generation, at temperature $T_0$, in a GaAs chip and the final sink for heat removal. The layer thicknesses are denoted $t_i$; $K_i$ is the media thermal conductivities; and $\Delta T_i$ is the temperature gradient between upper and lower surfaces of each layer. Not shown but often required is a metalization layer on the package base plate. This layer generally is very thin, would have high thermal conductivity, and should not introduce a significant thermal resistance (or $\Delta T$) in the calculation. *Note:* Gold plating per MIL-G-45204, Type III, Grade A, 100–150 $\mu$ in. thick with an electroless nickel underplate per MIL-C-26074, Class 1, 200–300 $\mu$ in. thick is typical.

A key difficulty immediately encountered with GaAs chips is the relatively poor thermal conductivity, $K_1$, of the material itself, typically one-third to one-fourth that of silicon, as compared in Figure 10.14. Because the sources of heat in a MMIC chip are in the gate-drain regions of the FET's top surface, the transfer of heat is through the 4 or 5 mil thickness of GaAs by conduction. Although there have been demonstrations of flip chipping (i.e., mounting the device face down to

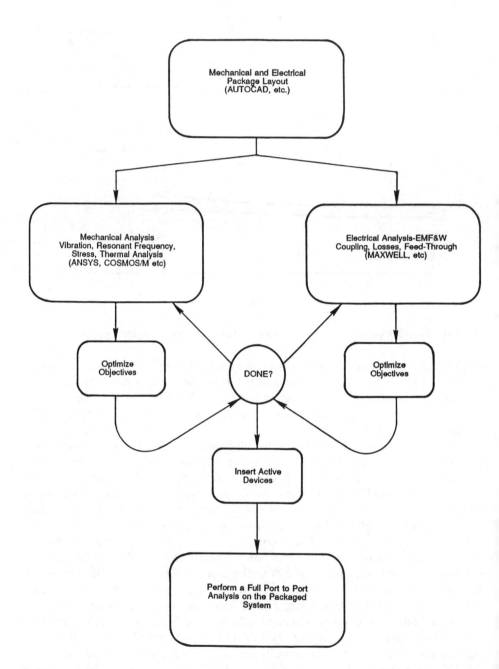

**Fig. 10.12** Integrated package modeling and analysis system.

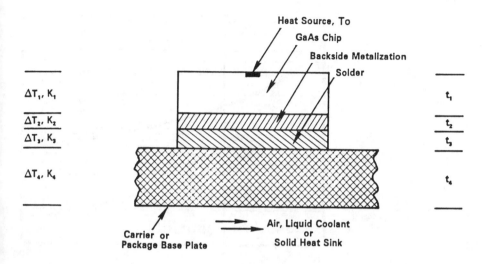

**Fig. 10.13** Cross section of MMIC chip thermal environment.

put the thermal generation regions as close to the heat sink as possible), the technique has not been developed into a practical production method. Thus, to remove heat in the current face-up configuration, the chip backside is metalized, $t_2$, $K_2$, generally with gold, and attached to a carrier or package base plate, $t_4$, $K_4$, by a low thermal resistant solder or eutectic, $t_3$, $K_3$. This die attach process has been studied extensively [5] with the evaluation of a wide variety of alloys in both paste and preform configuration.

A major concern for power MMICs is voids in the solder immediately below the heat sources on the chip. These voids arise from the entrapment of gas or improper cleaning or wetting of the chip-base plate interface during the die attach process. Both x-ray and sonic scan nondestructive diagnostic techniques have been employed to determine the presence and location of voids. In critical power chip assemblies, it may be necessary to inspect all chips by these means until a fully reliable die attach process is developed.

Assuming a reliable die attachment, the base plate or carrier material must be chosen to match the coefficient of thermal expansion of the GaAs chip. Table 10.1 lists the properties of several materials that may be used in microwave packaging. The availability of sintered composite powder metals of copper and tungsten in recent years has allowed a nearly ideal solution for mounting power MMICs. By varying the percentage of copper between 10 and 25% an almost perfect match in expansion characteristics can be achieved with GaAs over a

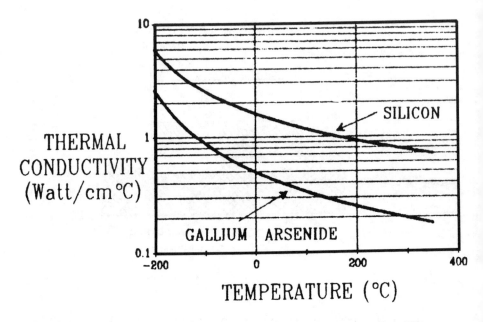

**Fig. 10.14** Thermal conductivity of silicon and GaAs as a function of temperature (from [4]).

**Table 10.1** Thermal Properties of MMIC Packaging Materials

| Material | Thermal conductivity (W/in.-°C) | Thermal expansion ($10^{-6}/°C$) |
|---|---|---|
| Alumina | 0.95 | 6.5 |
| Aluminum | 5.7 | 23.4 |
| Beryllia | 6 | 6.1 |
| Copper | 9.5 | 17 |
| GaAs | 1 | 5.8 |
| Invar | 0.28 | 1.6 |
| Kovar | 0.45 | 4.9 |
| Silicon | 3.31 | 7.6 |
| Silicon dioxide | 0.04 | 0.6 |

temperature range typical of military applications. Table 10.2 lists several properties of Thermkon materials, a copper-tungsten product of CMW, Inc. Figure 10.15 is a bar chart that compares thermal expansion coefficient of copper-tungsten and several other materials with GaAs and alumina substrates [6].

**Table 10.2** Typical Properties of Thermkon Materials (from CMW, Inc., Technical Data Brochure, 1987)

| Property | Thermkon 83 | Thermkon 76 | Thermkon 68 | Thermkon 62 |
|---|---|---|---|---|
| Thermal Conductivity Btu/h/ft/F | 110 | 104 | 97 | 91 |
| w/m/K | 190 | 180 | 167 | 157 |
| Thermal Expansion $\times$ $10^{-6}$ in/in/F | 4.6 | 4.2 | 3.6 | 3.2 |
| $\times$ $10^{-6}$ m/m/C | 8.3 | 7.6 | 6.5 | 5.7 |
| Modulus of Elasticity $\times$ $10^6$ psi | 34 | 35 | 36 | 37 |
| Tensile Strength $\times$ $10^3$ psi | 100 | 115 | 120 | 125 |
| Yield Strength $\times$ $10^{-3}$ psi | 76 | 89 | 105 | 116 |
| Flexual Strength $\times$ $10^3$ psi | 150 | 170 | 175 | 180 |
| Electrical Conductivity % IACS | 45 | 41 | 35 | 32 |
| Hardness | 98 HRB | 103 HRB | 25 HRC | 27 HRC |
| Density g/cm$^3$ | 14.84 | 15.56 | 16.60 | 17.17 |
| lb/ln$^3$ | .536 | .562 | .600 | .620 |

Thus, having established the basic thermal design model and material parameters of the MMIC package, the means of cooling remains to be defined. Forced air or liquid coolant are most commonly used in military system applications. The choice of fluid and flow rate often are dictated by the top-level system design approach. The present conversion efficiency from dc to RF of typical MMIC chips may constrain the operating environment of the system or require more innovative solutions to the temperature control problem. Fundamentally, high operating channel temperatures are associated with shortened median device life and, hence, poor reliability. Current industry activity is geared to develop stress test methodologies and reliability data bases on specific packaged MMIC chips to fully establish device performance. For further detail on calculation of thermal properties of packages, see such standard references as *The Handbook of Microelectronics of Packaging and Interconnections Technologies* [7].

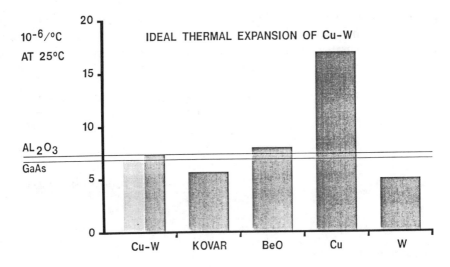

**Fig. 10.15** Comparison of thermal expansion coefficient for Cu-W and other materials with GaAs and alumina substrates (from [6]).

## 10.4 MECHANICAL DESIGN CONSIDERATIONS

Currently meeting specified electrical performance requirements is the dominant force in MMIC chip design for most applications. Mechanical design considerations tend to be of secondary importance, the popular view being that some sort of housing can always be found to accommodate the chip. The thrust of this section is to temper that view with the suggestion that a great deal of cost can be avoided by thinking about the package during the chip design layout. The basic mechanical design parameters that affect the MMIC chip package selection process are as follows:

Material: Metal, ceramic, or composite
Size: length, width, and height (including lid)
Lead count: bias, RF, control
Interconnect medium: coaxial, microstrip, stripline tabs, leadless
Weight: for space, airborne, or ground application
Die attach method: hot gas, scrubbing, reflow furnace
Lid seal: solder, seam weld, laser weld
Mounting: solder, epoxy, surface mount, screw down

Of equal importance are the constraints imposed by the system application into which the MMICs are to be inserted.

Possibly the first decision to be faced is whether the package will be metal, ceramic, or a combination of both. In the microwave industry, a plethora of custom metal housings has been developed for hybrid devices and components, which can be found in any number of vendors' catalogs. Although some attempt has been made to standardize package sizes within a given vendor's product line, there has been virtually no coordination among suppliers. The subject of standards is addressed further in Section 10.6.

Current package suppliers to the microwave industry would appear to have evolved in one of two ways. First, some have a "machine shop" origin, in which precision tolerance metal cutting, stamping, and forming are combined with brazing and hermetic glass-to-metal-seal technologies. In many cases this industry developed around the surge in hybrid microelectronics of the last two decades. A great deal of skill in producing high-quality packaging products has gone into creating this currently viable business area. The precision ceramicists are the second group of packaging specialists that has developed multilayer cofired packages for high-frequency applications. The needs of the silicon device and integrated circuit industry have fostered this technology for developing packages with dozens or hundreds of lead-ins.

The cofire multilayer process illustrated in Figure 10.16 begins by milling precise amounts of raw materials into a homogeneous slurry. This mixture, which has the consistency of latex paint, is principally alumina (aluminum oxide, $Al_2O_3$) and fluxes with small amounts of organic binders and solvents. Coloring oxides are added to an otherwise white slurry when opacity is required to protect light-sensitive devices or when cosmetic preference dictates a dark body. The opaque material will vary among manufacturers from blue-black to deep purple when fired, depending on the amount and types of oxides used. In general, coloring oxides degrade the performance of the body, and the clear finish should be our first choice.

The slurry is poured onto a mylar sheet and then passed under a doctor blade to produce a uniform strip of specified thickness. When dried, this strip will become a ceramic-filled plastic tape with the look and feel of thick vinyl. The sheets of "green" (unfired) ceramic tape are cut into cards. Alignment holes are punched in each card to ensure accurate layer-to-layer alignment during subsequent operations. Exact registration becomes more critical as circuit densities increase. Via holes, cavities, and notches are punched in the cards as required. The via holes are filled with tungsten paste, a mixture of metal powder and organics that becomes conductive when fired. Although other metals (e.g., copper) are better electrical conductors, their melting points are lower than the 1600°C necessary to cofire alumina ceramic. Tungsten combines a satisfactory level of electrical conductivity with a sintering point and shrink rate compatible with alumina. The standard filled via hole will carry the electrical signal from one metalized layer to the next. When densities dictate via hole diameters of .010 in. or less, process precision is essential. The bore-coated via hole, a large open via hole with a metalized ID coating, can function either as a castellation or receptacle

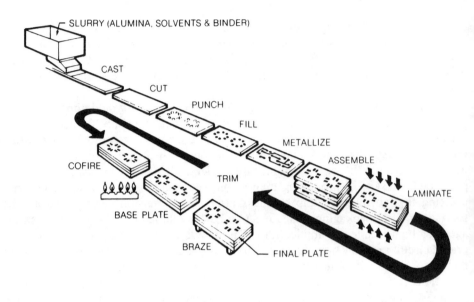

**Fig. 10.16** Key steps in the cofired multilayer ceramic package process (courtesy Interamics, San Diego, CA).

for interconnecting wires or pins. These via holes are punched in the same manner as standard ones and coated.

Conductive circuits are printed onto ceramic tape by forcing tungsten paste through an open mesh metal screen. The circuit pattern for each artwork layer is worked into a screen by the same photolithic process used in thick film and other screen-printing techniques. The artwork is adjusted to allow for shrinkage during firing. Screened layers are inspected, aligned, and laminated under high pressure and low temperature into one assembly. This assembly may be composed of a single unit or include multiple repetitions of the same product. The multiple units are processed as a single assembly from punching through lamination.

A high-precision scoring tool is used to trim single units or separate multiples. This tool also is used to partially score green ceramic structures to allow plating interconnects to be snapped off after firing. When conductor pads are required on the side of the product, additional screening operations are performed on the laminated, trimmed units. The ceramic structure is cofired at approximately 1600°C in a carefully controlled reducing atmosphere. During the firing process, the product shrinks approximately 20% (linear) for a total volume reduction of 40%. A subsequent high-temperature operation may be used to achieve .003 in. flatness or better. The base finish material, usually nickel, is electrochemically applied to all exposed tungsten metallization to improve performance, prepare the surface for subsequent braze operations, and prevent oxidation of conductor pads.

Brazing is required to join metal components, such as seal rings and leads, to the nickel-plated ceramic structure. The use and configuration of these components are determined by product design. Carbon or ceramic braze fixtures are utilized to properly position metal components in relation to the ceramic structure as it is processed through a braze furnace. Copper-silver eutectic (Cu-Ag) is the most widely used material for brazing. It forms a strong hermetic joint. When thermal-expansion mismatch between alumina ceramic and the metal component (Kovar or Alloy 42) becomes significant, an indium-copper-silver (In-Cu-Ag) eutectic may be used.

Either electroless or electrolytic plating processes may be used. Electrolytic plating requires that all exposed circuits be connected electrically by means of a lead frame, plating bus or tie bar, or a combination of the two. (When a plating bus is used, it must be removed after plating to isolate the circuitry.) The plating metal, its thickness, and process requirements are defined by the product application. Although nickel-plus-tin and nickel-only plating are available, gold is the most common final plating material. Gold facilitates wire bonding, protects the nickel from corrosion, maintains solderability for surface mounting, via-hole mounting, and socketing. A disassembled and complete view of a ceramic package suitable for a MMIC chip insertion is shown in Figure 10.17. In this case, a flange or metal base plate has been brazed to the ceramic to improve thermal performance.

One important characteristic in the design of the metal systems for ceramic packages is the choice of metals for transmission lines. Depending on the frequency of operation and the cost requirements of the package, thin film or thick film metalization may be used. Other issues that are similar to those in microwave hybrid circuits and the packaging of GaAs discrete devices include the choice of metals for transmission lines, bonding, and pull strength as well as the surface roughness of substrates on which chips must be mounted.

The package size clearly depends on the size of the chip or chips to be protected. A scattergraph of chip sizes shown in Figure 10.20 suggests that nearly 80% of current designs would fit in a .140 × .140 mil area. Package height should be sufficient to ensure no interaction with the circuit or allow excitation of volume modes or resonances.

The number of input/output leads depends heavily on the circuit design, with a minimum of at least one pair of RF I/O leads and a bias lead in. Both dc and RF grounds are common in a metal housing but may be separate in a ceramic walled package requiring an additional feedthrough. The need for control line feedthroughs, as for driving phase shifters, switches, or attenuators, creates additional I/O demands. Mechanical tolerances on bias and control line feedthroughs should be more relaxed than for RF I/Os and may help reduce the package cost.

The choice of the transmission medium for the interconnection I/Os is often dictated by the application and the next level of system integration of the device.

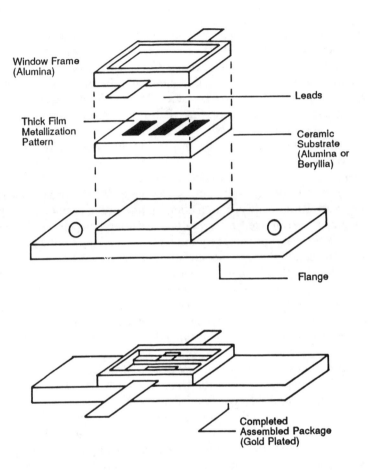

Window Frame
(Alumina)

Leads

Thick Film
Metallization
Pattern

Ceramic
Substrate
(Alumina or
Beryllia)

Flange

Completed
Assembled Package
(Gold Plated)

**Fig. 10.17** Ceramic package assembly.

Microstrip or stripline feedthroughs are compatible with the component (level 2) and module (level 3) integration indicated in Figure 10.1; however, a cascade of several MMIC chips in a package that must interconnect with coaxial lines or an air stripline medium may be simplified by the use of coaxial connectors or the design of a transition section. We should assess these interface requirements as early in the design as possible to minimize system losses and mismatches and satisfy MMIC device and component testability criteria.

The planned use of MMICs in space and airborne applications has emphasized the need to keep package weights to a minimum. Several new lightweight metal matrix and composite materials are being investigated that will help satisfy this need, some of which offer excellent thermal properties as well. Aluminum nitride, aluminum silicon carbide, and graphite fibers embedded in metal and ceramic matrix composites are just a few of the materials being studied.

The final device assembly process shown in Figure 10.18 illustrates the mechanical aspects of die attach, lid sealing, and mounting provisions that must be addressed in the package design. If the MMIC chip is to be scrubbed in, the entire package must be placed on a hot plate and the die with solder preform inserted. Attachment in a reflow furnace may require special fixtures and weights to ensure alignment of the chip with package interconnects. Considerable effort is being placed on developing automatic pick-and-place machines and robots to eliminate the high cost of manual assembly and improve reproducibility of performance.

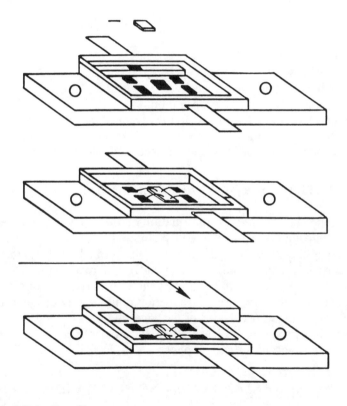

**Fig. 10.18** Final device assembly.

Methods of lidding and sealing have evolved over the years and become rather routine. In designing the package, a key engineering consideration should be to ensure compatibility of the lidding technique with the equipment available in the manufacturing facility. In today's environment of many GaAs foundries with excess capacity, it may be most cost effective to have the production performed out of house.

The means of mounting the package must be addressed early in the design, too. Providing bolt holes on flange mounts is used extensively and ensures a positive mechanical attachment that will survive any specified shock or vibration. Surface mounting of small MMIC packages also is proving popular, because of the momentum induced by the printed circuit industry in recent years. Surface mount techniques and new equipment that have been appearing in the trade journals and exhibitions feature high-throughput, reliable performance for both commercial and military applications.

Finally, the environment that the mechanical design must provide for the chip to optimize performance is as follows:

- Compatible 50 Ω MIC transmission lines and other passive circuits.
- Easy access to bias connections.
- Compatible mounting plates.
- Good heat transfer.
- Coefficient of expansion match.
- Suppression of waveguide propagation.
- Accommodation of MIC and striplines where higher $Q$ or non-GaAs is appropriate.
- Epoxies minimized and fluxes eliminated to avoid cleaning.

## 10.5 CHARACTERISTICS OF AVAILABLE MMIC PACKAGES

In the appendix, we present information on the types of packages available in 1988. This information was obained by circulating a survey to package users and producers in the industry. Each package is described in a table that discusses various performance considerations and includes, where available, either a photograph or an outline diagram of the package. Table 10.3 summarizes the data on the appendix tables in three key areas: frequency range, range of applications, and cost. Table 10.4 summarizes classes of packages in terms of their nominal performance, cost, and properties. Figure 10.19 shows photographs of a number of available packages and lists their suppliers.

## 10.6 MMIC PACKAGE STANDARDS AND STANDARDIZATION

The relatively early stage of development of MMIC packaging has two factors associated with it: lack of packaging standards and lack of package standardization. Both are key issues for the development of MMIC packaging and, ultimately, of MMICs themselves. The two factors need to be carefully distinguished. Standards refer to a means of determining and communicating package performance and characteristics. Standardization refers to limiting package types

**Table 10.3** Package Types and Characteristics.

| Table | Package type | Frequency range | Range of applications | Cost | Package level | Comments |
|-------|-------------|-----------------|----------------------|------|---------------|----------|
| 10.A1 | Cofired ceramic Cu-W base | + | + | − | 1,2 | Proven reliability |
| 10.A2 | Polyimide multilayer | = | − | + | 1,2 | Hermeticity is an issue |
| 10.A3 | Metal-cofired feed-throughs | + | + | − | 1,2 | Good impedance control on feedthroughs |
| 10.A4 | Metal-coaxial connectors | + | + | − − | 2,3 | Traditional Module Packaging Technology |
| 10.A5 | Molded microcoaxial | + | + | = | 2,3 | New development |
| 10.A6 | Thin film | + | = | + | 1,2 | Low tooling costs |
| 10.A7 | Surface mount | = | + | + | 1,2,3 | Cost + in volume |
| 10.A8 | Glass flatpack | = | + | + | 1,2,3 | Cost + in volume |
| 10.A9 | Ceramic flat | = | = | + | 1,2,3 | Cost + particularly considering # leads |
| 10.A10 | Glass feedthroughs | = | + | − | 1,2,3 | Seals can break if not handled properly |
| 10.A11 | T.O. cans | − | − | + | 1,2 | Problem is lack of commercial suppliers |
| 10.A12 | Chip carriers | = | + | + | 1 | Very cost effective |
| 10.A13 | Waffle line | + | + | − | 2,3 | High-priced custom application |
| 10.A14 | Hermetic metal or tray | + | + | = | 2,3 | Multichips are packaged |
| 10.A15 | Molded package | − | − | + | 1 | Nonhermetic |

**Table 10.3** cont.

to particular lead formats, dimensions, or properties. Standards can include the following:

- Lists of terms, definitions, or symbols applicable to packaging.
- Expositions of methods of measurement or tests of the parameters or performance of any device, apparatus, system, or phenomenon associated with MMIC packaging.
- Characteristics, performance, and safety requirements associated with the devices, equipment, and systems.
- Recommendations reflecting the current best application of engineering principles to packaging.

The adoption of packaging standards will speed the development of MMICs and enhance their acceptance by users. The purposes of MMIC packaging standards are as follows:

To develop accepted measurement techniques.
To develop accepted performance measures.
To identify material and process requirements.
To develop accepted design techniques.
To promote compatibility.
To simplify applications.
To help guide the development of the field in an orderly manner.

A major difficulty in meeting both military and commercial cost targets for MMIC packages is the plethora of package types, styles, and outlines used in the industry today; each in quite low volumes. Although the field may not be mature enough to allow a single standard or even family of standards, it is certainly *not* too early to consider the relevant issues:

Cost: Packaging will be more expensive than the chip itself. Cost is driven by volume; increase the volume, lower the cost.

**Table 10.4 Summarized Results of Packaging Survey (courtesy of Doug Mathews, Motorola, Inc.)**

| Package | Frequency | | Loss per I/O port | Isolation | Cost | | | Thermal resistance | Reliability | Adaptable |
|---|---|---|---|---|---|---|---|---|---|---|
| | 1.2:1 | 2.0:1 | | | NRE | 100 | 100K | | | |
| Cofired ceramic with Cu-W base | 20 GHz | 26 GHz | <.25 dB | >40 dB | $15–40K | <$50 | <$7 | Low | Excellent | Bolt or solder, coplanar or microside |
| Metal housing cofired feedthroughs | 20 GHz | 30 GHz | <.25 dB | >40 dB | $10–15K | $50–100 | $20–30 | Low | Excellent | Feedthroughs adapt to your housing design |
| Microwave glass flatpack | N/A | 20 GHz | <.25 dB | >25 dB | $5K | <$6 | <$4 | Low when configured with a Cu-W base | Good, usual handling precautions | Coplanar, microstrip version OK at lower frequency |
| Surface mount | | 12 GHz | <.5 dB | >30 dB | $1–2K | $10–15 | <$3 | High 35° C/W | Excellent | Poor, specific medium required |
| Metal housing glass feedthrough | | VSWR highly dependent on transition | <.2 | >40 | $1–5K | <$50 | <$15 | Low when configured with Cu-W base | Susceptible to mishandling | Requires motherboard |
| Microwave quality T.O. cans | 16 GHz | 18 GHz | .1 dB | >50 | $1–5K | Not known | Not known | Dependent on attachment method | Susceptible to mishandling | Very inflexible |
| Nonhermetic chip carrier | Circuit and cavity dependent | | Low | Topology dependent | <500 | <$2 | $<2 | Excellent, when made of Cu-W or moly | Susceptible to mishandling | Best |

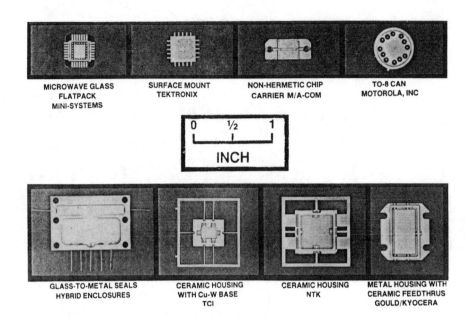

MICROWAVE GLASS
FLATPACK
MINI-SYSTEMS

SURFACE MOUNT
TEKTRONIX

NON-HERMETIC CHIP
CARRIER M/A-COM

TO-8 CAN
MOTOROLA, INC

GLASS-TO-METAL SEALS
HYBRID ENCLOSURES

CERAMIC HOUSING
WITH Cu-W BASE
TCI

CERAMIC HOUSING
NTK

METAL HOUSING WITH
CERAMIC FEEDTHRUS
GOULD/KYOCERA

**Fig. 10.19** MMIC package types (courtesy Doug Mathews, Motorola, Inc.).

Availability: "Company proprietary" limits industry use to a small number of off-the-shelf suppliers, and those available are performance or application limited.

Performance: Industry involvement in setting parameters for microwave circuitry will ensure that vendors produce those packages the industry needs.

Variability in chip size: Chip sizes are determined by function. Previous chip designs have shown that size variability is the rule, even though chip specific packaging is too costly.

Standardization: Common tooling, automated assembly equipment, test fixturing, and interconnection schemes can be developed that will be usable by the industry as a whole.

Figures 10.20 and 10.21 continue the case for an industry standard package. Figure 10.20 presents a scattergraph of chip sizes, showing the types most likely to be packaged. Figure 10.21 illustrates Mr. Mathews's proposal: the enhanced package. It would be available in either a coplanar or microstrip configuration and any of a variety of bases to suit different mounting needs. Whether this particular package design will evolve as the industry choice is far from clear, but it is certainly useful as a starting point for discussions of a standard package.

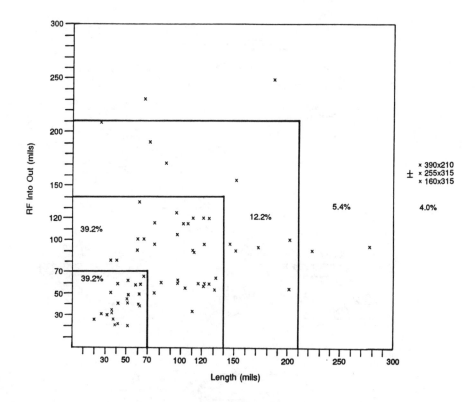

**Fig. 10.20** Scattergraph of chip sizes (courtesy Doug Mathews, Motorola, Inc.).

## 10.7 CONCLUSIONS AND ACKNOWLEDGMENTS

At present the status of MMIC packaging allows the user the following options:

1. The use of available catalog packages. In this case, users are likely to find packages that do not meet their basic needs but involve tradeoffs in a variety of important areas. What is possible, however, is good delivery, little or no tooling costs, and perhaps if some differences are tolerated, even alternative sources.

2. Custom packages can be procured, where users design the packages themselves, with help from the vendor. In addition to development costs, users have to bear tooling charges and high material package price costs. In return, users have a package that, within the capabilities of the designer and manufacturer, meets their requirements precisely.

3. For most companies, the in-house design and fabrication of packages is

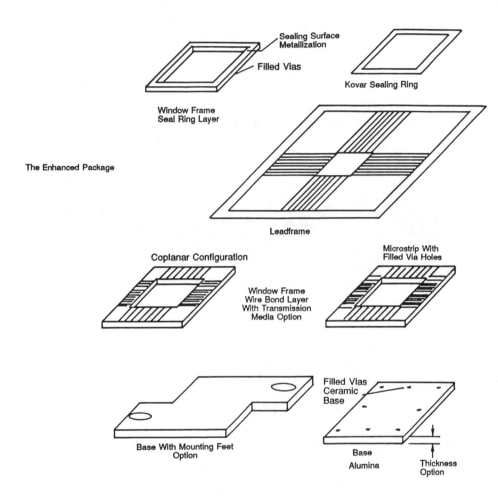

**Fig. 10.21** The enhanced package (courtesy Doug Mathews, Motorola, Inc.).

possible only for machined housings. The technology for ceramic packages, or even glass-to-metal seals, is not available within most companies. Even when these capabilities do exist, they are limited to prototyping capabilities. Many companies design and manufacture their own machined housings for in-house use. They often achieve as close to optimum performance as possible. However, this is at great expense in design, tooling, and fabrication.

Clearly, this choice between a package that does not quite do what it is supposed to do and one that is prohibitively expensive is not a happy one. We hope that, as the field progresses, more satisfactory options will appear, better performing and more versatile packages will become available, and better design tools will be developed. We hope that by summing up the present status of the field and suggesting some future directions, we may have contributed to accelerating this process.

Special thanks go to the following individuals and companies that provided materials and comments for this chapter: John Ellenberger, Hewlett-Packard; Takahashi Furutsuka, NEC; Phil Gerrou, Dow Chemical; Gary Holz, Holz Industries; Doug Maki, M/A-COM; Douglas Mathews, Motorola; Ray Pengelley, Tachonics; Thomas O. Perkins, Sanders; Leonard Schieber, Kollmorgan/Multiwire Div.; and David H. Smith, Tri Quint. In addition, we would like to thank Karen Krebser, without whose unflagging support this work would never have been completed.

# REFERENCES

1. E. Cohen, "The MIMIC Program—Key to Affordable MMICS for DOD Systems", *IEEE Monolithic Circuits Symp. Dig.,* May 1988, pp. 1–4.
2. F. Ishitsuka and N. Sato, "Low Cost, High-Performance Package for a Multi-Chip MMIC Module," *GaAs IC Symp. Dig.,* November 1988, pp. 221–224.
3. R. S. Pengelly and P. Schumacher, "High-Performance 20 GHz Package for GaAs MMICs," *Microwave Systems News,* January 1988, pp. 10–19.
4. W. J. Roesch, "Thermo-Reliability Relations of GaAs IC's," *IEEE GaAs IC Symp. Dig.,* November 1988, pp. 61–64.
5. J. S. Pavio, "Successful Alloy Attachment of GaAs MMIC's," *IEEE Trans. Microwave Theory Tech.,* Vol. MTT-S 35, December 1987, pp. 1507–1511.
6. T. O. Perkins, III, "MMIC/MIC Insertion," Microwave Hybrid Circuits Conf., Sedona, AZ, September 1988.
7. F. N. Sinnadurai, *The Handbook of Microelectronics of Packaging and Interconnections Technologies,* Electrochemical Publications, 1985.

# APPENDIX 10A SURVEY OF AVAILABLE PACKAGE TYPES AND THEIR PROPERTIES

**Table 10.A1** Cofired Ceramic Cu-W Base Package

Performance: Good up through 20 GHz; usable up through 26 GHz; low loss < 0.25 dB per feed-through.

Microstrip or coplanar configuration: Can be either.

Thermal resistance: Low.

Reliability: Proven.

Cost: Long tooling delay for new designs; high tooling charges, $10,000–40,000 for new designs; expensive in low volumes, $50 each in 100 quantities.

Mounting: Boltable or solderable.

Isolation: High input-output isolation, greater than 40 dB.

Production Costs: Reasonable cost in high volume, $5–7 in 100,000 quantities.

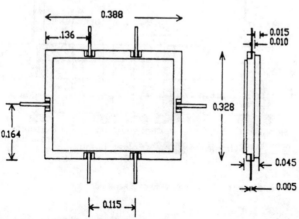

**Fig. 10.A1** Cofired ceramic Cu-W base package.

**Table 10.A2** Polyimide Multilayer Package

Performance: Acceptable but not outstanding.
Microstrip or coplanar configuration: Good interface.
Thermal resistance: Can be low.
Reliability: Questionable; more data needed.
Cost: Low.
Mounting: Restricted techniques.
Lead count: Good.
Controlled impedance: Good.
Die attach and wire bonding: Good with correct equipment.
Tooling: Varies depending on package company experience.
Dielectric constant: Can be low, leading to good propagation delays and phase reproducibility.
Loss: Good to variable.
Crosstalk: Good to questionable.
VSWR: Average.
Return loss: Average.
Hermeticity: Questionable.
Isolation: Good to questionable.
Adaptability: Good.
Pickup: Good.
Production Costs: Should be low.

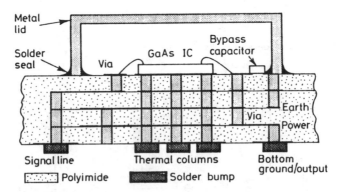

**Fig. 10.A2** Polyimide multilayer package (courtesy of Tachonics Corp.).

**Table 10.A3** Metal-Cofired Feedthrough Package

Performance: Good through 20 GHz; usable through at least 30 GHz.
Microstrip or coplanar configuration: Good in either.
Thermal resistance: Good.
Reliability: More work needed.
Cost: Shifts.
Mounting: Can be easy.
Lead count: Restricted, but depends on package size.
Controlled impedance: Good.
Die attach and wire bonding: Good.
Tooling: Expensive.
Dielectric constant: High (9.4).
Loss: Low, < .25 dB per feedthrough.
Crosstalk: Very good.
VSWR: Good.
Return loss: Good.
Hermeticity: Good.
Isolation: > 40 dB.
Adaptability: Poor.
Pickup: Poor.
Production costs: High.

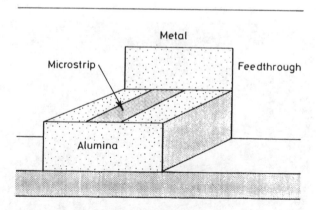

**Fig. 10.A3** Metal-cofired feedthrough package (courtesy of Tachonics Corp.).

**Table 10.A4** Metal Packages with Coaxial Connectors

Performance: Very good, both electrically and environmentally.

Microstrip or coplanar: Microstrip.

Thermal resistance: Very low, alumina soldered to Kovar soldered to aluminum housing. If housing has a heatsink, it will dissipate 500 W per in.[2]

Reliability: Very high, designed for 25–50 years nonoperating life.

Mounting: For this application, unit was fastened with adhesive to other assemblies and then potted in place in a hermetic container. Mounting feet could be added for other applications.

Lead count: One to three coaxial connectors per hybrid; one to five leads as needed.

Controlled impedance: 50 $\Omega$ on coaxial lines. > 15 dB return loss.

Die attach and wire bonding: Packaged semiconductors soldered to gold path or to metal backing plates; ribbon bonding where needed.

Tooling: Milled from solid block for small quantities; could use casting for larger quantities.

Dielectric constant: $Al_2O_3 = 9.8$.

Loss: Depends on frequency.

Crosstalk: < −60 dB from input to output of a hybrid; < −160 dB from cavity to cavity.

VSWR: 1.2 : 1 at coax connectors.

Return loss: > 15 dB.

Hermeticity: Not designed to be hermetic as it is installed in a hermetic package; however, it could be made hermetic by using hermetic filters and connectors.

Isolation: The case is ground.

Adaptability: A custom package, but the technology is versatile.

Production costs: The housing, as received from the vendor, cost in the order of $600–700 in small quantities and $300 in lots of 1000.

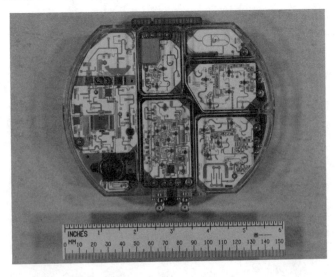

**Fig. 10.A4** Metal packages with coaxial connectors.

**Table 10.A5** Molded Microcoax

Performance: True coaxial interconnection.
Microstrip or coplanar configuration: Coaxial wires embedded in a circuit board.
Thermal resistance: $-65°C$ to $+125°C$.
Reliability: No data available.
Cost: Moderate.
Mounting: Standard circuit board mounting.
Lead Count: 76 leads/in.$^2$
Controlled impedance: 50 $\Omega$ + 3 $\Omega$.
Die attach and wire bonding: Surface mount or via hole.
Dielectric constant: 1.3 (or 2.2).
Loss: Dissipation factor 2 × 10 E-4.
Crosstalk: Nil.
Adaptability: Excellent.
Pickup: Nil.
Production costs: Moderate.
Characteristic impedance: 50 $\Omega$.
Velocity of propagation: 87% of C or 26.3 cm/ns.
Time delay: 1.2 ns/ft.
Dissipation factor: .0002.
Effective dielectric constant: 1.3.
dc resistance: 1.08 $\Omega$/ft.

**Fig. 10.A5** Thin film package (courtesy of Holz Industries, Inc.).

**Table 10.A6** Thin Film Packages

Performance: dc to 26 GHz

Microstrip or coplanar configuration: Microstrip or stripline.

Thermal conductivity: Alumina 25 w/mk; beryllia 240 W/mk; aluminum nitride 200 w/mk.

Reliability: Qualified to MIL STD 883 requirements.

Cost: High quantity less than $1.50 with lid.

Mounting: Epoxy, solder.

Lead count: Analog up to 48 I/O, digital up to 308 I/O.

Controlled impedance: Epoxy or gold eutectic; Au or Al wedge and ribbon.

Tooling: Generally less than $2500.

Dielectric constant: In alumina packages, 9.5; in BeO 6.5: in AlN 8.5.

Loss insertion: Less than 1 dB through 26 GHz (2 feedthroughs).

Crosstalk: At 20 GHz = 35 dB.

VSWR: Greater than 12 dB from dc to 26 GHz.

Return loss: Greater than 12 dB from dc to 26 GHz.

Hermeticity: Meets MIL-STD 883 Meth 5005 (fine leak at 10 E-8).

Isolation: Current products 30–60 dB.

Adaptability: With small mask change, up and down compatible.

**Fig. 10.A6** Surface mount packages.

**Table 10.A7** Surface Mount Packages

Performance: Up to 5 GHz for this package design.

Microstrip or coplanar configuration: Designed for microstrip due to customer needs.

Thermal resistance: Of lower importance, but still considered.

Reliability: Of highest importance.

Cost: Designed for potentially low cost (few piece parts, standardized assembly).

Mounting: Surface mount a must.

Lead count: New requirements are up to 28; only RF pins.

Controlled impedance: Designed into the lead shape and mounting processes.

Die attach and wire bonding: This package can use all but epoxy die attach.

Tooling: Designed so that packaging pinout can change with little tooling change.

Adaptability: High flexibility achieved through ability to change package size and pin count with little tooling and no process development.

Production costs: Designed in leadframe for future automation.

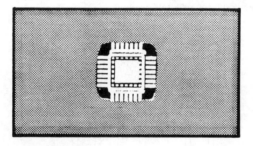

**Fig. 10.A7** Microwave glass flatpack package.

**Table 10.A8** Microwave Glass Flatpack Package

Performance: Performance to 20 GHz demonstrated with an insertion loss of .3 dB per I/O port and a return loss of 10 dB. The main problem with the unit is the metal seal ring: fields can couple to the ring, causing degraded performance. This can be alleviated by grounding the seal ring but is not easily accomplished in a cost-effective manner.

Microstrip or coplanar configuration: Originally configured as an all coplanar structure, the design currently is being revised to provide both coplanar and microstrip ports, any of which can be utilized as dc bias or grounds.

Thermal resistance: Normally configured with a Kovar base, but may have a Cu-W based brazed on, added cost for this feature.

**Table 10.A8**

Reliability: Very good, this technology has been utilized for a good number of years both in high-reliability space environment and in the extreme G forces of fuze environments.

Cost: There is no lower-cost package; in high volume, expect this package to be less than $5, with the potential to be around $2 or $3.

Mounting: Solder or epoxy in place; the base could be configured to have mounting ears for bolt-in applications.

Lead count: 28 leads, maximum of 12 coplanar microwave sets.

Controlled Z: 50 $\Omega$.

Die attach and wire bonding: Not a problem with proper metalization.

Tooling: A new design from scratch will cost about $5000.

Dielectric constant: 7052 glass with 25% alumina = 6.2.

Loss: .3 dB/port.

Crosstalk: 25 dB minimum port to port, adjacent or across.

VSWR: 2 : 1 maximum.

Hermeticity: Very good, seal with seam sealer or solder seal.

Adaptability: Coplanar only at this time.

Pickup: Glass walled, potential for signal intrusion.

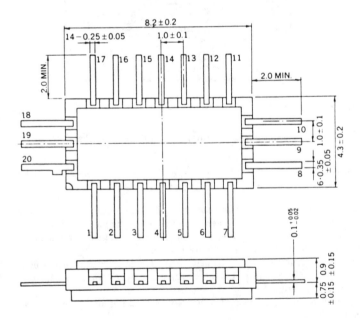

**Fig. 10.A8** Ceramic flatpack package.

**Table 10.A9** Ceramic Flatpack Package

Performance: Applicable up to 6 GHz.
Microstrip or coplanar configuration: Microstrip.
Thermal resistance: 5°C/W.
Reliability: Meets MIL-STD 883.
Mounting: AuSn is used as solder.
Lead count: 16,20.
Controlled impedance: 50 Ω.
Die attach and wire bonding: Die attach: AuSn; bonding: ball bond (Au wire).
Tooling: Fully automated.
Dielectric constant: 9–10.
Hermeticity: Hermetic.

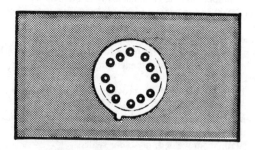

**Fig. 10.A9** T. O. can package (courtesy Motorola, Inc.).

**Table 10.A10** Glass Feedthrough Package

Performance: Highly dependent on frequency of use, the transition from the glass seal to the circuitry within the internal cavity, and the type of glass utilized. A proper design can result in
performance to 20 GHz with a VSWR or 1.2 : 1. A poor design will yield a radiation element
within the cavity exhibiting high VSWR and insertion loss.
Microstrip or coplanar configuration: Coaxial connection that may be connectorized or utilized as a
drop in microstrip; coplanar, as a drop in medium also could be used.
Thermal resistance: Glass-to-metal sealing requires matched thermal coefficients of expansion;
hence almost all of these packages are made from Kovar. Kovar has a thermal conductivity
similar to alumina and therefore is not a good thermal conductor. Thermal resistance may be
decreased by the addition of a heat sinking metal plug or base, which may be either Cu-W

**Table 10.A10**

(Thermkon, Elkonite), molybdenum or copper. Generally bases are Cu-W and plugs are copper. Added cost for this enhancement.

Reliability: Generally very good but, when people handle the parts, seal may crack; with proper care, problem becomes less severe.

Cost: Probably the most expensive package, depending upon the metal machining required; even in fairly large quantities (10,000) can exceed $20.

Mounting: By bolts generally.

Lead count: Limited by the size of the feedthroughs used and the package: the more leads, the larger the package.

Impedance control: By coaxial configuration and proper transition match, really the same as performance.

Die attach and wire bonding: If proper gold metalization is used, wire bonding and die attach are no problem. Die can be attached either by eutectic or conductive epoxy. Wire bonding generally is wedge bonding; ball bonding may also be used where applicable.

Tooling: On the order of $5000 and is needed for the graphite boats that hold the feedthroughs, glass beads, and metal housing during the firing process.

Dielectric constant: Depends upon which glass is used: 7052 is 4.7, 7070 is 4.1.

Loss: Dependent upon which glass used: 7070 has a loss tangent of .0025, which makes it desirable for low loss requirements; 7052 has a loss tangent in excess of .015. Also remember the better the transition, the less the mismatch loss.

Crosstalk: Isolation can be in excess of 50 dB, provided that the transition is not a radiating element.

VSWR: Highly dependent on the transition, if properly compensated no reason why it cannot be on the order of 1.05 to 1.

Sealing: May be laser sealed, seam welded, or solder sealed.

Adaptability: Very good, can be cabled or mounted into any of the various strip transmissions (microstrip, coplanar) usually bolted in place; tends to be the largest housing for MIMICs.

Pickup: All metal enclosed cavity; should be minimal.

Production costs: In high volume, with a coined housing, can be brought down to $20 each.

Approximate Size 6 mm Square

JEDEC Leads

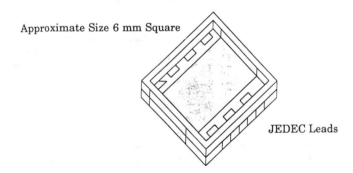

**Fig. 10.A10** Chip carrier (courtesy Tachonics Corp.).

**Table 10.A11** T.O. Can (microwave quality) Package

Performance: Relatively poor frequency performance: 16–18 GHz.
Reliability: Susceptible to human handling.
Mounting: Limited mounting options.
Tooling: Low tooling cost: $1000–5000.
Loss: < .1 dB.
Isolation: > 50 dB.
Production costs: Low.

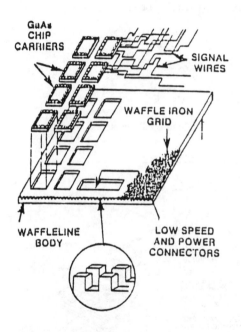

Harris' Waffleline subsystem
packaging scheme contains dielectric-
coated wires or grids with hollowed out
areas for GaAs chips.

**Fig. 10.A11** Waffle line package.

**Table 10.A12** Chip Carriers

Performance: Reasonable.
Microstrip or coplanar configuration: Good.
Thermal resistance: Low to good.
Reliability: Questionable with handling.
Cost: Low to very low (plastic).
Mounting: Sometimes inconvenient.
Lead count: Count high usually, but not for MIMICs.
Controlled impedance: Poor.
Die attach and wire bonding: Good.
Tooling: Low, < $500.
Dielectric constant: Epoxy board (EPIC), Low.
Loss: Poor to fairly good.
Crosstalk: Poor.
VSWR: Poor.
Return loss: Poor.
Hermeticity: Good.
Isolation: Poor.
Adaptability: Fairly good for MIMICs.
Pickup: Good.
Production costs: Very low, < $2.00 in small quantities.

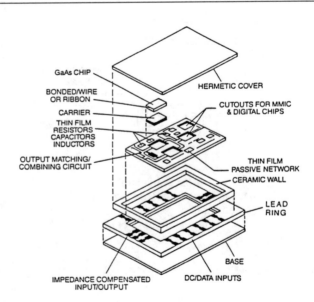

**Fig. 10.A12** Hermetic metal-tray package.

**Table 10.A13  Waffle Line Packages**

Microstrip or coplanar configuration: Controlled impedance.
Mounting: Flexible and adaptable to a variety of applications.
Crosstalk: Minimum cross coupling between lines.
Isolation: Minimum pickup from outside sources.
Production costs: Costs are relatively large.

**Fig. 10.A13**  Molded dip.

**Table 10.A14  Hermetic Metal-Tray Package**

Mounting: Slim profile for multiple phased array chips.
Controlled impedance: Compensated lines.
Adaptability: Minimizes GaAs real estate, hermetic.
Production costs: Planar wire bonding, production costs reasonable because multiple chips are
packaged.

## Table 10.A15 Molded Dip Package

Performance: Limited frequency response.
Thermal resistance: High.
Tooling: Low cost.
Hermeticity: Not hermetic.
Production costs: Low.

# Chapter 11
# MMIC Reliability

## W. J. Roesch and R. Goyal

## 11.1 INTRODUCTION

*Reliability* is commonly defined as the probability of a circuit performing a specific function, under certain environmental and operational conditions, for a specific period of time. These operating conditions are defined during design of the circuit and are based on the available information about process, design data, design rules, and reliability. From the user's viewpoint, *reliability* is simply defined as the length of trouble-free operation of the circuit in a functional system under actual conditions. With the technological advances in GaAs material and processing methodology, the manufacturing of GaAs MMICs for high-reliability applications has grown dramatically. Many of these applications in military and space systems that require rigorous reliability in the circuit performance. A survey of on-going reliability investigations clearly shows a large data base for reliability of GaAs discrete devices such as MESFETs, but scarce data on the reliability of GaAs MMICs. This is because of the limited commercial availability of manufacturable MMICs.

Although the performance of an individual MMIC chip can be predicted broadly, using computer-aided design tools, based on the prior information about process variations and in-process test data base, its reliability can be determined only by actual testing. Slight changes in chip layout at critical places, metal thickness, passivation, and packaging can result in very different results. There are three important aspects of any reliability study. First, extensive data must be collected, covering all aspects of operating conditions, that are sufficient to make useful projections of the failure rate of the chip under specified operating conditions. This needs to be done in conjunction with the characterization of the failure distribution and the identification of the activation energy of the failure mechanisms to predict MMIC life. Second, the failure mechanisms must be identified, understood, and possibly eliminated. Finally, if required, a screening procedure must be established, with an optimal number of tests performed, on a reasonable sample size, under certain operating conditions, for a specific amount of time, to screen reliable ICs from possibly failing devices. Proper documentation and analysis of the collected data is an essential part of this study.

Failure mechanisms depend greatly on the MMIC functional application, such as power, low noise, general purpose, or switching circuits, and thus their failure definition and criteria differ accordingly. Failures are broadly divided into three major classes: design related, material related, and process related. Design related failures generally emanate from the lack of a design data base; for example, electromigration of narrow interconnect lines carrying more than specified current, voltage breakdown, leakage current, and crosstalk between closely spaced devices in an MMIC. Such failures can be eliminated as the design data are collected on test structures or MMICs. Material-related failures depend on the technique used for growing the substrate and are neither well characterized nor understood for the GaAs material commonly used for manufacturing MMICs. Process-dependent failures occur during the processing and can vary drastically from one MMIC manufacturer to another. It is essential for individual manufacturers to collect extensive reliability data to ensure reliability of their MMICs. A consistent process, yielding a high throughput within certain specifications, is needed before process-dependent failure modes can be addressed. A simple change in a process step or design rule, if not performed with reliability in view, could significantly alter the reliability of an MMIC.

The evaluation of any IC reliability must begin at the very early stages of its development, including conceptual design, definition of specifications, operating conditions, circuit design voltages and currents, physical layout, assembly, and packaging. A reliability model is developed, based on the extensive data collected on past IC designs and reliability results for active and passive elements, such as FETs, diodes, resistors, inductors, capacitors, interconnect lines, crossovers, via holes, and other components used in MMIC design. This data base is complemented with in-process test results to evaluate the reliability of a new design. In a MMIC, the main reliability concerns are as follows:

- Ohmic contact metalizations, such as for MESFET source-drain.
- MESFET channel conduction integrity.
- Electromigration of gate, first-level interconnect, and airbridge metalizations.
- Thermal diffusion of deep-level traps and impurities in the substrate.
- Failures induced by external stress, such as electrostatic discharge and radiation.
- Short-term and long-term current and RF performance drifts.
- Device instability after turn on.

- Burn-out caused by large input pulses; this is particularly important for power ICs.
- Relatively short MTTF when operated under adverse channel temperatures.
- Mechanical failures due to die attach, wire bonding, and package sealing.

## 11.2 CIRCUIT FAILURE MECHANISMS

Whether made from gallium arsenide or silicon, integrated circuits are susceptible to the same classes of failure mechanisms, which generally are attributed to metalization, dielectric, semiconductor material, and the devices. These failures occur in different relative proportions depending upon the type of device, operating conditions, and more significantly the manufacturers. Recently many of these failure mechanisms were investigated in detail, and design and processing techniques were improved to prevent or reduce their occurrence. For example, several metalization steps associated with ohmic contacts, Schottky, and interconnect metalizations were studied in detail, and the proper combination of several metal layers of appropriate thickness were used to improve the reliability. Integration of active and passive devices on the same chip offers increased performance and reliability at lower cost. However, the trend toward greater integration increases the packing density, which introduces different failure mechanisms, such as coupling between close components and interconnects, electromigration in narrow interconnect lines, and increased power dissipation.

### 11.2.1 Metalization

Metalization failures result from degradation in several basic attributes. Interconnects are expected to be conductive, immune to electromigration, bondable, able to adhere to other circuit layers, resist corrosion, form good contacts, and be patternable into the desired structures.

Silicon technology uses aluminum, with small percentages of copper and silicon added, almost exclusively for metalizations. Processing has long been performed with the single material aluminum-type runs and contacts. The desirability of adding about 2% silicon was found to form direct ohmic contacts to the silicon active regions without interdiffusion and contacts spiking. The addition of small amounts of copper has become popular, mainly due to the increase of the aluminum's immunity to electromigration as metalization size decrease. Because of its maturity, the aluminum used in silicon processes generally meets all criteria for metalizations mentioned earlier. There have been persistent problems with corrosion, intermetallic formation, and electromigration, but those have been controlled.

Gallium arsenide MMIC processing generally involves specialized metalization for ohmic contacts, Schottky gate formation, thin-film resistors, and interconnects, including MIM capacitor electrodes. Naturally, each type of metal is designed to meet specific metalization properties, sometimes at the expense of other properties. Most commercial GaAs ICs employ gold-based metals, principally titanium and various other refractory metals. Gold-germanium ohmic contacts are used consistently. Aluminum has been used for gates of discrete power MESFETs and for some interconnects. However, gold has been proven superior to aluminum in side-by-side tests for many years. Gold-based connections avoid the possibility of intermetalic problems that an aluminum-gold metalization introduces. Gold-based metals therefore have dominated GaAs MMIC production. Performance of gallium arsenide metalizations in conductivity, bondability, and adherence are probably about equal to those of aluminum used in silicon processing. However, mixing gold and aluminum results in intermetallic problems such as purple plague.

Next, metalizations are expected to have an immunity to electromigration and resistance to corrosion. Tests on GaAs ICs have indicated susceptibility to both mechanisms, but not to the degree of aluminum metal problems for silicon devices. Electromigration and corrosion are not considered to be primary failure mechanisms for GaAs ICs.

The principal requirement of metalization for ohmic contact is to ensure that low series resistance by the metal-semiconductor interface and no rectifying property is present. Such contacts theoretically can be achieved by correctly matching the metal's work function to that of the semiconductor. However, in practice, due to the presence of semiconductor surface states and interdiffusion of metal problems, the choice of metal is not so simple. To achieve ohmic behavior, it is often necessary to heavily implant GaAs where ohmic contacts are to be formed. This ensures a very thin, if any, depletion region between the metal and the semiconductor, and electronics can tunnel through the barrier. Alloying the contacts further improves the ohmic behavior by controlled mixing of the metal and GaAs. In practical devices, formation of ohmic contact involves multilayer metalization. A reliable ohmic contact must have the following properties:

- The contact must present the least resistance to the flow of current.
- It must be stable against metal-semiconductor interdiffusion during operation.
- It must be resistant to metallurgical reactions, oxidation, and corrosion.
- It must be easily patternable in a desired and convenient shape.
- It must have good adherence to the substrate, thus requiring good thermal matching.

Au-Ge-Ni is widely used in multilayer metalization systems for making ohmic contacts on GaAs that is heavily doped with $Si^+$ before depositing the layers of metal.

Metalization also should be fairly inert with respect to other materials in the process. Silicon probably has the edge in this category. Even though gold-GaAs interdiffusion is essential in forming ohmic contacts, too much interdiffusion of gold, either on the ohmic contacts or the Schottky gates, is the primary failure mechanism in GaAs ICs. This interdiffusion may be at a "controlled" state, but the variety of existing metal schemes and process techniques would indicate there is no single, superior answer to the interdiffusion problem. Some gate metalizations are being produced from refractory metals, which can withstand high temperatures without interdiffusion, but Au-Ge ohmic contacts that are susceptible to interdiffusion after long high temperature exposures are still in widespread use.

### 11.2.2 Dielectric

For the most part, the materials used to form GaAs and Si dielectrics are quite similar but their requirements are different. Silicon can grow a stable, high-quality oxide that is used principally to form MOS gates. The silicon oxide is also used to isolate individual transistors in a high-performance silicon bipolar-MOS process and to form MOS capacitors. Above the surface of silicon, layers of nitride generally are used for interlevel dielectrics. Gallium arsenide has no native oxide; therefore, silicon oxide is often used in GaAs processing as a capping material or sometimes as an interlevel dielectric. Usually, silicon nitride is the dielectric used in MIM capacitors deposited on GaAs circuits. The usual form of deposition is plasma enhanced CVD (chemical vapor deposition). The silicon industry has devoted much effort to the development of dielectrics because of their importance in gate formation and isolation. Gallium arsenide IC manufacturers have taken advantage of this work, even though their dielectrics are much more forgiving, because Schottky gates are used and the substrate is self-insulating. Except for capacitors, GaAs circuits could be constructed without any dielectric. As we might expect, dielectric failure has not been reported as a degradation mechanism in gallium arsenide.

### 11.2.3 Substrate: Gate and Channel

The final general category of failure mechanisms involves the substrate material. The difference in resistance gives gallium arsenide an advantage in this area. The semi-insulating properties of GaAs, for the most part, have eliminated the problems with isolation and latch-up that are constant nuisances for silicon ICs. With the increasing density of GaAs circuitry and the application of E/D devices, a phenomenon of circuit interaction, called *back gating* or *side gating*, has been discovered. Semiinsulation may not be quite enough to prevent substrate conduction for MMICs and digital GaAs LSI devices. However, isolation implantation, substrate bias, and special circuit layout techniques offer possible answers to this potential problem. For the present time, the bulk properties of GaAs give it superior performance for microwave applications and low-complexity circuits.

The primary failure mechanisms of silicon technology generally involve gate oxide integrity, electromigration, and isolation, whereas gallium arsenide problems center around metalization interdiffusion. The MESFET itself is the center of attention in gallium arsenide circuits. Testing has revealed degradation of channel resistance and decreasing saturation current as clear failure modes of MESFETs at the end of their expected lifetimes.

Early GaAs MESFETs suffered from direct channel burn-out, as shown in Figure 11.1. Shorting in the channel from gate-to-drain, gate-to-source, or drain-to-source resulted from lack of consistent processing and poor substrate materials. More recent studies have found this shorting only when MESFETs were overstressed. The popularity of liquid encapsulated Czochralski growth of GaAs and processing facilities geared toward production (instead of just research) are two specific examples of enhancement that have helped reduce the incidence of burn-out.

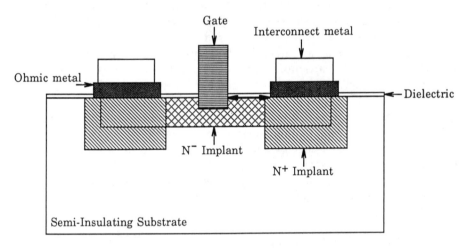

**Fig. 11.1** Burn-out mechanism.

Another reported failure mechanism that often could account for increased channel resistance and decreased saturation current was ohmic contact degradation, as shown in Figure 11.2. Interdiffusion of gallium and gold was a problem for manufacturers a few years ago. This interdiffusion often caused an increase in effective channel resistance, when the contacts degraded. As manufacturers standardized use of Au-Ge-Ni ohmic materials, instead of incorporating Cr or In, this problem subsided.

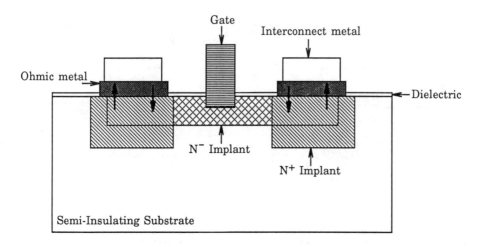

**Fig. 11.2** Ohmic degradation mechanism.

As GaAs IC development has progressed, the ohmic contacts have become less of a problem. Investigations of MESFET degradation on a device physics level has found that changes in pinch-off or transconductance often occur near the end of device life, which cannot be explained by ohmic contact degradation. Analysis generally also indicates a reduction of carrier concentration in the channel of the MESFET. At least three possible mechanisms could be responsible for this predominant failure mode.

Changes in the carrier concentration of the MESFET can be caused by compensation from gate atoms, as shown in Figure 11.3; diffusion of carriers out of the channel, as shown in Figure 11.4; or effective reduction of the channel depth by the encroachment of the gate, as shown in Figure 11.5. Compensation can be measured by *deep-level transient spectroscopy* (DLTS). Diffusion can be assessed by the effects of bias during lifetesting. Movement of the gate into the channel can be confirmed by auger analysis.

The effects of burn-out, ohmic contact degradation, carrier compensation, diffusion of carriers, or gate metal interdiffusion is specifically dependent on the manufacturer's process. The important failure modes for GaAs ICs are summarized in Table 11.1.

## 11.3 APPLICATION FAILURE MECHANISMS

### 11.3.1 Electrostatic Discharge

Military systems can be electrically overstressed by a variety of disturbances. Such overstress can cause either permanent damage to the hardware or

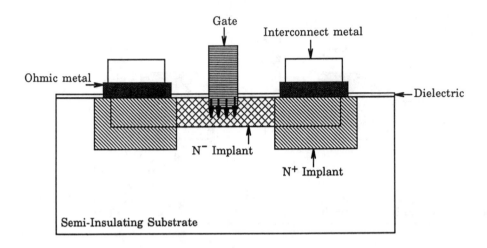

**Fig. 11.3** Channel compensation mechanism.

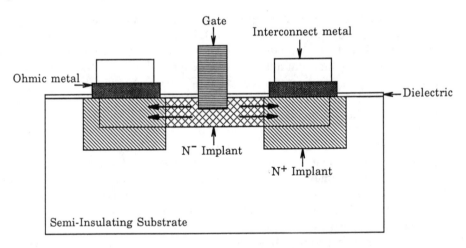

**Fig. 11.4** Diffusion mechanism.

transient upset of the constituent electronics. Some of the more common hazards which must be survived by deployed systems are enumerated below [1,2]:

- Electromagnetic Pulse (EMP)
- Electrostatic Discharge (ESD)
- Lightning
- Electronic Countermeasures (ECM)
- High Power Microwave (HPM)
- System Transients

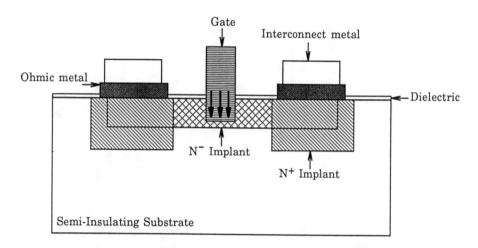

**Fig. 11.5** "Sinking" gate mechanism.

Attempts to protect electrical hardware from these threats can involve comprehensive shielding of the system or increasing the hardness of the individual components. Occasionally, protection techniques that defeat one of the hazards can also improve immunity against a similar threat. For example, the energy spectral density of an *electromagnetic pulse* (EMP) can be very similar to that of an *electrostatic discharge* (ESD) event. Thus, any technology developed to improve the ESD hardness of integrated systems might, in turn, increase their immunity to nuclear disturbances. In contrast, EMP fields generally reach their maximum intensity very rapidly compared to fields induced by lightning. Therefore, standard lightning protection methods are unlikely to provide the desired level of EMP protection [2,3].

All of the threats listed above can have a detrimental effect on electronic systems and solid-state ICs are particularly at jeopardy. Because of the severe effect of ESD on the ICs, considerable effort has been devoted to the development of technologies for simulating and defeating the effects of electrostatic discharge on silicon [4]. Extrapolation of this knowledge will help to elucidate analogous methods for ESD-hardening of gallium arsenide devices and circuits.

### 11.3.1.1 Implications of ESD for Gallium Arsenide

Overstress damage to a solid-state device is frequently induced by discharging a triboelectrically generated static charge through the pins of a packaged semiconductor. Such an event is called electrostatic discharge or ESD. One very common ESD event occurs when an electrostatically charged person or object is accidentally discharged through the pins of a solid-state component. Such an

**Table 11.1 Important Failure Modes in GaAs Integrated Circuits**

| Device Feature | Failure Modes | Failure Process | Acceleration Factors | Comments |
|---|---|---|---|---|
| Schottky barrier | Junction degradation | Gate metal diffusion | Temperature Voltage | Apparent failure mode for "good" MESFETs |
| | | Gate region sputter damage | Temperature Voltage | Fixed by use of nonsputter gate recess process |
| | | Gate metal used as interconnect | Temperature Voltage | Fixed by confining gate metal to gate implant region |
| Ohmic contacts | Increased resistance | Interdiffusion, electromigration | Temperature Voltage | Present designs have been established empirically |
| Airbridge, interconnect, metalization | Shorts/opens due to warp | Differential expansion | Temperature | Fixed by geometric design rule for standard temperature ranges |
| | Opens due to high current | Electromigration | Temperature Current | Requires maximum current density design rule |
| Wire bond and other mechanical interfaces | Intermetallic growth | Impurities, bond strength | Temperature | Constraints similar to those seen in silicon ICs |
| | Fatigue | Temperature cycling, bond strength | Extremes in temperature cycling | Constraints similar to those seen in silicon ICs |
| Hermeticity | Seal leaks | Pressure differential | High pressure | Constraints similar to those seen in silicon ICs |

event is analogous to the shock delivered when you walk across a carpet and then touch a doorknob. Although the doorknob generally escapes unscathed, the corresponding spark can obviously be very damaging to something as sensitive as a solid-state device. For example, just touching a component can discharge pulses of 50 kv through the packaged part [5]. The silicon semiconductor industry has suffered both yield losses and field failures from these types of events, and has spent millions of dollars trying to prevent ESD damage. A few examples illustrating the magnitude of the problem are presented in Table 11.2 [5]. Immunizing chips from such aggressive overstress is a necessary, but formidable, undertaking.

The magnitude of an ESD pulse is a function of the environment in which, and the methodology by which, the semiconductor device is handled and deployed. Improvements in static control have reduced the severity of commonly encountered ESD pulses to tolerable levels. Nonetheless, some static is always present, and hence the design engineer must provide a system with a nominal level of ESD immunity. The military's current strategy is to encourage whatever layout and circuit techniques are required to make its chips ESD-hard for discharge levels of up to several thousand volts, at least.

A popular methodology for evaluating ESD compliance is to test the part to ensure that it is initially within specification, to apply high voltage pulses with both positive and negative polarities to various pin combinations and then to test the part again to see if it is still within specification [4]. Many of the parts being manufactured today do not pass this ESD screening test.

Recently, some research involving circuit layout variations, wafer processing improvements, and a menu of input protection clamps has resulted in improved ESD immunity of silicon integrated circuits [6]. Most of this work has been directed toward input protection devices. These clamping structures attempt to defeat ESD pulses by routing the transients through less sensitive portions of the chip. Such techniques are often called *input protection*, although the general methodology can also be applied to outputs or devices internal to the circuit. The ideal protection network should be invisible (absent) to normal signals, but provide a low-impedance shunt for ESD pulses. In addition, fast switching between the "absent" and the "shunt" modes is desired. Of course, such ideality is not possible to achieve in practice. Physical realizations of input protectors often use some type of clamp to prevent excessively high voltages from appearing across critical devices and a series impedance to limit the ESD current.

Placing protection elements on the chip adds to the die size and degrades the frequency response of the circuit. For example, locating a current-limiting resistor between an output driver and a package pin will slow the characteristic delay time of the part. Such performance degradation may be particularly detrimental when driving capacitive loads. Generally, when input protection is used, design tradeoffs must be made between the desired amount of protection and the tolerable

**Table 11.2** A Selection of ESD Case Histories [5]

| Company-Project | Technology | Remarks |
|---|---|---|
| McDonnell-Douglas Aircraft | Low-Powered Schottky Logic | Many soft failures in the equipment; $\frac{1}{3}$ of these were traced to ESD degradation. |
| Hewlett-Packard | Various Integrated Circuits | Between 5% and 25% of all component failures are caused by ESD. |
| Bell Telephone | Various Integrated Circuits | More than half of the early-life operating failures are ESD induced; also, 50% of the incoming inspection failures are caused by ESD. |
| JPL Galileo | Sandia CMOS RAM | Between 80% and 95% of the chip failures were attributed to ESD damage. |
| Titan III | Junction Field-Effect Transistors | Ongoing failures in the flight control computer, many of which have been attributed to ESD damage. |
| Viking Lander | Hex Inverter | Two checkout failures at NASA; both of these were caused by ESD damage. |
| Delco Electronics-General Motors | Various Integrated Circuits | ESD-related problems cost Delco up to $22 million for a given automobile model year; ESD causes 40% of all failures in on-board computers and 44% of the failures in radios. |

performance penalty. These protection methodologies, however, have significantly increased the ESD hardness of integrated circuits made from silicon.

Unfortunately, very little complementary research has addressed ways of increasing the static damage threshold of devices made from compound semiconductors such as gallium arsenide (GaAs) [7–13]. This situation should be remedied because many of GaAs MMICs being considered for new electronic hardware are very ESD-sensitive in comparison to their silicon counterparts. In fact, numerous ESD-related failures of GaAs microcircuits have been reported in the literature and a summary of the corresponding damage types is presented below [7–24]:

- Decrease in bipolar transistor beta;
- Decrease in FET transconductance;
- Increase in FET gate leakage current;
- Increase in the ideality factor of a Schottky diode;
- Increase in noise figure of a device or circuit;
- Open-circuit of the gate of a FET;
- Short-circuit of the gate-source of a FET;
- Open-circuit of interconnect metalization;
- Other catastrophic electrical failures of a device or circuit;
- Visual physical damage to a device or circuit;
- Catastrophic physical damage to a device or circuit.

These contemporary GaAs devices are often handicapped with extremely low ESD-failure thresholds, which thus limit the yield and cause a concomitant increase in deployment cost. Furthermore, without additional protection, the next generation of circuits is likely to be even more ESD-sensitive as a consequence of the projected MMIC architectures, enumerated as follows:

- Small lateral dimensions;
- Shallow active layer and junction;
- Low-melting-point substrates;
- Nonplanar semiconductor surfaces;
- Deposited, rather than thermally grown, passivation layers;
- Reverse-biased circuitry.

### 11.3.1.2 ESD Protection for MMICs

Several manufacturers of GaAs MMICs are placing ESD protection networks on their chips. These clamps are generally located between the bonding pads and the corresponding MESFET gate to which the pad is ultimately connected. Figure 11.6(a) illustrates a typical ESD pad clamp. Components $R_1$ and $R_2$ are current-limiting resistors while elements $D_1$ and $D_2$ are Schottky diodes. The clamping diodes terminate at tie points 1 and 2 as shown on the figure. Frequently, tie point 1 is the $V_{dd}$ power supply rail and tie point 2 is the $V_{ss}$ rail. In normal operation, $D_1$ and $D_2$ are reverse biased and the signal appearing at the bonding pad is passed to the interior of the circuit as shown in Figure 11.6(b). However, a positive ESD pulse turns on $D_1$ and routes this transient energy to the robust $V_{dd}$ line. At the same time, the positive transient tends to break down diode $D_2$ and this augments $D_1$'s ability to shunt the ESD pulse. Similarly, negative ESD stress forward biases $D_2$ and breaks down $D_1$.

Figure 11.6(c) presents an equivalent circuit for the pad clamp when a positive ESD pulse is applied to the bonding pad. Resistor $R_{df1}$ is the forward resistance of diode $D_1$, resistor $R_{dr2}$ is the reverse breakdown resistance of diode $D_2$ and voltage $BV_{d2}$ is the breakdown voltage of $D_2$. Figure 11.6(d) shows the analogous case when the bonding pad is stressed with a negative ESD transient.

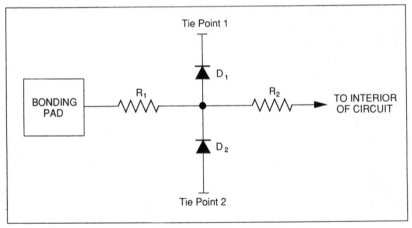

**Fig. 11.6(a)** Typical ESD pag clamp.

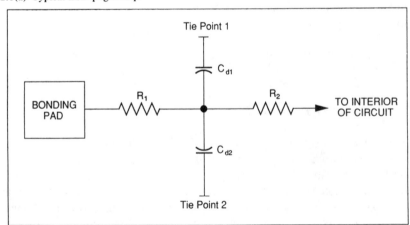

**Fig. 11.6(b)** Equivalent circuit for ESD pad clamp under normal operation.

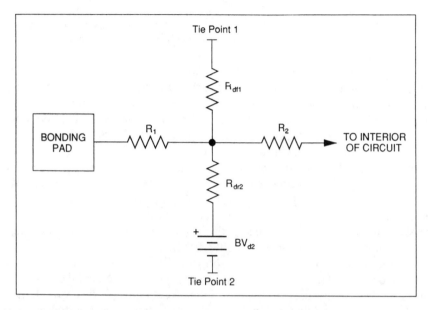

**Fig. 11.6(c)** Equivalent circuit for ESD pad clamp under positive ESD stress.

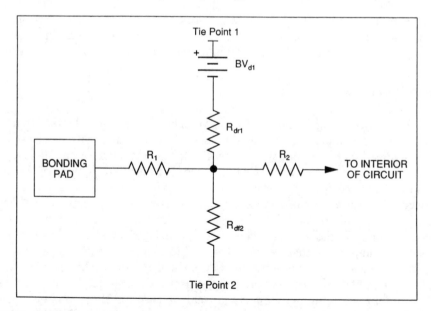

**Fig. 11.6(d)** Equivalent circuit for ESD pad clamp under negative ESD stress.

The Schottky diodes utilized in this clamping arrangement turn on very rapidly, compared to the rise time of typical ESD threats, because the Schottky devices have no minority carrier charge storage.

The pad clamping arrangement illustrated in Figure 11.6(a) is currently employed on a number of GaAs ICs. Apparently adding the associated impedance to various nodes may degrade the frequency response of the circuit. As a consequence, design trade-offs are often needed to optimize both the frequency response and the ESD hardness of a GaAs chip. At one extreme, to incorporate no ESD pad protection on the GaAs IC may be desirable. In this case, the part suffers no performance degradation, but a GaAs chip with minimum geometry input transistors will perhaps suffer from an ESD damage threshold similar to that of unprotected CMOS [7–13]. Alternatively, a MMIC with large input devices may need no additional ESD protection. In fact, the large input transistors often used in MMICs for their superior transconductance may be more ESD-immune than the tiny diode clamps purported to protect them.

Protection networks like the one shown in Figure 11.6(a) can be successfully employed on control inputs, power supply pins and selected digital circuitry. Such networks can also be used for some higher frequency applications if they are carefully designed into the original circuit, rather than incorporated as an afterthought. What is recommended, of course, is to develop the pad clamps as an integral part of the MMIC–protection-network ensemble. In such an arrangement, the circuit components can be selected and the corresponding chip can be tuned to account for the additional impedance inherent in the protection networks.

### 11.3.2 Assembly

The MMIC chip is mounted on a metallic or ceramic substrate that is an integral part of the package. Commonly used mounting materials are low-melting point solders, such as gold-germanium or gold-tin eutectic alloy, or silver epoxy. The temperature must be controlled to ensure the complete alloying of the eutectic solder and the thermal curing of the epoxy in order to avoid a weak bond between the mounting substrate and the chip, especially on chips with substrate via holes. Most often, during the backside metalization of the chip with via holes, metal does not fill the via holes completely, causing a nonplanar structure at the back. It is important that, when such a chip is mounted on substrate in the package, the eutectic metal or the epoxy is sucked into the via hole void so that no air is trapped between the via hole and the package substrate. The primary cause for reliability failure in mounting is the weakening of the alloy bond between the backside metalization on the chip and the package, due to the solid-state diffusion reaction forming brittle intermetallic compounds. However, failures of chip-to-package bond constitute only a minor reliability hazard.

The bonding pads on the chip are connected to the package leads using thermocompression or ultrasonic wire bonding. Both wedge bonding and ball bonding methods are widely used for this purpose. In a gold-based metalization system for the MMICs, using a gold wire for bonding inhibits the intermetallic reaction and produces a reliable bond. Bonding failures in IC assembly are usually because of over- or underbonding, which results in weak bonds. These bonds are easily screened either optically or by centrifuge testing, which stresses the bond wire leading to a clean break and thus a failed device. Another reason for bonding failure is the poor adhesion of pads to the substrate. Bond-pull testing techniques commonly are used to evaluate such a failure.

### 11.3.3 Packaging

The package provides electrical and mechanical protection for the device as well as easy access to the device through its lead connections. The package also provides the IC an inert, dry atmosphere and protection against mechanical and thermal shock. Three types of packages commonly are used for MMICs: metal cans, ceramics, and plastic capsule. Plastic packages are only used in commercial or consumer applications at lower frequencies.

The package seals are made as hermetic as possible to isolate the IC from the atmosphere. Leak testing of the seals is commonly performed by measuring the rate at which helium escapes from the package. A gold-plated header is welded to the metal can package. The seal is made in nitrogen ambient with a moisture content less than 10 ppm. The hermeticity of the can depends on the integrity of the welded seal and the glass-to-metal seals often used to isolate the electrical leads from the metal can itself.

The reliability of the seal depends upon the close matching of the thermal coefficient of glass and metal leads. In ceramic packages, the chip and lead frame assembly are sandwiched between two ceramic slices, held together by glass sealant. The thermal coefficient of the glass and ceramic are carefully matched to prevent thermal stress cracking. Such a package is quite strong, although brittle and susceptible to torsional stress and thermal shock.

## 11.4 RELIABILITY TEST STRATEGY

One drawback of the rapid development of GaAs IC technology is the lack of available circuits to test. The GaAs IC industry is still in its infancy and lacks the large volume TTL-like circuit families to learn from. Standard components have not been universally popular. Most of the circuits have been custom-made or application-specific, generally produced in low volumes and difficult to test by nature. The combination of evolving device complexity with electrical measurement problems at GHz frequencies has resulted in relatively small sample sizes in comparison to the large historical data base developed for silicon device manufacturers.

A small sample size is treated with skepticism by the reliability professionals, so new approaches to reliability testing have to be developed while conventional data is amassed over time. Millions of device-hours are being counterbalanced with specific failure mechanism studies and acceleration factor examinations. Failure mechanism data imply that failures must be obtained, however. A well-characterized failure distribution becomes essential.

### 11.4.1 Failure Distributions

The most commonly used failure distribution curve in silicon ICs is the bathtub curve shown in Figure 11.7. This curve is composed of three failure rate zones during a product's life; infant mortality, useful life and wearout as shown in Figure 11.8. When devices are manufactured in volume, there is always a possibility of certain failure mechanisms, which cause it to fail under moderate operating or stress conditions. Such devices fail very easily in the useful life and are termed *infant mortality rejects*. These devices generally have gross defects or have been damaged in processing due to mishandling. Such failures can be reduced significantly by proper manufacturing controls and screening procedures. The use of burn-in procedures is designed to screen such devices. Moreover, because of the low volume demand for MMICs, the wafers are processed carefully with extensive in-process testing on larger numbers of PCMs, thus rejecting the marginal and low yielding wafers at early stages of processing.

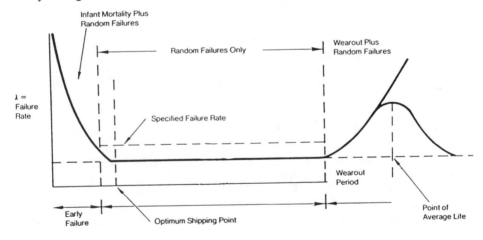

**Fig. 11.7** Bathtub curve used in silicon ICs.

The second curve in Figure 11.8 defines the predominant zone of a well-designed product life cycle and represents the period of time when the product population is undergoing failures at a low constant rate. Failures in this part of the curve are random in nature and depend strongly on the technology and its maturity. The rising tail of the curve or final phase is the "wearout," indicating the median active life of the device. This wearout is due to the intrinsic degradation of the device. It is possible to extend the period at which such wearout occurs by using proper design techniques such as interconnect line widths, thermal management on the chip, appropriate supply voltages, and use of available reliability data base collected on similar products.

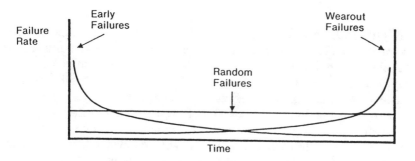

**Fig. 11.8** The three curves making up the bathtub curve.

Both the infant mortality and random failure rate regions can be described by the same calculations. During these periods of failure, the probability of having no failures at a specific time can be expressed by

$$P_0 = \exp(-\lambda t) \tag{11.1}$$

where $\lambda$ = failure rate
$t$ = time

The value of $\lambda$ changes rapidly during infant mortality. Hence, Equation (11.1) is not particularly useful until the random failure period, when the value of $\lambda$ becomes relatively constant. The failure rate, $\lambda$, is usually expressed in percent failures per thousand hours. Alternatively, $\lambda$ is expressed as *failure in time* (FIT), defined as failure in $10^9$ hours and *mean time to failure* (MTTF) or *mean time between failures* (MTBF), both being equal to $1/\lambda$.

Typically, reliability evaluations, such as extended life tests at accelerated test conditions, involve only samples of an entire population of devices. Hence, the statistical concepts of the central limit theorem apply, and $\lambda$ is calculated using the $\chi^2$ distribution:

$$\lambda \leq \frac{\chi^2(\alpha,\ 2r\ +\ 2)}{2nt} \tag{11.2}$$

where $\alpha = (100 - CL)/100$

$CL$ = confidence limit in percent

$r$ = number of rejects

$n$ = number of devices

$t$ = duration of test

The confidence limit is the degree of conservatism desired in the calculation of failure rate. The central limit theorem states that the values of any sample of units out of a large population will produce a normal distribution. A 50% confidence limit is considered the best estimate and is the mean of this distribution. A 90% confidence limit is a very conservative value and results in a higher value of $\lambda$, which represents the point at which 90% of the area of the distribution is to the left of that value, as shown in Figure 11.9. It is common practice in the semiconductor IC industry to predict failure rates at a 60% confidence limit. The term $(2r + 2)$ is called the *degrees of freedom* and expresses the number of rejects in a form suitable to the $\chi^2$ distribution shown in Table 11.3.

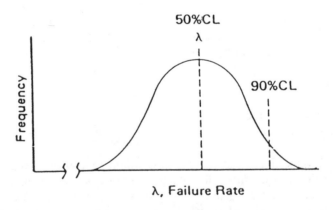

**Fig. 11.9** Confidence limits and the distribution of sample failure rates.

As the sample size and test time are decreased, the probability of the test not being a true representative of the entire population increases. Calculating $\chi^2$ using such test results on a limited sample gives high values of $\lambda$, and hence large failure rates, even though the true long-term failure rate may be quite low. For this

**Table 11.3** $\chi^2$ Distribution (confidence level)

| Degees of Freedom | 50% | 60% | 70% | 80% | 90% | 95% |
|---|---|---|---|---|---|---|
| 2 | 1.39 | 1.83 | 2.41 | 3.22 | 4.61 | 5.99 |
| 4 | 3.36 | 4.04 | 4.88 | 5.99 | 7.78 | 9.49 |
| 6 | 5.35 | 6.21 | 7.23 | 8.56 | 10.6 | 12.6 |
| 8 | 7.34 | 8.35 | 9.52 | 11.0 | 13.4 | 15.5 |
| 10 | 9.34 | 10.5 | 11.8 | 13.4 | 16.0 | 18.3 |
| 12 | 11.3 | 12.6 | 14.0 | 15.8 | 18.5 | 21.0 |

*Note:* Degrees of freedom = (2 × no. of rejects) + 2.

reason relatively large amounts of data must be gathered to evaluate the real long-term failure rate. Since this would require years of testing on thousands of devices, methods of accelerated testing are developed.

The failure rate is expressed in an alternate form known as *lognormal distribution*. In such a distribution, the cumulative failures are plotted as a function of the logarithm of the time-to-failure. This distribution is expected, because it has been well established for silicon ICs. All reported GaAs IC failure mechanisms result from the diffusion of atoms. In other words, GaAs IC failure mechanisms have been found to be chemical and physiochemical processes that interact multiplicatively and therefore should result in log normally distributed populations. Infant and freak populations have not been reported for GaAs ICs. Some indication of early failure has been made by a few companies that perform burn-in, but failure analysis data is not reported. More early failures may be discovered as production volumes increase. However, thus far, the bathtub curve and exponential failure rate have not been proven for gallium arsenide IC reliability analysis.

## 11.4.2 Acceleration Methods

Most integrated circuits have lifetimes in the order of a few million hours at normal operating temperatures. To demonstrate the lifetime of an IC with reasonable accuracy, either an enormously large sample size or life tests over long periods, close to lifetime of the device, are required. Accelerated temperature testing is a commonly used method to overcome this anomaly.

As circuit lifetimes improve, failure distributions become more difficult to characterize. Methods of accelerating failure mechanisms or artificial "aging" are applied using accelerated stress testing procedures to determine potential sources of failure. These tests include operation of ICs at rated currents and voltages at elevated temperatures; temperature, humidity, and bias testing at extreme conditions; and temperature cycling to introduce mechanical stress. These tests are performed upon many devices over considerable periods of time to obtain any statistically significant failure distribution.

To obtain any meaningful results using this method, two basic rules must be observed. First, the failure mechanisms achieved in the accelerated environment must be the same as those observed under the normal operating conditions of the device. Second, the results from accelerated life tests must be extrapolated to the normal operating conditions with a reasonable degree of assurance. One of the most commonly used models relating IC lifetime with temperature is the Arrhenius model. The basis for this model assumes that the performance degradation leading to component failure is governed by chemical and physical processes with a reaction rate given by

$$R(t, T) = R_0(t) \exp(-E_A/kT) \tag{11.3}$$

where $R(t, T)$ = reaction rate as a function of time and temperature
$R_0(t)$ = constant with respect to temperature
$E_A$ = activation energy in eV
$k$ = Boltzmann's constant
$T$ = temperature in K
$t$ = time

To provide time-temperature equivalents, Equation (11.3) is applied to failure rate calculations in the following form:

$$t = t_0 \exp(E_A/kT) \tag{11.4}$$

where $t$ = time
$t_0$ = constant

Combining Equations (11.3) and (11.4), a more useful relationship is obtained:

$$\text{Acceleration factor} = \exp\left(E_A/k \left(\frac{1}{To} - \frac{1}{Ta}\right)\right) \tag{11.5}$$

where $To$ = operating temperature in K
$Ta$ = accelerated temperature in K

The most crucial parameter in this model is the activation energy, $E_A$, which is determined experimentally by collecting time-to-failure life data on different batches of devices at different temperatures of operation.

The activation energy is a measure of how effectively a failure mechanism utilizes thermal energy. In short, the higher is the activation energy, the greater the accelerating effect of temperature in causing failures. There is no particular advantage to high or low activation energies, because the Arrhenius equation determines only an acceleration factor not overall reliability. Devices with poor reliability are not necessarily "saved" by a high activation energy, and the reliability of good devices is not jeopardized because of a low activation energy.

The failure mechanisms in semiconductor ICs can be attributed to physical phenomenon related to metalization, substrate material, and dielectric, as described in Section 11.2. Hence, it is possible to accelerate the onset of such failures by elevating the temperature of the IC. Si and GaAs devices have very different activation energies. Most Si MOS devices fail because of time-dependent dielectric breakdown of thin-gate oxides. Dielectric breakdown is affected more by the electric field across the gate oxide than the temperature at which the stress is applied. As we might expect, dielectric breakdown has a low activation energy, of about 0.3 eV.

Most other Si failure mechanisms have similarly low activation energies, below 1.0 eV. However, the primary failure mechanisms in GaAs involve atomic interdiffusions in MESFET circuit elements, which have relatively higher activation energies. Ohmic, gate, and channel degradation mechanisms have reported activation energies between 1.4 eV and 2.6 eV.

The activation energy of a failure mechanism is determined simply by conducting two or more life tests at various temperatures. It is essential that only a single failure mechanism occur during these tests or if multiple mechanisms occur they can be distinguished and that we know which half of the devices fail. This point is called the *median life*. Reliability comparisons based on median lives is convenient, because most failure rates are not constant during the life of a semiconductor. Assuming $E_A$ to be constant, a plot of log $R$ *versus* $1/T$ results in a straight line with its slope equal to $E_A$, which may be interpreted as the energy threshold of a particular failure mechanism. For different values of activation energy, the device lifetime is plotted as a function of channel temperature, as shown in Figure 11.10. Using such a plot, failure rate data collected at one channel temperature can be translated to another temperature with reasonable accuracy. The Arrhenius equation is used to predict the life time of ICs by (1) conducting life

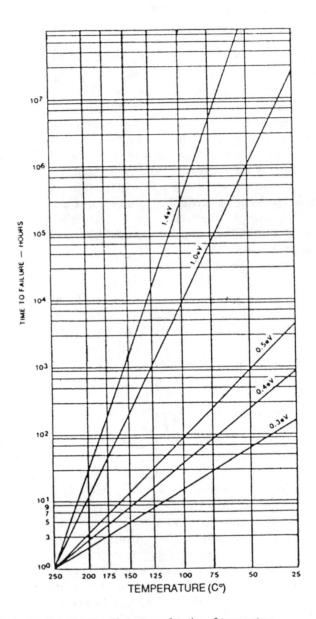

**Fig. 11.10** Arrhenius plot of device lifetime as a function of temperature.

tests on sample lots at multiple temperatures, (2) plotting the median lives of identical failure mechanisms as a function of reciprocal of temperature providing the activation energy for the specific failure mechanism; (3) conducting life tests at specific temperatures on a sample of devices on which data is required, (4) calculating the median life at elevated temperatures, and finally (5) extrapolating the median life at normal operating temperatures using the Arrhenius plot shown in Figure 11.10.

### 11.4.3 Environmental Testing

GaAs MMICs are very fragile compared to silicon devices, particularly because the GaAs material is brittle. Moreover, some design considerations require the substrate to be thinned down to 4 mil or even less, essentially for power devices or the requirement of substrate via holes. Larger chip areas are generally associated with MMIC power devices resulting in significant stress within the chip as well as between chip and eutectic or epoxy mounting material and mounting substrate. This occurs because of the dissimilar thermal expansion rates of the materials. During the normal life cycle, events of thermal cycling, shock, vibration, *et cetera* can cause the bonded chip to degrade by developing cracks through the GaAs substrate, fracturing of the mounting material, or worst, the separation of the chip from the mounting substrate. Although these catastrophic failures may take several thousands of hours, some MMIC performance parameters are affected earlier. For example, the RF performance of a power amplifier strongly depends on the device channel temperature. With the degradation of GaAs structures or the adhesion between the IC and mounting substrate, the thermal impedance increases rapidly. This increases the channel temperature of the device and thus adversely affects the gain, power output, and efficiency.

Environmental screening tests and procedures for GaAs MMICs are adopted from the standards developed for testing silicon ICs and microwave hybrid ICs. Most of these tests are directly applicable to MMICs with some changes in test conditions due to different materials used in the assembly and packaging of MMICs. Commonly used environmental and mechanical tests include the following:

1. *High temperature storage life test:* This is an environmental test, where temperature is the only stress. Temperature and test duration must be specified. Usually temperature is the maximum storage temperature of the devices under test. Failure mainly is due to metalization, bulk GaAs, and corrosion.

2. *High humidity, high temperature, and bias:* A combined environmental-electrical stress test in which devices are subjected to an elevated ambient temperature and high humidity, simultaneously biased for a period of time. Normally, it is performed on a sample basis (qualification) on nonhermetic devices. The most common condition is 85°C and 85% relative humidity. More extreme conditions generally are very destructive to the chambers used. Time, temperature, humidity, and voltage must be specified. This accelerated test mainly detects corrosion risks.

3. *Steady-state operating life:* An electrical stress test in which devices are biased at full rated power for prolonged duration. The test normally is performed at 25°C ambient and power is 100% of full rated. Duration, power, and ambience, if other than 25°C, must be specified. Accelerated failure mechanisms mainly are metalization, bulk GaAs, and passivation.

4. *Dynamic operating life:* An electrical stress test in which devices are alternately subjected to forward bias at full rated power or current and reverse bias. Duration, power, duty cycle, reverse voltage ambience, and frequency must be specified. Failure mechanisms essentially are the same as in the steady-state operating life test.

5. *Intermittent operating life (power cycling):* An electrical stress test in which devices are turned on and off for a period of time. During the "on" time, the devices are fed a power that causes the junction temperature to reach its maximum rating. During "off" cycle, the devices return to 25°C ambient. Duration, power, or duty cycle must be specified individually. Main accelerated failure mechanisms are die bonds, wire bond, metalization, bulk GaAs, and passivation.

6. *Temperature cycling:* An environmental stress test in which devices are alternately subjected to low and high temperatures with or without a dwell time to stabilize the devices to 25°C ambience—the medium is usually air. Temperatures, dwell times, and cycles must be specified. Failure mechanisms essentially are die bonds, wire bonds, and package.

7. *Thermal shock (glass strain):* An environmental stress test in which the devices are subjected to a low temperature, stabilized, then immediately transferred to a high temperature. The medium is usually liquid. Failure mechanisms essentially are the same as in the temperature cycling test.

8. *Mechanical shock:* A mechanical stress test in which the devices are subjected to high impact forces normally in two or more of the six orientations: X1, Y1, Z1, X2, Y2, Z2. Tests are used to verify the physical integrity of the devices. G forces, pulse duration, and number of shocks and axes must be specified.

9. *Vibration variable frequency:* The vibration frequency is logarithmically var-

ied from 100 Hz to 1 kHz and back, normally over four cycles. Cycle time, amplitude, and total duration must be specified. Failure mechanisms mainly are package and wire bond.

### 11.4.4 Reliability Philosophy

General reliability studies should encompass three areas. First, the possible failure mechanisms must be identified and characterized. This is useful not only for subsequent reliability tests but is necessary to set meaningful limits for design and circuit layout. Without this elemental characterization, design for reliability would be hit or miss. Once all of the failure mechanisms have been identified, their interaction must be investigated. Reliability of entire circuits must be determined. Predictions made by mechanism studies must be confirmed. This step can be performed by lifetesting circuits and establishing failure distribution characteristics. This data is used to project failure rates. After the first two steps have been completed, a monitoring procedure is necessary to maintain a level of confidence about the results. As gallium arsenide processing matures and process control improves, changes will be likely in the reliability of the devices produced. When these changes occur, a complete recharacterization may be required. Over time, monitoring may discover the existence of infant failure mechanisms. Screening procedures eventually might be required and screening techniques would be developed at this stage.

This reliability philosophy differs from that of silicon devices. The historical approach for silicon has been to test hundreds of devices and establish the length of failure-free operation. This type of testing has been driven by MIL-STD-883 requirements for Group C testing and naturally has resulted in very good characterization of infant and random failures. In some cases, constant failure rates apply and exponential distributions are appropriate. The key parameter of this kind of testing has been the mean-time-between-failures number. Large samples are necessary to build confidence in the time span of a low failure rate. Unfortunately, this MTBF-mentality does not fit especially well with the early results of GaAs reliability testing. Available samples sizes are much smaller and activation energies are much higher for GaAs devices.

While building a respectable data base, the GaAs IC industry must find alternatives. One alternative is to test in ways that establish failure distributions through accelerated methods to cause failure during the test period. This technique may require fewer devices, as long as the distribution can be established with certainty. The actual results of both philosophies are identical, but one method seeks no failures during testing and the other requires them.

Every manufacturer has a significantly different method of generating reliability data, particularly for GaAs ICs, where no standards of reliability qualification criteria, test conditions, and procedures are universally adopted. Most of the standards used by MMIC manufacturers either are borrowed from silicon technology or selected at their own convenience. In such a situation it is difficult to compare reliability from one manufacturer to another. As a guideline, the following parameters are important to evaluate for any meaningful comparison:

- Confidence limit as defined in Section 11.4.1.
- Reject criteria, such as degradation, functional, catastrophic, out-of-specifications, and the specific mechanisms.
- Temperature of test, this includes channel or case temperature; method of measurement; and average or hot spot temperature measurement.
- Activation energy and whether the activation energy is empirical or assumed.
- Biasing conditions.

## 11.5 RELIABILITY DATA

### 11.5.1 Element Studies

High temperature studies are commonly used to evaluate the reliability of GaAs MMICs. These can be performed either at rated or stressed bias conditions or with no bias at all. When relegated to temperature variations, separation of failure mechanisms on an integrated circuit becomes time consuming. Moreover, it becomes difficult to identify the specific cause of failure of an MMIC during reliability test because of the presence of several different components, such as monolithic and thin film resistors, MESFETs, and MIM capacitors. Degradation of one component can degrade the overall performance of the MMIC. One alternative to lengthy failure mechanism studies for MMICs is to examine each circuit element separately. This not only allows each mechanism to be identified independently but also provides for verification of design rules. Handling of the element data is usually simplified, because single mechanisms are identified and interaction between elements does not occur.

A typical life test procedure to determine the reliability of elemental building blocks of MMIC is shown in Table 11.4. In the absence of characterization standards in terms of channel temperature, biasing conditions, sample size, and criterion for failure, MMIC manufacturers follow their own procedures and standards. Table 11.3 is just an example of the test conditions, device types to be tested, and failure criterion to achieve any meaningful reliability results. By no means are these the standards or suggested parameter values. The objectives of such a life test study is to (1) determine failure modes in different circuit elements; (2) evaluate the electrical parameter stability significant to an element; (3) determine MTTF (mean time to failure) for the element and evaluate how temperature, bias conditions, voltage, and current affect MTTF; and finally (4) correlate the element

**Table 11.4 Life Test Procedures of PCM Elements**

| Type of Devices | Description | Burn-in $T_{ch}$ Temperature (°C) | Burn-in Bias Condition | Accelerated Life Test $T_{ch}$ (°C) | Bias Condition | Sample Size | Parameter Monitored and Failure Criteria |
|---|---|---|---|---|---|---|---|
| PCM FET | 0.5 $\mu$m × 100 $\mu$m FET | 150 | $V_{ds} = 8$ V $I_{ds} = 1/2{*}I_{dss}$ | 175, 250, 275, 300 | $V_{ds} = 5$ V $I_{ds} = 1/2{*}I_{dss}$ | 50 at each temperature | $I_{dss}$, $V_p$ Detect 10% change in $I_{dss}$ |
| PCM Schottky diode | Width = 100 $\mu$m | 150 | Forward bias at 150 mA/mm | 250, 275, 300 | Same as burn-in | 100 | Reverse leakage, $V_{on}$ at 150 mA/mm 10% change in $V_{on}$ |
| PCM transmission line contact resistance test patterns | Spacing = 5, 10, 15, 20, and 25 $\mu$m, width = 100 $\mu$m | 150 | Bias voltage of 3 V/$\mu$m spacing | 250, 275, 300 | Bis voltage of 3 V/$\mu$m spacing | 100 at each temperature | 10% change in contact resistance |
| Implanted and annealed 3″ GaAs wafer | Whole 3″ GaAs wafers with $R_\square$ = 1000 Ω/sq. | 150 | | 250, 275, 300 | Unbiased | 5 wafers | Change in $R_\square$, monitored by M-Gauge® |
| PCM NiCr resistor | Evaporated NiCr, $R = 50$ 100 $\mu$m × 100 $\mu$m | 150 | $J = 2 \times 10^6$ amp/cm$^2$ | 225 | $J = 2 \times 10^6$ amp/cm$^2$ | 100 | 5% change in resistance |
| PCM airbridge | TiW-Au plated Au air-bridge 8 $\mu$m wide | 150 | $J = 5 \times 10^5$ amp/cm$^2$ | 225 | $J = 5 \times 10^5$ amp/cm$^2$ | 100 | 10% change in resistance |

*Note:* M-Gauge is an instrument made by Tencor for monitoring sheet resistance without contact. It offers better than 1% in reproducibility.

reliability and MTTF to the MMIC itself. Three tests commonly used to accomplish such objectives are the step stress test, the life test, and the accelerated life test. During the step stress test the device is subjected to ramping stress of temperature, voltage, and current until it fails. This type of test validates the test fixturing and measurement method, estimates the maximum stress level for subsequent tests, and finally indicates the failure mechanism that can be expected during life tests.

Life tests are performed at relatively low temperatures for a longer period of time. Due to the lower temperatures, no catastrophic failures are expected, but the test identifies low-activation energy failure mechanisms, which can not be distinguished from mechanisms that cause failure at high temperatures. The lifetime of the device cannot be determined from such testing. An accelerated life test is performed at several stress levels of temperature, voltage, and current or their combination. Temperature stress is used universally for all elements besides current stress testing on interconnects and voltage stress testing on MIM capacitors. This is the most useful testing for determination of lifetimes and identification of failure mechanisms.

GaAs IC suppliers have benefited from the experience of the discrete and power MESFET suppliers with elements such as interconnect metal layers, gate metals, ohmic contacts, dielectric passivation layers, diode structures, die attach methods, and several types of MESFETs. Specific testing of GaAs elements was initiated as early as 1978 and has been adopted for additional GaAs IC elements, adding data on resistor and capacitor elements. The following reliability results were obtained on different circuit elements manufactured by TriQuint's foundry process.

Ohmic contact reliability is characterized by the transmission line contact resistance pattern (as described in Chapter 3) or by characterizing the monolithic resistors. Implant degradation is tested using the M-gauge resistivity measurement technique on implanted and annealed wafers. The main failure mechanism in ohmic contacts is the movement of ohmic metal both on the surface and within the contact. Median life as a function of current density is plotted for ohmic contact at 203°C in Figure 11.11 indicating a current density exponent of 3.5. These results on ohmic contacts are obtained from reliability tests on monolithic resistors of different sizes and, thus, isolate the ohmic contact characteristics from the implanted bulk. The results obtained by the other MMIC manufacturers may be quite different from these results based on several fabrication factors, such as metalization, its composition and thickness, doping levels in the substrate under the ohmic contact, and the alloying temperatures.

The reliable life of the gold-based airbridge interconnect metal was found to exceed 1 million hours and exhibit very minimal resistance drift (less than 3%) under the maximum rated current density ($5 \times 10^6$ amps/cm$^2$) at 150°C. Complete and uniform nitride passivation may increase the life of metal runs, as shown in

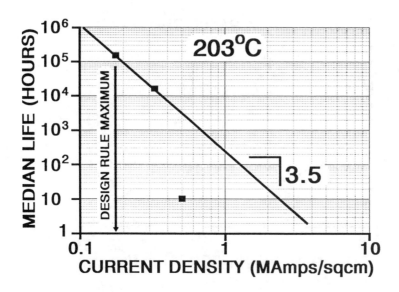

**Fig. 11.11** Ohmic current exponent.

Figure 11.12 provided the current density is kept low. No studies have indicated any particular susceptibility to corrosion.

The activation energy for airbridge metalization was found to be between 0.38 eV and 0.45 eV, as shown in Figure 11.13, where the median life *versus* temperature is plotted for different types of passivation. The current exponent for airbridge metalization is shown in Figure 11.14 to be between 3 and 5, where the median life *versus* current density is plotted as a function of types of passivation.

The main failure mode of airbridges was the formation of voids and mechanical cracks in protected passivation layers with metalization openings. The eventual failure mechanism was the fusing of airbridges due to mechanical deformation and molten gold. The reliability results may vary significantly for different manufacturers, based on the airbridge metal composition, its thickness, and the processing method used, such as E-beam evaporated, sputtered, or plated. The different metalization techniques may produce significantly different metal densities in the airbridge and thus affect the reliability. The first-level metal interconnect formed by Ti-Pd-Au is found comparatively less reliable than airbridge metalization. During the accelerated life test, the first-level metal changed its resistance continuously during the test procedure, making it difficult to select a failure criterion based on change in resistance value. Table 11.5 lists the activation energy and current density exponent values for each failure criterion.

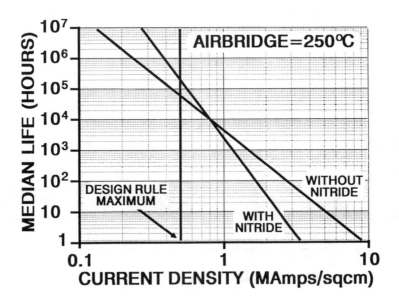

**Fig. 11.12** Passivation effects.

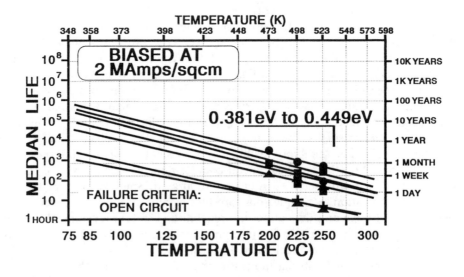

**Fig. 11.13** AB activation energy.

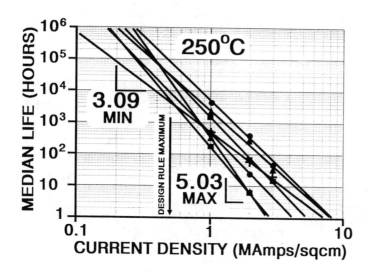

**Fig. 11.14** AB current exponent.

**Table 11.5** Failure Criteria (percent of change in resistance)

|  | *50%* | *100%* | *250%* |
|---|---|---|---|
| Unbiased activation energy | 2.18 eV | 1.86 eV | 1.81 eV |
| Biased at $2.7 \times 10^6$ amps/cm$^2$ activation energy | 2.24 eV | 2.42 eV | 2.55 eV |
| Current density exponent | −0.49 | −0.94 | −1.5 |

Figure 11.15 shows the median life at different temperatures as a function of failure criteria when the interconnect metal was unbiased. When the interconnect metal was biased at $2.7 \times 10^6$ amps/cm$^2$, the activation energy was found to be comparatively higher than the unbiased activation energy, as shown in Figure 11.16. The main failure mechanism obtained from accelerated life tests was the

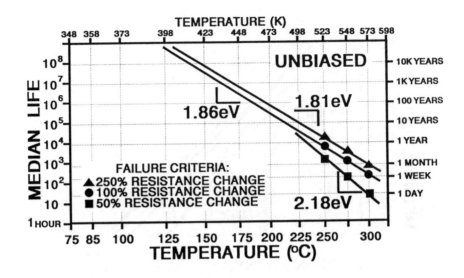

**Fig. 11.15** 1-ME activation energy, unbiased.

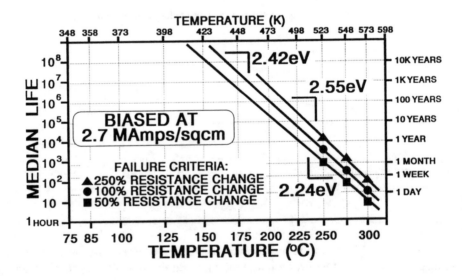

**Fig. 11.16** 1-ME activation energy, biased at 2.7 MAmps/cm².

change from sandwich layers of Ti-Pd-Au, which are E-beam deposited, to a nearly homogeneous metal, as shown in Figure 11.17. This interdiffusion of metals explains the change in resistance and contributes to the highest activation energy during the first 50% change in resistance value. After the metals were interdiffused, failure due to electromigration was the prominent effect responsible for final failure and eventual open circuits in the metal lines. Similar to airbridge metal, the reliability of the first-level metal depends significantly on the metalization constituents, their thickness, and the process used.

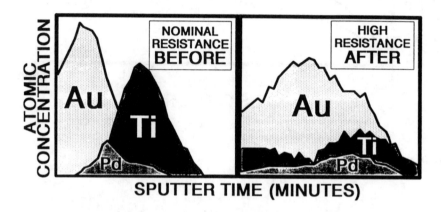

**SPUTTER TIME (MINUTES)**

**Fig. 11.17** 1-ME failure mechanism.

Thin-film NiCr resistors, which are passivated, show better results than metal layers, even to the point of not being able to predict their lifespans at normal use. Extremely high stress conditions are required to cause degradation in thin-film resistors. Figure 11.18 shows the change in the value of NiCr resistors *versus* time as a function of temperature. Figure 11.19 plots the median life *versus* temperature at bias condition of maximum current density, resulting in the activation energy of 1.03 eV. The reliability of thin-film resistors depends strongly on the thin-film material, the thickness of the metal, the annealing procedure to stabilize its resistivity, and the protected passivation. Hence, the reliability results on thin-film resistors could be significantly different for different MMIC manufacturers.

Reliability results on monolithic resistors and MIM capacitors indicate their lifetimes to far exceed that of any other component in MMIC. However, MIM capacitor reliability depends strongly on the quality of the insulator sandwiched between the two metal layers. The main parameters affecting this reliability are insulator thickness, pin holes, and the insulator material density.

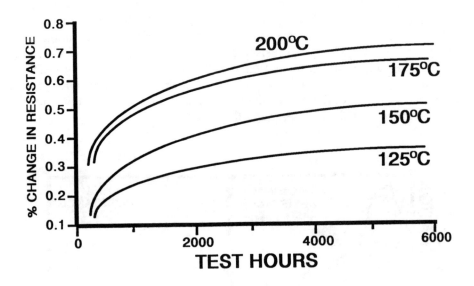

**Fig. 11.18** NiCr AF tests.

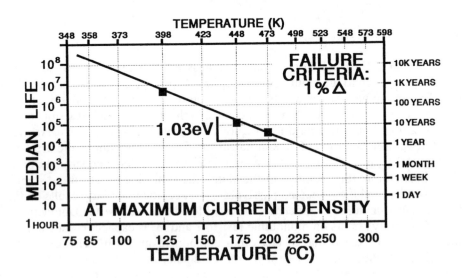

**Fig. 11.19** NiCr activation energy at maximum current density.

MESFETs have proven to be the key in MMIC element reliability studies. Some studies, especially on discrete MESFETs, have centered on the elements of the FET itself, specifically the formation of the Schottky gate and ohmic contact. Literature on power MESFETs and IC MESFETs tends to conflict. The abundant data from the 1970s on power and discrete MESFETs suggest a primary failure mode of decreasing channel current and a corresponding mechanism of ohmic contact degradation. Studies in the 1980s similarly have found that channel current decreases but that pinch-off voltage magnitude also decreases. Although some disagreement exists as to the precise mechanism (probably caused by differences in IC processes), failure analyses in most of the recent IC MESFET studies have indicated that the mechanism is a reduction in effective channel thickness due to the Schottky interface moving into the channel. Such a difference in failure mechanism between old and new reliability studies on MESFETs may come from the technology for ohmic contact metalization, which has improved over time.

MESFETs are composed of gate Schottky metal, ohmic contact, lightly doped channel implants, first-level metal, and optional airbridge metalization. Because of the complexity of the structure, composed of several basic processes and elements, MESFETs are the most difficult elements to test and estimate their lifetimes. Recent acceleration factor tests on IC MESFETs over a range of temperatures of 245–310°C have determined an activation energy of 2.6 eV and a median life over 1 billion hours at 150°C. This appears to be higher than other reported activation energy studies, covering a range of 1.4 to 1.8 eV, and is strongly dependent upon the device processing. The activation energy is a measure of how effectively a failure mechanism utilizes thermal energy. In short, the higher the activation energy, the greater is the acceleration effect of temperature. Failure percentage *versus* test time is shown in Figure 11.20 for different temperatures. The failure criterion is 20% change in $I_{dss}$. We emphasize that such a criterion is set completely arbitrarily by the manufacturer, which is important to consider when comparing reliability of FETs from different manufacturers. Figure 11.21 shows temperature *versus* median life data for standard depletion FETs (BFETs) and lightly doped channel FETs (DFETs), biased at $I_{dss}$ and using the failure criterion described earlier. All these FETs have 1.0 $\mu$m gate lengths. Table 11.6 presents a summary of the results of individual element testing. Each failure mechanism is listed with its empirically determined activation energies and current density exponents. Median lives are predicted for 150°C operation at the maximum current density limits for individual elements. The median life of different MMIC elements is plotted as a function of temperature in Figure 11.22.

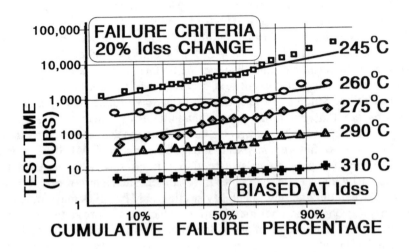

**Fig. 11.20** MESFET failure distribution, biased at $I_{dss}$.

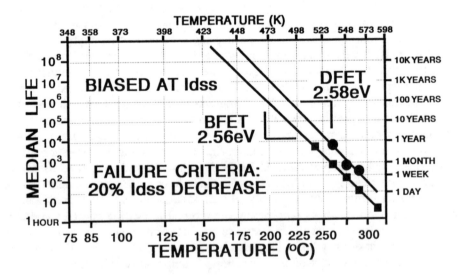

**Fig. 11.21** FET activation energy, biased at $I_{dss}$.

**Table 11.6** Element Accelerated Life Testing Data Summary

| Element | Failure Mechanism | Failure Criteria | Activation Energy | Current Density Exponent | Median Life at 150°C and Max. Current |
|---|---|---|---|---|---|
| Implanted resistors | Ohmic degradation | 6% resistance |  | −3.5 | 3 million hours |
| Nichrome resistors | Intermetallic degradation electromigration | 1% resistance | 1.0 eV |  | 500 thousand hours |
| First-layer metal interconnect | Intermetallic degradation electromigration | 50% | 2.18 eV | None | 80 million hours |
| Airbridge metal interconnect | Electromigration intermetallic degradation | 250% resistance | 2.55 eV | −1.5 | 255 million hours |
|  | Stress mechanical deformation | 25% resistance | .43 eV | −3 to −5 | 200 thousand hours to 240 million hours |
| MESFETs | Sinking gate | 20% channel current | 2.56 eV |  | 1 billion hours to 10 billion hours |

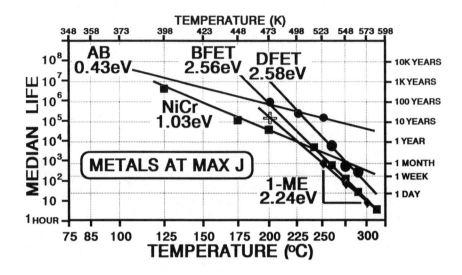

**Fig. 11.22** Element summary of metals at maximum J.

## 11.5.2 MMIC Reliability

Reported data on GaAs integrated circuits started to appear in 1979, proving that GaAs ICs could be produced reliably for benign operating conditions. But the testing did not run long enough to generate sufficient failures for lifespan predictions. Recent studies also have had difficulty in identifying failure mechanisms during life tests lasting at least 2000 hours but have predicted respectable lifetimes. The MMIC reliability studies based on element tests have clearly identified failure mechanisms, mainly around the MESFETs. These recent IC studies have indicated other "silicon-type" reliability trends. Comparisons of level of integration have borne out an increasing reliability level on a per MESFET basis. Improvements in the quality of bulk material, the application of automated process equipment, increasing volumes, and better cleanliness are specific examples of contributions to a general process and, hence, the reliability improvement. GaAs MMIC manufacturers claim to have expected median lifetimes on the order of 100 years. Failure rates have been below 1000 FITs (i.e., 1000 failures per billion hours). Once the reliability data and median lifetimes are available for individual MMIC components, it will be possible to predict the lifetime of MMIC. As shown in the following example, the predictions of IC life based on modeling the elements under highest stress are very close to actual IC results.

A typical MMIC widely used in various microwave applications is an amplifier. For the purpose of this study reliability data is gathered on distributed amplifiers whose design includes four 300 $\mu$m $\times$ 1/2 $\mu$m MESFETs, several spiral inductors fabricated from airbridge metal, nichrome resistors, and silicon nitride MIM capacitors. The total die size is 54 $\times$ 54 mil and the unit under normal bias condition provides 8 dB gain over 1–8 GHz and dissipates 1 W at 12 V supply. As with all GaAs ICs, special measurement hardware was required to test and measure the MMIC amplifiers. For easy biasing and measuring, the devices were mounted on small teflon evaluation boards. These boards had a large aluminum heatsink, connected to a network analyzer that was used to measure the devices. The test board was made of high-temperature polymide and employed spring-loaded "pogo" pins to make dc bias connections to the evaluation boards for steady-state tests. Thermal analysis of the device determined the hottest element at normal bias was a 100 $\times$ 150 $\mu$m nichrome resistor. This resistor dissipated roughly half of the power for the device (i.e., 0.5 W) and ran at 0.9 of the maximum rated current for nichrome. Initially, the amplifier MMICs were step-stress tested. The step stress involved a constant 15 V dc bias with temperatures that increased 25°C per 24 hour interval. The test started at a channel temperature of 125°C and ended at 250°C, when the MMIC packages and fixturing began to fail. All S-parameters were measured from 1–8 GHz. The power supply current was also measured. The step-stress test was performed for several reasons. First, it was used to verify the fixturing and measurement methods. Second, it gave an indication of the types and distributions of failure mechanism that could occur during subsequent lifetimes. Last, it provided information on the appropriate temperatures for life testing. Based on the step stress results, a life test was performed at 12 V supply voltage. During the life test, the temperature of this resistor was 290°C, the MESFETs were at 225°C, and the remaining circuitry was much cooler. Based on one definite failure (i.e., a nichrome resistor fused open) and the degradation of other amplifiers during the life test, a 3157 hour median life was established for this IC at 290°C. A median life of 1681 hours would be predicted for nichrome at 290°C under maximum rated current from element modeling parameters in Table 11.6. The comparison of accumulated failures for the IC life test and the element prediction is made in Figure 11.23. If the nichrome resistor in the MMIC operated at the maximum current density instead of 90%, the two curves probably would match even better. Element modeling predicts that these amplifiers would have a median life greater than 60 years when the MMIC is operating with the nichrome resistor at 150°C.

## 11.6 RADIATION HARDENING

A number of the GaAs semiconductor devices manufactured today must ultimately function in an environment fraught with radiation hazards. Certainly,

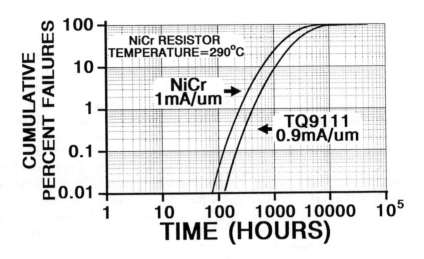

**Fig. 11.23** Element and MMIC comparison.

components which must be successfully deployed in outer space, in nuclear reactors, or in proximity to detonations from nuclear weapons need to feature some nominal level of radiation hardness. The purpose of this section is to review the radiation hazards to which GaAs devices may be exposed, examine the effects of typical irradiations on such semiconductors, and propose some guidelines for the hardening of the corresponding electronics.

### 11.6.1 Radiation Hazards

Emissions from typical radiation hazards can include neutral particles, charged particles, or photons, and a categorization of the common radiation products that can detrimentally interact with GaAs is presented in Table 11.7.

The amount of energy absorbed by an irradiated semiconductor is conventionally expressed in terms of the *radiation absorbed dose,* or rad. One rad is defined as the radiation dose which deposits an energy density of 100 ergs per gram in the absorbing material. Obviously, different materials have different energy absorption affinities, and thus it is important to specify the material being irradiated. Also note the radiation absorbed dose can be used to express the total, cumulative energy absorbed by the semiconductor over its entire lifetime of operation.

We emphasize that the effect of radiation on a semiconductor sample is not solely a function of the radiation absorbed dose. The dose rate, the type of irradi-

**Table 11.7** Common Particles and Photons Emitted from Radiation Sources

| Category of Radiation | Type of Radiation | Charge Polarity | Remarks |
|---|---|---|---|
| Neutral Particles | Neutrons | None | Very penetrating. |
| Charged Particles | Alpha Particles | + | |
| | Beta Particles | − | |
| | Electrons | − | |
| | Ions | + or − | |
| | Protons | + | |
| Photons | Gamma Rays | None | Very penetrating. For example, $60_{Co}$ radiation requires 5 cm of lead to reduce intensity by $10\times$. |
| | X-Rays | None | Very penetrating. |

ating particle or photon, and the energy spectrum of the incident radiation is also important [1–7].

When the transient radiation dose is of interest, this can be described as, for example, a dose rate of $N$ rads (GaAs)/s. Sometimes, however, a description is desired of the actual irradiating particle flux. Such a flux is often expressed in units of particles/cm$^2 \cdot$ s and is thus independent of the material upon which this flux is incident. Finally, the time integral of the flux is called the *fluence* and it commonly has units of particles/cm$^2$. These definitions are summarized in Table 11.8.

### 11.6.2 Types of Radiation Damage

The particles and photons elucidated in Table 11.7 can cause either permanent damage or temporary upset of the semiconductor device. Permanent effects include displacement, charge transfer, and mechanical damage. In contrast, temporary effects induce only a transient upset. Each of these four types of damage will be discussed in detail below.

*Displacement Damage*

In nuclear environments, energetic neutrons can impinge on a GaAs crystal

**Table 11.8** A Summary of Radiation Nomenclature

| | |
|---|---|
| *Dose* | Radiation absorbed dose (rad); |
| | A rad is the dose which deposits 100 ergs/gram in the absorbing material; |
| | A function of the absorbing material, which should thus be specified; |
| | Often used to express the total, cumulative energy absorbed by the semiconductor over its entire lifetime. |
| *Dose Rate* | Rads (GaAs)/second; |
| | Often used to describe the transient rate. |
| *Flux* | Irradiating particles/cm$^2$. second; |
| | Independent of material upon which this flux is incident. |
| *Fluence* | Time integral of the flux; |
| | Irradiating particles/cm$^2$. |

and actually knock lattice atoms out of position. These displaced atoms normally occupy interstitial sites and as a result leave behind vacancies in the lattice. These vacancy-interstitial pairs are called *Frenkel defects* and are assumed to be point disturbances, which, by definition, do not preferentially occur at any particular location in the lattice. Previous work on GaAs has shown that the group III (Ga) atom has a lower displacement energy threshold than the group V (As) atom [8].

The net result of such displacements is the formation of defect centers, which, in turn, produce traps in the energy bandgap. Obviously, such traps decrease the effective majority carrier concentration and mobility in addition to compromising the minority carriers lifetime. Many of these traps have been found to be near the center of the bandgap [9]. Hence, at extremely high neutron fluences, the GaAs material has a tendency to become intrinsic.

Not surprisingly, high energy electrons, protons, and gamma rays can also cause displacement damage by elastic collision mechanisms, similar to the neutron impact event described above [10]. Barring annealing phenomena, displacement damage is believed to be permanent, and thus the effect of subsequent irradiation is cumulative.

## Charge Transfer Damage

Another type of radiation damage considered to be permanent is that caused by the transfer of charge from the incident radiation to the semiconductor device. The magnitude of such an effect is a function of the total radiation dose received by the sample. For example, such irradiation might result in charge injection into an oxide designed for use as a MOSFET gate dielectric. This oxide charging results in modification of inversion potentials, and thus is extremely detrimental to MOS circuitry. In contrast, GaAs technologies often employ unpassivated MES-FETs, which are expected to be relatively immune to dielectric charging. However, placement of a passivation layer superficial to the MESFET gate region increases the likelihood of trapping radiation-supplied charges, and hence disturbing the parameters of the corresponding transistor.

## Mechanical Damage

In the limit, a neutron fluence in excess of $10^{15}/cm^2$ is likely to cause changes in the mechanical properties of the substrate [8]. Presumably, catastrophic electrical damage will occur well before the onset of mechanical damage. Therefore, effects induced by such elevated fluences are considered beyond the scope of this section.

## Transient Damage

In contrast to the various types of permanent damage discussed above, radiation can also cause the temporary upset of a semiconductor device. Such effects are usually associated with ionization tracks left behind by the passage of particles or photons through the crystal [1]. The resulting electron-hole pairs are then routed around in the electronics by the various electric fields. If, for instance, these excess carriers are collected by p-n junctions, voltage and current perturbations are caused in the corresponding circuit. In a digital system, such perturbations can induce logic errors, and in analog electronics the fidelity of signal swings is disturbed [9]. Regardless of the details, ionizing radiation impinging on a semiconductor can force both the reverse leakage current and the surface state charge to increase [8].

Transient effects are temporary disturbances, and are therefore a function of the rate at which the sample is irradiated. In extreme cases, the resulting transient currents can induce a permanent failure by encouraging the system to latch or otherwise to burn out. Although some IC architectures are prone to latching, GaAs MESFET structures normally do not possess the required number of layers to sustain a latch.

Radiation products which can cause transient effects include alpha particles, protons, heavy ions, cosmic rays, x-rays, high energy neutrons, and gamma rays [5,9,11]. Interactions between such radiation and the semiconductor might induce transient damage, displacement damage, or both, depending on the circumstances. For instance, some portion of a flux of protons may elastically scatter with atoms in the semiconductor, and hence cause displacement damage. Other protons might undergo inelastic scattering or nuclear reactions with the lattice atoms. The resulting recoil nuclei and secondary alpha particles can, in turn, produce ionization tracks [9]. Space radiation, however, with its complement of light charged particles is much more likely to cause simply an ionization upset than to displace atoms in the crystal [12]. Independent of the detailed mechanisms, the transient currents formed directly as a consequence of the ionization track are often called *primary photocurrents*. If they are subsequently amplified by a transistor or otherwise augmented, a *secondary photocurrent* is said to have been generated [9].

Perhaps the best known type of transient effect is the soft error suffered when, for example, an alpha particle toggles the state of a dynamic memory cell [13]. As a memory cell becomes smaller, the number of electron-hole pairs required to toggle its state is reduced, and hence the likelihood of a soft error increases [14]. However, the probability of the alpha particle striking the smaller cell at all is somewhat diminished. Thus, the optimal cell size for a radiation-hardened memory is not immediately obvious.

Another common type of ionizing disturbance occurs when the radiation-induced electron-hole pairs perturb critical voltages and currents in analog circuitry. An important figure of merit is the amount of time the circuit takes to recover from this transient. Initially, GaAs is expected to enjoy very fast recovery becaue the minority carrier lifetime is so short. This is a consequence of the rapid radiative recombination characteristic of direct gap semiconductors. However, the GaAs circuits being produced today are loaded with energy levels and interfaces where radiation-produced carriers can be trapped. These traps empty by various mechanisms, and provide the impetus for persistent currents to exist. In fact, measurements on various GaAs circuits reveal transient recovery times ranging from nanoseconds to seconds [9,15–18]. Unfortunately, some advanced devices, such as the HEMT, contain a multiplicity of important interfaces. Thus, these semiconductors may tend to suffer from very long transient recovery times [19]. In contrast, structures with a minimum number of interfaces and which are built on high quality material can be used to reduce these trapping effects [20–22].

An important piece of information for the systems designer is the effect of radiation on the electrical properties of semiconductor devices and circuits. As illustrated in Figure 11.24 the transconductance, pinch-off voltage, and drain saturation current all decrease monotonically as a MESFET is bombarded with gamma rays [23]. Other specific examples of the parametric changes caused by various types of radiation are presented in Table 11.9 [1,8,9,23–31]. Note that, in general, nearly all important device parameters are degraded by the respective irradiation.

### 11.6.3 Suggestions for Improving Radiation Hardness

Proper design, layout, and processing technologies are required to achieve

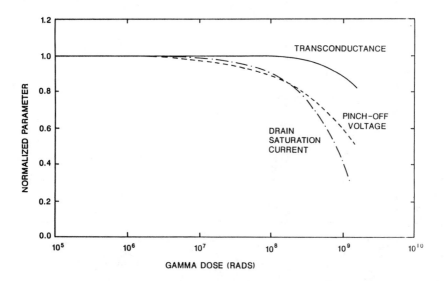

**Fig. 11.24** Gamma radiation-induced changes in MESFET parameters (from Kadowski *et al.* [23]).

the radiation hardness potential of GaAs. Unfortunately, very little rigorous research has been performed on the relationship between these important variables and the physics of radiation-induced failures. Nevertheless, heuristic arguments allow us to propose some interim solutions. A compilation of these suggested guidelines is enumerated below.

### Starting Material Considerations

- Use high quality, undoped substrates in an attempt to reduce the transient recovery time [15,16,22,23]. Avoid, in particular, chromium-doped wafers [33]. Liquid-encapsulated Czochralski (LEC) material appears at present to be an attractive option [9].
- Use the most heavily doped active layer possible, while still meeting the minimum breakdown voltage requirement [5]. Consider, for example, gradient doping of the active layer to permit more highly doped material to achieve reasonable breakdown potentials [5].

### Device Layout Considerations

- Minimize the interaction between the active devices and the substrate by placing the bonding pads and the metal interconnections superficial to an insulating layer [9]. Dielectrically isolated material, if available, can be used

**Table 11.9 Radiation-Induced Damage to Gallium Arsenide Semiconductors**

| Item Being Irradiated | Parameter of Interest | Type of Radiation | | | | |
|---|---|---|---|---|---|---|
| | | Neutrons | Alpha Particles | Electrons | Gamma Rays | Ions |
| MESFET | Transconductance | Decreases [9] | | | Decreases [23,24] | |
| | Drain Saturation Current | Decreases [9] | Decreases [25] | Increases [26] | Decreases [23,24] | Decreases [25] |
| | Pinch-off Voltage | Decreases [9] | | | Decreases [23] | |
| | Saturation Resistance | Increases [27] | | | | |
| | Noise Figure | Increases [9] | | | Increases [23,24,27] | |
| | Cut-off Frequency | Decreases [9] | | | | |
| MESFET Amplifier | Power Output | Decreases [28] | | | | |
| | Small-Signal Gain | Decreases [28] | | | | |
| | Power-Added Efficiency | Decreases [28] | | | | |
| Bipolar Transistor | Current Gain | Decreases [9] | | | | |
| | Reverse Leakage Current | Increases [9] | | | | |
| Dielectric Capacitor | Capacitance | | | | Decreases [24] | |
| GaAs Wafer | Carrier Lifetime | Decreases [8] | | Decreases [29] | | |
| | Carrier Mobility | Decreases | | Decreases [30] | | |
| | Carrier Concentration | Decreases | | Decreases [30] | | |
| | Resistivity | Increases [8] | | | | |
| | Breakdown Voltage | Increases [8] | | | Decreases [31] | |
| | Surface State Charge Density | | | | Increases [8] | |

to further insulate the active layer from substrate effects [5,32].

- Reduce radiation-induced charge collection by employing thin active layers and small device geometries [9,34,35]. Recall, however, that a lesser number of electron-hole pairs are ultimately required to upset smaller devices [5,9,32].
- Nondestructively absorb radiation-induced charges or localize their effect. Several ways of accomplishing this include the use of a high conductivity layer buried under sensitive devices or surrounding these devices with $p^+$ guard rings [9,18,32–34,36].

*Circuit Design Considerations*

- Dissipate radiation-induced photocurrents by using resistors or diodes in shunt with sensitive devices [5].
- Use thin-film resistors instead of diffused or implanted resistors [5].
- Employ *N*-channel devices because the corresponding structure has a smaller carrier removal rate than similar, *P*-channel components [5].
- Minimize radiation-induced charge collection volumes by operating reverse-biased junctions at the lowest possible voltage [5].
- Use high drive levels and conservative design margins [9,17].
- Select transistors which feature a high transconductance and a high drain saturation current [5].
- Bias transistors near the peak of their transconductance *versus* drain current characteristic to enjoy improved gain margins [5,16]. Similarly, use *N*-channel, rather than *P*-channel, MESFETs because the N-channel transistors have more available gain [5].
- Where appropriate, operate output stages in a saturated mode. Then, radiation-induced photocurrents simply increase saturation [5].
- Design circuits with conservative fan-out to minimize the effect of radiation-degraded gain in the driver transistors [5].
- Employ active, rather than passive, bias networks [12].
- Design circuits to have terminal characteristics that are dependent on the ratio of internal device parameters rather than their absolute values. This technique prevents radiation damage of similar components from degrading the performance of the overall circuit.
- Operate the electronics at an elevated temperature to anneal some of the radiation-induced damage [20,24,37]. Trade this improved radiation tolerance with the degradation in the conventional reliability suffered by operating at higher temperatures.
- Consider the use of auxilliary circuitry that senses the onset of a radiation pulse and then places the electronics in a less vulnerable state [5]. Design a

power-up and power-down protocol such that the system resets to its desired mode after the radiation hazard has passed.

### 11.6.4 Radiation Shielding Guidelines

In addition to the wafer processing and design guidelines presented above, employing shielding to attenuate the offending radiation may also be possible. Of course, auxiliary shields can significantly increase the mass of a system, and thus introduce a particularly large handicap for airborne and spaceborne electronics. As a result, judicious use of shielding is recommended only after all other radiation-hardening methods have been tried.

Unfortunately, to shield against a number of even the most common threats is not a simple matter. For example, although shielding is effective against electrons and low frequency x-rays, it is generally much more difficult to block gamma rays and fast neutrons [5]. In addition, the radiation incident on the shield can itself generate second-order problems, as illustrated in Figure 11.25 [38]. The figure suggests that an aluminum shield is incapable of reducing this particular radiation dose to less than approximately $4 \times 10^3$ rads (silicon). This is because in the process of slowing the incident radiation the aluminum releases bremsstrahlung x-rays as a byproduct [12,22,38].

Sometimes multilayer shields can be devised so that each succeeding layer absorbs the secondaries generated by the preceding layer. However, even with these advanced geometries, arrangement of the different shielding materials in the correct order is important. The shield with the highest atomic number is best placed next to the incident radiation, whereas the lighter material should be located adjacent to the electronics [5]. Other techniques useful for improving the shielding afforded to the underlying components are summarized below:

- Employ dense circuitry featuring high levels of integration. Then, more functions can be protected for a given mass of shielding material [38].
- Avoid placing particularly sensitive components on the top layer of circuit boards [38].
- Collocate sensitive elements for their mutual protection. For instance, that planar arrays of packaged integrated circuits can provide edge protection for each other has been demonstrated [38].
- Site sensitive circuitry to take advantage of the inherent shielding afforded by structural elements in either the electronics bay or the vehicle [12].

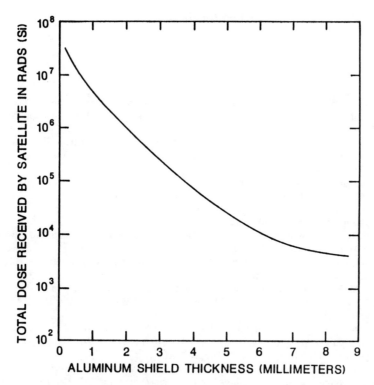

**Fig. 11.25** Total radiation dose received by a satellite operating for seven years in geosynchronous orbit (from Holmes-Siedel and Freeman [38]).

## ACKNOWLEDGEMENTS

The editor wishes to thank the following for his contributions to this chapter: Dr. David W. Hughes of Georgia Tech Research Institute, Atlanta, Georgia for contributing two of the sections.

# REFERENCES

## Section 11.3

1. C-L. Chen and W. D. Peele, "Federal Aviation Administration-Electromagnetic Pulse (EMP) Protection Study: A Reexamination and Update," Federal Aviation Administration Technical Report FAA-RD-80-11, November 1979.
2. D. G. Pierce and D. L. Durgin, "An Overview of Electrical Overstress Effects on Semiconductor Devices," *Electrical Overstress/Electrostatic Discharge Symposium Proceedings*, EOS-3, pp. 120–131, September 1981.
3. E. F. Vance and M. A. Uman, "Differences Between Lightning and Nuclear Electromagnetic Pulse Interactions," *IEEE Transactions on Electromagnetic Compatibility*, Vol. 30, No. 1, pp. 54–62, February 1988.
4. MIL-STD-883C, Test Method 3015.6, Notice 7.
5. H. Domingos and R. Walker, *Electrostatic Discharge Protective Design*, Static Awareness Reliability Associates, Rome, New York, 1985.
6. D. Hughes, "Guidelines for Prevention of ESD Damage to Integrated Circuits," *Semiconductor International*, pp. 188–193, May 1986.
7. S. Nambu, M. Hagio, A. Nagashima, K. Goda, G. Kano and I. Teramota, "A Low-Noise Dual-Gate MESFET for UHF TV Tuner," *IEEE Journal of Solid-State Circuits*, Vol. SC-17, No. 4, pp. 648–653, August 1982.
8. M. Hagio, K. Kanazawa, S. Nambu, S. Tohmori and S. Ogata, "Monolithic Integration of Surge Protection Diodes Into Low-Noise GaAs MESFETs," *IEEE Transactions on Electron Devices*, Vol. ED-32, No. 5, pp. 892–895, May 1985.
9. A. Fraser and D. Ogbonnah, "Reliability Investigation of GaAs IC Components," *IEEE Gallium Arsenide Integrated Circuit Symposium Technical Digest*, pp. 161–164, 1985.
10. A. L. Rubalcava, D. Stunkard and W. J. Roesch, "Electrostatic Discharge Effects on Gallium Arsenide Integrated Circuits," *Electrical Overstress/Electrostatic Discharge Symposium Proceedings*, EOS-8, pp. 159–165, September 1986.
11. T. E. Paquette, C. S. Bhasker, J. Cyr and R. Rosenberry, "GaAs Integrated Circuit Reliability Studies at Ford Microelectronics, Inc.," *GaAs Reliability Workshop Program and Abstracts*, pp. II-3, October 1986.
12. W. J. Orvis, A. L. Rubalcava, E. W. Chase, R. G. Taylor and R. H. Morrow, "Gallium Arsenide Protection Devices," *Electrical Overstress/Electrostatic Discharge Symposium Proceedings*, EOS-9, p. 297, September/October 1987.
13. A. L. Rubalcava, "Electrostatic Discharge Thresholds of Gallium Arsenide Integrated Circuits," *GaAs Reliability Workshop Program and Abstracts*, pp. 29–30, October 1987.
14. S. P. Bellier, R. F. Haythornthwaite, J. L. May and P. J. Woods, "Reliability of Microwave Gallium Arsenide Field Effect Transistors," *IEEE International Reliability Physics Symposium Proceedings*, pp. 193–199, 1975.
15. D. A. Abbott and J. A. Turner, "Some Aspects of GaAs MESFET Reliability," *IEEE Transactions on Microwave Theory and Techniques*, Vol. MTT-24, No. 6, pp. 317–321, June 1976.
16. T. Irie, I. Nagasako, H. Kohzu and K. Sekido, "Reliability Study of GaAs MESFETs," *IEEE Transactions on Microwave Theory and Techniques*, Vol. MTT-24, No. 6, pp. 321–328, June 1976.
17. R. Lundgren, "Reliability Study of GaAs FET," Rome Air Development Center Technical Report RADC-TR-78-213, October 1978.

18. C. L. Huang, F. Kwan, S. Y. Wang, P. Galle and J. S. Barrera, "Reliability Aspects of 0.5 um and 1.0 um Gate Low Noise GaAs FETs," *IEEE International Reliability Physics Symposium Proceedings*, pp. 143–149, 1979.

19. E. D. Cohen, A. C. MacPherson and A. Christou, "Reliability of Power GaAs FETs—Au Gates and Al-Au Linked Gates," *IEEE Transactions on Microwave Theory and Techniques*, Vol. MTT-29, No. 7, pp. 636–642, July 1981.

20. A. Christou, "Report on the 1982 GaAs Device Reliability Workshop," *IEEE International Reliability Physics Symposium Proceedings*, pp. 276–277, 1982.

21. C. Canali, F. Castaldo, F. Fantini, D. Ogliari, M. Vanzi, M. Zicolillo and E. Zanoni, "Power GaAs MESFET: Reliability Aspects and Failure Mechanisms," *Microelectronics and Reliability*, Vol. 24, No. 5, pp. 947–955, 1984.

22. A. A. Immorlica and J. R. Michener, "High Voltage Screening of GaAs Power FETs: Effect on Burn-In Yield and Modes of Catastrophic Device Failure," *IEEE International Reliability Physics Symposium Proceedings*, pp. 49–53, 1985.

23. W. T. Anderson and E. W. Chase, "Electrostatic Discharge Thresholds for GaAs FETs," *GaAs Reliability Workshop Program and Abstracts*, pp. II-4, October 1986.

24. W. T. Anderson and E. W. Chase, "Electrostatic Discharge Effects in GaAs FETs and MOD-FETs," *Electrical Overstress/Electrostatic Discharge Symposium Proceedings*, EOS-9, pp. 205–207, September/October 1987.

## *Section 11.6*

1. R. J. Chaffin, *Microwave Semiconductor Devices: Fundamentals and Radiation Effects*, John Wiley and Sons, New York, 1973.

2. R. J. Wilks, *Principles of Radiological Physics*, Churchill Livingstone, Edinburgh, 1981.

3. J. I. Vette, "The Space Radiation Environment," *IEEE Transactions on Nuclear Science*, Vol. NS-12, No. 5, pp. 1–17, October 1965.

4. P. G. Kase, "The Radiation Environments of Outer-Planet Missions," *IEEE Transactions on Nuclear Science*, Vol. NS-19, No. 6, pp. 141–146, December 1972.

5. L. W. Ricketts, *Fundamentals of Nuclear Hardening of Electronic Equipment*, Wiley-Interscience, New York, 1972.

6. W. Rosenzweig, "Space Radiation Effects In Silicon Devices," *IEEE Transactions on Nuclear Science*, Vol. NS-12, No. 5, pp. 18–29, October 1965.

7. R. J. Boucher and J. B. Pearson, "Transient Radiation Environment of Nuclear Weapon Detonations," *Journal of Applied Physics*, Vol. 36, No. 9, pp. 2722–2730, September 1965.

8. V. A. J. van Lint, T. M. Flanagan, R. E. Leadon, J. A. Naber and V. C. Rogers, *Mechanisms of Radiation Effects in Electronic Materials*, Vol. 1, John Wiley and Sons, New York, 1980.

9. M. Simons, "Radiation Effects In GaAs Integrated Circuits: A Comparison With Silicon," *IEEE Gallium Arsenide Integrated Circuit Symposium Technical Digest*, pp. 124–128, 1983.

10. J. H. Cahn, "Irradiation Damage in Germanium and Silicon due to Electrons and Gamma Rays," *Journal of Applied Physics*, Vol. 30, No. 8, pp. 1310–1316, August 1959.

11. R. M. Gilbert, G. K. Ovrebo, J. Schifano and T. R. Oldham, "Charge Collection in N-Type GaAs Schottky-Barrier Diodes Struck By Heavy Energetic Ions," *IEEE Transactions on Nuclear Science*, Vol. NS-31, No. 6, pp. 1570–1573, December 1984.

12. M. H. Gibson and I. Thomson, "Radiation Considerations in the Design of Linear Microwave Transistor Amplifiers for Space Applications," *IEEE Transactions on Microwave Theory and Techniques*, Vol. MTT-26, No. 10, pp. 779–788, October 1978.

13. T. C. May and M. H. Woods, "Alpha-Particle-Induced Soft Errors in Dynamic Memories," *IEEE Transactions on Electron Devices*, Vol. ED-26, No. 1, pp. 2–9, January 1979.

14. R. Zuleeg, J. K. Notthoff and G. L. Troeger, "Latch-Up in GaAs ICs During Ionizing Radiation Due to Surface Potential Shift," *IEEE Gallium Arsenide Integrated Circuit Symposium Technical Digest*, pp. 123–126, 1982.

15. M. Simons and E. E. King, "Long-Term Radiation Transients in GaAs FETs," *IEEE Transactions on Nuclear Science*, Vol. NS-26, No. 6, pp. 5080–5086, December 1979.

16. M. Simons, E. E. King, W. T. Anderson and H. M. Day, "Transient Radiation Study of GaAs Metal Semiconductor Field Effect Transistors Implanted in Cr-Doped and Undoped Substrates," *Journal of Applied Physics*, Vol. 52, No. 11, pp. 6630–6636, November 1981.

17. S. I. Long, F. S. Lee and P. Pellegrini, "Pulsed Ionizing Radiation Recovery Characteristics of MSI GaAs Integrated Circuits," *IEEE Electron Device Letters*, Vol. EDL-2, No. 7, pp. 173–176, July 1981.

18. J. G. Castle, M. G. Armendariz, D. R. Smith and G. R. Schuster, "Transient Response of a Small-Signal Microwave GaAs FET to X-Rays," *IEEE Transactions on Nuclear Science*, Vol. NS-31, No. 6, pp. 1596–1598, December, 1984.

19. H. Morkoc and P. M. Solomon, "The HEMT: A Superfast Transistor," *IEEE Spectrum*, Vol. 21, No. 2, pp. 28–35, February 1984.

20. R. Zuleeg, "Radiation Effects in GaAs Integrated Circuits," in N. G. Einspruch and W. R. Wisseman (editors). *VLSI Electronics Microstructure Science*, Vol. 11, Academic Press, New York, 1985.

21. K. Sleger, I. Mack, C. Scott and F. Buot, "Compound Semiconductor Digital Integrated Circuits," *Microwave Journal*, Vol. 29, No. 8, pp. 85–86, August 1986.

22. A. Firstenberg and S. Roosild, "GaAs ICs for New Defense Systems Offer Speed and Radiation Hardness Benefits," *Microwave Journal*, Vol. 28, No. 3, pp. 145–146, March 1985.

23. Y. Kadowaki, Y. Mitsui, T. Takebe, O. Ishihara and M. Nakatani, "Effects of Gamma Ray Irradiation on GaAs MMICs," *IEEE Gallium Arsenide Integrated Circuit Symposium Technical Digest*, pp. 83–86, 1982.

24. K. Aono, O. Ishihara, K. Nishitani, M. Nakatani, K. Fujikawa, M. Ohtani and T. Odaka, "Gamma Ray Radiation Effects on MMIC Elements," *IEEE Gallium Arsenide Integrated Circuit Symposium Technical Digest*, pp. 139–142, 1984.

25. W. T. Anderson, A. B. Campbell, A. R. Knudson, A. Christou and B. R. Wilkins, "Degradation in GaAs FETs Resulting from Alpha Particle Irradiation," *IEEE Transactions on Nuclear Science*, Vol. NS-31, No. 6, pp. 1124–1127, December 1984.

26. D. S. Newman, D. K. Ferry and J. R. Sites, "Measurement and Simulation of GaAs FETs Under Electron-Beam Irradiation," *IEEE Transactions on Electron Devices*, Vol. ED-30, No. 7, pp. 849–855, July 1983.

27. J. M. Borrego, R. J. Gutman and S. B. Moghe, "Radiation Effects on Signal and Noise Characteristics of GaAs MESFET Microwave Amplifiers," *IEEE Transactions on Nuclear Science*, Vol. NS-26, No. 6, pp. 5092–5099, December 1979.

28. S. B. Moghe, R. J. Gutmann and J. M. Borrego, "Radiation Effects on Distortion Characteristics of Power GaAs MESFET Amplifiers," *IEEE Transactions on Nuclear Science*, Vol. NS-29, No. 6, pp. 1545–1550, December 1982.

29. L. W. Aukerman, M. F. Millea and M. McColl, "Diffusion Lengths of Electrons and Holes in GaAs," *Journal of Applied Physics*, Vol. 38, No. 2, pp. 685–690, February 1967.

30. H. J. Stein, "Electrical Studies of Low-Temperature Neutron- and Electron-Irradiated Epitaxial N-Type GaAs," *Journal of Applied Physics*, Vol. 40, No. 13, pp. 5300–5307, December 1969.

31. G. E. Brehm and G. L. Pearson, "Gamma-Radiation Damage in Epitaxial Gallium Arsenide," *Journal of Applied Physics*, Vol. 43, No. 2, pp. 568–574, February 1972.

32. J. Spadaro, "Rad-Hard Product Mix Grows as GaAs Joins Silicon," *Electronic Products*, Vol. 28, No. 9, pp. 39–43, October 1985.

33. W. T. Anderson and S. C. Binari, "Radiation Effects in GaAs Devices and ICs," *IEEE International Reliability Physics Symposium Proceedings,* pp. 316–319, 1983.

34. J. G. Fossum, H. H. Sander and H. J. Gerwin, "The Effects of Ionizing Radiation on Diffused Resistors," *IEEE Transactions on Nuclear Science,* Vol. NS-21, No. 6, pp. 315–322, December 1974.

35. P. J. McNulty, W. Abdel-Kader, A. B. Campbell, A. R. Knudson, P. Shapiro, F. Eisen and S. Roosild, "Charge Collection in GaAs Test Structures," *IEEE Transactions on Nuclear Science,* Vol. NS-31, No. 6, pp. 1128–1131, December 1984.

36. Y. Umemoto, N. Masuda and K. Mitsusada, "Effects of a Buried P-Layer on Alpha-Particle Immunity of MESFETs Fabricated on Semi-Insulating GaAs Substrates," *IEEE Electron Device Letters,* Vol. EDL-7, No. 6, pp. 396–397, June 1986.

37. I. Thomson, M. H. Gibson, M. B. Christensen and G. J. G. Janssens, "The Sensitivity of Microwave Bipolar Transistors and Amplifiers to Ionizing Radiation," *IEEE Transactions on Nuclear Science,* Vol. NS-26, No. 3, pp. 4298–4306, June 1979.

38. A. Holmes-Siedle and R. F. A. Freeman, "Improving Radiation Tolerance in Space-Borne Electronics," *IEEE Transactions on Nuclear Science,* Vol. NS-24, No. 6, pp. 2259–2265, December 1977.

---

## EDITOR'S NOTE

In addition to the references and citations associated with the text, figures, and tables throughout this book, we wish to acknowledge the following:

Fig. 2.14: Horizon House/Microwave, Inc.
Figs. 3.1, 3.2, 3.3, 3.5: Artech House, Inc.
Figs. 3.10: © 1983 IEEE. Reprinted with permission.
Fig. 3.11: Courtesy of Eaton Corp.
Figs. 3.33, 3.37, 3.39, 3.44, 351, 354, 6.67: Courtesy of Anadigics, Inc.
Figs. 8.11, 8.12, 8.13: © 1987 Semiconductor International. Reprinted with permission.

Also note that the cross-reference of Fig. 3.49 (p. 110) in Chapter 3 should be to Fig. 3.44; the citation of reference [28] as the source of Fig. 4.46 (p.300) in Chapter 4 is erroneous; the description of Fig. 4.3.6 (p. 308–309) is erroneous, Figs. 4.3.5 and 4.3.6 respectively being the forward and reversed biased I-V conditions of a Schottky diode. We apologize to the reader for any confusion resulting from these errors.

# INDEX

Accelerated testing, 801–805
Activation, 138–139
Admittance, 63–64, 91
Airbridge,
  crossover connections, 270, 272
  metal, 341, 810–811
  removal, 463–467
Amplifier (microwave) design,
  circuit performance, 456
  combining techniques, 456–462
  considerations for, 393–399
  cost, 399
  performance, 399
  procedure for, 405–406
  reliability, 399
Amplifier design examples,
  distributed, 427–431
  feedback, 410–419
  feedback gain module, 431–439
  lossy match, 406–410
  low–noise, 440–456
  power amplifier, 419–427, 456–462
Amplifier noise, 402–404
Amplifier nonlinear behavior, 404–405
Amplifier power gain, 400
Amplifier stability, 400–402
ATE (automated test systems), 684
  architecture, 704–707
Auger analysis, 109

Backside via etch, 165–166
Balun topology, 512
Bathtub curve, 798–799
Biasing techniques, 387–392
  for a GaAs FET, 387–392

Calibration, computer aided, 146–138
  using GATES, 148–151
CAD, (Computer-aided design), 529–530, 580, 636
CAM (computer-aided manufacturing), 196–201, 573, 636
  CAMEO, 198
  features of, 197
  software packages, 198
Capacitance-voltage (C–V), 110
Chebyschev response, 480
Chemical vapor deposition (CVD), 123–124
Characteristic impedance, 41–42, 45–46, 57–63
Circuit simulation program, 593–594
Combining techniques,
  off-chip, 460–462
  on-chip, 458–460
Computer-Aided process design,
  device modeling, 140–144
  process modeling, 128–140
CPW (coplanar waveguide) system, 608, 691, 693–694
  grounded (GCPW), 694–695, 700

Deembedding, 684
Design languages, 634–636
Design verification,
  back-end, 628–634
  software for, 629, 630, 632–634
  techniques, 631–633
Device analysis,
  dc measurements, 215
  large-signal measurements, 220–221
  RF measurements, 219–220
Device modeling, 212–214, 655–678

computer aided, 140–146
empirical, 213–214
large-signal, 244–245
of GaAs MESFETs, 244–269
physical, 213–214, 254–255
Dielectric deposition, 123
CVD (Chemical vapor deposition), 123–124
pyrolytic deposition, 124
sputtering, 123
Dielectric layer, 162
Dielectric resonating oscillator (DRO), 568
Diffusion, 137–138
Dislocation density effects, 134–137
Distributed amplifier, 9
DLTS (deep-level transient spectroscopy), 787
Drain lag, 263
Drain resistance, 142–143
DRC (design-rules checker), 628, 633
DRO (dielectric resonating oscillator), 568
DSB (direct broadcast satellite), 23
Dual-gate FET, 510
applications of, 299–303
as mixer, 474–477, 479, 510
basic operation, 286–288
dc transfer characteristic, 286–287
lumped element equivalent circuits, 289–298
modeling of, 289–298
DVS (design verification software system), 628

EDA (electronic design automation), 574
EDIF (electronic design interface format), 635
Electrolytic profiling, 111
Electron beam deposition, 122–123
Electro-optic coupling, 696–698
EMP (electromagnetic pulse), 789
ENR (excess noise ratio), 722
EPC (electrical parameters checker), 629
Epitaxial growth, 101–108
Epitaxial layers, 140
ERC (electrical rules checker), 629–633
ESD (electrostatic discharge), 787–783
protection for MMICs, 793–796
ESM (electronic support measures), 23
Etching, 124–125
of GaAs, 789–793

Faraday's law, 35
Failure distributions, 798–801
Failure mechanisms, 807
dielectric, 785

electrostatic discharge, *see* ESD
metalization, 783–785
substrate, 785–787
Fermi level, 305
Final passivation, 163
FIT (failure in time), 799
Fourier analysis, 38
Fourier series, 472, 610
Fourier transform, 623, 624
discrete (DFT), 613, 617
fast (FFT), 617, 619
multidimensional (MFT), 621
multiple Fourier transform method, 620–622

GaAs (gallium arsenide)
doping profile, 110–111
fabrication of, 125–128
growth of, 101–108
mobility of, 111
resistivity, 109–110
GaAs MESFET, 151–152
basic operation, 208–211
device modeling of, 244–269
low-frequency anomolies, 262–269
performance analysis, 211–215
GaAs MMIC, 2–3,4, 151–152
applications, 18–27
cost, 17
disadvantages of, 108
reliability of, 16
*versus* Si, 5–6
yield, 16
GaAs MMIC processing, 97, 117–125, 151–167
active layer formation, 153–155
computer aided, see Computer-aided process
design
device isolation, 155–158
dielectric layer, 162
first-level metalization, 162
ohmic contacts, 158–159
resistor etch, 161–162
Schottky metalization, 159–161
thin film resistors, 158
GaAs-on-Si technology, 208
Gain block, 393
GATES (Gallium Arsenide Transistor
Engineering Models), 127, 128
Graphics processing, 632–633
Guassian law, 35

Hall effect, 111
Harmonic components, 613
Hartley transform, 617
HB (harmonic balance)
  analysis, 622–625
  errors, 624–625
  in mixer design, 470
  piecewise technique, 622. *See also* PHB.
  simulator, 611–626
HDTV (high definition television) receivers, 24
HEMT (high electron mobility transistor), 6–7,
  17, 111–113, 114–116
HJBT (heterojunction bipolar transistor),
  114–116, 396
Horizontal Bridgeman technique, 97, 98–99, 127
Hybrid MIC (HMIC), 1
  miniaturized (MHMIC), 8
  *versus* MMIC, 575
Hyperbolic tangent model, 245, 251
Hysteresis, 263

IEEE, 31, 33
IM (intermodulation) products, 405, 612
IMD (intermodulation distortion), 238
Implanted layer resistivity, 142
Ion implantation, 119–121, 128–133

Junction capacitance, 256
Junction capacitance models, 256–261

Kirchoff's laws, 36–37, 598

LAN (local area network), 23
Lange coupler, 395, 461
Large-signal parameter extraction, 237–244
LBE (liquid phase epitaxy), 101, 102
LEC (liquid encapsulated Czochralski
  technique), 97, 99–101, 127, 153, 786, 827
Line-to-line coupling, 359–361
Linear frequency domain simulation, 595–596
  Monte Carlo analysis, 595–596
Loss (in transmission lines), 42
Lumped element equivalent circuits,
  for dual-gate FETs, 289–298
  for Schottky diodes, 310–311
  for single-gate FETs, 289
LVS (layout versus schematics), 629, 633

MAG (maximum available gain), 15
Marquadt algorithm, 243
Mask fabrication, 636–646

Materials analysis, 109–111
  Auger analysis, 109
  resistivity, 109–110
  secondary ion mass spectrometry (SIMS)
    analysis, 109
Maximum channel current, 178
Maxwell's equations, 35, 38
MBE (molecular beam epitaxy), 7, 15
Metalization,
  airbridge, 163, 164, 341
  first-level, 162, 341
  Schottky, 159–161
  vacuum deposition, 121–123
MIC, (microwave integrated circuit), 1, 573
  hybrid (HMIC), 1
Microwave, 31
Microwave Amplifier Design, see Amplifier
  design
Microwave frequencies, 31, 32
Microwave probes, 684–698
  limitations of, 693–694, 716–723
Millimeter waves, 31
MIM (metal-insulator-metal) structures, 7
  capacitors, see Planar capacitors
MIMIC (microwave-millimeter wave monolithic
  integrated circuit), 735–736
Miniaturized HMIC (MHMIC), 8
Mixers,
  double-double balanced, 519–521
  single balanced, 525, 27
Mixer circuit design (monolithic),
  MIRM (image rejection), 479–483
  MIRMA, 481
  single-ended broadband, 483–489
Mixer circuit designs,
  diodes, 477
  dual-gate FET, 474–477, 479, 510
  single-gate FET, 470–474
MMIC (monolithic microwave integrated
  circuit), 318-319
  distributed elements, 319
  GaAs, see GaAs MMIC
  lumped elements, 319
  passive components of, 318
  performance of, 8–15
  reliability, 782, 820–821
  Si, 3–4
  *versus* HMIC, 9–11, 575
  *versus* digital-analog, 576
  wafer testing, 711–716, 724–733

MMIC applications,
    commercial, 23–24
    consumer, 24
    military, 18–23
MMIC design, 576–580
    physical layout, 626–628
MMIC devices, 269
MMIC foundries, 181–190
MMIC packaging, 736–738, 767–780, 797
    design of, 743, 752–758
    testing, 745–746
    thermal testing, 732–746
    characteristics of, 758, 759
    standards, 758–763
MMIC process control, 167–170
MOCVD (metal-organic chemical vapor
    deposition), 7
MODFET (modulation doped FET), 6–7
MOMBE (metal organic molecular beam
    epitaxy), 7
MTBF (mean time between failures), 799, 807
MTTF (mean time to failure), 799
Mushroom gate formation, 160

NCC (network consistency checker), 629
Network parameters, 70–82, 647–655
    ABCD (chain parameter), 72–73, 81
    h, 71, 82
    S, 70, 73–82, 221, 647–655
    Y, 71, 82
    z, 71, 82
Newton algorithm, 625
Newton iteration, 140, 625
Noise,
    in a 2-port network, 84–91
    shot (Schottky), 84
    temperature, 93
    thermal (Johnson), 83–84
    voltage, 94
Noise factor (F), 88, 90, 91, 94
Noise figure (NF), 88, 90, 92, 94, 278, 283, 402,
    449
Noise modeling, 278–285
Noise parameters, 82–95
    in a 2-port network, 84–91
    on the Smith chart, 92
Nonlinear-time domain simulator, 596–610
    SPICE program, 599–609
    SPICE 2, 602
Norton equivalent circuit, 83, 85

Ohmic contacts, 158–159, 176, 177, 308–309,
    312–313
    reliability, 810
    resistance, 313
OMVPE (organometallic vapor phase epitaxy),
    101, 105–106
On-chip tuning, 462–467
On-wafer calibration, 699–704
    SOLT, 699, 703–704
    TRL, 703–704
On-wafer microwave testing, 707–724
On-wafer probe testing,
    in-circuit, 686, 696–697
    terminal, 686
OPAMP (operational amplifier), 662, 665

Parameter extraction, 212, 214–221, 633
    and equivalent circuits, 221–244
    large-signal, 237–244
Parasitic capacitance, 341
Parasitic coupling, 358–359
Parasitic resistance, 224
Phase shifter design, 491–504
    lumped element, 649–652
Phase velocity, 40
PHB (piecewise harmonic balance) technique,
    622
    *versus* time domain analysis, 625–626
Photolithography, 117–119
    deep UV, 118
    electron beam, 119
    optical, 117–118
    x-ray, 118–119
Piezoelectric charges, 143–144, 146
Pinch-off voltage, 140–142
Planar capacitors, 329–341
    interdigitated, 329–331
    MIM (overlay), 331–340, 552, 815
    Q-factor of, 338–339
Planar lumped elements, 318–329
Planar resistors, 342–347
    fabrication of, 343
    GaAs, 343
    impedance of, 345–346
    metal film, 345
    single-gate FET, 347
Plating, 125, 755
PM (process monitor), 193, 195
Poisson's equation, 140, 144
Power spectral density, 279

Probe card technique, 684–685. See also
    Microwave probes
Process control monitors (PCM), 110
Process modeling, 128–140
Propagation constant, 39–40
Propagation modes, 385–386
Purple plague, 784

Radiation,
    damage, 823–826
    hardness, 826–830
    hazards, 822–823
    sheilding, 830–832
Reactive ion etching, 124–125
Recess etching, 138
Reflection coefficient, 46–47, 48, 51, 64, 91
RFOW test, 707–708
Resistor etch, 161–162
Reliability, 781, 807–821
RTL (register transfer level), 635

S-parameters (scattering parameters), see
    Network parameters
Schematic entry, 591–593
Schottky diodes, 304–318
    applications of, 316–318
    forward biased, 305–306
    GaAs, 304
    layout, 316–318
    lumped equivalent circuit, 310–311
    reverse biased, 305
    semidistributed element equivalent circuit of,
        312–316
    SPICE model of, 310–311
Sensitivity analysis, 450–456, 499, 595–596
Short-channel effects, 144, 208, 276
Shot noise (Schottky noise), 84
SIMS, see Materials analysis
Skin depth, 43
Small-signal equivalent circuits, 227–237
Smith Chart, 54–70
    and admittance, 63–64, 66–67
    and characteristic impedance, 57–63
    and noise parameter, 91
    components of, 59–61
    effects of frequency variation, 68–70
    features of, 57
    lossy circuits, 70
    reflection coefficient, 64
    VSWR, 64

Solid-state device, 383
SOLT (short-open-load-through) technique,
    699, 703–704
Source resistances, 142–143
SPDT switch, 489
SPICE, 252, 283, 470, 599–610, 647–655
    ac analysis, 608–610
    dc analysis, 606
    device modeling, 655–677
    transient analysis, 606–608
Standing wave, 49–54
Standing wave ratio (SWR), 51–54, 64
    power, 51
    voltage, see Voltage SWR
Substrate material effects, 133–134
SUPREM (Stanford University Process
    Engineering Models), 127, 128
Surface acoustic wave (SAW) filters, 24, 64
SUXES (Stanford University extractor model
    parameters, 243–244

TDR (time domain reflectometry), 689–690
Testability, 562–563, 568
    dc, 562
    RF, 562
Thermal noise, 83–84
Thevenin equivalent circuit, 85
Thin film resistors, 162
Threshold voltage, 246
Transconductance, 216, 241
Transmission coefficient, 47–48
Transmission line, 347–372
    ac solution for, 38–39
    coplanar (waveguide), 350–351, 352–355,
        373–375, 739
    discontinuities, 369–372
    flat line, 49
    frequency dispersion of, 356–358
    impedance limits of, 355–356
    in packaging, 739, 755–756
    losses, 42–45, 365–369
    microstrip, 351–352, 739, 756
    nonresonant line, 49
    properly terminated line, 49
    resonant line, 50
    terminated, 49–50
Transmission line theory, 35–54, 319, see also
    Transmission line
Transmit-receive (T/R) module, 18

TRL (through-reflection-line) calibration,
    703–704
TVRO (television receive-only), 551–562

Vacuum deposition,
    electron beam deposition, 122–123
    sputtering, 122
    thermal evaporation, 122
Van per Paw probe structure, 175, 181
VHDL (VHSIC hardware description
    language), 635–636
VHSIC (very high speed integrated circuit), 117,
    635
VPE (vapor phase epitaxy), 101, 102–104
VSATs(Very small aperture terminals ), 23, 64
VSWR ( Voltage standing wave ratio), 51, 54, 64

Wafer,
    process qualifications, 190–196
    sawing, 167
    testing, 564–567, 711–716
    thinning, 167
Wafer probing,
    and MMIC product testing, 724–733
    dc, 567–568
    electro-optic, 697–698
    RF, 562
White noise, 84, 95, 279
Wilkinson divider-combiner, 460
Work function, 305
Workstation, 582–591